Crop Production and Yield Management

Crop Production and Yield Management

Editor: Lennon Carlson

RCALLISTO
REFERENCE

www.callistoreference.com

Callisto Reference,
118-35 Queens Blvd., Suite 400,
Forest Hills, NY 11375, USA

Visit us on the World Wide Web at:
www.callistoreference.com

ISBN: 978-1-64116-200-5 (Hardback)

Cataloging-in-Publication Data

Crop production and yield management / edited by Lennon Carlson.
   p. cm.
Includes bibliographical references and index.
ISBN 978-1-64116-200-5
1. Crop yields. 2. Crops. 3. Crop science. 4. Farm management.
5. Agricultural productivity. I. Carlson, Lennon.
SB91 .C76 2019
631.558--dc23

Table of Contents

Preface

Crop refers to any plant that is grown and harvested for subsistence or to earn a profit. Crops are generally harvested to function as food for humans or fodder for livestock. Agricultural, horticultural, floricultural and industrial crops are the primary categories of crop. Crop yield is defined as the quantity of the harvested crop per unit area of land along with its seed generation. Crop rotation, intercropping and multiple cropping are some common techniques for improving and increasing crop yield. Yield management is a pivotal part of optimizing revenue and profit from crop production. It is a complex process, which includes understanding revenue streams management, rate management and distribution channel management. This book explores all the important aspects of crop production and yield management in the present day scenario. It presents some path-breaking studies in this field. The extensive content of this book provides the readers with a thorough understanding of the subject.

This book is a result of research of several months to collate the most relevant data in the field.

When I was approached with the idea of this book and the proposal to edit it, I was overwhelmed. It gave me an opportunity to reach out to all those who share a common interest with me in this field. I had 3 main parameters for editing this text:

1. Accuracy – The data and information provided in this book should be up-to-date and valuable to the readers.

2. Structure – The data must be presented in a structured format for easy understanding and better grasping of the readers.

3. Universal Approach – This book not only targets students but also experts and innovators in the field, thus my aim was to present topics which are of use to all.

Thus, it took me a couple of months to finish the editing of this book.

I would like to make a special mention of my publisher who considered me worthy of this opportunity and also supported me throughout the editing process. I would also like to thank the editing team at the back-end who extended their help whenever required.

Editor

Molecular Analysis of Biofield Treated Eggplant and Watermelon Crops

Mahendra Kumar Trivedi[1], Alice Branton[1], Dahryn Trivedi[1], Gopal Nayak[1], Mayank Gangwar[2] and Snehasis Jana[2]*

[1]Trivedi Global Inc., Henderson, USA
[2]Trivedi Science Research Laboratory Pvt. Ltd., Bhopal, Madhya Pradesh, India

Abstract

Eggplant and watermelon, as one of the important vegetative crops have grown worldwide. The aim of the present study was to analyze the overall growth of the two inbreed crops varieties after the biofield energy treatment. The plots were selected for the study, and divided into two parts, control and treated. The control plots were left as untreated, while the treated plots were exposed with Mr. Trivedi's biofield energy treatment. Both the crops were cultivated in different fields and were analyzed for the growth contributing parameters as compared with their respective control. To study the genetic variability in both plants after biofield energy treatment, DNA fingerprinting was performed using RAPD method. The eggplants were reported to have uniform colored, glossy, and greener leaves, which are bigger in size. The canopy of the eggplant was larger with early fruiting, while the fruits have uniform shape and the texture as compared with the control. However, the watermelon plants after the biofield treatment showed higher survival rate, with larger canopy, bright and dark green leaves compared with the untreated plants. The percentage of true polymorphism observed between control and treated samples of eggplant and watermelon seed samples were an average value of 18% and 17%, respectively. Overall, the data suggest that Mr. Trivedi's biofield energy treatment has the ability to alter the plant growth rate, and can be utilized in better way as compared with the existing agricultural crop improvement techniques to improve the overall crop yield.

Keywords: *Solanum melongena; Citrullus lanatus*; Biofield energy; Plant growth attributes; DNA Fingerprinting; Polymorphism

Introduction

The eggplant (*Solanum melongena* L.) is considered as one of the most important fruit vegetable crops all over the World [1]. In South Asia, Southeast Asia and South Africa, eggplant is commonly known as brinjal of family *Solanaceae*. The fruit grown are utilized for vegetables, which contributes all the essential nutrients in our diet [2-4]. The yield of eggplant fruit is dependent on several factors such as its flowering rate (anthesis), pest attack, and diseases infections, soil nutrient status, its fertility, and application of fertilizers [5]. Eggplant is considered as heavy feeder, which occupies the ground for long time, so at least two dressings for fertilizers are required [6]. The low level of soil fertility was linked with the poor prevailing climatic conditions, that results in low final yield.

Watermelon (*Citrullus lanatus*) belongs to the family of *Cucurbitaceae* [7], grown as a cash crop. It is mainly grown for its edible fruit that is a special kind of berry named as pepo. This plant is originally from Southern Africa, while its center of origin is between Kalahari and Sahara deserts in Africa [8]. These areas has been regarded as the point of diversification to other parts of the World [7]. For better nutrient status, the soil fertility factors must meet the criteria for better yield of fruit crop. Some methods has been prescribed for better yield of soil is to boost it with the use of organic materials, like animal waste, poultry manure, and use of compost or with the use of inorganic fertilizers [9]. This crop is considered as heavy feeder of nitrogen that required a high application of NPK fertilizers before sowing, followed by nitrogenous fertilizers till flowering stage [10]. The most important source of nitrogen is the inorganic fertilizers, which yield the vigorous vegetative growth, dark green leaves, and high photosynthetic rates. It was reported that extensive use of fertilizers will delay the ripening, reduce fruit setting and its number [11]. Therefore, some alternative approach besides the use of fertilizers, which could improve crop yield, overall plant growth, and its vegetative growth.

Phenotypic characters are based on the genetic identification, which affect the morphological characters of plant. DNA polymorphism identification is independent of environmental conditions using different molecular markers. Molecular markers of randomly amplified polymorphic DNAs (RAPD) analysis shows variation in the genome, which might expressed or not, while morphological markers reflect variation in expressed regions [12]. Using RAPD analysis, maximum genetic relatedness among plant genome can be identified, due to their simplicity, speed and low-cost [13].

Apart from these traditional approaches to improve the crop yield, recent research suggest that treatment of seeds with electric and magnetic field can improve the growth and yield of agricultural crops [14-16]. National Center for Complementary and Alternative Medicine (NCCAM) recommended the use of energy treatment as an alternative integrative medicine to promote human wellness [17]. Biofield is a type of electromagnetic field that permeates and surrounds the living organisms. Scientifically, it can be defined as biologically produced electromagnetic and subtle energy field within the organism. The objects always receive the energy and responding to the useful way that is called biofield energy treatment. Mr. Trivedi's unique biofield treatment is known as The Trivedi Effect®. Mr. Trivedi having the unique biofield energy, which has been reported in several research areas [18-21]. On the basis, of present literatures, present study was designed to evaluate the biofield treatment on selected plots (control and treated) for the seeds of eggplant and watermelon crop. Genetic variability parameters of both the crops were studied using RAPD (DNA fingerprinting).

***Corresponding author:** Snehasis Jana, Trivedi Science Research Laboratory Pvt. Ltd., Hall-A, Chinar Mega Mall, Chinar Fortune City, Hoshangabad Rd, Bhopal- 462026, Madhya Pradesh, India, E-mail: publication@trivedisrl.com

Materials and Methods

Eggplant (*Solanum melongena*) and watermelon (*Citrullus lanatus*) were selected for the present study. Both the plants were selected from inbreed variety for all the experimental parameters. The biofield treated plot size for eggplant was 64 × 8 feet, while the control plot size was 47 × 12 feet. The treated plot size for watermelon was 64 × 16 feet, while the control plot size was 35 × 25 feet. Both the plots were have same number of plants, and were compared with respect to respective control. The control plots were left untreated, while the treated plots of eggplant and watermelon was subjected to Mr. Trivedi's biofield energy treatment. The seeds from each crop were cultivated for analysis. However, the control plants were given standard cultivation parameters such as proper irrigation, fertilizers, pesticides and fungicides; while the treated plots were given only irrigation, without any supportive measure. DNA fingerprinting of both the plants were performed using RAPD techniques using Ultrapure Genomic DNA Prep Kit; Cat KT 83 (Bangalore Genei, India) to study the genetic relationship before and after treatment.

Biofield treatment strategy

The treated plots were subjected to Mr. Trivedi's biofield energy treatment. Mr. Trivedi provided the unique biofield treatment through his energy transmission process to the selected treated plots of both the crops. The plant samples of treated plots were assessed for the growth attributes with respect to control. Variability in different growth contributing parameters and genetic relatedness using RAPD of control and treated crops were compared [18].

Analysis of growth and related parameters of crops

The seeds of eggplant and watermelon were cultivated under similar conditions. The vegetative growth of the crops with respect to plant canopy, the shape of leaves, flowering conditions, infection rate, etc. were analyzed and compared with respect to the plants of control plots [22].

DNA fingerprinting isolation of plant genomic DNA using CTAB method

The leaves disc of both plants were harvested after germination, as it reached the appropriate stage. The genomic DNA from both plant leaves was isolated according to the standard cetyl-trimethyl-ammonium bromide (CTAB) method [23]. Approximately 200 mg of plant tissues were grinded to a fine paste in approximately 500 µL of CTAB buffer. The mixture (CTAB/plant extract) was transferred to a microcentrifuge tube, and incubated for about 15 min at 55°C in a recirculating water bath. After incubation, the mixture was centrifuged at 12000 g for 5 min and the supernatant was transferred to a clean microcentrifuge tube. After mixing with chloroform and iso-amyl alcohol followed by centrifugation the aqueous layers were isolated which contain the DNA. Then, ammonium acetate followed by chilled absolute ethanol were added, to precipitate the DNA content and stored at -20°C. The RNase treatment was provided to remove any RNA material followed by washing with DNA free sterile solution. The quantity of genomic DNA was measured at 260 nm using spectrophotometer [22].

Random amplified polymorphic DNA (RAPD) analysis

The RAPD analysis was performed on the each treated plot plants using RAPD primers, which were label as RPL 6A, RPL 13A, RPL 16A, RPL 18A, and RPL 19A for eggplant, while RPL 2A, RPL 7A, RPL 12A, RPL 14A, RPL 18A, and RPL 23A for watermelon. The DNA concentration was considered about 25 ng/µL using distilled deionized water for polymerase chain reaction (PCR) experiment. The PCR mixture including 2.5 µL each of buffer, 4.0 mM each of dNTP, 2.5 µM each of primer, 5.0 µL (approximately 20 ng) of each genomic DNA, 2U each of *Thermus aquaticus* (*Taq*) polymerase, 1.5 µL of MgCl$_2$ and 9.5 µL of water in a total of 25 µL with the following PCR amplification protocol. The PCR cycle condition for eggplant and watermelon includes initial denaturation at 94°C for 5 min, followed by 40 cycles of annealing at 94°C for 1 min, annealing at 36°C for 1 min, and extension at 72°C for 2 min, while final extension was carried out at 72°C for 10 min. Amplified PCR products (12 µL of each) from control and treated samples were loaded on to 1.5% agarose gel and resolved by electrophoresis at 75 volts. Each fragment was estimated using 100 bp ladder (Genei™; Cat # RMBD19S). The watermelon sample was analyzed with help another ladder of 500 bp ladder (Genei™; Cat # RMBD13S). The gel was subsequently stained with ethidium bromide and viewed under UV-light [24]. Photographs were documented subsequently. The following formula was used for calculation of the percentage of polymorphism.

Percent polymorphism = A/B × 100

Where, A = number of polymorphic bands in treated plant; and B = number of polymorphic bands in the control plant.

Results and Discussion

Effect of biofield treatment on growth contributing parameters of eggplant

The eggplant crop in control plots showed the survival as less than 60-65%. The growth of eggplants was much less in the control group as well. The eggplants had a small canopy in control plot. The leaves were small in size, and their color was light green in eggplants of control plot. The fruit of eggplant in control plot did not have uniform shape and most of the fruit was diseased even after being sprayed with pesticides and fungicides.

On the other hand, the biofield treated plot for eggplants showed the leaves were thick, glossy, more green in color and bigger in size. The canopy of the eggplant in biofield treated plot was also larger as compared with the control crop. The budding was early, that suggest early fruiting as compared with the control. In biofield treated plots, all eggplants fruits had uniform shape and the texture as compared with the control. Further, no disease in the treated plants were observed as compared with the control (Figure 1).

Research study suggest that both mineral fertilizers and organic manures have their own roles in soil fertility management, however none can completely provide all the nutrients and conditions, which may enhance the growth of eggplant [25]. Biofield treatment on soil selected for eggplant crops, showed enhanced growth in the absence of chemical fertilizers, and suggested the alternate method to improve the crop yield.

Effect of biofield treatment on growth contributing parameters of watermelon

In the control plot, the survival rate of the watermelon plants was less than 60 to 65%. The canopy was small, and the color of the leaves were pale green in plants of control plot. Many of the plants were reported as diseased and even the fruits were infected at an early stage, with small fruit size in plants of control plot.

The biofield energy treatment on plots with watermelon plants showed high survival rate i.e., more than 99%. The canopy of watermelon plants was much larger in treated plots than in the control plot plants.

Figure 1: The Trivedi Effect® on eggplant and watermelon.
(a) leaf of control eggplant was reported with less growth and infection,
(b) biofield treated leaves and flowers of eggplant are healthy and infection free,
(c) control watermelon plants showed infection in fruits and leafs, while
(d) biofield treated watermelon showed leaves were free from any kind of disease with healthy growth and fruits in high yield.
C: Control; T: Treated

RAPD analysis of eggplant and watermelon

DNA fingerprinting using RAPD molecular markers have been widely accepted technique to study the changes in vegetable crops [29]. Using different RAPD markers, important information for genetic diversity can be evaluated for different species of plants. Besides genetic diversity, population genetics study, pedigree analysis and taxonomic discrimination can also be correlated [24]. However, RAPD is considered as a powerful tool to evaluate the differences between inter- and intra-population of plants [30]. Biofield energy treatment was reported with high genetic variability among species using RAPD fingerprinting [18]. However, the effect was also reported in case of biofield treated ginseng, blueberry [31], and lettuce, tomato [32] with an improved overall agronomical characteristics.

Effect on plants genetic characters from control and biofield treated plots were compared and analyzed for their epidemiological relatedness and genetic characteristics. Genetic similarity or mutations between the two groups were analyzed using RAPD. Both the samples required short nucleotide random primers, which were unrelated to known DNA sequences of the target genome.

Random amplified polymorphic-DNA fragment patterns of control and treated eggplant samples were generated using five RAPD primers, with 100 base pair DNA ladder. The results of DNA polymorphism in control and treated samples are presented in Figure 2. The DNA profiles of treated group were compared with their respective control.

The color of the leaves was bright and dark green. The watermelons were bigger in size, and the texture of the fruit was different. The treated watermelon plants were absolutely free from disease. Although the growth of watermelon plants has been reported to show vast variation due to different seasons such as effect of light, heat, temperature, etc. [26] After biofield treatment, the growth characters were reported with huge change as compared with the control in similar conditions, so results suggest that biofield treatment might alter some basic physiological character of plants responsible for the overall growth (Figure 1).

Therefore, it can be suggested that biofield treatment on land could be a new approach to improve the yield of eggplant and watermelon as compared with the control. Mahajan et al. reported that on exposure of plant seeds to electric and magnetic field, seeds become polarized and can retain the change in polarization. However, the polarized seeds when come in contact with water, significant interaction takes place between water dipole and seed dipoles, which results in better water uptake. This phenomenon might be responsible for better yield of crops [15]. Our experimental results suggest better growth of plants that might be due to the higher water retention in biofield treated land plants may be due to better dipole interaction.

The different environmental factors somehow contribute to the growth of the plant. A report on the effect of magnetic field on plant seeds with respect to the growth of the plant was measured, and suggest the improved growth of roots and shoots [27]. Further, the effect was also reported to have improved level of photosynthesis, stomatal conductance and chlorophyll content after magnetic field treatment under stress conditions [28]. Biofield energy treatment is a type of complementary and alternative energy medicine, which involves low-level energy field interactions. Overall results assumed that the biofield energy might provide energy to the plant that change the paramagnetic behaviors of the tested plants, which might help in improved growth of eggplant and watermelon.

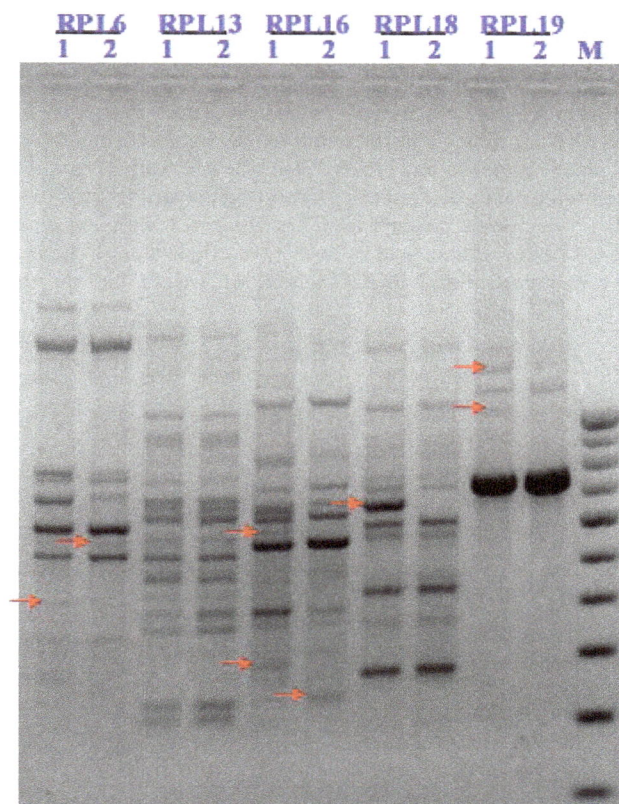

Figure 2: Random amplified polymorphic-DNA fragment patterns of eggplants of biofield treated plots generated using five RAPD primers, RPL 6A, RPL 13A, RPL 16A, RPL 18A, and RPL 19A. M: 100 bp DNA Ladder; Lane 1: Control; Lane 2: Treated.

The polymorphic bands observed using different primers in control and treated samples were marked by arrows. The results of RAPD patterns in biofield treated eggplant showed some unique, common and dissimilar bands as compared with the control. The DNA polymorphism analyzed by RAPD analysis, showed different banding pattern in terms of total number of bands, and common, and unique bands, which are summarized in Table 1. The percentage of polymorphism between samples was varied in all the primers, and were ranged from 0 to 40% between control and treated samples. However, level of polymorphism was maximum using the primer RPL 19A and minimum with RPL 18A, while RPL 13A primers did not show any level of polymorphism.

On the other hand, watermelon also showed high level for polymorphism using five primers with respect to the control. Different banding pattern was observed using RAPD DNA polymorphism in terms of total number of bands, and common, and unique bands, which are summarized in Table 2. The polymorphic bands observed using six different primers in control and treated samples of watermelon were marked by arrows in Figure 3. The level of polymorphism percentage in watermelon samples were varied in all the primers, and were ranged from 8 to 100% between control and treated samples. However, level of polymorphism was detected as 7%, 16%, 18%, 12%, and 33% using the primer RPL 2A, RPL 7A, RPL 12A, RPL 18A, and RPL 23A respectively. Highest level of polymorphism was detected using primer RPL 23A, 33%, while RPL 14A did not shown any polymorphism.

RAPD analysis using different primers explains the relevant degree of genetic diversity among the tested samples. Overall, RAPD showed that polymorphism was detected between control and treated samples. The percentage of true polymorphism observed between control and treated samples of eggplant and watermelon sample was an average value of 18% and 17%, respectively.

However, RAPD is a tool which will detect the potential of polymorphism throughout the entire tested genome. Biofield treated plot plants eggplant and watermelon showed varied number of polymorphic bands that indicated that the genotypes selected possess a higher degree of polymorphism as compared with the control. Molecular analyses and genetic diversity of eggplant and watermelon have been reported using RAPD analysis. Mr. Trivedi's biofield energy

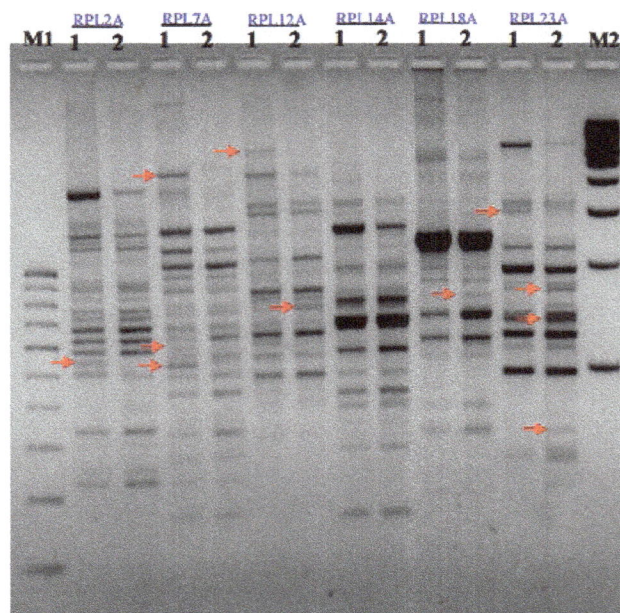

Figure 3: Random amplified polymorphic-DNA fragment patterns of watermelon in biofield treated plot generated using six RAPD primers, RPL 2A, RPL 7A, RPL 12A, RPL 14A, RPL 18A, and RPL 23A. M1: 100 bp, M2: 500 bp DNA Ladder; Lane 1: Control; Lane 2: Treated.

treatment on plots showed different level of polymorphism in eggplant and watermelon that suggested that biofield energy treatment might have the capability to alter the genetic character of plants, which might be useful in terms of productivity.

Conclusions

In summary, biofield energy treatment on the eggplant and watermelon showed improved growth characteristics such as fruits, leaves and free from pest attack. The canopy of plant and fruits of eggplant and watermelon was reported as large compared to their respective control. Biofield treated eggplant and watermelon plants showed strong and uniform colored leaves, with high survival rate, which suggest higher immunity of plant as compared with the control. Further, the watermelons were bigger in size, and the texture of the fruit was different from untreated fruits. It is assumed that after biofield treatment, the polarization of seeds might be affected that changed the interaction between water and seed during germination. Besides, the percentage of true polymorphism observed between control and treated samples of eggplant and watermelon seed sample was an average value of 18% and 17%, respectively. Overall, the experimental results suggested that Mr. Trivedi's biofield energy treatment might be used to improve the overall crop productivity with the capability to alter at genetic level.

Acknowledgments

Authors thanks to Bangalore Genei Private Limited, for conducting DNA fingerprinting using RAPD analysis. Authors are grateful to Trivedi science, Trivedi testimonials and Trivedi master wellness for their support throughout the work.

References

1. Thompson HC, Kelly CW (1977) Vegetable Crops, New York: McGraw Hill Book company.

2. Norman JC (1974) Egg plant production in Ghana. Ghana Farmer 17: 25–27.

S.No.	Primer	Primer Sequence	Band Scored	Common bands	Unique band	
					Control	Treated
1.	RPL 6A	TGGACCGGTG	11	8	1	1
2.	RPL 13A	CCTACGTCAG	16	16	-	-
3.	RPL 16A	AGGCGGGAAC	14	8	2	1
4.	RPL 18A	GAACGGACTC	11	10	1	-
5.	RPL 19A	CACACTCCAG	5	2	2	-

Table 1: DNA polymorphism of eggplant analyzed after biofield treatment using random amplified polymorphic DNA (RAPD) analysis.

S.No.	Primer	Primer Sequence	Band Scored	Common bands	Unique band	
					Control	Treated
1.	RPL 2A	CAGGCCCTTC	19	18	1	-
2.	RPL 7A	GTGATCGCAG	19	17	3	-
3.	RPL 12A	AGGACTGCCA	18	16	1	1
4.	RPL 14A	ACGGATCCTG	13	13	-	-
5.	RPL 18A	GAACGGACTC	13	13	-	1
6.	RPL 23A	CAGCACCCAC	14	14	1	3

Table 2: DNA polymorphism of watermelon analyzed after biofield treatment in plot using random amplified polymorphic DNA (RAPD) analysis.

3. Langer RA, Hill GD (1976) Agricultural Plants. London: Cambridge University Press.

4. Siemonsma JS (1981) A survey of indigenous vegetables in Ivory Coast Proc. (6th edn) African Symposium on Horticultural crops, Ibadan, Nigeria.

5. Huth CJ, Pellmyer D (1977) Nutrient requirements of solanaceous vegetable crops. Indian journal of agricultural sciences 58: 668-672.

6. Mc Collum JP (1980) Producing Vegetable Crops. Interstate printers and publishers Inc. pp: 518-522.

7. Schippers RR (2000) African Indigenous Vegetable. An overview of the cultivated species. Chatthan, U.K, N.R/ACO, EU.

8. Jarret B, Bill R, Tom W, Garry A (1996) Cucurbits germplasm report. Watermelon National Germplasm System, Agricultural Service, U.S.D.A.

9. Dauda SN, Aliyu L, Chiezey UF (2005) Effect of variety, seedling age and poultry manure on growth and yield of garden egg (Solamun gilo L.). Nigerian Acad. Forum 9: 88-95.

10. Rice RP, Rice LW, Tindal HD (1986) Fruit and Vegetable Production in Africa. Macmillan Publications.

11. Aliyu L (2000) the effect of organic and mineral fertilizer on growth, yield and composition of pepper (Capsicum annum L). Biol Agric Hort 18: 29-36.

12. Thormann CE, Ferreira ME, Camargo LE, Tivang JG, Osborn TC (1994) Comparison of RFLP and RAPD markers to estimating genetic relationships within and among cruciferous species. Theor Appl Genet 88: 973-980.

13. Rafalski JA, Tingey SV (1993) Genetic diagnostics in plant breeding: RAPDs, microsatellites and machines. Trends Genet 9: 275-280.

14. Maffei ME (2014) Magnetic field effects on plant growth, development, and evolution. Front Plant Sci 5: 445.

15. Mahajan TS, Pandey OP (2015) Effect of electric and magnetic treatments on germination of bitter gourd (Momordica charantia) seed. Int J Agric Biol 17: 351-356.

16. Alexander MP, Doijode SD (1995) Electromagnetic field, a novel tool to increase germination and seedling vigour of conserved onion (Allium cepa L.) and rice (Oryza sativa L.) seeds with low viability. Plant Genet Resour Newslett 104: 1-5.

17. NIH (2008) National Center for Complementary and Alternative Medicine. CAM Basics. Publication 347. [October 2, 2008].

18. Lenssen AW (2013) Biofield and fungicide seed treatment influences on soybean productivity, seed quality and weed community. Agricultural Journal 8: 138-143.

19. Nayak G, Altekar N (2015) Effect of biofield treatment on plant growth and adaptation. J Environ Health Sci 1: 1-9.

20. Trivedi MK, Patil S, Shettigar H, Bairwa K, Jana S (2015) Phenotypic and biotypic characterization of Klebsiella oxytoca: An impact of biofield treatment. J Microb Biochem Technol 7: 203-206.

21. Trivedi MK, Patil S, Nayak G, Jana S, Latiyal O (2015) Influence of biofield treatment on physical, structural and spectral properties of boron nitride. J Material Sci Eng 4: 181.

22. Shinde VD, Trivedi MK, Patil S (2015) Impact of biofield treatment on yield, quality and control of nematode in carrots. J Horticulture 2: 150.

23. Green MR, Sambrook J (2012) Molecular cloning: A laboratory manual. (3rd edn), Cold Spring Harbor, Cold Spring Harbor Laboratory Press, NY.

24. Welsh J, McClelland M (1990) Fingerprinting genomes using PCR with arbitrary primers. Nucleic Acids Res 18: 7213-7218.

25. Suge JK, Omunyin ME, Omami EN (2011) Effect of organic and inorganic sources of fertilizer on growth, yield and fruit quality of eggplant (Solanum Melongena L). Arch Appl Sci Res 3: 470-479.

26. Ufoegbune GC, Fadipe OA, Belloo NJ, Eruola AO, Makinde AA, et al. (2014) Growth and Development of Watermelon in Response to Seasonal Variation of Rainfall. J Climatol Weather Forecasting 2: 117.

27. Florez M, Carbonell MV, Martinez E (2007) Exposure of maize seeds to stationary magnetic fields: Effects on germination and early growth. Environ Exp Bot 59: 68-75.

28. Anand A, Nagarajan S, Verma AP, Joshi DK, Pathak PC, et al. (2012) Pre-treatment of seeds with static magnetic field ameliorates soil water stress in seedlings of maize (Zea mays L.). Indian J Biochem Biophys 49: 63-70.

29. Raj M, Prasanna NKP, Peter KB (1993) Bitter gourd Momordica ssp. In: Berg BO, Kalo G (eds) Genetic improvement of vegetable crops. Pergmon Press, Oxford.

30. Archak S, Karihaloo JL, Jain A (2002) RAPD markers reveal narrowing genetic base of Indian tomato cultivars. Curr Sci 82: 1139-1143.

31. Sances F, Flora E, Patil S, Spence A, Shinde V (2013) Impact of biofield treatment on ginseng and organic blueberry yield. AGRIVITA J Agri Sci 35: 22-29.

32. Shinde V, Sances F, Patil S, Spence A (2012) Impact of biofield treatment on growth and yield of lettuce and tomato. Aust J Basic Appl Sci 6: 100-105.

Effect of Nitrogen Levels and Plant Population on Yield and Yield Components of Maize

Shahzad Imran[1], Muhammad Arif[1], Arsalan Khan[2]*, Muhammad Ali Khan[2], Wasif Shah[1] and Abdul Latif[1]

[1]The University of Agriculture, Peshawar, Khyber Pakhtunkhwa, Pakistan
[2]Agriculture Research Institute (ARI) Tarnab Peshawar, Pakistan

Abstract

Field experiment was conducted to study the effect of nitrogen levels and plant population on maize. Maximum number of days to tasseling (71), silking (76) and maturity (108) were recorded with the application of nitrogen at 210 kg ha^{-1}. Higher plant height (202 cm), leaf area plant^{-1} (2757 cm^2), leaf area index (2.16), ear length (18.0 cm), ear weight (150 g), grains ear^{-1} (548), thousand grain weight (258 g) and grain yield (2673 kg ha^{-1}) were recorded with application of 210 kg N ha^{-1} which was statistically similar to 180 and 150 kg N ha^{-1}. Higher biological yield (7189 kg ha^{-1}) was recorded from 150 kg N ha^{-1} which was similar to 210 kg N ha^{-1}. Plant population of 95000 plants ha^{-1} took more number of days to tasseling (70), silking (75) and maturity (107). Taller plants (197 cm) were measured for plant population of 95000 plants ha^{-1}. Maximum number of leaves plant^{-1} (10.45) was recorded for plant population of 80000 plants ha^{-1}. Higher leaf area plant^{-1} (2585 cm^2) and leaf area index (2.59) were recorded for 65000 plants ha^{-1} which was statistically at par with 80000 plants ha^{-1}. Higher ear length (17.71 cm), ear weight (145 g), grains ear^{-1} (515) and thousand grain weight (252 g) were recorded from 65000 plants ha^{-1} which was similar to 80000 plants ha^{-1}. Plant population of 95000 plants ha^{-1} produced maximum biological yield (7276 kg ha^{-1}) while plant population of 80000 plants ha^{-1} produced maximum grain yield (2551 kg ha^{-1}) and harvest index (35.95%). It is concluded from the study that application of 150 kg N ha^{-1} produced maximum grain yield and plant population of 80000 plants ha^{-1} produced higher grain yield.

Keywords: Nitrogen level; Plant population; Grain yield

Introduction

Maize is the most important cereal crop of the world after wheat and rice, growing everywhere in the irrigated as well as in rain-fed areas. Botanically, it is known as *Zea mays* L. and belongs to family Poaceae. It is an annual cross pollinated crop. Its stem or stalk is thick and strong. The stalk bears leaf at each node. The leaf consists of a sheath and a broad, large blade. The sheath covers the stem. The male or terminal inflorescence is called tassel. The female inflorescence is known as ear in the middle [1]. The internodes are straight and nearly cylindrical in the upper part of the plant, but alternatively grooved on the lower part [2]. In spite of high yielding potential of maize, its yield per unit area is very low in Pakistan as compared to advanced countries of the world. In Pakistan, maize was cultivated on an area of 1042.0 thousand hectares with the annual production of 3109.6 thousand tons with an average yield of 2984 kg ha^{-1}. In KP, it was grown on about 492.2 thousand ha with production of 782.4 thousand ton annually. The average yield of this crop was 1590 kg ha^{-1} [3]. Maize is grown as food as well as fodder crop and is the second most important crop after wheat in Khyber Pukhtunkhwa (KP). It is staple food of rural population in Pakistan. As a grain crop, maize is a rich source of food and it is also used on large scale in industries for manufacturing of corn oil, corn flakes and corn sugar [4]. Nitrogen fertilization plays significant role in improving soil fertility and increasing crop productivity. Nitrogen fertilization results in increased grain yield (43-68%) and biomass (25-42%) in maize [5]. Nitrogen fertilization increase corn yield when N supply by soil is low [6]. Chemical fertilizer application could not be avoided completely since they are the potential sources of high amount of nutrients in easily available forms and maize is more responsive to it [7]. An increase in yield of maize with increasing rate of nitrogen has been reported by many researchers [8] primarily due to its favorable effect on yield components of maize [9]. Plant population is another factor which affects the plant yield. Yield was increased by 4% with increasing plant density [10]. Higher plant population produce 25%

more grain yield and 38% more biomass as compared with low plant population and early sown crop produce 19% more grain yield and 11% more biomass than late planted crop [11]. Keeping in view the importance of plant density and nitrogen fertilization, the study was conducted to find out optimum plant population and appropriate level of nitrogen for obtaining higher yield of maize.

Materials and Methods

A field experiment was conducted at New Developmental Farm, the University of Agriculture, Peshawar during spring 2014. The experiment was conducted in split pot randomized complete block design having four replications. The plot size of 3 m by 5 m (15 m^2) with row to row distance of 75 cm was used. For cultivation of maize crop field was ploughed two times using cultivator. Azam variety was sown at a higher seed rate of 60 kg ha^{-1} using drill. The required plant population i.e. 65000, 80000 and 95000 plants ha^{-1} was maintained by thinning after emergence. Number of plants per row was determined according to the respective plant population and then each row was thinned to achieve the respective plants per row (26 plants per row for 65,000 plants ha^{-1}, 32 plants per row for 80,000 plants ha^{-1}, 38 plants per row for 95,000 plants ha^{-1}). Nitrogen levels (0, 120, 150, 180 and 210 kg ha^{-1}) was applied in three split doses i.e. 1/3 at the time of sowing, 1/3 at knee height (5-6 leaf stage) and 1/3 at pre-tasseling stages. Urea was used

---

*Corresponding author: Arsalan khan, Agriculture Research Institute Tarnab Peshawar, Khyber PukhtoonKhwa Pakistan, E-mail: arsalankhan.fst@gmail.com

as source of N. Phosphorus was applied at the rate of 100 kg ha^{-1} before sowing. SSP was used as source of phosphorus. Weeds were removed through herbicides and hand howing. Irrigation was done through canal water as per crop water demand. Insecticide 'chloropirophide' was applied for controlling stem borer. All the other agronomic and cultural practices were done uniformly.

Statistical analysis

The data were statistically analyzed using analysis of variance technique appropriate for split plot randomized complete block design. Means were compared using LSD test at 0.05 level of probability, if the F-values are significant [12].

Results

Emergence m^{-2}

Data regarding emergance^{-2} are presented in Table 1. Analysis of the data revealed that nitrogen levels and plant densities had non-significant effect on emergence m^{-2} of maize. Likewise, the nitrogen x plant density interaction was also non-significant.

Days to tasseling

Data regarding days to tasseling are reported in Table 2. Analysis of the data showed that nitrogen levels and plant densities had significant effect on days to tasseling of maize. However, interaction of nitrogen x plant density was non-significant. Mean values of the data showed that increasing nitrogen level consistently increased days to tasseling. Plots received 210 kg N ha^{-1} took higher numbers of days to tasseling (71) while control plots took lower numbers of days to tasseling (66). The data further revealed that higher plant density of 95000 plants ha^{-1} took more numbers of days to tasseling as compared to lower plant density of 65000 plants kg ha^{-1}.

Days to silking

Data regarding days to silking are given in Table 3. Analysis of the data revealed that nitrogen levels and plant densities significantly affected days to silking of maize. However, interaction of nitrogen x plant density was non-significant. Mean values of the data revealed that each increment in nitrogen level consistently delayed days to silking. Maximum number of days to silking (75.92) was recorded for 210 kg N ha^{-1} while control treatment took minimum numbers of days to silking (71.50). In case of planting densities, higher plant density of 95000 plants ha^{-1} took maximum numbers of days to silking (74.65). Minimum numbers of days to silking (72.80) was recorded for lower plant density of 65000 plants kg ha^{-1}.

Plant height (cm)

Data on plant height are given in Table 4. Statistical analysis of the data revealed that nitrogen levels and plant densities had significant effect on plant height of maize. However, interaction of nitrogen x plant density was non-significant. Application of nitrogen at the rate of 210 kg ha^{-1} produced taller plants (202 cm) which were statistically at par with 180 and 150 kg N ha^{-1}(201 and 198 cm, respectively). Minimum plant height (181 cm) was recorded from control plots. Similarly, higher planting density of 95000 plants ha^{-1} produced taller plants (197 cm) as compared to lower plant density of 65000 plants ha^{-1}.

Number of leaves plant^{-1}

Analysis of the data revealed nitrogen levels had non-significant effect on number of leaves plant^{-1} however planting density had significant effect on number of leaves plant^{-1} (Table 5). Interaction of N x PD was non-significant. Mean values of the data revealed maximum number of leaves plant^{-1} (10.45) was recorded for planting density of 80000 plants ha^{-1}. While minimum number of leaves plant^{-1} (10.20) was recorded for planting density of 65000 plants ha^{-1}.

Days to maturity

Data on days to maturity of maize are given Table 6. Analysis of data revealed that nitrogen levels and planting densities had significant effects on days to maturity of maize. Interaction was found non-

N-levels (kg ha^{-1})	Plant density (plants ha^{-1})			Mean
	65000	80000	95000	
0	139	131	65.5	112
40	128	136	145	137
50	136	146	133	139
60	70.0	146	72.0	96
70	142	141	119	135
Mean	123	140	107	

Table 1: Emergence m^{-2} of maize as affected by nitrogen levels and plant population.

N-levels (kg ha^{-1})	Plant density (plants ha^{-1})			Mean
	65000	80000	95000	
0	66	67	66	66 e
120	67	68	69	68d
150	68	69	70	69c
180	70	70	72	71b
210	71	71	72	71a
Mean	68c	69b	70a	

LSD$_{(0.05)}$ for Nirogen=0.5527; LSD$_{(0.05)}$ for Plant density=0.4281.
Means of the same category followed by different letters are significantly different at 0.05 level of probability using LSD test.

Table 2: Days to tasseling of maize affected by nitrogen levels and plant population.

N-levels (kg ha^{-1})	Plant density (plants ha^{-1})			Mean
	65000	80000	95000	
0	70	71	72	71e
120	72	73	73	73d
150	72	74	75	74c
180	73	75	76	75b
210	76	75	76	76a
Mean	73c	74b	75a	

LSD$_{(0.05)}$ for Nitrogen=0.5478; LSD$_{(0.05)}$ for Plant density=0.4243.
Means of the same category followed by different letters are significantly different at 0.05 level of probability using LSD test.

Table 3: Days to silking of maize as affected by nitrogen and plant population.

N-levels (kg ha^{-1})	Plant density (plants ha^{-1})			Mean
	65000	80000	95000	
0	102	103	104	103e
120	104	105	106	105d
150	104	106	107	106c
180	105	107	108	107b
210	108	107	108	108a
Mean	105c	106b	107a	

LSD$_{(0.05)}$ for Nitrogen=0.55; LSD$_{(0.05)}$ for Plant density=0.42.
Means of the same category followed by different letters are significantly different at 0.05 level of probability using LSD test.

Table 4: Days to maturity of maize affected by nitrogen levels and plant population.

N-levels (kg ha⁻¹)	Plant density (plants ha⁻¹)			Mean
	65000	80000	95000	
0	176	186	183	182c
120	189	193	198	193b
150	199	195	201	198a
180	200	200	203	201a
210	200	202	204	202a
Mean	193b	195ab	198a	

LSD$_{(0.05)}$ for Nitrogen=3.337; LSD$_{(0.05)}$ for Plant density=2.585.
Means of the same category followed by different letters are significantly different at 0.05 level of probability using LSD test.

Table 5: Plant height (cm) of maize as affected by nitrogen and plant population.

N-levels (kg ha⁻¹)	Plant density (plants ha⁻¹)			Mean
	65000	80000	95000	
0	10.05	10.50	10.50	10.35
120	10.20	10.00	10.00	10.07
150	10.25	9.75	10.25	10.08
180	10.00	11.75	10.00	10.58
210	10.50	10.25	10.50	10.42
Mean	10.20b	10.45a	10.25b	

LSD$_{(0.05)}$ for Plant density=0.29.
Means of the same category followed by different letters are significantly different at 0.05 level of probability using LSD test.

Table 6: Number of leaves plant⁻¹ of maize affected by nitrogen and plant population.

significant. Mean values of the data showed constant increase in days to maturity with each increment in nitrogen level. Maximum days to maturity were taken by nitrogen level of 210 kg ha⁻¹ followed by nitrogen level of 180 and 150 kg ha⁻¹ with days to maturity of 106 and 105, respectively. Minimum days to maturity were recorded in control treatment. Maturity delayed with increasing plant density. The highest planting density of 95000 plants ha⁻¹ took more days to maturity (107) while minimum days to maturity (105) were observed in planting density of 65000 plants ha⁻¹.

Leaf area plant⁻¹

Data on leaf area plant⁻¹ are presented in Table 7. Nitrogen levels and planting densities significantly affected leaf area plant⁻¹. However, interaction of nitrogen x plant density was non-significant. Maximum leaf area plant⁻¹ (2757 cm²) was recorded with application of nitrogen at the rate of 210 kg ha⁻¹ which is statistically at par with 180 and 150 kg N ha⁻¹ with leaf area of 2523 and 2544 cm², respectively. Likewise, leaf area decreased with increasing plant density. Higher leaf area plant⁻¹ (2585 cm²) was recorded for 65000 plants ha⁻¹. Minimum leaf area plant⁻¹ (2316 cm²) was recorded from planting density of 95000 plants ha⁻¹.

Leaf area index

Data on leaf area index (LAI) of maize are presented in Table 8. Nitrogen levels and planting densities had significant effects on leaf area index. However, interaction of nitrogen x plant density was non-significant. Mean values of data revealed that application of nitrogen at the rate of 210 kg ha⁻¹ produced maximum leaf area index (2.76) which is statistically at par with 180 and 150 kg N ha⁻¹ with the LAI of 2.52 and 2.54, respectively. Lower planting density (65000 plants ha⁻¹) produced maximum LAI (2.59) and minimum LAI (2.32) was recorded from higher plant density 95000 plants ha⁻¹.

Ear length (cm)

Data on ear length of maize are shown in Table 9. Statistical

analysis of the data showed that nitrogen levels and planting densities had significant effects on ear length of maize. However, interaction of nitrogen x plant density was non-significant. Higher ear length (18.20 cm) was recorded from 180 kg N ha⁻¹ which is statistically at par with 210 and 150 kg N ha⁻¹ with the ear length of 17.28 cm and 17.09 cm, respectively. In case of planting densities, maximum ear length (17.71 cm) was recorded from lower plant density (65000 plants ha⁻¹) which is at par with higher plant density of 95000 plants ha⁻¹. Minimum ear length (16.84 cm) was recorded from 80000 plants ha⁻¹.

Ear weight (g)

Data regarding ear weight of maize as affected by nitrogen and plant densities are showed in Table 10. Analysis of the data showed significant effect of nitrogen levels and planting densities on ear weight. However, interaction of nitrogen x plant density was non-significant. Mean values of the data revealed that higher ear weight (150 g) was recorded from the treatment of nitrogen at the rate of 210 kg ha⁻¹ which is statistically at par with 180 and 150 kg N ha⁻¹ with the ear weight of 148 and 138 g, respectively. Similarly, maximum ear weight (145 g) was recorded from lower plant density which is statistically at par with

N-levels (kg ha⁻¹)	Plant density (plants ha⁻¹)			Mean
	65000	80000	95000	
0	2112	1943	1865	1973c
120	2410	2409	2550	2456b
150	2561	2716	2355	2544ab
180	2787	2505	2277	2523ab
210	3055	2683	2532	2757a
Mean	2585a	2451ab	2316b	

LSD$_{(0.05)}$ for Nitrogen=199.78; LSD$_{(0.05)}$ for Plant density=154.75.
Means of the same category followed by different letters are significantly different at 0.05 level of probability using LSD test.

Table 7: Leaf area plant⁻¹ (cm²) of maize as affected by nitrogen levels and plant population.

N-levels (kg ha⁻¹)	Plant density (plants ha⁻¹)			Mean
	65000	80000	95000	
0	2.11	1.94	1.87	1.97c
120	2.41	2.41	2.55	2.46b
150	2.56	2.72	2.36	2.54ab
180	2.79	2.51	2.28	2.52ab
210	3.06	2.68	2.53	2.76a
Mean	2.59a	2.45ab	2.32b	

LSD$_{(0.05)}$ for Nitrogen=0.1995; LSD$_{(0.05)}$ for Plant density=0.1546.
Means of the same category followed by different letters are significantly different at 0.05 level of probability using LSD test.

Table 8: Leaf area index of maize as affected by nitrogen levels and plant population.

N-levels (kg ha⁻¹)	Plant density (plants ha⁻¹)			Mean
	65000	80000	95000	
0	16.5	16.6	17.1	16.7b
120	16.3	16.5	17.6	16.8b
150	17.8	16.9	16.5	17.0ab
180	19.4	17.2	17.9	18.2a
210	18.4	16.7	16.6	17.2ab
Mean	17.7a	16.8b	17.1ab	

LSD$_{(0.05)}$ for Nitrogen=0.8637; LSD$_{(0.05)}$ for Plant density=0.6690.
Means of the same category followed by different letters are significantly different at 0.05 level of probability using LSD test.

Table 9: Ear length (cm) of maize as affected by nitrogen and plant population.

N-levels (kg ha⁻¹)	Plant density (plants ha⁻¹)			Mean
	65000	80000	95000	
0	130	121	116	123c
120	135	130	131	132bc
150	144	138	133	139ab
180	158	143	143	148a
210	158	149	143	150a
Mean	145a	136ab	133b	

LSD$_{(0.05)}$ for Nitrogen=10.912; LSD$_{(0.05)}$ for Plant density=8.4526.
Means of the same category followed by different letters are significantly different at 0.05 level of probability using LSD test.

Table 10: Ear weight (g) of maize affected by nitrogen levels and plant population.

80000 plants ha⁻¹ with the ear weight of 136 g. Minimum ear weight (133 g) was recorded from higher plant density (95000 plants ha⁻¹).

Number of plants at harvest ha⁻¹

The effect of nitrogen and plant population on plant at harvest ha⁻¹ is reported in Table 11. The effect of N had non-significant while plant population had significantly affected plants at harvest. Its interaction was also non-significant. Highest number of (92984) plant ha⁻¹ were observed at 95000 plant population fallowed by 77037 plant ha⁻¹ at 80000 plant population while lowest number of (59986) plant ha⁻¹ was observed at 65000 plant population.

Grains ear⁻¹

Grains ear⁻¹ of maize as affected by nitrogen and plant densities are presented in Table 12. Statistical analysis of the data showed that nitrogen levels and plant densities had significant influence on grains ear⁻¹ of maize. However, interaction of nitrogen x plant density was non-significant. Maximum grains ear⁻¹ (548) was recorded from nitrogen application at the rate of 210 kg ha⁻¹ which is at par with 180 kg N ha⁻¹ with 531 grains ear⁻¹, respectively. Minimum grains ear⁻¹ was recorded from control plots. In case of planting densities, maximum grains ear⁻¹ (515) was recorded from lower plant density (65000 plants ha⁻¹) which is statistically at par with 80000 plants ha⁻¹ with 497 grains ear⁻¹. Minimum grains ear⁻¹ (470) was recorded from 95000 plants ha⁻¹.

Thousand grain weight (g)

Data on thousand grain weight of maize as affected by nitrogen and planting density are presented in Table 13. Analysis of the data revealed significant effect of nitrogen levels and planting density on thousand grain weight. Interaction of nitrogen and planting density was non-significant. Maximum thousand grain weight of 259 g was recorded from 210 kg N ha⁻¹ which is statistically at par with 180 and 150 kg N ha⁻¹ with thousand grain weights of 258 and 250 g, respectively. Thousand grain weights decreased with increasing planting density. Maximum thousand grain weight of 253 g was recorded from lowest plant density of 65000 plants ha⁻¹ which is at par with 80000 plants ha⁻¹ with thousand grain weight of 250 g. Minimum thousand grain weight of 242 g was recorded from the highest planting density of 95000 plants ha⁻¹.

Thousand grain weight (g)

Data on thousand grain weight of maize as affected by nitrogen and planting density are presented in Table 13. Analysis of the data revealed significant effect of nitrogen levels and planting density on thousand grain weight. Interaction of nitrogen and planting density was non-significant. Maximum thousand grain weight of 259 g was recorded from 210 kg N ha⁻¹ which is statistically at par with 180 and 150 kg

N ha⁻¹ with thousand grain weights of 258 and 250 g, respectively. Thousand grain weights decreased with increasing planting density. Maximum thousand grain weight of 253 g was recorded from lowest plant density of 65000 plants ha⁻¹ which is at par with 80000 plants ha⁻¹ with thousand grain weight of 250 g. Minimum thousand grain weight of 242 g was recorded from the highest planting density of 95000 plants ha⁻¹.

Biological yield (kg ha⁻¹)

Biological yield of maize as affected by nitrogen levels and planting density are given in Table 14. Statistical analysis of the data showed that nitrogen levels and planting density had significant effects on biological yield of maize. Interaction of nitrogen and planting density was significant. Mean values of the data revealed that application of nitrogen at the rate of 150 kg ha⁻¹ produced maximum biological yield (7189 kg ha⁻¹) which is statistically at par with 120 and 210 kg N ha⁻¹ with biological yields of 6517 and 7863 kg ha⁻¹, respectively. Minimum biological yield (5884 kg ha⁻¹) was recorded from control plots. Higher biological yield of 7276 kg ha⁻¹ was recorded for planting

N-levels (kg ha⁻¹)	Plant density (plants ha⁻¹)			Mean
	65000	80000	95000	
0	60006	77007	93004	76672
120	60006	77006	93004	76672
150	59993	77244	92991	76743
180	60000	77001	92998	76667
210	59923	76924	92921	76590
Mean	59986 c	77037 b	92984 a	

LSD$_{(0.05)}$ for Plant density=84.34.
Means of the same category followed by different letters are significantly different at 0.05 level of probability using LSD test.

Table 11: Number of plants at harvest ha⁻¹ of maize as affected by nitrogen levels and plant population.

N-levels (kg ha⁻¹)	Plant density (plants ha⁻¹)			Mean
	65000	80000	95000	
0	448	439	411	433c
120	494	482	418	465bc
150	523	484	467	492b
180	545	528	519	531a
210	561	551	531	548a
Mean	515a	497ab	470b	

LSD$_{(0.05)}$ for Nitrogen=37.317; LSD$_{(0.05)}$ for Plant density=28.906.
Means of the same category followed by different letters are significantly different at 0.05 level of probability using LSD test.

Table 12: Grains ear⁻¹ of maize as affected by nitrogen levels and plant population.

N-levels (kg ha⁻¹)	Plant density (plants ha⁻¹)			Mean
	65000	80000	95000	
0	228	245	223	232c
120	253	222	253	243bc
150	247	260	242	250ab
180	267	258	250	258a
210	270	267	240	259a
Mean	253a	250a	242b	

LSD$_{(0.05)}$ for Nitrogen=9.0032; LSD$_{(0.05)}$ for Plant density=6.9739.
Means of the same category followed by different letters are significantly different at 0.05 level of probability using LSD test.

Table 13: Thousand grain weight (g) of maize affected by nitrogen and plant population.

N-levels (kg ha⁻¹)	Plant density (plants ha⁻¹)			Mean
	65000	80000	95000	
0	5246	6070	6338	5884c
120	5370	6981	7200	6517ab
150	6949	7045	7572	7189a
180	6724	7154	7394	7091 b
210	7319	8392	7879	7863 ab
Mean	6321 b	7128 a	7276 a	

LSD$_{(0.05)}$ for Nitrogen=646.20; LSD$_{(0.05)}$ for Plant density=500.54.
Means of the same category followed by different letters are significantly different at 0.05 level of probability using LSD test.

Table 14: Biological yield (kg ha⁻¹) of maize as affected by nitrogen levels and plant population.

density of 95000 plants ha⁻¹ which is statistically at par with highest planting density of 80000 plants ha⁻¹ with biological yield of 7128 kg ha⁻¹. Minimum biological yield of 6321 kg ha⁻¹ was recorded for 65000 plants ha⁻¹.

Grain yield (kg ha⁻¹)

Data on grain yield of maize as influenced by nitrogen levels and planting density are reported in Table 15. Nitrogen levels and planting density had significant effects on grain yield of maize. Interaction of nitrogen and planting densities was non-significant. Mean values of the data revealed that application of nitrogen at the rate of 210 kg ha⁻¹ produced maximum grain yield of 2673 kg ha⁻¹ which is statistically at par with 180 and 150 kg N ha⁻¹ with grain yield of 2475 and 2461 kg ha⁻¹, respectively. Minimum grain yield of 1803 kg ha⁻¹ was recorded in control plots. The plant density of 80000 plants ha⁻¹ produced maximum grain yield of 2551 kg ha⁻¹ while minimum grain yield of 2143 kg ha⁻¹ was recorded from 95000 plants ha⁻¹.

Harvest index (%)

Data on harvest index of maize are presented in Table 16. Statistical analysis of the data showed that planting density had significant effects on harvest index of maize. Maximum harvest index (35.95%) was recorded from planting density of 80000 plants ha⁻¹ which is at par with planting density of 65000 plants ha⁻¹ with harvest index of 34.52%. Minimum harvest index (29.54%) was recorded from 95000 plants ha⁻¹.

Discussion

Influence of nitrogen levels and planting density was found non-significant for emergence⁻² of maize. These results are in line with Le Gouis who reported that nitrogen had little or no effect on emergence m⁻² [13]. Nitrogen and planting densities significantly affected days to tasseling of maize. Plots received 210 kg N ha⁻¹ took higher numbers of days to tasseling in comparison to control plots. These results are in line with Amanullah et al. who stated that delay in days to tasseling was observed with increase in N rate and number of N splits [14]. The data further revealed that higher plant density of 95000 plants ha⁻¹ took more numbers of days to tasseling as compared to lower plant density of 65000 plants kg ha⁻¹. Our results agreed with Shafi et al. who reported that higher plant population took more numbers of days to tasseling compared to optimum and lower plant population [15]. Influence of nitrogen levels and planting density significantly affected days to silking of maize. Maximum number of days to silking was recorded from the treatment of nitrogen at 210 kg ha⁻¹ in comparison to control treatment. These results are consistent with the finding of Amanullah et al. who stated that increasing N application delay silking in maize [14]. In case of planting densities, higher plant density of 95000 plants

ha⁻¹ took more numbers of days to silking. Minimum numbers of days to silking was observed in lower plant density of 65000 plants kg ha⁻¹. These results are similar with Bhatt delayed silking were observed at more dense population as compare less dense population [16]. Nitrogen levels and planting densities had significant effect on days to maturity of maize. Maximum days to maturity were observed in the treatment of nitrogen at 210 kg ha⁻¹ followed by treatment of nitrogen at 180 and 150 kg ha⁻¹. Shrestha observed delay maturity with increase in nitrogen rate because nitrogen delay vegetative growth and as a result delay maturity [17,18]. Minimum days to maturity were recorded in control treatment. The data further revealed that highest planting density of 95000 plant ha⁻¹ took more number of days to maturity. Minimum days to maturity were observed in planting density of 65000 plants ha⁻¹. Our results are in line with Bhatt who stated that optimum plant population completes their life cycle earlier due to the enough water and nutrients availability [16].

Influence nitrogen levels and plant densities had significant effect on plant height of maize. Maximum plant height was recorded with the application of nitrogen at the rate of 210 kg ha⁻¹. Minimum plant height was recorded from control plots. These results agreed with Wajid et al. who investigated that higher nitrogen level influence plant height [18]. In case of planting density maximum plant height was recorded from 95000 plants ha⁻¹ while minimum plant height was recorded from 65000 plants ha⁻¹. Our results are supported by Malaviarachchi et al. who reported higher plant height with increase in plant population [19]. The planting density had significant effect on number of leaves plant⁻¹. Maximum number of leaves plant⁻¹ was recorded from planting density of 80000 plants ha⁻¹. Minimum number of leaves plant⁻¹ was recorded from planting density of 65000 plants ha⁻¹. These results are in agreement with Zandi who observed highest number of leaves plant⁻¹ at optimum planting density. The effect of nitrogen levels and planting densities had significant effect on leaf area index. Mean values of data revealed that application of nitrogen at the rate of 210 kg ha⁻¹ produced

N-levels (kg ha⁻¹)	Plant density (plants ha⁻¹)			Mean
	65000	80000	95000	
0	1726	1929	1755	1803 c
120	1678	2467	2158	2101 b
150	2383	2917	2083	2461 a
180	2325	2675	2425	2475 a
210	2957	2767	2295	2673 a
Mean	2214 b	2551 a	2143 b	

LSD$_{(0.05)}$ for Nitrogen=335.31; LSD$_{(0.05)}$ for Plant density=259.73.
Means of the same category followed by different letters are significantly different at 0.05 level of probability using LSD test.

Table 15: Grain yield (kg ha⁻¹) of maize as affected by nitrogen levels and plant population.

N-levels (kg ha⁻¹)	Plant density (plants ha⁻¹)			Mean
	65000	80000	95000	
0	32.90	32.11	27.72	30.91
120	30.86	36.06	30.14	32.35
150	33.87	41.36	27.59	34.27
180	34.35	37.33	32.93	34.87
210	40.64	32.89	29.32	34.28
Mean	34.52 a	35.95 a	29.54 b	

LSD$_{(0.05)}$ for Plant density=3.09.
Means of the same category followed by different letters are significantly different at 0.05 level of probability using LSD test.

Table 16: Harvest index (%) of maize as affected by nitrogen and plant population.

higher leaf area index which is statistically at par with 180 and 150 kg N ha^{-1} with the LAI of 2.52 and 2.54, respectively [20]. These results are similar with Jasemi et al. who reported that higher LAI associated with nitrogen treated plants have been probably due to increased leaf production and leaf area duration [21]. In case of plant density the lower planting density (65000 plants ha^{-1}) produced higher LAI. Lower LAI was recorded from higher plant density (95000 plants ha^{-1}). These results are similar with the finding of Maddoni et al. who stated that lower plant population got more nutrients and water compared to higher population and in turn increased growth and LAI [22]. It can be inferred from the data showed that nitrogen levels and planting densities had significant effect on ear length of maize. Maximum ear length was recorded from 180 kg N ha^{-1} which is statistically at par with 210 and 150 kg N ha^{-1} with the ear length of 17.28 cm and 17.09 cm, respectively. These results are similar with the results of Akram et al. who reported that cob length increases with increase in nitrogen level [23]. In case of planting densities, higher ear length was recorded from lower plant density (65000 plants ha^{-1}) which is at par with higher plant density of 95000 plants ha^{-1}. Minimum ear length was recorded from 80000 plants ha^{-1}. Similar results were also obtained by Khah et al. reported that ear length reduced with increasing plant population [24]. Influence of nitrogen levels and planting densities had significant effects on ear weight. Mean values of the data revealed that higher ear weight was recorded from the treatment of nitrogen at the rate of 210 kg ha^{-1} which is statistically at par with 180 and 150 kg N ha^{-1} with the ear weight of 148.06 and 138.52 g, respectively. These results are in line with Bhatt In case of planting density maximum ear weight was recorded from lower plant density which is statistically at par with 80000 plants ha^{-1} with the ear weight of 136.33 g [16]. Minimum ear weights was recorded from higher plant density (95000 plants ha^{-1}). Our results are supported by Hoshang who concluded that ear weight decreased with increasing plant population [25]. The effect of N had non-significant while plant population had significantly affected plants at harvest. Its interaction was also non-significant Highest number of (92984) plant ha^{-1} were observed at 95000 plant population fallowed by 77037 plant ha^{-1} at 80000 plant population while lowest number of (59986) plant ha^{-1} was observed at 65000 plant population. It may be the higher plants population having higher plant at harvest as compare to other plant population. Influence of nitrogen levels and plant densities had significant effects on grains ear^{-1} of maize. Higher grains ear^{-1} was recorded from the treatment of nitrogen at the rate of 210 kg ha^{-1} which is at par with 180 kg N ha^{-1} with 531 grains ear^{-1}. Lower grains ear^{-1} was recorded from control plots. These results are in line with Rizwan et al. who observed that number of grains per cob increased significantly with increasing nitrogen rates [26]. In case of planting densities, maximum grains ear^{-1} was recorded from lower plant density (65000 plants ha^{-1}) which is statistically at par with 80000 plants ha^{-1} with 497 grains ear^{-1}. Minimum grains ear^{-1} was recorded from 95000 plants ha^{-1}. These results are further endorsed by Abuzar et al. who reported that increase in plant population decreased grains ear^{-1} [27]. Thousand grain weight was significantly affected by nitrogen levels and planting density. Maximum thousand grain weight was recorded from 210 kg N ha^{-1} which is statistically at par with 180 and 150 kg N ha^{-1}. These results are in line with Arif et al. who reported maximum thousand grains weight (254.1 g) with 160 kg N ha^{-1} [28]. In case of planting densities, maximum thousand grain weight was recorded from lowest plant density of 65000 plants ha^{-1} which is at par with 80000 plants ha^{-1}. Minimum thousand grain weights was recorded from highest planting density of 95000 plants ha^{-1}. These results are in agreement with the finding of Radma and Dagash who reported that thousand grain weight increases with the increase in nitrogen level [29]. Influence of nitrogen

levels and planting densities had significantly affected biological yield of maize. Application of nitrogen at the rate of 150 kg ha^{-1} produced maximum biological which is statistically at par with 210 kg N ha^{-1} with biological yield of 7863. Minimum biological yield was recorded from control plots. These results are in line with Arif et al. who found that increase in nitrogen levels increased biological yield [28]. The data further revealed that higher biological yield was produced in the planting density of 95000 plants ha^{-1} which is statistically at par with highest planting density of 80000 plants ha^{-1} with biological yield of 7128 kg ha^{-1}. Minimum biological yield was recorded from 65000 plants ha^{-1}. These results are in line with Bhatt who reported higher biological from higher planting density [16]. Influence of nitrogen levels and planting densities had significant effect on grain yield of maize. Mean values of the data revealed that application of nitrogen at the rate of 210 kg ha^{-1} produced maximum grain yield which is statistically at par with 180 and 150 kg N ha^{-1} with grain yield of 2475 and 2461 kg ha^{-1}, respectively. These results are in line with Sharifi et al. who reported that increase in nitrogen significantly increased grain yield [30]. Minimum grain yield was recorded in control plots. The data further revealed that plant density of 80000 plants ha^{-1} produced maximum grain yield. Minimum grain yield was recorded from 95000 plants ha^{-1}. These results are supported by Aziz et al. who stated that increase in grain yield at optimum planting densities may be due to the availability of more nutrients which led to more growth and higher assimilates translocation to grains [31]. The influence of planting densities had significant effect on harvest index of maize. Maximum harvest index was recorded from planting density of 80000 plants ha^{-1}. Minimum harvest index was recorded from 65000 plants ha^{-1} which is at par with planting density of 95000 plants ha^{-1} with harvest index of 29.54%. These results are supported by Bahadar et al. who reported higher harvest index with optimum plant population [32].

Conclusion and Recommendation

From the results of the study it is concluded that, application of nitrogen at the rate of 150 kg ha^{-1} produced highest grain yield while plant population of 80000 plants ha^{-1} produced maximum yield and yield components. Application of 150 kg N ha^{-1} on the basis of higher grain yield is recommended. Plant population of 80000 plants ha^{-1} is recommended on the basis of higher grain yield and other parameter.

References

1. Khalil IA, Jan A (2002) Textbook of cropping technology (1stedn) National Book Found, Pakistan: 204-224.

2. Arnon I (1972) Crop production in dry regions. Leonard Hill Book, London. Nicholas Polunin 163: 1-2

3. Minfal (2006) Govt. of Pakistan, Ministry of Food, Agric. and Liv. Econ. Wing, Islamabad, 18-19.

4. Harris D, Rashid A, Miraj G, Arif M, Yunas M (2007) 'On-farm' seed priming with zinc in chickpea and wheat in Pakistan. Plant and Soil 306: 3-10.

5. Ogola JBO, Wheeler TR, Harris PM (2002) Effects of nitrogen and irrigation on water use of maize crops. Field Crop Res. 78: 105-117.

6. Ayeni LS, Adetunji MT (2010) Integrated Application of Poultry Manure and Mineral Fertilizer on Soil Chemical Properties, Nutrient Uptake, Yield and growth components of maize. J. Nature and Sci 60-67.

7. Obi ME, Ebo PO (1995) The effect of organic and inorganic amendments on soil physical properties and maize production in a several degraded sandy soils. Bioresource Tech 51: 117-123.

8. Khan A, Sarfraz M, Ahmad N, Ahmad B (1994) Effect of N dose and irrigation depth on nitrate movement in soil and N-uptake by maize. Agric. Res 32: 47-54.

9. Alvi MI (1994) Effect of different levels of NPK on growth and yield of maize. M.Sc. Thesis, Deptt, Agron, Univ Agric. Faisalabad.

10. Shapiro CA, Wortmann CS (2006) Corn response to nitrogen rate, row spacing and plant density Eastern Nebraska. Agron J 98: 529-535.

11. Abdul A, Rehman H, Khan N (2007) Maize culivar response to population density and planting date for grain and biomass yield. Sarhad J Agric 23: 25-30.

12. Steel RGD, Torrie JH (1997) Principles and procedures of statistics. A Biometrical approach (3rdedn) McGraw Hill book Co. NY. USA.

13. LeGouis J, Delebarre O, Beghin D, Heumez E, Pluchard P (1999) Nitrogen uptake and utilization efficiency of two-row and six-row winter barley cultivars grown at two N levels. Eur J Agron 10: 73-79.

14. Amanullah, Khattak RA, Khalil SK (2009) Effects of plant density and N on phenology and yield of maize. J Plant Nutr 32: 246-260.

15. Shafi MJ, Bakht S, Ali H, Khan MA, Khan, et al. (2012) Effect of planting density on phenology, growth and yield of maize. Pak J Bot 44: 691-696.

16. Bhatt PS (2012) Response of sweet corn hybrid to varying plant densities and nitrogen levels. African J of Agri Res 7: 6158-6166.

17. Shrestha J (2013) Effect of nitrogen and plant population on flowering and grain yield of winter maize. Sky J Agri Res 2: 64-68.

18. Wajid A, Ghaffar A, Maqsood M, Hussain K, Nasim W (2007) Yield response of maize hybrids to varying nitrogen rates. Pak J Agri Sci.

19. Malaviarachchi MAPWK, Karunarathneand KM, Jayawardane SN (2007) Influence of plant density on yield of hybrid maize under supplementary irrigation J Agri Sci 3: 58-66.

20. Zandi P (2012) Effect of plant density on yield new hybrids of maize in the region. Master Thesis of Agronomy, Faculty of Agriculture, Islamic Azad University (Isfahan).

21. Jasemi M, Darab F, Naser R (2013) Effect of Planting Date and Nitrogen Fertilizer Application on Grain Yield and Yield Components in Maize. American-Eurasian J Agric & Environ Sci 13: 914-919.

22. Maddoni GA, Otegui ME, Cirilo AG (2001) Plant population density, row spacing and hybrid effects on maize canopy architecture and light attenuation. Field crop Res 71: 183-193.

23. Akram M, Ashraf MY, Waraich EA, Hussain M, Hussain N (2010) Performance of autumn planted maize (Zea mays L.) hybrids at various nitrogen levels under salt affected soils. Soil & Environ. 29: 23-32.

24. Khah MN, Kheibari, Khorasani SK, Taheri G (2012) Effects of plant density and variety on some of morphological traits, yield and yield components of baby corn (Zea mays L). Int Res J of Applied and Basic Sci 3: 2009-2014.

25. Hoshang R (2012) Effect of plant density and nitrogen rates on morphological characteristics grain maize. J Basic Appl Sci Res 2: 4680-4683.

26. Rizwan M, Maqsood M, Rafiq M, Saeed M, Ali Z (2003) Maize (Zea maysL.) Response to Split application of Nitrogen. Int J Agri Biol 1560-8530.

27. Abuzar MR, Sadozai GU, Baloch MS, Baloch AA, Shah IH, et al. (2011) Effect of plant population densities on yield of maize. J Ani and Plant Sci 21: 692-695.

28. Arif M, Amin I, Jan MT, Munir I, Nawab K, et al. (2010) Effect of plant population and nitrogen levels and methods of application on ear characters and yield of maize. Pak J Bot 42: 1959-1967.

29. Radma IAM, Dagash YMI (2013) Effect of different nitrogen and weeding levels on yield of five maize cultivars under irrigation. Univ J Agric Res 1: 119-125.

30. Sharifi RS, Taghizadeh R (2009) Response of maize (Zea mays L.) cultivars to different levels of nitrogen fertilizer. J F Agric Envi 7 3: 518-521.

31. Aziz A, Rehman H, Khan N (2007) Maize cultivar response to population density and planting date for grain and biomass yield. Sarhad J Agric 23: 25-31.

32. Bahadar MM, Zaman MA, Chowdry MF, Shaidullah SM (1999) Growth and yield component responses of maize as affected by plant population. Pak J of Bio Sci 2: 1092-1095.

3

Combining Ability of Elite Highland Maize (*Zea mays* L.) Inbred Lines at Jimma Dedo, South West Ethiopia

Amare Seyoum[1]*, Dagne Wegary[2] and Sentayehu Alamerew[1]

[1]College of Agriculture and Veterinary Medicine, Jimma University, PO Box 307, Jimma, Ethiopia
[2]CIMMYT–Ethiopia, ILRI Campus, PO Box 5689, Addis Ababa, Ethiopia

Abstract

Breeding efforts to develop high yielding and improved maize varieties for highland areas (altitude of 1700-2400 m) has been recently, launched in Ethiopia. In Ethiopia, national average maize yield under farmer condition is far below attainable. Thirty-four crosses (seventeen inbred lines and two testers) along with two popular standard checks were evaluated for 17 traits in alpha lattice design at Jimma, Dedo. The objectives of this study were to evaluate top cross performance and to estimate combining abilities for grain yield and related traits. The inbred lines and crosses differ significantly for all of the studied traits except ASI, NPP and EPP. Among the crosses L5 × T1 and L16 × T2 showed higher grain yield, crosses L14 × T1 (158.13 cm), L6 × T2 (168.54 cm) expressed short plant height, crosses L4 × T1 (L16 × T1 (231.04 cm) and L14 × T1 (258.13 cm) expressed higher plant height. Crosses L5 × T1 (102.63 day) and L2 × T2 (107.95 day) displayed lowest an thesis date and crosses L5 × T1 (112.67 day), L2 × T2 (114.29 day) displayed lowest silking date. GCA mean squares due to lines were highly significant for most of the traits, while SCA mean squares were significant for some traits. The higher the percentage relative contribution of GCA sum square over SCA sum of square in all studied trait indicated the predominance of additive gene effect in controlling the inheritance of these traits. For grain yield inbred lines L5, L6, L16 and L17 were the top general combiner. Lines with positive and significant GCA effects for grain yield were generally considered as good general combiner for improvement of grain yield. L2, L5, L17 were the best general combiners for days to anthesis. L2, L5, L17, L12, L14 for days to silking and L11 for tallness, while L2, L6, L12, L13, L8, L5, L14 for shortness in plant height were the top general combiner. Future studies should explore the possibility of separating the inbred lines used in this study into distinct heterotic groups using divergent tester. By and large, the information from this study could be useful for researchers who need to develop high yielding varieties of maize adapted to the highland area of Ethiopia.

Keywords: Combining ability; Line × Tester; SCA; GCA; Maize inbred lines; Ethiopia

Introduction

Maize (*Zea mays* L.) is a diploid (2n=20) crop and one of the oldest food grains in the world. It is a member of order *Oales*, family *Poaceae*, and sub family *Panicoideae* tribe *maydeae*. It is believed that the crop is originated in Mexico and introduced to West Africa in the early 1500s by the Portuguese traders [1]. Maize is one of the most important cereal crops with a high rate of photosynthetic activity leading to high grain and biomass yield potential called C4 grain crop. It is predominantly a cross-pollinating species, a feature that has contributed to its broad morphological variability and geographical adaptability. Maize is currently produced on nearly 100 million hectares in 125 developing countries and is among the most widely grown crops in 75 of those countries [2]. Between now and 2050, the demand for maize in the developing world will double, and by 2025, maize production is expected to be highest globally, especially in the developing countries [3]. Production may not be able to meet out the demands without strong technological and policy interventions [3].

The average yield of maize in developed world is high (7.2 t ha^{-1}), the national average yield in Ethiopia is still as low as 3.2 t ha^{-1} [4] and thus, increasing maize productivity is a high national priority. The wide gap in the yield is attributed to an array of abiotic and biotic stresses, besides other factors. However, in spite of its wide adaptation and efforts made to develop improved maize technologies for different maize agro-ecological zones, still many biotic and a biotic constraint limit maize production and productivity in different maize producing area of Ethiopia [5]. However, breeding efforts to develop high yielding and improved maize varieties for highland areas (altitude of 1700-2400 m) has been recently, launched in Ethiopia, Jimma. Most of the varieties grown in south western highland part of Ethiopia are low yielding local cultivars with very tall in plant and ear height that result into root and stalk lodging, and also are late in physiological maturity and susceptibility to various foliar diseases, mainly grey leaf spot (GLS) (*Cercospora zeae maydis*), Northern corn leaf blight (NCLB) (*Exserohilum turcicum*) and common leaf rust (*Puccinia polysora*) which are the most important diseases [6]. It is estimated that the high altitude covers 20 % of land devoted annually to maize cultivation, and more than 30 % of small-scale farmers in the area depend on maize production for their livelihood [7].

Enhancement of maize production and productivity can be achieved by identifying elite parent materials, which could be used to develop high yielding varieties, and by forming broad based source population serving the breeding program. Line × tester mating design is a modified form of the top cross scheme proposed by Davis in 1927 for inbred lines evaluation [8]. Line × tester is useful in deciding the relative ability of female and male lines to produce desirable hybrid combinations. Kruvadi [9] reported Knowledge of GCA and SCA combining abilities influencing yield and its components has become increasingly important to plant breeders in the choice of suitable parents for developing potential hybrids in many crop plants.

Corresponding author: Amare Seyoum, College of Agriculture and Veterinary Medicine, Jimma University, PO Box 307, Jimma, Ethiopia
E-mail: seyoumamare99@gmail.com

Large numbers of elite highland maize inbred lines and their F1 were developed by Ethiopian high land maize improvement project. Currently, these are available at different centers such as Ambo, Holetta, Kulumsa, Adet and Jimma Agricultural Research Center. To enhance hybrid formation and open pollinated variety development information on the magnitude of combining abilities is extremely important. However, there is little or no information on the magnitude of combining abilities of the 17 inbred lines used in this study. Therefore; the objectives of this study were: To evaluate top cross performance and estimate combining abilities for grain yield and other agronomic traits of seventeen elite highland maize inbred lines using line × tester mating design.

Materials and Methods

The experimental material for the study consisted of thirty-four lines by test crosses (L1 × T1, L1 × T2, L2 × T1, L2 × T2... L17 × T1 and L17 × T2) and three standard checks. The parents were crossed in Line × Tester mating design to generate 34 F1 hybrids at Ambo Agricultural Research Center, Ethiopia at main cropping season 2012. The experiment was conducted during the main cropping season of 2013 at the Jimma Dedo on F1 (Line × Tester crosses). Early generation highland maize lines originally introduced from CIMMYT-Mexico were advanced to S_5 stage through selection at Ambo highland maize breeding program, Ethiopia. Elite lines at S_5 stage of inbreeding were crossed to a well-adapted local inbred line, F215. Line development was re-initiated from the F2 populations of these crosses, and has been advanced to S3 level of inbreeding. The inbred lines were developed from germplasm collection of CIMMYT-Mexico and were bred for resistance to various biotic and abiotic stresses. The most important stresses against which the inbred lines were selected include susceptibility to various foliar diseases, mainly grey leaf spot (GLS) (*Cercospora zeae maydis*), Northern corn leaf blight (NCLB) (*Exserohilum turcicum*) and common leaf rust (*Puccinia polysora*) All 34 test crosses were evaluated along with two most popularly grown highland maize hybrids, Wenchi and Jibat. The two checks are released by Ambo agricultural research center recently.

Experimental design and field managements

The experimental design used for the field evaluation was 12 × 3

alpha- lattice design replicated twice. Design and randomization of the trial was generated using CIMMYT's computer software known as Field book Bindiganavile et al. [10]. Spacing was 75 cm between rows and 25 cm between plants within the row. Each entry consisted of one rows of 5.25 m long. Two seeds were planted per hill to ensure uniform and enough stand and then thinning was performed at the three to five leaf stages to attain a final plant density of 53,000 plants hectar. As recommended by AARC, 100 kg DAP ha^{-1} and 75 kg Urea ha^{-1} applied at planting and additional 75 kg N hectarside dressed at 45 days after planting. Urea and dominium phosphate (DAP) used as sources of N and P_2O_5, respectively. Other crop management practices such as land preparation three times before sowing, weeding once per month and slashing in 15 days interval was applied following research recommendations for the site.

Statistical analysis and procedures

Data collection and analysis of variance (ANOVA): Data were recorded on seventeen quantitative characters. Data related to days to 50 % anthesis, 50 % days to silking, 50 % days to Maturity, 1000-kernel weight, grain yield and anthesis silking interval were recorded on the plot basis while data related to other characters were recorded on five randomly selected plants leaving border plants of each row. The mean values were subjected to line × tester analysis. Analyses of variances (ANOVA) were computed for grain yield and other agronomic traits by using SAS 9.2 software.

Combining ability analysis: Line × tester analysis was done for traits that showed statistically significant differences among the crosses using the adjusted means based on the method described by Kempthorne. General combining ability (GCA) and specific combining ability (SCA) effects for grain yield and other agronomic traits were calculated using the line × tester model. The F-test of mean square due to lines, testers and their interactions were computed against mean square due to error Significances of GCA and SCA effects of the lines and hybrids were determined by an F - test using the standard errors of GCA and SCA effects.

Results and Discussion

Analysis of variance

Analyses of variances were computed and presented in Table 1.

							Mean squares											
Source of Variation	DF	GY (t/ha)	AD (days)	SD (day)	ASI (day)	PH (cm)	EH (cm)	ED (cm)	EL (cm)	NPP (#)	EPP (#)	KPR (#)	RPE (#)	GM (%)	EA (#)	PA (#)	TKWT (gm)	MD (day)
Rep	1	10.31**	159.01**	105.13**	4.01	16.15	41.24	0.40**	0.33	34.72*	0.01	66.51**	0.11	14.05**	0.00	0.03	112.50	29.38**
Block	4	0.91	16.61	14.37	0.31	138.02	90.02	0.07*	4.14**	2.37	0.01	11.48**	0.18	5.68**	0.16	0.03	906.25	11.97*
Entry	35	3..56**	19.73**	18.62**	3.47	991.95**	451.37**	0.10**	1.85**	6.09	0.02	10.66**	1.34**	8.41**	0.24**	0.26**	5419.88**	13.46**
Crosses(Cr)	33	3.45**	19.32**	17.88**	4.17	951.28 **	383.91**	0.13**	2.38**	6.78	0.01	17.00**	1.69**	10.37**	0.24**	0.27**	5871.30**	11.68**
GCA line	16	4.39**	30.59**	29.62**	4.53	620.43**	443.12**	0.12**	2.57**	6.74	0.02	21.24**	1.09**	12.59**	0.33**	0.30**	8190.81**	13.53**
GCA tester	1	11.07**	41.31*	14.13	4.25	19111.76**	7.00**	1.60**	4.96**	31.12**	0.02	120.71**	28.99**	54.90**	0.94**	1.62**	52.94	91.77**
SCA	16	2.04**	6.68	6.38	3.81	147.11	2.00	0.04	2.04	5.31	0.01	6.28	0.58**	5.36**	0.12	0.17*	3915.44	4.84
Ck	1	1.55	0.00	1.00	1.00	3306.25	30.00	0.00	0.06	4.00	0.02	0.04	0.16**	0.20	0.06	0.25**	2025.00	56.25
Ck vs Cr	1	1.90	19.33	16.88	3.77	281.25	26.00	0.13	2.32	2.78	0.02	16.96	1.53**	10.17	0.19	0.02**	3846.00	14.57
Error	31	0.64	9.38	6.73	3.97	166.72	165.89	0.02	0.68	5.47	0.01	4.77	0.03	1.66	0.11	0.07	2487.50	4.66
% Cont. GCA		71.34	83.23	82.69	55.7	92.50 7.50	84.29	81.52	58.60	62.08	72.3	82.07 17.93	83.27	74.91	76.43	70.50	67.67 32.33	79.92
% Contr. SCA		28.66	16.77	17.31	44.3		15.71	17.78	41.40	37.92	24.7		16.73	25.09	23.57	29.50		20.08

**=Significant at P<0.01 level of probability; *=Significant at P<0.05 Level of probability; DF=Degrees of freedom; Rep= Replication; GY= Grain yield; AD=Number of days to anthesis; ED=Ear diameter; EH= Ear height; EL=Ear length; EPP= Number of ears per plant; NP=Number of plant per plot; KPR=Number of kernels per row; PH=Plant height; KRPE=Number of rows per ear; SD=Number of days to silking; TKWT=Thousand kernels weight; ASI=Anthesis silking interval; GM=Grain moisture; MD=Maturity date; PA=Plant aspect and EA=Ear aspect

Table 1: Analysis of variance for grain yield and other agronomic traits of line by tester crosses involving 17 lines and 2 testers evaluated at Jimma Dedo, south west Ethiopia in 2013 cropping season.

Highly significant differences *(P<0.01)* were obtained among the genotypes (entries) and crosses for all traits except ASI, NPP and EPP. In addition, mean squares due to checks and check vs. cross were highly significant (P<0.01) for RPE and PA. The rest of the traits had non-significant. Significant differences observed among the entries (genotypes) for most of the traits studied, indicating the presence of genetic variation among the materials, which makes possible for improvement of the traits. In consistence with this finding Dagne et al.[11], Teshale [12], Jemal [13], Amiruzzaman et al.[14], Hadji [15], Gudeta [16], Alamenesh [17] reported the presence of significant differences among genotypes for grain yield and other traits in different sets of maize parental inbred lines.

Mean performance of F1 crosses (Line × Tester)

The overall mean performance of the 36 entries i.e., thirty-four crosses and two standard check evaluated for grain yield and related agronomic traits presented in Table 2. In current study, the overall mean grain yields (GY) of the entries were 5.61 t/ha ranging from 3.46 t/ha to 8.36 t/ha. Cross L5 × T1 (8.36 t/ha) and WENCHI (8.02 t/ha) expressed higher grain yield, while cross L1 × T2 (3.46 t/ha) and L7 × T2 (3.76 t/ha) showed lower GY. Overall mean for number of kernels per row (KPR) were 31.86 ranging from 26.86 to 36.80. Cross L15 × T1 (36.80), cross L1 × T1 (36.15) and L14 × T1 (35.88) displayed higher number of kernels per row. Cross L11 × T2 (26.68), cross L10 × T2 (27.75) and L16 × T2 (28.05) displayed lower number of KPR. Thousand kernels weight (TKW) ranging from 267.08 g cross (L2 × T1) to 492.08 g cross (L10 × T1) gm with overall mean of 387.71. Cross L10 × T1 (492.8 g), cross L13 × T1 (477.08 g) and cross L13 × T2 (452.08 g) expressed higher

TKW, while cross L2 × T1 (267.08 g) and cross L7 × T1 (295.83 g) displayed lower TKW (Tables 2-5). In agreement with this study Dagne et al. [18], Zerihun [19], Alemenesh [17] in their studies reported that experimental varieties showed better performance than the best check for most of yield and other traits.

The overall mean values of number kernels per ear (KRPE) were 11.46 ranging from 10.03 to 13.07. Cross L3 × T1 (13.07), cross L5 × T1 (12.85) and WENCHI (12.80) had higher number of kernels per row, while cross L7 × T1 (10.03), cross L4 × T2 (10.05) and cross L8 × T2 (10.13) showed lower number of KRPE. Ear diameter (ED) were ranged from 3.47 cross (L4 × T2) to 4.55 cross (L10 × T1) cm, with overall mean of 4.07 cm. Cross L10 × T1 (4.55) cm, cross L16 × T1 (4.43) cm and cross L1 × T1 (4.38) cm were displayed higher ED, while cross L4 × T2 (3.47) cm and cross L7 × T2 (3.59) cm expressed lower ED. Mean value for ear length (EL) were 16.33 cm ranged from 14.19 cm cross (L7 × T2) to 17.85 cm WENCHI. For ear length standard check (WENCHI) displayed (17.85) cm (Tables 6-8). In agreement with this study Dagne et al. [18], Zerihun [19], Alemenesh [17] in their studies reported that experimental varieties showed better performance than the best check for most of yield and other traits.

Cross L14 × T1 (258.13) cm, cross L16 × T1 (231.01) cm and cross L10 × T1 (225.00) cm were showed higher plant height while cross L14 × T1 (158.13) cm, WENCHI (168.13) cm and cross L6 × T2 (168.54) cm expressed lower PH. Ear height (EH) ranged from 70.00 standard check (WENCHI) to 135.83 (L13 × T1) with a mean height of 101.72 cm. Cross L13 × T1 (135.83), cross L14 × T1 (134.79) and cross L16 × T1 (127.50) cm expressed higher ear height, while WENCHI (70.00), cross L7 × T2 (73.13) cm and cross L2 ×T2 (80.83) cm expressed lower

Genotypes	GY (t/ha)	AD (days)	SD (day)	ASI (day)	PH (cm)	EH (cm)	ED (cm)	EL (cm)	NPP (#)	EPP (#)	KPR (#)	KPE (#)	GM (%)	EA (#)	PA (#)	TKW (gm)	MD (day)
L1 × T2	3.46	111.37	117.54	6.13	177.08	96.67	3.76	14.53	18.46	1.2	29.68	11.13	16.28	3.00	1.87	322.08	218.42
L2 × T2	5.05	107.95	114.29	6.33	184.37	80.83	4.03	16.84	20.54	1.2	32.88	11.38	19.65	2.46	1.92	372.08	220.29
L3 × T2	4.46	117.71	124.17	6.42	203.13	114.58	4.08	15.77	17.63	1.1	31.54	11.87	14.74	2.25	2.13	327.08	219.58
L4 × T2	6.31	111.13	117.67	6.54	175.83	110.42	3.47	16.02	19.37	1.2	28.22	10.05	17.38	2.71	2.16	432.08	221.62
L5 × T2	4.97	112.67	117.67	5.04	181.04	92.50	4.18	16.86	16.00	1.3	32.24	11.88	15.77	2.04	2.37	420.08	220.29
L6 × T2	5.97	118.67	124.67	6.04	168.54	100.00	3.88	15.91	20.50	1.1	32.94	11.08	21.52	2.04	1.75	410.83	221.29
L7 × T2	3.76	115.08	120.54	5.33	186.25	73.13	3.59	14.19	14.17	0.9	28.38	10.03	16.64	2.87	1.63	317.08	218.75
L8 × T2	3.77	111.37	117.54	6.13	169.58	84.17	3.83	16.29	15.46	1.1	31.58	10.13	17.78	3.00	2.29	392.08	220.42
L9 × T2	5.43	113.13	120.67	7.54	185.83	90.42	3.68	17.27	17.37	1.3	30.82	10.25	19.24	1.71	2.37	367.08	224.63
L10 × T2	5.75	117.58	122.54	4.83	178.75	88.13	4.24	16.55	15.17	1.1	27.75	11.63	22.89	2.37	2.00	362.08	219.75
L11 × T2	5.05	117.46	124.29	6.83	186.87	88.33	3.88	15.84	17.04	1.3	26.68	10.58	18.45	2.21	2.25	312.08	221.29
L12 × T2	3.87	111.58	117.04	5.33	171.25	83.13	3.95	16.55	16.17	1.3	31.75	10.63	20.59	2.37	1.87	447.08	217.75
L13 × T2	4.19	118.21	123.17	4.92	158.13	87.08	4.04	16.92	16.63	1.2	30.14	10.47	18.09	2.75	1.92	452.08	217.08
L14 × T2	5.38	111.13	117.17	6.04	180.83	90.42	3.78	17.72	19.87	1.1	32.57	10.25	19.94	1.96	2.13	427.08	220.62
L15 × T2	4.04	114.96	120.29	5.33	184.37	98.33	4.01	17.29	18.04	1.4	33.68	10.38	15.89	1.96	2.16	387.08	218.29
L16 × T2	7.89	114.08	119.04	4.83	191.25	95.63	4.17	14.95	19.17	1.3	28.05	10.23	21.04	2.13	2.37	417.08	225.25
L17 × T2	7.49	111.21	116.17	4.92	193.13	104.58	4.18	16.07	20.63	1.2	34.24	10.87	21.69	2.25	1.75	407.08	224.58
L1 × T1	5.99	112.04	119.91	8.04	216.67	110.21	4.38	15.56	19.87	1.2	30.66	11.83	18.37	2.37	1.63	425.83	223.83
L2 ×T1	5.98	110.33	116.41	5.92	217.50	91.87	3.79	16.82	17.25	1.3	33.52	12.72	19.44	2.67	2.29	267.08	224.04
L3 × T1	6.58	114.71	119.67	4.92	213.13	127.08	4.30	15.57	20.13	1.0	34.74	13.07	20.34	2.00	2.37	367.08	220.08
L4 × T1	5.05	111.54	121.41	10.04	219.17	127.08	4.23	16.60	17.37	1.0	29.26	12.43	19.77	2.13	2.00	385.83	222.83
L5 × T1	8.36	102.63	112.67	10.04	203.33	105.42	4.33	14.62	21.37	1.3	30.12	12.85	20.28	1.71	2.25	377.08	225.13
L6 × T1	7.74	115.37	123.04	7.63	192.08	99.17	4.10	15.75	19.56	1.2	33.48	11.33	20.18	2.00	1.87	342.08	222.42
L7 × T1	4.36	113.67	121.17	5.54	208.54	102.50	4.15	14.67	17.50	1.1	32.14	11.48	19.32	2.54	1.92	295.83	220.29
L8 × T1	6.47	112.33	117.42	4.42	205.00	101.87	4.13	17.62	20.75	1.2	32.92	12.32	19.44	2.17	2.13	412.08	219.54

GY=Grain yield; AD=Number of days to anthesis; ED=Ear diameter; EH=Ear height; EL=Ear length; EPP=Number of ears per plant; NPP=Number of plant per plot; KPR=Number of kernels per row; PH=Plant height; KRPE=Number of rows per ear; SD=Number of days to silking; TKWT=Thousand kernels weight; ASI=Anthesis silking interval; GM=Grain moisture; MD=Maturity date; PA=Plant aspect and EA=Ear aspect

Table 2: (Continued) Estimates of mean values for grain yield and related traits at Jimma Dedo, south west Ethiopia during 2013 cropping season.

Line	GY (t/ha)	AD (days)	SD (day)	PH (cm)	EH (cm)	ED (cm)	EL (cm)	NPP (#)	KPR (#)	KRPE (#)	GM (%)	EA (#)	PA (#)	TKWT (gm)	MD (day)
L1	-0.78**	-1.10	-0.28	3.08	4.03	-0.02	-0.86**	0.93	-1.15*	0.12	-2.71**	0.53**	0.37**	-13.97	-0.84
L2	0.00	-4.35**	-4.28**	4.34	-10.97**	-0.18**	0.49	0.43	1.45*	0.62**	0.19	0.28**	0.24**	-71.47**	0.91
L3	0.01	2.89**	2.47**	18.08**	22.78**	0.02	-0.13	0.43	2.25**	1.02**	-2.33**	-0.09	-0.26**	-51.47**	-1.08
L4	0.17	-1.60	0.47	0.58	-13.47**	-0.22**	-0.11	0.17	-3.45**	-0.08	-1.48**	0.28**	-0.26**	26.03*	-0.34
L5	1.15**	-5.35**	-4.03**	-8.41*	0.28	0.24**	-0.71*	0.43	-1.00	1.02**	-1.48**	-0.35**	-0.38**	18.53	0.41
L6	1.34**	4.15**	4.72**	-16.91**	-1.97	-0.03	-0.11	1.93**	1.89**	-0.18	1.37**	-0.22*	0.12	-8.97	0.16
L7	-1.45**	0.39	0.72	-4.41	-10.97**	-0.12**	-1.96**	-2.32**	-1.65**	-0.58**	-0.91*	0.40**	0.37**	-73.97**	-2.34**
L8	-0.39	-0.60	-1.28	-8.16*	-7.22*	-0.07	1.04**	-0.57	1.09	-0.28*	-0.78	0.28**	0.12	8.53	-0.84
L9	-0.00	0.65	0.97	-1.66	0.28	-0.13**	0.36	-0.57	-0.65	-0.28*	-1.11**	-0.35**	-0.13	3.53	1.91**
L10	0.19	3.65**	1.22	1.84	4.03	0.37**	0.54	-2.57**	-3.05**	0.32*	2.64**	-0.22*	-0.26**	38.53**	-1.84**
L11	-0.87**	3.65**	4.72**	6.84*	0.28	-0.22**	-0.96**	-1.07	-4.05**	0.32*	-0.71	0.03	0.49**	-78.97**	0.16
L12	-0.98**	-1.35	-2.03*	-13.16**	-5.97**	-0.04	0.29	-0.57	0.89	-0.48**	2.42**	-0.09	0.12	58.53**	-1.34**
L13	-1.31**	1.39	0.72	-16.91**	-9.72**	-0.03	0.36	-1.07	-0.50	-0.88**	0.49	0.15	0.24**	71.03**	-1.34**
L14	-0.55*	-1.60	-2.03*	-8.16*	-5.97**	-0.07	0.51	1.17*	1.92**	-0.28*	1.07**	-0.35**	-0.26**	8.53	-0.84
L15	-0.41*	2.15**	2.47**	1.84	1.53	0.02	1.01**	0.17	3.05**	-0.08	-1.48**	-0.35**	-0.13	6.03	-0.84
L16	2.07**	-0.35	-1.03	9.34**	12.78**	0.31**	-0.58*	1.17*	-0.90	-0.28*	2.54**	-0.09	-0.26**	36.03**	4.41**
L17	1.81**	-2.60**	-3.53**	31.84**	20.28**	0.14**	0.82**	1.93**	3.87**	0.02	2.29**	0.15	-0.13	23.53	3.66**
SE	0.19	0.81	0.70	3.08	2.87	0.04	0.26	0.55	0.54	0.12	0.37	0.08	0.07	12.06	0.60
SED	0.27	1.14	0.99	4.36	4.06	0.06	0.37	0.78	0.77	0.17	0.53	0.12	0.10	17.06	0.85

**=Significant at P<0.01 level of probability; *=Significant at P<0.05 Level of Probability; GY= Grain yield; AD=Number of days to anthesis; ED=Ear diameter; EH=Ear height; EL=Ear length; NPP=Number of plant per plot; KPR=Number of kernels Per row; PH=Plant height; KRPE=Number of rows per ear; SD=Number of days to silking; TKWT=Thousand kernels weight; GM=Grain moisture; MD=Maturity date; PA=Plant aspect and EA=Ear aspect

Table 3: Estimates of General combining ability effects (GCA) for line by tester crosses of maize inbred lines evaluated at Jimma, Dedo 2013.

Cross	GY (t/ha)	KRPE (#)	GM (%)	PA (#)
L1 × T1	0.86	-0.15	0.13	0.03
L1 × T2	-0.86	0.15	-0.13	-0.03
L2 × T1	0.06	-0.05	-0.42	-0.09
L2 × T2	-0.06	0.05	0.42	0.09
L3 × T1	0.66	-0.05	1.90*	-0.09
L3 × T2	-0.66	0.05	-1.90*	0.09
L4 × T1	-1.03	-0.65*	0.30	-0.59**
L4 × T2	1.03	0.65*	-0.30	0.59**
L5 × T1	1.29	-0.25	0.80	-0.22
L5 × T2	-1.29	0.25	-0.80	0.22
L6 × T1	0.48	-0.65*	-2.09**	0.03
L6 × T2	-0.48	0.65*	2 09**	-0.03
L7 × T1	-0.10	0.15	0.37	0.28
L7 × T2	0.10	-0.15	-0.37	-0.28
L8 × T1	0.94	0.45	0.55	0.03
L8 × T2	-0.94	-0.45	-0.55	-0.03
L9 × T1	-0.32	0.25	-1.17	-0.22
L9 × T2	0.32	-0.25	1.17	0.22
L10 × T1	-0.45	-0.55*	-2.32**	0.15
L10 × T2	0.45	0.55*	2.32**	-0.15
L11 × T1	-0.82	0.45	-0.12	0.15
L11 × T2	0.82	-0.45	0.12	-0.15
L12 × T1	0.25	-0.35	-0.25	0.03
L12 × T2	-0.25	0.35	0.25	-0.03
L13 × T1	-0.38	-0.55*	1.37	0.15
L13 × T2	0.38	0.55*	-1.37	-0.15
L14 × T1	-0.82	0.25	0.30	0.15
L14 × T2	0.82	-0.25	-0.30	-0.15
L15 × T1	0.67	0.25	1.65*	0.03
L15 × T2	-0.67	-0.25	-1.65*	-0.03
L16 × T1	0.72	0.25	-0.57	0.15
L16 × T2	-0.71	-0.25	0.57	-0.15
L17 × T1	-0.57	-0.05	-0.42	0.03
L17 × T1	0.57	0.05	0.42	-0.03
SE	0.75	0.24	0.75	0.14
SED	1.06	0.35	1.06	0.20

**=Significant at P<0.01 level of probability; *=Significant at P<0.05 Level of probability; KRPE=Number of rows per ear; GM=Grain moisture; MD=Maturity date and PA=Plant aspect

Table 4: Estimates of specific combining ability effects (SCA) for line by tester crosses of maize inbred lines evaluated at Jimma, Dedo 2012/13.

GY (t/ha)		TKW (gm)		KPE(#)		KPR(#)		ED(cm)		EL(cm)	
Bottom five mean values from smallest to largest											
L1 × T2	3.46	L2 × T1	267.08	L7 × T2	10.03	L11 × T2	26.68	L4 × T2	3.47	L7 × T2	14.19
L7 × T2	3.76	L7 × T1	295.83	L4 × T2	10.05	L10 × T2	27.75	L7 × T2	3.59	L1 × T2	14.53
L8 × T2	3.77	L11 × T1	300.83	L8 × T2	10.13	L16 × T2	28.05	L9 × T2	3.68	L5 × T1	14.62
L12 × T2	3.87	L11 × T2	312.08	L16 × T2	10.23	L4 × T2	28.22	L1 × T2	3.76	L7 × T1	14.67
L15 × T2	4.04	L7 × T2	317.08	L9 × T2	10.25	L7 × T2	28.38	L14 × T2	3.78	L11 × T1	14.90
Top five mean values from smallest to largest											
L17 × T2	7.49	L12 × T2	447.08	L11 × T1	12.63	L17 × T2	34.24	L3 × T1	4.30	L15 × T1	17.57
L6 × T1	7.74	L17 × T2	447.08	L2 × T1	12.72	L3 × T1	34.74	L5 × T1	4.33	JIBAT	17.60
L16 × T2	7.89	L13 × T2	452.08	WENCHI	12.80	L14 × T1	35.88	L1 × T1	4.38	L8 × T1	17.62
WENCHI	8.02	L13 × T1	477.08	L5 × T1	12.85	L1 × T1	36.15	L16 × T1	4.43	L14 × T2	17.72
L5 × T1	8.36	L10 × T1	492.08	L3 × T1	13.07	L15 × T1	36.80	L10 × T1	4.55	WENCHI	17.85
Mean	5.61		387.71		11.46		31.86		4.07		16.33
LSD (5 %)	0.96		59.58		0.57		2.61		0.20		0.98
CV	14.35		12.87		4.17		6.86		3.93		5.07

Table 5: Top and bottom five mean values of genotypes for GY, TKW, KPE, KPR, ED and EL location Jimma, Dedo 2012/13.

PH(cm)		EH(cm)		PA (1-5)		EA (1-5)		NPP(#)		EPP(#)	
Bottom five mean values from smallest to largest											
L14 × T1	158.13	WENCHI	70.00	L4 × T1	1.54	L15 × T1	1.63	L7 × T2	14.17	L3 × T1	0.90
WENCHI	168.13	L7 × T2	73.13	L5 × T1	1.68		1.71	L10 × T2 & WENCHI	15.17	L4 × T1	1.00
L6 × T2	168.54	L2 × T2	80.83	L3 × T1	2.02	L5 × T1	1.71	L8 × T2	15.46	L10 × T2	1.00
L8 × T2	169.58	L10 × T2	83.13	L9 × T1	2.04	L14 × T1	1.75	L5 × T2	16.00	L12 × T1	1.00
L12 × T2	171.25	L8 × T2	84.17	JIBAT	2.04	L9 × T1	1.87	L12 × T2	16.17	L15 × T1	1.00
Top five mean values from smallest to largest											
L4 × T1	219.17	L3 × T1	127.08	L2 × T2	2.94	L4 × T2	2.71	L6 × T2	20.50	L5 × T2	1.30
JIBAT	221.67	L4 × T1	127.08	L4 × T2	2.94	L13 × T2	2.75	L2 × T2	20.54	L9 × T2	130
L10 × T1	225.00	L16 × T1	127.50	L1 × T2	3.02	L7 × T2	2.87	L17 × T2	20.63	L13 × T1	1.30
L16 × T1	231.04	L14 × T1	134.79	L7 × T1	3.04	L1 × T2	3.00	L8 × T1	20.75	L14 × T1	1.30
L14 × T1	258.13	L13 × T1	135.83	L11 × T1	3.04	L8 × T2	3.00	L5 × T1	21.37	L15 × T2	1.40
Mean	197.02		101.72		2.50		2.23		17.99		1.17
LSD (5%)	0.17		0.20		0.34		0.39		2.79		14.1
CV	6.28		17.39		11.27		15.13		12.76		15.38

Table 6: Top and bottom five mean values of genotypes for PH, EH, PA, EA, NPPP and NPPE location Jimma, Dedo2012/13.

AD (day)		SD (day)		ASI (day)		MD (day)		GM (%)	
Bottom five mean values from smallest to largest									
L5 × T1	102.63	L5 × T1	112.67	L10 × T1	2.42	L13 × T2	217.08	L3 × T2	14.74
L2 × T2	107.95	L2 × T2	114.29	L8 × T1	4.42	L12 × T2	217.75	L5 × T2	15.77
L14 × T1	108.87	L14 × T1	114.54	L10 × T2	4.83	L15 × T2	218.29	L15 × T2	15.89
L14 × T2	111.13	L14 × T2	114.79	L16 × T2	4.83	L1 × T2	218.42	L1 × T2	16.28
WENCHI	111.00	WENCHI	115.29	L17 × T2	4.92	L10 × T1	218.54	L7 × T2	16.64
Top five mean values from smallest to largest									
L11 × T1	117.54	L13 × T2	123.17	L6 × T1	7.63	JIBAT	225.00	L12 × T1	21.84
L10 × T2	117.58	L3 × T2	124.17	L15 × T1	7.75	L5 × T1	225.13	L14 × T1	22.28
L3 × T2	117.71	L11 × T2	124.29	L1 × T1	8.04	L16 × T2	225.25	L13 × T1	22.69
L13 × T2	118.21	L6 × T2	124.67	L4 × T1	10.04	L14 × T1	226.67	L10 × T2	22.89
L6 × T2	118.67	L11 × T1	124.91	L5 × T1	10.04	L16 × T1	227.29	L14 × T1	22.97
Mean	112.97		119.12		6.06		221.76		19.58
LSD (5%)	3.67		3.11		2.38		2.59		1.54
CV	2.71		2.18		32.85		0.97		6.85

Table 7: Top and bottom five mean values of genotypes for AD, SD, ASI, MD, and GM location Jimma, Dedo 2012/13.

EH. The overall mean values for plant aspect (PA) of entries were 2.50 % ranging from 1.54 % to 3.04 %. Cross L11 × T1 and cross L7 × T1 (3.04) % and cross L1 × T2 (3.02) % showed higher % of plant aspect, while cross L4 × T1 (1.54 %) and cross L5 × T1 (1.68.54) % showed lower PA. The overall mean for ear aspect (EA) of the entries were 2.23

% ranging from 1.63 % to 3.00 %. Cross L8 × T2 and L1 × T2 (3.00) % and cross L7 × T21 (2.87) %) expressed higher % of ear aspect while cross L15 × T1 (1.63) % and cross L9 × T2 (1.71) % showed lower EA. Mean value for number of plant per plot (NPP) were 17.99 ranged from 14.17 (L7 × T1) to 21.37 (L5 × T1). Cross L5 × T1 (21.37), cross L8 × T1

Value	GY (t/ha)	AD (days)	SD (day)	ASI (day)	PH (cm)	EH (cm)	ED (cm)	EL (cm)	NPP (#)	EPP (#)	KPR (#)	KPE (#)	GM (%)	EA (#)	PA (#)	TKWT (gm)	MD (day)
Location Jimma, Dedo																	
Cr mean	5.51	113.09	119.27	6.09	197.06	101.97	4.07	16.25	18.34	1.17	31.77	11.39	19.60	2.24	2.20	386.84	221.77
Ck mean	7.24	111.11	116.61	5.65	194.90	99.00	4.16	17.43	18.04	1.17	33.23	12.46	19.67	2.00	2.29	395.83	223.00
Mean	5.61	112.97	119.12	6.06	197.02	101.72	4.07	16.33	17.99	1.17	31.86	11.46	19.58	2.23	2.50	387.71	221.76
Min	3.46	102.63	114.29	2.42	158.13	72.79	3.47	14.19	14.17	0.90	27.75	10.13	14.74	1.63	1.54	267.08	217.08
Max	8.36	118.67	124.91	10.04	258.13	135.83	4.55	17.72	21.37	1.40	36.80	13.07	22.97	3.00	3.04	492.08	227.29
LSD (5%)	0.96	3.67	3.11	2.38	0.17	0.20	0.20	0.98	2.79	0.30	2.61	0.57	1.54	0.39	0.34	59.58	2.59
CV	14.35	2.71	2.17	32.85	6.28	17.39	3.93	5.07	12.76	14.1	6.86	4.17	6.58	15.13	11.27	12.87	0.97

Table 8: Mean, maximum and minimum values of entries (genotypes) for grain yield and yield related traits Jimma, Dedo 2012/2013.

(20.75) and cross L17 × T2 (20.63) expressed higher number of plants per plot, while cross L7 × T2 (14.17), cross L10 × T2 and WENCHI (15.17) showed lower NPP. Mean value for EPP were 1.17 ranged from 0.90 cross (L3 × T1) to 1.40 cross (L15 × T2). Days to anthesis (AD) were ranged between 102.63 cross (L5 × T1) and 118.67 cross (L6 × T2) with overall mean of 112.97. Cross L6 × T2 (118.67), cross L13 × T2 (118.21) and cross L3 × T2 (117.71) showed higher anthesis date while cross L5 × T1 (102.63), cross L2 × T2 (107.95) and L14 × T1 (168.87) showed lower AD (Tables 6-8). In agreement with this Dagne et al. [18], Zerihun [19], Alemenesh [17] in their studies reported that experimental varieties showed better performance than the best check for most of yield and other traits.

Overall, mean values of days to silking (SD) of the entries (genotypes) were 119.12 with a range of 112.67 cross (L5 × T1) to 124.91 cross (L11 × T1). Cross L11 × T1 (124.91), cross L6 × T2 (124.67) and cross L11 × T2 (124.29) expressed higher silking date while cross L5 × T1 (112.67), cross L2 × T2 (114.29) and cross L14 × T1 (114.54) displayed lower SD. Overall mean values of days to maturity (MD) of the entries (genotypes) were 221.76 with a range 217.08 (L13 × T2) to 227.29 (L16 × T1). Cross L16 × T1 (227.29), cross L14 × T1 (226.67) and JIBAT (226.33) showed higher record for days to maturity while cross L13 × T2 (217.08), cross L12 × T2 (217.75) and cross L15 × T2 (218.29) showed lower record for days to maturity. Grain moisture at harvest (GM) ranged between 14.74 % (L3 × T2) and 22.97 % (L14 × T1) with over all mean of 19.58%. Cross L14 × T1 (22.97) %, cross L10 × T2 (22.89) % and cross L13 × T1 (22.69) % expressed higher grain moisture at harvest, while cross L3 × T2 (14.74) %, cross L5 v T2 (15.77) and cross L15 × T2 (15.87) displayed lower grain moisture.

A number of crosses showed better performances for more than one trait. Therefore, crosses that had high grain yield could be used in further across location breeding program to improve grain yield and other traits of interest. Hence, a hybrid expressed earlier in anthesis and silking, medium in ear and plant heights, better performance in ear and plant aspect could be used as sources of genes for development of high yielder, early maturing varieties. This study gives a clue for highland maize breeding program to design appropriate breeding strategies. In agreement with this study Dagne et al. [18], Zerihun [19], Alemenesh [17] in their studies reported that experimental varieties showed better performance than the best check for most of yield and other traits.

Combining ability analysis

In this study the contribution of general combining ability variance was much greater than those of specific combining ability variance for all the characters studied. The higher percentage relative contribution of GCA sum of square over SCA sum of square showed the predominance role of additive gene action over non-additive action in the inheritance of all traits studied. For the evaluation of an inbred line in the production

of hybrid maize two factors are considered i.e., characteristics of the line itself and behavior of the line in a particular hybrid combination. As revealed from the present study results, the inbred lines displayed superior performance in their GCA effects especially for grain yield and other prominent traits contributing towards grain yield i.e., AD, SD, PH, EH, ED, EL, KPR, RPE, EA, PA and TKWT.

Grain yield (GY)

Line by tester analysis of variance (ANOVA) of combining ability for grain yield and related traits is given in Table 1. Line GCA, tester GCA and SCA means squares were highly significant ($p < 0.01$) for grain yield. Significant GCA and SCA mean square indicated the importance of both additive and non-additive gene actions in governing grain yield. In agreement with the present study, Hadji found highly significant mean squares due to GCA and SCA for grain yield in diallel study of quality protein maize inbred lines. In addition, both the role of additive and non-additive gene actions in governing grain yield in maize was reported by other works Mandefro and Habtamu [20], Dagne et al. [11], Demissew et al. [21] On the other hand, Bayisa [22] found non-significant GCA effects for grain yield in line × tester study of transition highland inbred lines at Kulumsa [16] carried out line × tester analysis of QPM versions of early generation highland maize inbred lines and reported significant GCA mean squares due to lines at Holeta and Kulumsa but non-significant GCA mean squares at Ambo and Haramaya.

Thousand kernels weight (TKWT)

Mean squares due to line GCA for thousand kernels weight were highly significant ($P < 0.01$) Table 1. In this study the result observed indicates the predominant role of additive gene effect in governing this trait. However, tester GCA and SCA mean square were no significant for the same trait. In agreement with this study, reported non-significant SCA mean squares for TKWT. In addition, Joshi et al. [23] observed importance of additive genetic variance for this trait.

Anthesis and silking days (AD and SD)

For number of days to anthesis (AD) and silking (SD), mean squares due to line GCA were highly significant($p<0.01$). Mean squares due to tester GCA for number of days to anthesis (AD) were significant ($p<0.05$) (Table 1). In line with this finding, Gudeta [16] reported significant GCA effects due to testers at Ambo. The predominance effect of GCA mean of squares over SCA mean squares for these traits indicates the relative importance of additive gene action to non-additive gene action for these traits. Similar to this study Shewangizaw [24], Leta et al. [25] reported the highest contribution of GCA than SCA for days to silking. In line with this study, Ahmad and Saleem [26] reported the preponderance of additive gene action in the inheritance of days to anthesis and days to silking. Also Legesse et al. [27] reported the predominance role of additive gene action in inheritance of days to silking.

Kernel rows per ear (KRPE)

For number of kernel rows per ear mean squares due to line GCA and tester GCA and SCA were highly significant *(p<0.01)* (Table 1). In this study, the effects of additive and non additive gene action for the inheritance of KRPE were identified. In agreement with this study, Hadji [15], Dagne et al. [11] reported both additive and non-additive were important for this trait. In disagreement with this finding Petrovic [28], Mathur et al. [29] observed a significant GCA effect and the predominance role of additive gene action for KRPE. On the other hand, Pal and Prodhan [30], Dehghanapour et al. [31], Kumar et al. [32] reported the more importance of non-additive gene effects in the inheritance of this trait.

Kernels per row (KPR)

For number of kernels per row mean squares due to line GCA and tester GCA were highly significant *(p<0.01)* (Table 1). Significant GCA mean square were implied the importance of additive gene action in controlling the inheritance of KPR. In line with this study, Dagne [33], Gudeta [16] reported the importance of additive gene action for controlling (number of kernels per row) in maize. In addition, Mathur et al. [29] observed the predominance of additive genetic in the inheritance of this trait unlike Dehghanapour et al. [31] was reported the more importance of non-additive component effects for this trait.

Ear length and ear diameter (EL and ED)

Line GCA and tester GCA mean squares were highly significantly *(P < 0.01)* for ear diameter and ear length found (Table 1). In this finding the predominant role of additive gene effect in the inheritance of both EL and ED were observed. Similarly, Mandefro [34] reported no importance of non- additive gene action for ear length. As opposite to this result Dagne [33], Hadji [15], Gudeta [16] observed the importance of both additive and non-additive gene effects in the inheritance of ear diameter.

Plant height and ear height (PH and EH)

Combining ability analysis revealed highly significant *(p<0.01)* mean squares due to line GCA and tester GCA effects of lines for plant and ear height (Table 1) while, SCA mean squares were not significant for both traits. Plant height and ear height are important morphological traits affecting the final yield of maize crop. Extremely dwarf varieties have the problem of crowded canopy, aeration and transmission of sun light to the lower parts resulting in drastic reduction in yield while the high stature plants are highly susceptible to lodging. Greater ear height is undesirable because the ear placement at a greater height from the ground level exerts pressure on plant during grain filling and physiological maturity and causes lodging, which could ultimately affect the final yield. In this study the result obtained indicated the predominance role of additive gene effect in controlling the inheritance of both PH and EH. In line with these findings, Leta et al. [25] found significant GCA effect and non-significant SCA effect for plant- and-ear height. On the other hand, Gudeta [16] reported significant GCA and non-significant SCA mean squares for plant height.

Plant aspect and ear aspect (PA and EA)

Mean squares due to line GCA and tester GCA were highly significant *(p<0.01)* for both plant and ear aspect (Table 1) while, SCA mean square was significant *(p<0.05)* for plant aspect only. In this study

only the relevance of additive gene effect in controlling the inheritance of EA was observed. The role of additive and non additive gene action showed the presence of variation among lines and crosses. Nevertheless, Significant GCA and SCA mean squares implied the importance of additive and non additive gene action in controlling PA in maize. Based on the view of the farmers at Jimma, Dedo the farm wants maize hybrid having good physical appearance such as medium height, strong and thick stalk, upward leave branching, free and resistance to wards any pest and disease, variety having good husk cover, good kernel per ear and row and variety having two ear per plant.

Grain moisture at harvest (GM)

For grain moisture at harvest mean squares due to line GCA, tester GCA and SCA were highly significant *(p<0.01)* (Table 1). In this study the importance of both additive and non-additive gene action were found in controlling grain moisture at harvest. Therefore, the result obtained indicated the improvement of this trait through exploitation of both additive and non-additive gene action. In contrary to this study, Saad et al. [35] in his finding observed highly significant GCA mean squares for trait grain moisture while SCA mean square were not significant for grain moisture at harvest and reported the importance of both of additive and non additive gene action.

Number of plant per plot (NPP)

For trait number of plant per plot mean square due to tester GCA were highly significant (p<0.01) (Table 1) while, line GCA and SCA mean square were not significant. In the current finding the role of additive gene action in controlling the inheritance of NPP was indicated. In contrary to this finding Alemnesh [17] found no significant mean square due to tester GCA.

Days to maturity (DM)

For maturity date mean squares due to line GCA were significant at *(p<0.05)* and tester GCA were highly significant at *(p<0.01)* while, non significant SCA was observed (Table 1). In this study additive gene actions are important in governing this trait. Similarly, the predominant roles of additive gene effect in controlling maturity dates were reported by Dagne [33], Hadji [15] Absence of significant SCA mean squares at this location makes the current finding similar with the previous work of Mandefro [34], Jemal [13] in which non-significant SCA effect for days to maturity was reported. Also, Legesse et al. [27] reported highly significant (P < 0.01) line GCA mean squares for days to maturity.

Conclusion and Recommendation

The present study consisted of 34 entries(crosses) along with two popular standard check were evaluated at Jimma, Dedo Ethiopia during the 2013 cropping season with the objectives of evaluating top cross performance and estimating combining abilities for 17 characters. The analysis of variance showed highly significantly (p<0.01) differences for all the characters except for ASI, NPP and EPP. Further, significant differences were not recorded among the checks and checks vs crosses for most traits. Line GCA means squares were highly significant for the studied traits except ASI, NPP and EPP. Testers GCA mean squares were significant for most of studied traits except SD, ASI, EPP and TKWT. SCA mean squares were significant mainly for GY, RPE, GM and PA. Significant GCA mean squares for all traits indicated the predominant role of additive gene actions in determining the inheritance of these traits. Generally, GCA sum of squares component was greater than SCA sum of squares for all of the studied traits, suggesting that variations among crosses were mainly due to additive rather than non-additive gene effects; and hence, selection would be effective in improving grain yield and other agronomic traits.

Based on GCA analysis L5, L6, L16 and L17 were the top general combiners for grain yield and these inbred lines can be used for variety development in the future highland maize improvement program. Inbred lines L2, L5, and L17 were the best general combiners for days to anthesis and silking, respectively, indicating these lines had favorable allele frequency for earliness and could be used to develop early maturing varieties. Inbred lines L2, L3, L5, L10 and L11 and L6, L14, L15 and L17 were the best general combiners for number of rows per ear and number of kernel per row, respectively. These lines had favorable allele are to improve RPE and KPR to enhance grain yield. For ear diameter L5, L10, L16, and L17 were good general combiner, indicating these lines had the tendency to increase ear diameter. For ear length L8, L15, and L17 were good general combiner indicating these lines had the tendency to increase ear length. For thousand-kernel weight L4, L10, L12, L13 and L16 were the top general combiners as such line had the tendency to increase thousand kernel weights.

An inbred line L2, L5, L6, L8, L12, L13 and L14 were top general combiners for shorter plant height, which are desirable for lodging resistance. On the other hand, L11, L13, L16 and L17 were the top general combiners for increased plant height. An inbred line L2, L4, L7, L12 and L13 was top general combiners for enhancing shortness of ear height. On the other hand, L3, L16 and L17 were the top general combiners for increased ear height. Among the lines L1, L2, L4, L7, L11 and L13 were the best general combiner for plant aspect since these line indicated a tendency to improve this trait in future hybrids development program. An inbred line L1, L2, L3, L7 and L8 were top general combiners for enhancing better ear aspect since these line indicated a tendency to improve this trait in future hybrids development program. Among the lines L7, L10, L12 and L13 were the top general combiners for enhancing early maturity. Among the crosses L5 × T1 (8.36 t/ha) and L16 × T2 (7.89 t/ha) were showed higher grain yield (t/ha). These hybrids could be included in further investigation for grain yield and related traits and could be possible candidates of future release.

For plant height crosses L14 × T1 (158.13 cm), L6 × T2 (168.54 cm), L8 × T2 (169.25 cm) and L12 × T2 (171.25 cm) expressed short plant height and these crosses were the best specific combiner for shortness whereas crosses L4 × T1 (219.17 cm), L10 × T1 (225.00 cm), L16 × T1 (231.04 cm) and L14 × T1 (258.13 cm) expressed higher plant height indicated these crosses where the best specific combiner for tallness. Among crosses L5 × T1 (102.63 day), L2 × T2 (107.95 day), L14 × T1 (108.87 day) and L14 × T2 (102.63 day) displayed lowest an thesis date. These crosses were the best crosses for development of early matured hybrids. Among Crosses L5 × T1 (112.67 day), L2 × T2 (114.29 day), L14 × T1 (114.54 day) and L14 × T1 (114,79 day) displayed lowest silking date. These crosses were the best specific crosses for development of early matured hybrid. For number kernels per ear, only four crosses (L4 × T2, L6 × T2, L10 × T2 and L13 × T2) showed positive and significant (p<0.05) SCA effects in desired direction. This result indicates these crosses were good specific combiner for the improvement of this trait. For trait grain moisture crosses (L6 × T1 and L0 × T1) displayed negative and highly significant SCA effect (p<0.01) for this trait. For improvement of this trait, these crosses were appropriate specific crosses combiner. For plant aspect cross (L4 × T2) showed positive and significant SCA affect and this cross was the good specific combiner for this trait. For plant aspect cross (L4 × T2) expressed positive and highly significant (p<0.05) SCA effect was displayed and this cross is the best specific combiner for this trait.

From these finding better performing testcrosses, inbred lines with desirable GCA effects for grain yield and other grain yield related traits were successfully identified. These germplasm constitute a source of valuable genetic material that could be used for future highland maize improvement program. Generally, the results of this study could be useful for researchers who need to develop high yielding variety of maize adapted to highland areas of Ethiopia.

However, the present study was conducted at one location and the result is only an indication and we cannot reach at definite conclusion. Therefore, it is advisable to continue with this study over many years and locations. Moreover, future studies should explore the possibility of separating the inbred lines used in this study in to distinct heterotic groups by using divergent tester.

References

1. Galinat WC (1988) The origin of corn. In Sprague GF and Dudley JW (Eds.). Corn and corn improvement. Agronomy 18: 1-31.

2. FAOSTAT (2010) Statistical databases and data sets of the Food and Agriculture Organization of the United Nations.

3. Rosegrant MR, Ringler C, Sulser TB, Ewing M, Palazzo A, et al. (2009) Agriculture and food security under global change: Prospects for 2025/2050. International Food Policy Research Institute.

4. CSA (Central Statically Agency) Federal Democratic Republic of Ethiopia (2014) Agricultural sample survey report on area and production of crops, Addis Ababa, Ethiopia.

5. EARO/CIMMYT (2002) Ethiopian Agricultural Research Organization (EARO), Research strategy plan for maize. Addis Ababa, Ethiopia.

6. Salasya BDS, Mwangi W, Verkuijl H, Odendo MA, Odenya JO (1998) An assessment of the adoption of seed and fertilizer package and role of credit in smallholder maize production in Kakamega and Vihiga districts, Kenya. AGRIS. p: 35.

7. Twumasi A, Habtamu Z, Kassa Y, Bayisa A, Sewagegn T (2002) Development and improvement of highland maize in Ethiopia. Proceedings of the Second National Maize Workshop of Ethiopia, Agricultural Research organization (EARO) and CIMMYT. p: 31-38.

8. Prashanth M (2008) Isolation and early generation evaluation of inbred lines derived from yellow pool population of maize (Zea mays L.). PhD Thesis in Genetics and plant Breeding, Dharwad University of Agricultural Sciences, Dharwad, India.

9. Kruvadi S (1991) Diallel analysis and heterosis for yield and associated characters in durum wheat under upland conditions. Turrialba Publ Canada 41: 335-338.

10. Bindiganavile S, Vivek V, Joseph K, Simbarashe C, Cosmos M (2007) Fieldbook: Software For Managing A Maize Breeding Program: A Cookbook For Handling Field Experiments, Data, Stocks and Pedigree Information. CIMMYT.

11. Dagne W, Habtamu Z, Labuschagne MT, Hussien T, Singh H (2007) Heterosis and combining ability for grain yield and its component in selected maize inbred lines. S Afr J Plant Soil 24: 1.

12. Teshale A (2001) Analysis of tropical Highland Maize (Zea mays L.) Inbred lines top crossed with three east African Populations. MSc Thesis School of Graduate studies, Alemaya University, Ethiopia.

13. Jemal A (1999) Heterosis and combining ability for yield and related traits in maize. MSc Thesis. School of Graduate Studies, Alemaya University, Ethiopia.

14. Amiruzzaman M, Islam MA, Hassan L, Rohman MM (2010) Combining ability and heterosis for yield and component characters in maize. Academic Journal of Plant Sciences 3: 79-84.

15. Hadji T (2004) Combining Ability Analysis for yield and yield related traits in quality Protein maize (QPM) inbred lines. MSc Thesis. School of Graduate studies, Alemaya University.

16. Gudeta N (2007) Heterosis and combining abilities in QPM versions of early generation highland maize (Zea mays L.) inbred lines. MSc Thesis. Alemaya University of Agriculture, Ethiopia.

17. Alemnesh A (2012) Test cross performance and combining ability studies of elite maize (Zea mays L.) inbred lines in the central rift valley of Ethiopia. MSc Thesis. School of Graduate studies, Jimma University, Ethiopia.

18. Dagne W, Vivek BS, Birhanu T, Koste A, Mosisa W, et al. (2010) Combining ability and heterotic relationships between CIMMYT and Ethiopan inbred lines. Ethiop J Agric Sci 20: 82-93.

19. Zerihun T (2011) Genotype x Environment interaction and yield stability analysis of maize (Zea mays L.) in Ethiopia. MSc Thesis. School of Graduate Studies, Jimma University, Ethiopia.

20. Mandefro N, Habtamu Z (2001) Heterosis and combining ability in 8 x 8 diallel crosses of drought resistant maize (Zea mays L.) populations. African Crop Science Journal 9: 471-479.

21. Demissew A, Habtamu Z, Kanuajia KR, Dagne W (2011) Combining ability in maize lines for agronomic traits and resistance to weevil. Ethiop J crop scil 2: 1.

22. Bayisa A (2004) Heterosis and Combining Ability of Transitional highland maize (Zea mays L.). MSc Thesis. School of Graduate Studies, Alemaya University, Ethiopia.

23. Joshi VN, Pandiya NK, Dubey RB (1998) Heterosis and combining ability for quality and yield in early maturing single cross hybrids of maize (Zea mays L.). Indian Journal of Genetics and Plant Breeding 58: 519-524.

24. Shewangizaw A (1983) Heterosis and Combining ability in 7 x 7 diallel crosses of selected inbred lines of maize (Zea mays L.). MSc Thesis. School of Graduate Studies, Addis Ababa University, Ethiopia.

25. Tulu L (1999) Combining ability of some traits in seven- parent diallel crosses of selected maize (Zea mays L.) Populations. AGRIS. pp: 78-80.

26. Ahmad A, Saleem M (2003) Combining ability analysis in Zea mays L. International journal of agriculture & biology 5: 239-244.

27. Legesse BW, Pixely KV, Botha AM (2009) Combining ability and heterotic grouping of highland transition maize inbred lines. Maydica 54: 1-9.

28. Petrovic Z (1998) Combining abilities and mode of inheritance of yield and yield components in maize (Zea mays L.). AGRIS, p: 85.

29. Mathur RK, Chunilal C, Bhatnagar SK, Singh V (1998) Combining ability for yield, phonological and ear characters in white seeded maize. Indian Journal of Genetics and Plant Breeding 58: 117-182.

30. Pal AK, Prodhan HS (1994) Combining ability analysis of grain yield and oil content along with some other attributes in maize (Zea mays L.). Indian J Genet and Plant Breeding 54: 376-380.

31. Dehghanpour Z, Ehdaie B, Moghaddam M, Griffinh B, Hayman B (1997) Diallel analysis of agronomic characters in white endosperm corn. Journal of Genetics and Plant Breeding 50: 357-365.

32. Kumar A, Ganshetti MG, Kumar A (1998) Gene effects in some metric traits of maize (Zea mays L.). Ann Agric Biological Res 3: 139-143.

33. Dagne W (2002) Combining ability analysis for traits of agronomic importance in Maize (Zea mays L.) inbred lines with different levels of resistance to grey leaf spot (Cercospora Zea maydis). MSc Thesis. School of Graduate studies, Alemaya University, Ethiopia.

34. Mandefro N (1999) Heterosis, Combining ability and correlation in 8 x 8 diallel crosses of drought tolerant maize (Zea mays L) population. MSc Thesis, School of Graduate studies, Alemaya University, Ethiopia.

35. Saad IM, Haq NM, Nasir MM, Muhammad M (2004) General and Specific Combining Ability Studies in Maize Diallel Crosses. International Journal of Agriculture and Biology 5: 856-859.

Degradation and Downward Movement of Lindane in Soil Under Cultivated Field Conditions

S.K. Sahoo* and B. Singh

Department of Entomology, Punjab Agricultural University, Ludhiana-141004, Punjab, India

Abstract

Transport of pesticides in soil is important because it determines the extent to which pesticides reach groundwater. Many investigators have studied in a qualitative manner the tendency of insecticides to move by leaching through the soil by developing a mathematical expression. Little attention has been given to the actual formulation of such a model particularly under cultivated field conditions. The investigations described in this paper presented quantitative data which has indicated the relative importance of the factors which must be considered in predicating pesticide movement under cultivable field conditions. The data on residues of lindane in soil under cropped conditions showed the highly persistence nature of the pesticide. The movement of lindane being a matter of inches rather than feet holds good with the present experimental findings though the experiment was carried out in a sandy loam soil with very little organic carbon and clay contents.

Keywords: Degradation; Downward movement; Field conditions; Lindane

Introduction

The accumulation of organ chlorine insecticides in soils has been reported with increasing frequencies since decades. The accumulation of these insecticides in soil, will in part depend upon the adsorption characteristics of these compounds and subsequent movement through soil profile [1]. Transport of these insecticides in soil is important because it determines the extent to which insecticides reach groundwater and transport may also be important for agricultural purpose: for soil applied insecticides some redistribution in the root zone is essential for good efficacy [2]. Information on the environment fate of pesticides has been generated mostly from studies in the temperate environment, because use of pesticides in agriculture and public health has been more extensive in temperate countries than in tropics and subtropics. However, a steady increase in use of pesticides, insecticides in particular, in tropics in recent years prompted studies on the fate and significance of pesticide residues in the tropical environment [3]. Surfaced applied or soil incorporated pesticides, after entering the agricultural system, may be translocated into plants, volatized into atmosphere, leached downward below the root zone, sorbed onto soil constituents, transported while being adsorbed on soil particles, or degraded to nontoxic molecules [4]. These pesticides also may affect the next crop, as well as non-target species. The transport of pesticides in soil, and their rate of disappearance from soil, is of considerable importance as down ward movement of pesticides may result in the contamination of ground water, yet the composite behavior of pesticides in sub-surface is almost impossible to determine accurately. Many investigators have studied in a qualitative manner the tendency of insecticides to move by leaching through the soil by developing a mathematical expression. Little attention has been given to the actual formulation of such a model particularly under cultivated field conditions [5]. The investigations described in this paper presented quantitative data which has indicated the relative importance of the factors which must be considered in predicating pesticide movement under cultivable field conditions.

Materials and Methods

A field experiment was conducted under tomato crop at Entomological Research Farm, Punjab Agricultural University, Ludhiana, Punjab, India during the year 2005-06 to study the rate of degradation and downward movement of lindane in soil. Tomato (var. Punjab Chhuhara) was transplanted in February following good agricultural practices that includes field preparation, intercultural operations like weeding, timely irrigation etc.

The important soil characteristics which were noted were clay content, organic matter content, pH (Table 1). Water solubility, molecular weight, polarity and biodegradability were defined as the principal pesticide properties of importance.

Lindane (99.9% gamma isomer of 1, 2, 3, 4, 5, 6 hexachloro cyclohexane) stable to light, air, acid and temperature up to 180°C with molecular weight of 290.8 and water solubility of 7.3 mg lit^{-1} at 25°C was used as an experimental insecticide. Environmental factors which must influenced the degradation of insecticides are temperature, rainfall and relative humidity are presented in (Table 2).

The field experiment was set up to study the rate of degradation and downward movement of lindane in soil. PVC tubes (18" × 4") were vertically pushed into the field soil without disturbing the soil texture. One kg of top layer soil form each tube was removed and mixed with

Soil characteristics	
Texture	Sandy loam
Sand (%)	15.1
Silt (%)	79.7
Clay (%)	5.2
pH	7.5
Organic Carbon (%)	0.18

Table 1: Properties of Field Soils used for the experiment.

*Corresponding author: S.K. Sahoo, Assistant Residue Analyst, Department of Entomology, Punjab Agricultural University, Ludhiana-141004, Punjab, India E-mail: sksahoo_2006@rediffmail.com

Month	Maximum* Temp. (°C)	Minimum* Temp. (°C)	Relative Humidity* (%)	Rainfall* (mm)
February	18.7	8.5	79	47.4
March	25.8	13.6	72	42.2
April	34.6	16.6	45	6.1
May	37.9	22.2	43	Nil
June	39.3	26.5	49	48.1

*Mean of the month

Table 2: Weather conditions during the experimental period (Feb 2005 to June 2005).

lindane (Kanodane 20 EC) to give concentrations of 100, 200 and 500 mg a.i. kg^{-1} of soil. The scheduled cultural practices like weeding, irrigation, earthing up were carried out. For residue analysis samples were taken from depths of 0-15, 15-30 and 30-45 cms up to 90 days to study the dissipation pattern with timely movement of lindane under field conditions whereas samples from the depths of 0-15, 15-30, 30-45, 45-60, 60-75 and 75-90 cms at 150 days (after the crop was harvested) were taken in order to study the rate of downward movement of lindane in soil under a cropped area.

Fifty-gram soil sample was dipped in 100 ml mixture of methanol and water (2:1 v/v) for overnight. The contents were filtered into one litre separatory funnel, diluted with 500 ml of 5 per cent sodium chloride solution and partitioned twice into hexane (100+50 ml, each) separately. The combined hexane layers were concentrated and transferred to hexane for estimation by gas liquid chromatography (GLC) equipped with electron capture detector (ECD) and a glass column (1 m × 2 mm i.d) packed with ready to use 1.5% SP-2250+1.95% SP2401 on 80-100 mesh supelco port. The operating conditions of GLC were as follows: detector temperature: 270°C, oven (column) temperature: 200°C, injector temperature: 240°C and carrier gas (N$_2$) flow rate: 3.5 kg cm^{-2}. Under these operating conditions, lindane gave peak with retention time of 0.97 minutes. The average recoveries of lindane from soil sample spiked with concentrations ranging from 0.2, 0.5 and 1.0 mg kg^{-1} were found to be more than 80 percent.

Results and Discussion

Degradation of lindane under field conditions

The quantitative estimates of lindane residues in soils at different depths drawn at varying intervals, after applications @ 100, 200 and 500 mg a.i. kg^{-1} of soil are presented in tables 3 to 5. The residues of lindane in control were found to be below the detectable limit of 0.01 mg kg^{-1}.

The mean initial deposit of lindane in 0-15 cm layer was found to be 104.51 mg kg^{-1}, which dissipated to a mean level of 29.09 mg kg^{-1}, thus showing a loss of 72.16 percent after 30 days of application. After 150 days of application, the residues were found to be dissipated by 99.75 percent (Table 3).

Following application of lindane @ 200 mg a.i. kg^{-1} of soil, the mean initial deposit of 194.24 mg kg^{-1}, were found to be dissipated by 80.46 percent, after 30 days of application. The levels of lindane residues were found to be 1.99 mg kg^{-1} after 150 days of application, thereby, showing a degradation of 98.97 percent (Table 4).

When lindane was applied at the dose of 500 mg kg^{-1}, the mean initial deposit of 438.52 mg kg^{-1} was dissipated to 145.63 mg kg^{-1} after 30 days of application, thus showing a loss of 66.79 percent. Residues levels, at the end of 150 days were 4.44 mg kg^{-1} which were only 1.01 percent of the initial deposit (Table 5).

Downward movement of lindane in soil

The quantitative estimation of residues of lindane at different depth drawn at 150 days (crop period) after application of insecticide are presented in table 3 to 5. The data on residues of lindane in soil under cropped conditions show that there was no downward movement of lindane though the experiment was carried out in a sandy loam soil with very little organic carbon and clay content. This results deviates

Days after treatment	Depths (cms)	Mean ± S.D.	Per cent dissipation
0	0-15	104.51± 0.69	
7	0-15	78.53 ± 0.70	24.86
	15-30	BDL	
	30-45	BDL	
15	0-15	37.15 ± 1.24	64.45
	15-30	BDL	
	30-45	BDL	
30	0-15	29.01 ± 0.30	72.16
	15-30	BDL	
	30-45	BDL	
45	0-15	6.51 ± 0.62	93.77
	15-30	BDL	
	30-45	BDL	
60	0-15	4.99 ± 0.12	95.22
	15-30	BDL	
	30-45	BDL	
90	0-15	1.51 ± 0.02	98.55
	15-30	BDL	
	30-45	BDL	
150	0-15	0.26 ± 0.01	99.75
	15-30	BDL	
	30-45	BDL	
	45-60	BDL	
	60-75	BDL	
	75-90	BDL	

T$_{1/2}$ = 17.10 days; BDL= <0.01 mg kg^{-1}

Table 3: Residues of lindane (mg kg-1) in soil following its application @ 100 mg kg-1 of soil.

Days after treatment	Depths (cms)	Mean ± S.D.	Per cent dissipation
0	0-15	194.24 ± 2.66	
7	0-15	127.93 ± 1.29	34.75
	15-30	BDL	
	30-45	BDL	
15	0-15	73.04 ± 0.54	62.40
	15-30	BDL	
	30-45	BDL	
30	0-15	37.96 ± 0.97	80.46
	15-30	BDL	
	30-45	BDL	
45	0-15	17.61 ± 0.68	90.93
	15-30	BDL	
	30-45	BDL	
60	0-15	7.36 ± 0.22	96.21
	15-30	BDL	
	30-45	BDL	
90	0-15	4.98 ± 0.11	97.44
	15-30	BDL	
	30-45	BDL	
150	0-15	1.99 ± 0.03	98.97
	15-30	BDL	
	30-45	BDL	
	45-60	BDL	
	60-75	BDL	
	75-90	BDL	

T$_{1/2}$ = 22.46 days; BDL= <0.01 mg kg^{-1}

Table 4: Residues of lindane (mg kg^{-1}) in soil following its application @ 200 mg kg^{-1} of soil.

Days after treatment	Depths (cm)	Mean ± S.D.	Per cent dissipation
0	0-15	438.52 ± 2.78	
7	0-15	329.48 ± 1.64	24.86
	15-30	BDL	
	30-45	BDL	
15	0-15	182.29 ± 0.97	58.42
	15-30	BDL	
	30-45	BDL	
30	0-15	145.63 ± 2.28	66.79
	15-30	BDL	
	30-45	BDL	
45	0-15	76.57 ± 0.22	82.54
	15-30	BDL	
	30-45	BDL	
60	0-15	25.85 ± 0.21	94.10
	15-30	BDL	
	30-45	BDL	
90	0-15	11.13 ± 0.63	97.46
	15-30	BDL	
	30-45	BDL	
150	0-15	4.44 ± 0.35	98.99
	15-30	BDL	
	30-45	BDL	
	45-60	BDL	
	60-75	BDL	
	75-90	BDL	

$T_{\frac{1}{2}}$ = 21.66 days; BDL= <0.01 mg kg^{-1}

Table 5: Residues of lindane (mg kg^{-1}) in soil following its application @ 500 mg kg^{-1} of soil.

from the equation:-

$$K_{OC} = \frac{K_d}{F_{OC}}$$

(K_{OC}=soil organic carbon sorption coefficient; K_d=soil sorption partition constant and F_{OC}=organic carbon fraction of the specific soil).

The soil organic matter is the only sorbing material in the solid phase and that soil organic matter in all soils has the same affinity for solutes. However, soil/water/pesticide systems exhibit much more complex behaviour under field conditions than that of mathematical model developed in laboratory conditions. More polar solutes, surfaces of other materials in soils can also become important sorbents, particularly in soils where the organic matter fraction is low may holds good in this case [6-8]. The important soil parameters in the transport of pesticide is the sorption coefficient. This was measured in laboratory experiments in which a suspension of soil was shaken for about half a day or a long

term sorption process. But under field conditions, sampling dates 5 months (total crop period) after application of pesticides associated with irrigation at regular intervals resulted in a considerable effect on the transport of pesticides. Extreme sorption is associated with extreme persistence [9-11]. May be one of the reason regarding the movement of lindane in soil under field conditions. The experimental findings of Key and Elrick [1] regarding the failure of mathematical model to predict lindane elution for different low rate and conclusion made by him that under field conditions it is expected that the leaching of lindane will be limited, the movement being a matter of inches rather than feet holds good with the present experimental findings. However, to know accurately the degree of sorption of a specific pesticide in a specific soil, the approach is still empirical- a measurement must be made.

Acknowledgments

The authors are grateful to the Indian Council of Agricultural Research, New Delhi for the research grant. The authors thanks to the Head, Department of Entomology, Punjab Agricultural University, Ludhiana, Punjab, India for providing the necessary facilities during the course of investigations.

References

1. Kay BD, Elrick DE (1967) Adsorption and movement of lindane in soils. Soil Sci 104 (5): 314-322.

2. Boesten JJTI, Van der Pas LJT, Smelt JH (1989) Field test of a mathematical model for non-equilibrium transport of pesticides in soil. Pestic Sci 25: 187-203.

3. Ramanand K, Panda, S, Sharmila M, Adhya TK, Sethunathan N (1998) Development and Acclimatization of cabofuran degradaing soil enrichment cultures at different temperatures. J Agric Food Chem 36: 200-205.

4. Singh G, Spencer WF, Genuchten MTV, Kookana RS (1992) Predicting Pesticide Transport in Soil. Pestic Res J 4: 1-10.

5. King PH, Mc Carty PL (1968) a chromatographic model for predicting pesticide migration in soils. Soil Sci 106 (4): 248-261.

6. Helling CS, Kearney PC, Alexander M (1971) Behaviour of pesticides in soils. Adv Agron 23: 147-239.

7. Karickhoff SW (1984) Organic pollutant sorption in aquatic systems. Hydraul Engng 110: 707-735.

8. Hance RJ (1969) Influence of p^H, exchangeable cation and the presence of organic matter on the adsorption of some herbicides by montmorillonite. Canad J Soil Sci 49: 357-364.

9. Walker A, Welch SJ (1989) the relative movement and persistence in soil of chlorsulfuron, metsulfuron-methyl and trisulfuron. Weed Res 29: 337-383.

10. Gamerdinger AP, Achin RS, Traxler RW (1997) Approximating the impact of sorption on biodegradation kinetics in soil-water systems. Soil Sci Soc Am J 61: 1618-1626.

11. Guo L, Wagenet RJ, Jury WA (1999) Adsorption effects on kinetics of aldicarb degradation: equilibrium model and application to incubation and transport experiments. Soil Sci Soc Am J 63: 1637-1644.

5

Development of Mathematical Model for Repair and Maintenance of Some of the Farm Tractors of JNKVV, Jabalpur, India

Avinash Kumar Gautam* and Shrivastava AK
Department of Farm Machinery and Power Engineering, College of Agricultural Engineering, JNKVV, Jabalpur, Madhya Pradesh, India

Abstract

An experiment was conducted to studies on Development of mathematical model for repair and maintenance of some of the farm tractors JNKVV Jabalpur, the repair and maintenance data of the farm tractor were taking from the breeder soybean production farm, biotechnology, groundnut farm and horticulture farm. The data collect yearly working hours; yearly repair and maintenance costs included spare part and repairable part, lubricant, wages and others. A study was conducted to modelling of accumulated repair and maintenance costs of JNKVV farm tractors as percentage of initial purchase price (Y) based on accumulated usage hours (x). Recorded data were used to determine regression model(s). Exponential, logarithmic, linear, polynomial. The Prediction of cumulative repair and maintenance costs the power model is better than the models that is linear, polynomial, logarithmic and exponential, among the various alternatives power model was found ($Y=ax^b$) most suitable to predict accumulated repair and maintenance costs of tractor. The service life of the tractor near 1000 hours the power model $Y=1.910x^{1.64}$ (where x in 1000) with $R^2=0.989$ to predict accumulated repair and maintenance costs of JNKVV tractors can be strongly recommended. The repair and maintenance cost consist of spare parts, lubricant, wages and other. The average spare parts, wages and others and lubricants costs is 49.32, 17.24 and 12.15%.

Keywords: Exponential; Logarithmic; Linear and polynomial model

Introduction

Worldwide Tractor is the main source of power on the farm, and represents a major component of farm fixed costs. With due field maintenance tractors can operate for long period and do great deal of work before major repairs are required [1]. Tractor break down can be of a high cost not only from expenditure point of view, but also because of the disastrous effect on crop productivity, and the fact that idle staff must still be paid. The extent of the problem of tractor failure in developing countries is more serious as compared to developed countries. This is due to acute shortage of genuine spare parts, preventive maintenance, absence of future planning for integrated maintenance management and programs that strive to identify incipient faults before they become critical to enable more accurate planning of preventive actions. As such system performance can be improved by developing optimal maintenance prediction models that minimize overall maintenance cost or maximize system performance measures [2].

A repairable mechanical system (as agricultural tractor) is subjected to deterioration or repeated failure. The system is subjected to periodic inspection that identifies the condition of deterioration. Based on the degree of deterioration (system condition), preventive maintenance is performed or no action is taken. At each inspection of failure, the system status is classified into partial, combined and complete. According to this Condition-Based Maintenance classification the level of maintenance is determined and performed to restore the system to "as good as new" state.

Machinery ownership (fixed) and operational (variable) cost represent substantial portion of total production experiences. Machinery ownership costs usually include charge for depreciation, interest on investment, taxes, insurance and housing facilities. Operational costs include repair and maintenance costs of farm machinery which is necessary to restore or maintain technical soundness and reliability of the machine. The accurate prediction of repair and maintenance costs trends is critical for determination of optimum economical life of machine and to make appropriate decision for machinery replacement. The prediction of these costs at an acceptable level can be made by

fitting of linear, logarithmic, polynomial, and exponential and power equation.

The repair and maintenance cost of tractors is essential for both owner and manager to achieve information on overall cost to control financial and production economy. It is small but relatively important portion of owning and operating farm machinery, repair cost is generally 10 to 15% of the total cost. Since it tends to increase with machine usage, repair costs become important for replacement policy. Hence, five performance measures of the model process are used to find the optimal algorithm parameters that maximize the system availability. The model decision variables are working hours, and repair and maintenance costs. Thereafter, the model is used to predict the expected repair and maintenance cost [3].

Appropriate mathematical model for the maintenance costs of farm tractors provide planners and policy makers and farmers an opportunity to evaluate the machine economics.

Materials and Methods

This study is carried out at JNKVV Jabalpur Madhya Pradesh. The repair and maintenance data of farm tractors were collected from breeder soybean production unit, biotechnology and groundnut unit, and horticulture farm, the Major activities are Tillage, seed bed preparation, sowing, spraying, harvesting, threshing, and transportation. Major crop in that farms are Paddy, maize, soybean, wheat, gram, berseem, pea and tuar, potato, ground nut, cauliflower, chilli, bottle guard, cucumber, etc.

*Corresponding author: Avinash Kumar Gautam, Department of Farm Machinery and Power Engineering, College of Agricultural Engineering, JNKVV, Jabalpur, Madhya Pradesh, India, E-mail: avipavan75@gmail.com

All the selected farms are situated at JNKVV campus around the college of agricultural engineering Jabalpur, India, which lies in between 22° 49' N and 24° 8' N latitude; 78° 21' E and 80° 85' E longitude. The data sample was taken four tractors; the study data available from the first-year life of the tractors and tractor horse power 55 hp. The selected two tractors name are Hindustan HWD50, John-deer 5310D name like T1, T2 and these tractors data available up to 10 year or more.

The Information on yearly repair and maintenance cost data of the tractors such as use of tractor each year, repair and maintenance costs of major part, lubricants, wages etc. was collected [4]. Some variations were apparent between individual tractors for the service hours. As hours of annual usage for each tractor were needed for data analysis study. To determine tractors at any point of service life, accumulated hours of use for each year were added up to previous usage hours are independent variable (x) of the model (s). Then, repair and maintenance costs as percentage of initial purchase price which was dependent variable (Y) obtained through dividing the total accumulated repair and maintenance costs by initial purchase price of the tractors.

Regression analysis

In any kind of mathematical relationship, one value of the variable is known and the value of another variable can be determined exactly. But it is possible case of statistical relationship the value of one variable from that of the other variable cannot be determined exactly. In this case, the estimation of the other variable is made with the help of known by using regression analysis. It is an important statistical technique used in science, business and economics.

To determine regression model(s) for predicting repair and maintenance costs of these tractors at any point of service life, accumulated hours of use for each year were added up to previous usage hours and the sum was independent variable x (where x in 1000) of the model(s). Then, repair and maintenance costs as percentage of initial purchase price which was dependent variable (Y) obtained through dividing the total accumulated repair and maintenance costs by initial purchase price of the tractor. To acquire information (i.e., repair and maintenance costs, hours of service and initial purchase price) for all tractors, cumulative of data was employed for analysis. Regression analysis of data for all tractors was done. The regression model Exponential, Linear, Logarithmic, Polynomial and Power equation were tried. The regression model(s) having the highest coefficient of determination (R^2) was selected as the best model(s) for predicting actual repair and maintenance costs trends (Table 1).

Results and Discussion

Tractor is used for tractive as well as stationary work in the farm. To perform the work timely. For better performance of a tractor, repair and maintenance is done by the farm in charge. The repair and maintenance data for the tractors of the farms under this study were collected from the records. The data and the analysed result are below.

S. No.	Model	Equations
1	Exponential	$Y=ae^{bx}$
2	Linear	$Y=a+bx$
3	Logarithmic	$Y=a+\ln bx$
4	Polynomial	$Y=a+bx+cx^2$
5	Power	$Y=ax^b$

Table 1: Five models are used to perform regression analysis.

Determination of appropriate mathematical model to predict repair and maintenance cost for JNKVV tractors

Determination of appropriate mathematical model for cumulative repair and maintenance for JNKVV farm tractor. Five regression models namely polynomial, linear, logarithmic, polynomial and power are applied. For the determination of equation Y=crm/pu in percentage, x=cumulative working hours. Whereas 1, 2, 3…….10 shows the age of the tractor represent cumulative hours [5,6] (Figures 1 and 2) (Tables 2 and 3).

Considering R^2 values, there is a significant correlation between x and Y variables in all five models. However, R^2 values indicate that the power and polynomial models have higher conformity with actual data trend in comparison with the linear, exponential and logarithmic models, for prediction of accumulated repair and maintenance cost.

It was found that the power model is best for prediction of cumulative repair and maintenance costs and the polynomial model of second order also predict good cumulative repair and maintenance but power model is used to calculate the cumulative repair and maintenance because of the value of R^2 is 0.99 as well as simple structure and ease in calculation.

Repair and maintenance costs fractions

The average spare parts, wages and others and lubricants costs is 49.32, 17.24 and 12.15%. Among the cost of spare parts, wages, lubricant and other; spare parts cost is more than the other costs which varies from 41.37 to 50.23% of the total repair and maintenance costs [7] (Figures 3 and 4).

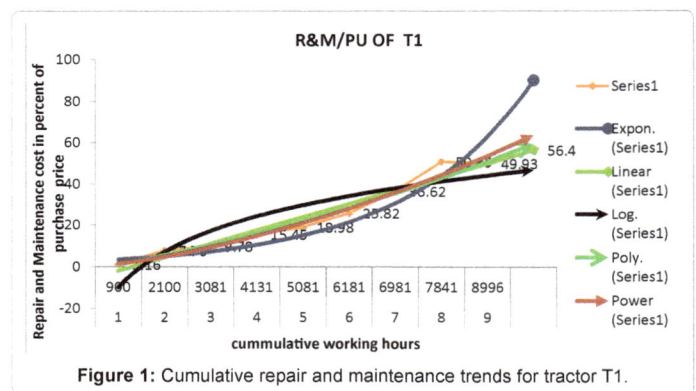

Figure 1: Cumulative repair and maintenance trends for tractor T1.

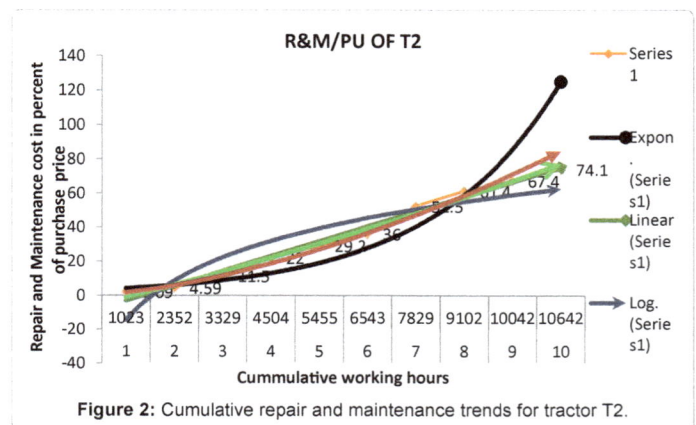

Figure 2: Cumulative repair and maintenance trends for tractor T2.

Figure 3: Fraction of repair and maintenance cost (in percentage of purchase price).

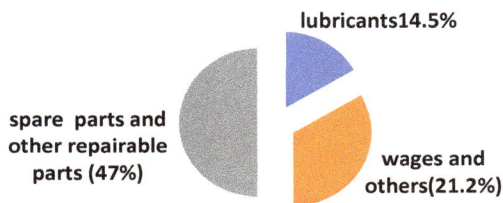

Figure 4: Fraction of repair and maintenance cost (in percentage of purchase price).

T1	Model Summary		Parameter		
Model	Equation	R^2	a	b	c
Exponential	$Y=ae^{bx}$	0.901	2.47	0.375	
Linear	$Y=a+bx$	0.917	8.378	−15.01	
Logarithmic	$Y=a+b\ln x$	0.709	30.43	−14.9	
Polynomial	$Y=a+bx+cx^2$	0.991	0.942	−1.991	5.727
Power	$Y=ax^b$	0.997	1.701	1.614	

Table 2: The coefficient and coefficient of determination (R^2) of the five-regression model obtained for tractor T1.

T2	Model Summary		Parameter		
	Equation	R^2	a	b	c
Exponential	$Y=ae^{bx}$	0.881	2.897	0.376	
Linear	$Y=a+bx$	0.985	8.704	−11.78	
Logarithmic	$Y=a+b\ln x$	0.850	33.39	−14.35	
Polynomial	$Y=a+bx+cx^2$	0.988	0.167	6.866	−8.110
Power	$Y=ax^b$	0.989	1.910	1.647	

Table 3: The coefficient and coefficient of determination (R^2) of the five-regression model obtained for tractor T2.

Conclusion

Results of this study indicated that average repair and maintenance costs per hour increased with tractor age. For Prediction of cumulative repair and maintenance costs the power ($Y=ax^b$) model is better than the models that is linear, polynomial, logarithmic and exponential. The value of "b" varies from 1.61 to 1.96 in power model and the value of R^2 is between 0.997 to 0.993. The values of R^2 indicate that the power, polynomial and linear models have higher conformity with actual data trend in comparison with exponential and logarithmic models. For exponential, logarithmic model there is no significant correlation between cumulative working hours and cumulative repair and maintenance costs.

References

1. Elbashir E, Ali H (1996) Tillage for sugar cane in Sudan with special reference to Kenana. PhD Thesis, University of Khartoum, Sudan.

2. Hunt DR (2001) Farm Power Machinery Management. 3rd edn. Iowa State University Press, Ames, Iowa, USA.

3. Ashid M, Ranjbar I (2010) Modeling of repair and maintenance costs of John Deere 4955 tractors in Iran. American-Eurosian J Agric and Environ Sci 9: 605-609.

4. Adekoya LO, Otono PA (1990) Repair and Maintenance Costs of Agricultural Tractors in Nigeria. Tropical Agriculture, Trinidad and Tobago 67: 119-122.

5. Bowers W, Hunt DR (1970) Application of mathematical formulas to repair cost data. Transactions of ASAE 13: 806-809.

6. Almassi M, Yeganeh HR (2002) Determination of a suitable mathematical model to predict the repair and maintenance costs of farm tractors in Karoon Agro-industry Company. Iranian J Agric Sci 33: 707-716.

7. Beppler DC, Hummedia MA (1985) Maintaining and repairing agricultural equipment in developing nations. Agricultural Engineering 66: 11-13.

6

An Appraisal of Governments Youth Empowerment Programme through Agriculture: The Need for a New Approach

Kator P E*[1] and Adaigho. P[2]

[1]Department of Agricultural Technology, Delta State Polytechnic, P.M. B. 005, Ozoro, Delta State, Nigeria
[2]Department of Agricultural Extension Management, Nigeria

Abstract

This paper appraised governments youth empowerment programme through agriculture with respect to the selection process and training programme. The multi-stage random sampling technique was used in selecting two hindered and seventy five trained youths(11 per local government area of the state), two hundred and fifty youths not trained(10 per local government), sixteen coordinators and two hundred and fifty other interested persons (10 per local government). Questionnaires and personal interview were used in collecting data. The findings show that age and personal interest constitutes the major criteria for selection of youths which is regarded as fair and equitable. The training programme for youths is perceived to have more practical content (60-90%)than theoretical content, but the 4-week period of training was considered inadequate and also not in line with the National Board for Technical Education provisions contained in the curriculum for technical and vocational Education. The study recommends that additional criteria in selection of youths such as good physical strength and good health as well as selection from the wards in each local government areas of the state should be embraced. The target group (youths) should be involved in the design and introduction of a youth empowerment programme through agriculture and more youths should be accommodated for a more intensive training programme lasting for longer period and going through different tiers amongst others.

Keywords: Agriculture; Appraisal; Approach; Government; Youth empowerment

Introduction

Vavrus and Elecher [1] define youth empowerment as an attitudinal, structural and cultural process whereby young people gain the ability, authority and agency to make decisions and implement change in their own lives and the lives of other people including youths and adult. Oluwasanjo [2] enumerated that youth empowerment simply means all positive efforts taken deliberately towards improving the lives of young people; efforts channeled towards developing the capacities of young ones to draw out the best out of them bearing in mind that they are truly leaders of tomorrow. Youth empowerment may involve five types [3]: (i) financial empowerment may involve offering cash to youths to start up or solve problems; (ii) academic empowerment entailing giving academic support to youths through scholarships; (iii) moral empowerment that entails empowering youths morally in families, schools and community gatherings to behave well; (iv) skill acquisition empowerment which encompasses training youths on skill development to support them in earning a living through acquisition of skills in areas such as plumbing, welding, mechanics and writing skills amongst others and (v) agricultural empowerment which empowers youths to be engaged in agriculture to feed the nation, earn income and make agriculture a major driver in the economy.

Youth empowerment constitutes a central tenet of the United Nations Convention on the Rights of the Child which every country in the world has signed into law [4]. Hence government youth empowerment programmes is in vogue globally to create and support the enabling conditions under which young people can act on their own behalf, and on their own terms, rather than at the direction of others. Uzochukwu [5] likened youth empowerment to saving in the bank that cannot be affected by any theft of any kind and further outlined some of the benefits to include, crime reduction, poverty eradication, national growth, security, technological development, employment creation, good education standard and good governance. Ali [6] also elucidated that youth empowerment leads to nations prosperity for its next

generations, innovation, transparent social and political systems, less dependability on government by youth communities, opportunity for voice on youth issues as well as advocate them at local and international level effectively.

Amongst the essence of governments youth empowerment programme through agriculture has been to battle unemployment through agriculture; remodel the agricultural sector and correct the problem of ageing farming population. One of the prominent programmes put in place by the Delta State government is the Youth Empowerment through Agriculture [7] (YETA). Ogbogu [8] noted that the programme was an initiative of the Uduaghan-led administration (2007-2015) that was introduced as part of the human capital development agenda to; (i) encourage youths to embrace farming (ii) make youths enjoy a good life devoid of poverty, ignorance, violence and other vices that reduce the capacity to complement governments efforts (iii) build around active and result oriented participation to secure their future and turn them into owners of their own businesses. The programme which was inaugurated on 27th November, 2008 commenced with the selection of eighteen youths from each of the twenty five local government areas of the state. The selection process was based on interest and devoid of any political interference. The youths were made to go through a 4-week intensive training at the Songhai-Delta Integrated Agricultural centre Amukpe-Sapele in Delta

***Corresponding author:** Kator P E, Department of Agricultural Technology, Delta State Polytechnic, P.M. B. 005, Ozoro, Delta State, Nigeria
E-mail: Katorpe@yahoo.com

State with a promise of start-up capital, implements and input materials. Several youth empowerment schemes have been in place, but much of them if not all have failed to achieve it desired objectives. Hence an appraisal is necessary to ascertain how the programme was operated with a view to addressing identified "grey areas" in any subsequent government youth empowerment programme. This paper therefore examines the Delta State governments youth empowerment through Agriculture programme with respect to; (i) the selection process (ii) the training programme and (iii) the strategies needed to address the shortcomings in future programmes.

Hypothesis

Ho_1: There is no significant relationship between the perception of the trained youths and other interested persons on the extent of fairness and equity in the selection process of the YETA programme.

Ho2: There is no significant relationship between the perception of trained youths and the other youths on the adequacy of the period of training needed to acquire practical hand-on skills under the YETA programme.

Materials and Methods

The study was carried out in Delta State, Nigeria. The state covers an area of 16,842 square kilometers. It lies in between Longitude 5°00' and 6°45' East and Latitude 5°00' and 6°30'North. The state is bounded in the North and West by Edo state the east by Anambra, Imo and Rivers states, South east by Bayelsa state and on the Southern flanks is Bight of Benin. The state comprises of twenty five local government areas contained in the three senatorial zones of Delta North, Delta South and Delta Central. The state has an average annual rainfall of 266 cm coastal area and 190.5 cm northern fringes running between March and November. Agriculture constitutes a dominant occupation in the state with crop production, poultry production, fisheries and animal husbandry being the specialized areas. The population of the study comprised of two categories (i) direct beneficiaries made up of youths trained as part of the YETA programme and (ii) indirect beneficiaries made up of other youths, co-ordinators/officers of the programme and other interested persons. The multi-stage random sampling technique was adopted by identifying the category of the beneficiaries, determining their total number/ number of interest and choosing convenient sample population. Two hundred and seventy five youths trained (eleven from each local government area), 16 co-ordinators, 250 other youths not trained and 250 interested persons were chosen for the study. Primary and secondary data were used. Primary data were generated using questionnaires and personal interview (structured). Secondary data were sourced from internet materials, journals, books, newspaper and periodicals amongst others. Data was analyzed using frequency, percentage and mean. The hypothesis was analysed using Z-statistics at 5% level of significance.

Results (Tables 1-7)

Discussion

The selection process for youths in an empowerment programme requires defined criteria and a logical process to guarantee transparency, equity and to avoid bias. The YETA programme is this regard is perceived to be based on the criteria of age and interest in agriculture. This does not fall below the expectation of all the categories of respondents, hence it is considered fair and equitable.

Transmission of more practical hand on practical skills with lesser

Criteria	Categories/Frequency/Percentage			
	TY	OY	CO	IP
* Age	-	04 (1.66%)	-	10
* Edu. Qual	-	04 (1.66%)	-	10 (4.00%)
* Interest	-	-	-	-
Political Affiliation	-	63 (125.20%)	-	-
Social Class	-	08 (3.02%)	-	-
Parental Background	-	-	-	-
Institutions Attended	-	-	-	-
Age and Interest	209 (76.00%)	123 (49.20%)	16 (100%)	63 (25.20%)
i, iii & iv	66 (24.00%)	48 (19.20%)	-	148 (59.20%)

Table 1: Perception of Criteria for Selection of Beneficiaries of the YETA Programme.

Extent	Category/Frequency			
	TY	OY	CO	IP
To large extent (5)	243 (1215)	160 (800)	14 (70)	133 (665)
To considerate extent(4)	07 (28)	41 (164)	02 (08)	48 (192)
To middle extent (3)	-	30 (90)	-	10 (30)
To small extent (2)	-	17 (34)	-	50 (100)
To no extent (1)	-	02 (02)	-	09 (09)
Total	1243	1090	78	996
N	275	250	16	250
X	4.58	4.36	4.88	3.98

Table 2: Perception of Extent of Fairness and Equity In the Selection Process of Beneficiaries of the YETA Programme.

%	Categories/Frequency/Percentage							
	TY		OY		CO		IP	
	No.	%	No.	%	No.	%	No.	%
0-29	-	-	08	3.20	-	-	10	4.00
>30-59	46	16.73	142	56.80	-	-	131	52.40
>60-90	229	83.27	100	40.00	16	100.00	109	43.60
> 90	-	-	-	-	-	-	-	-

Table 3: Perception of Practical Content of Training Programme for Beneficiaries.

%	Categories/Frequency/Percentage							
	TY		OY		CO		IP	
	No.	%	No.	%	No.	%	No.	%
0-30	208	75.64	64	25.60	16	100.00	106	42.40
>30-59	67	24.36	186	74.40	-	-	143	57.20
>60-90	-	-	-	-	-	-	01	0.40
> 90	-	-	-	-	-	-	-	-

Table 4: Perception of Theoretical Content of Training Programme for Beneficiaries.

theoretical skills is the focus of any youths empowerment programme. Hence, the perceived greater practical content over the theoretical is capable of producing youths that can practice agriculture. As indicated in the National Board for Technical Education Curriculum, practical hand on skills on an area of specialization should be rendered over a period of 16 weeks involving theory and practicals [9]. Thus, the 4 week

Item Statement	Response Options Options	Categories /Frequencies			
		TY	OY	CO	IP
*The 4-week training Programme was inadequate to acquire the needed hand on practical skills.	Strongly Agree (5)	186(930)	214 (1070)	16 (80)	211 (1055)
	Agree (4)	73(292)	29 (116)	-	28 (112)
	Undecided (3)	01 (03)	07(21)	-	11(33)
	Disagree (2)	10(20)	-	-	-
	Strongly Disagree(1)	-	-	-	-
	Total	1245	1207	80	1200
	N	275	250	16	250
	X	4.53	4.83	5.00	4.80
*The training was carried out in line with specified curriculum by the National Business and technical Education Provisions	Strongly Agree (5)	-	-	-	-
	Agree (4)	-	-	-	-
	Undecided (3)	09(27)	10(30)	-	28(84)
	Disagree (2)	50(150)	80(160)	06(12)	83(166)
	Strongly Disagree(1)	216(2160)	160(160)	10(10)	139(139)
	Total	393	350	22	389
	N	275	250	16	250
	X	1.43	1.40	1.38	1.56

Table 5: Perception of Period of Training and Use of Specified Curriculum.

Category	X	Standard Deviation (SD)	Variance (S²)	n	df	Zcal	Zcrit
Trained Youths	4.88	0.1032	0.3212	275	523	10.82	1.645
Other Interested Persons	3.96	1.7031	1.3050	250			

Level of Significance=0.05

Since Zcal (10.52) is greater than critical Zcrit (1.645) the null hypothesis is rejected and the alternative hypothesis which states that there is significant relationship between the perception of trained youths and other interested persons on the extent of fairness and equity in the selection process under YETA is validated

Table 6: Z – test of difference between the mean scores of Trained Youths and other Interested Persons on the extent of Fairness and Equity in the Selection process of YETA.

Category	X	Standard Deviation (SD)	Variance (S²)	n	df	Zcal	Zcrit
Trained Youths	4.53	0.69	0.47	275	523	-6.52	1.645
Other Interested Persons	4.83	0.45	0.20	250			

Level of Significance=0.05

Since Zcal (-6.52) is greater than critical Zcrit (1.645) the alternative hypothesis Ha is rejected and the null hypothesis which states that there is significant relationship between the perception of trained youths and other youths on the adequacy of the period of training needed to acquire practical on skills under the YETA programme is validated

Table 7: Z- test of Difference between mean scores of trained Youths and other youths on the adequacy of Period of training needed to Acquire Practical Hand-on Skills under YETA.

training programme embarked upon by the YETA programme was grossly inadequate and may have resulted in skipping certain relevant aspects of the training that could make it less effective. It is therefore, imperative that the training was carried out using an ad-hoc curriculum as against the specified curriculum in the area of specialization in agriculture as against the specified curriculum by the National Board for Technical Education provision. However, this does not imply that the youths that were engaged in the YETA programme would have not acquired the needed skills as attitude towards the training programme as well as instructors ability to disseminate the needed information and exposure amongst others could be ultimate determinants of the end resultant skills.

Conclusion and Recommendations

Based on the findings of the study, it can be concluded that the selection process of youths under the YETA programme was largely adjudged to be adequate in terms of being equitable and fair. The practical content of the training dominated the theoretical content, but the 4-week training period and non-adherence to the NBTE provisions for technical and vocational training was considered inadequate. The following recommendations are proffered.

1. The selection of youths for agricultural empowerment programmes should include the criteria of satisfactory physical and health conditions. This is because agricultural activities require some amount of good physical strength and good health status. In addition the selection should start from the wards in each local government area to avoid selection of youths from only a few wards in each local government.

2. The target group (youths) and others should be consulted. Local youth associations should be involved in setting up the youth empowerment programme. This is because the programme is meant for them and they will benefit from it such that they are in a better position to stipulate how the programme should be run to provide the needed impact.

3. The training programme should be put up in batches to accommodate more youths. One set should be made to go through training for a 5-months period. This training programme should be put up to comprise three tiers;

* Attitude re-orientation to be embarked upon for a one week period to prepare the minds of the trainees to expose them to leadership skills needed, discipline and direction of expectations.

* Vocational/Technical training where each of them will be required to select an area of specialization in agriculture that is available and undergo an 16-week training in line with the NBTE provisions Technical and Vocational. Education curriculum should be used as it structure training leading to the acquisition of applied skills and well as scientific knowledge (National Policy on Education, 2004). The length of training is important as it makes the training more intensive covering best practices and other essentials. Kalu [10] in describing the length of training of youths under the Lagos state government Agric.

Yes Scheme noted that a six-month intensive training would expose the participants to the rigours and varying aspects of the specialized area making them more skillful and ready to practice on their own.

* Business and Development training which should last for one week. Each of the participants would in the course of the period be taught how to set up and run an agricultural venture to achieve success.

References

1. Vavrus J, Fletcher A (2006) Guide to Social Change led by and with Young people. The Freechild Project.

2. Oluwasanjo AR (2013) Future Empowerment: Equipping the Future Leaders.

3. Uzochukwu M (2015) Youth Empowerment: Types of Youth Empowerment Hub pages Politics and Social Science. August 29, 2015.

4. Wikipedia the Free Encyclopedia (2015) Youth Empowerment. A Wikimedia Project, Powered by Mediawiki.

5. Uzochukwu M (2015) Importance of Youth Empowerment Hubpages. Politics and Social Science September 1st 2015.

6. Ali Q (2012) Benefits of Youth Empowerment, First thought, Friday September 07, 2012.

7. Youth empowerment through Agriculture (2013) Delta State Nigeria, March 14 2013.

8. Ogbogu A (2015) Delta: Giving Priority to Youth Empowerment. The Pointer Newspaper. 8th October, 2015.

9. National Board for Technical Education (2012) Curriculum and Course Specifications.

10. Kalu S (2015) Nigeria: Apply to the Agriculture youth Empowerment Scheme (Agric.Yes), Lagos Kalusami's Blog.

Effects of Polyvinyl Pyrrolidone and Activated Charcoal to Control Effect of Phenolic Oxidation on *In Vitro* Culture Establishment Stage of Micropropagation of Sugarcane (*Saccharum Officinarum L.*)

Dereje Shimelis[1]*, Kassahun Bantte[2] and Tilaye Feyissa[3]

[1]Ethiopian Sugar Corporation Research and Training Division, Variety Development Directorate, Biotechnology Research Team, Wonji Research Center, P. O. box 15,Wonji, Ethiopia.
[2]Jimma University College of Agriculture and Veterinary Medicine, Jimma, Ethiopia
[3]Addis Abeba University, Science and Technology faculty, Addis Abeba, Ethiopia

Abstract

Sugarcane isgrown in Ethiopia as both cash and industrial crop. Although it plays a great role in the economy of the country, there are no enough sugarcane varieties under production and it is not an ideal crop for conventional breeding. Hence, it takes around ten years for its breeding and extra years to scale up the planting material for commercial by vegetative propagation. To circumvent these limitations, biotechnological tool (Plant tissue culture) was born as best alternative. Though the tissue culture (micro propagation) plays a reliable role, culture establishment stage of sugarcane micro propagation has a serious phenolic oxidation problem which can kill the whole culture. Hence, these experiments were conducted to optimize the appropriate concentration of Polyvinylpyrrolidone (0.0, 0.1, 0.2, 0.3, 0.4 and 0.5 gL^{-1}) and Activated charcoal (0.0, 0.1, 0.2, 0.3, 0.4 and 0.5 gL^{-1}) on (C86-12, C86-56) genotypes in completely randomized design with 2 × 5 factorial treatment combinations arrangements to control the effect of phenolic oxidation. Analysis of variance showed that interaction effect of Polyvinylpyrrolidone and genotypes. Activated charcoal and genotypes on percentage of survived and dead explants due to phenolic oxidation were highly significant at (P=0.001). Murashige and Skoog medium supplemented with 0.2 gL^{-1} and 0.3 gL^{-1} of Polyvinylpyrrolidone has gave 100% and 80% survived explants of C86-56 and C86-12genotypes respectively while 0.4 gL^{-1} and 0.3 gLI^{-1}of activated charcoal resulted in 46% and 40% of survived explants of C86-56 and C86-12 genotypes respectively, after 30 days of culturing. Thus, these optimized concentrations of Polyvinylpyrrolidone and activated charcoal are useful to control the effect of phenolic oxidation on culture establishment of micro propagation, which helps to have enough starter culture for further multiplication stage of micro propagation.

Keywords: Sugarcane; Phenolic oxidation; Polyvinylpyrrolidone; Activated charcoal; Survived explant

Introduction

Sugarcane (*Saccharum officinarum L.*) belongs to the *Saccharum*genus of the Andropogoneae tribe of the Poaceae (Gramineae/grass) family with an octaploid 2n=8 × =80 number of chromosomes [1]. It is a perennial cash cropgrown in tropical and subtropical regions of the world, which accounts about 75% of the sugar in the world [2]. The commercially cultivated crops of sugarcane have two geographic centers of origin; in Newguinea and Northern India [3]. It is an important cash crop and the main source of raw material for the production of sugar [4]. It is cultivated as a commercial crop in nearly around 60 countries [5]. In Ethiopia sugarcane isgrown both as cash and industrial crop but there is no well documented reference on how, where and when it was introduced, although some records claim its introduction during the early 18th century [6]. Its properties such as efficient photosynthesis and biomass production make it an excellent target for industrial processing, valuable alternative for animal feed and production of bi-products such as ethanol production from molasses [7]. In Ethiopia the annual yield of sugar was around 300,000 tons from three established sugar factories but the annual domestic demand close to 450,000 tons [8]. However, the country has abundant water resource coupled with a vast fertile land favorable for sugarcane cultivation, suitable agro-ecological conditions, cheap labour and huge domestic and foreign demand for sugar and its by-products. So, by realizing these potentials and opportunities, the Ethiopian government has planned to establish ten sugar factories with 370,000 ha of plantation area.

Though, the edaphic and climatic factors are suitable for sugarcane cultivation, there are limited varieties under production and the country has no breeding facilities to release new high yielding and disease resistance varieties. Furthermore sugarcane is not an ideal candidate crop for conventional plant breeding; because its flowering is not-synchronised, it has low sexual seed viability and it is a perennial crop [9]. Hence, its improvement takes up to ten years from initial crosses to final agronomic assessment of elite varieties [10,3]. In addition it's vegetative propagation by cutting for commercial production takes extra years and favors dissemination of disease and pest from generation to generation [11,12]. To circumvent these limitations of conventional breeding and vegetative propagation of sugarcane, biotechnological tool (plant tissue culture) was born as best alternative.

So far in Ethiopia, new varieties have been imported and propagated vegetatively by cutting for commercial production. But currently, after importing the new varieties, to supplement the vegetative propagation, we have already started *in vitro* propagation of sugarcane varieties. Though sugarcane tissue culture is the best alternative technology

*Corresponding author: Dereje Shimelis, Ethiopian Sugar Corporation, Research and Training Division, Variety Development Directorate, Biotechnology Research Team, Wonji Research Center, P. O. box 15,Wonji, Ethiopia
E-mail: d.shimelis@yahoo.com.

or option, its operation is very difficult; because sugarcane has high content of phenols, especially in the apical meristem and leaf sheaths which poses browning problem during tissue culture process [13]. It is a serious problem associated with the culture establishment stage of micro propagation and the oxidation of phenolic substances leaches out from the explant resulting in browning of the explant, the medium and at last killing the whole culture [14]. Pretreatment of the explants with solutions of ascorbic acid, citric acid, Polyvinylpyrrolidone (PVP) and activated charcoal (AC) or culturing them with these antioxidants could solve the problem [14]. By using 0.3 gL^{-1} PVP [15] Shukla RK et al. found that all their cultures were survived. Around 80% browning free cultures (explants) were obtained by using activated charcoal of 0.2-0.5 gL^{-1}[16]. Hence, this research was conducted to optimize appropriate concentrations of PVP and AC to control the effect of phenolic oxidation on culture establishment stage of micro propagation which helps to have enough starter culture for the subsequent multiplication stage of micro propagation.

Materials and Method

The experiment was done at Plant tissue culture laboratory of Jimma University College of Agriculture and Veterinary Medicine (JUCAVM), Ethiopia. Two sugarcane genotypes, (C86-12 and C86-56), were used for the study. They were imported from Cuba in 2006. After agronomic performance evaluation; these genotypes were among the selected ones to be commercialized. The sets of these genotypes were prepared, treated with hot water and taken to JUCAVM green house and planted. After two to three months of growing, shoot tip explants were taken from the sugarcane plants. The explants were prepared according to Jalaja NC et al. procedures [12]. The surrounding leaf sheaths of sugarcane tops were carefully removed one by one until the inner white sheaths were exposed. The explants were sized to about 10 cm length by cutting off at the two ends, locating the growing point somewhere near to top. They were washed under running tap water and liquid detergents. They were socked in fungicide solution (0.3% kocid) for 30 minutes under laminar flow cabinet containing three drops of tween-20. After the kocid was properly washed off from the explants, they were rinsed three times with distilled water and disinfected with 70% ethanol for one minute. The ethanol was poured off and the explants were rinsed again with sterile distilled water. Disinfection of explants was done with 0.1% of HgCl$_2$ for 10 minutes [17] followed by 3-4 washing with sterile distilled water. The required amounts of all stock solutions of MS [18] medium, 20 gmL^{-1} sucrose [19], different concentrations of PVP and AC and 3 mgL^{-1} of BAP [20] were mixed in a beaker and the pH was adjusted to be 5.8.This was followed by addition of 0.8% agar for solidifying the media. Then, it was heated to melt the agar and then 30 ml media was dispensed in to culture jars. Finally, it was autoclaved at temperature of 121°C for 20 minutes with 15 psi of pressure.

About 1.5 cm explants were cultured under laminar flow hood aseptically and transferred to the growth room at which growth conditions were adjusted to be 16 hr of light and 8hr of dark with 25 μmolm^{-2}s^{-1} photosynthetic photon flux intensity and 26 ± 2°C of temperature. The experiments were laid down in complete randomized design with two factor factorial treatment combinations arrangements and each treatment was replicated three times. Data of survival and death of explants due to phenolic oxidation (browning) were collected after 30 days of culturing and subjected to two-way analysis of variance (ANOVA) using SAS statistical software version 9.2 (SAS Inc., 2008). Treatments means were separated by using REGWQ (Ryan-Einot-Gabreil-Welsch Multiple range test) mean separation method.

Results and Discussions

The analysis of variance showed a significant interaction effects of PVP and genotypes (p=0.001) for percent of survived and dead explants due to phenolic oxidation. Data of unchanged (neither survived nor dead) explants was not included in the analysis. Among the different concentrations of PVP, the highest percentages; 100% and 80% of survived explants were obtained at PVP concentration of 0.3 gL^{-1} for C86-12 and 0.2 gL^{-1} for C86-56 respectively.

As the concentration of PVP increased (Table 1) from 0.0 gL^{-1} to 0.3 gL^{-1}, the percentage of survived explants of C86-12 significantly increased from 0% to 100% then declined to 40% as the PVP concentration increased from 0.3 gL^{-1} to 0.5 gL^{-1}. However, the percentage of dead explants was declined from 80% to 0% then increased to 60%. For C86-56, the percentages of survived explants increased from 6.67% to 80% and then dropped to 40% as the concentration of PVP increased from 0.0 gL^{-1} to 0.2 gL^{-1} and 0.3 gL^{-1} to 0.5 gL^{-1} respectively. Hence, increasing the concentration of PVP beyond 0.2 gL^{-1} and 0.3 gL^{-1} is not economical for C86-56 and C86-12 respectively.

The best result of this experiment, i.e., PVP at a concentration of 0.3 gL^{-1} for C86-12 is in agreement with the result obtained by Shukla RK et al. [15] cultured shoot tips explants on MS media containing 0.5 mgL^{-1} IBA, 2 mgL^{-1} BAP+0.5 mgL^{-1} IAA, and 0.5 mgL^{-1} BAP+0.5 mgL^{-1}g A$_3$ with 0.3 gL^{-1} of PVP. They indicated that though the cultures showed browning after five days of culturing, most of the explants were survived. In this study, the best results obtained for both genotypes are in contrast to the results reported by Huang CM et al. [16] and Michael PS et al. [21]. Huang CM et al. [16] found 60% and 40% browning free explants for two different genotypes at a PVP concentration of (0.5-1) gL^{-1}. This could be due to genotypic differences among the materials used. MS medium supplemented with 3 mgL^{-1} of 2,4-D+0.5 gL^{-1} PVP+0.5 mgL^{-1} BAP+100 mlL^{-1} coconut water+8% agar resulted in successful initiation of large embryogenic callus ranging from 80 to 90% which were free of browning [21]. This difference may be happened due to the difference in genotypes and the type of *in vitro* regeneration path used (Table 2).

The ANOVA revealed that the interactions between concentrations of AC and genotypes were significantly influenced the percentage of survived and dead explants (p=0.001). Data of unchanged (neither dead nor survived) explants was not included in the analysis. The least percentages of survived explants were observed on MS medium without application of activated charcoal for both genotypes. The highest percentages of survived explants were observed on MS medium

Antioxidant (gL^{-1})	Genotype			
	C86-12		C86-56	
Levels of PVP	% Survived ± SD	% Dead(B) ± SD	% Survived ± SD	% Dead (B) ± SD
0	0.0^f ± 0.00	80.00^a ± 0.00	6.67^f ± 0.84	86.67^a ± 0.84
0.1	20.0^e ± 0.00	80.00^a ± 0.00	20.00^e ± 0.00	80.0^a ± 0.00
0.2	40.0^d± 0.017	60.00^b ± 0.00	80.00^b ± 0.00	0.0^e ± 0.00
0.3	100.0^a ± 0.00	0.00^e ± 0.00	80.00^b ± 0.00	20.0^d ± 0.00
0.4	60.0^c ± 0.00	40.00^c ± 0.00	40.00^d ± 0.00	40.0^c ± 0.00
0.5	40.0^d± 0.58	60.00^c ± 0.00	40.00^d ± 0.017	40.0^c ± 0.58
CV%	7.6	6.81	7.6	6.81

PVP=Poly vinyl pyrrolidone; B=Browning (phenolic oxidation).
*Values for percent of explants survived and dead given as mean ± SD.
*Numbers within the same column with different letter(s) are significantly different from each other according to REGWQ mean separation method at p ≤ 0.05

Table 1: Effects of PVP in preventing browning.

supplemented with 0.4 gL⁻¹ and 0.3 gL⁻¹ of AC for C86-12 and C86-56 respectively. For C86-12, increment of AC concentrations from 0.3 to 0.5 gL⁻¹ showed non-significant effect on percentage of survived explants. However, concentrations of 0.3 gL⁻¹ and 0.gL⁻¹ of AC resulted in 60% death of explants. At 0.4 gL⁻¹of AC only 40% of explants were died due to the effect of phenolic oxidation which is by far better than that of 0.3 gL⁻¹ and 0.5 gL⁻¹ of AC. Even though the percentage of survived explants is the same with 0.3 gL⁻¹ and 0.4 gL⁻¹ of AC for C86-56, the percentage of dead explants is 53.4% at 0.3 gL⁻¹ and 60% at 0.4 gL⁻¹ of AC respectively.

As the concentration of AC increased from 0.0 *gL⁻¹* to 0.4 *gL⁻¹* for C86-12, percentage of survived explants increased from zero to 46.6% whereas for C86-56 the increment was from zero to 40% as the concentration of AC increased from 0.0 *gL⁻¹*to 0.3 *gL⁻¹*. Increasing the concentrations of AC beyond 0.4 *gL⁻¹* for C86-12 and 0.3 *gL⁻¹* for C86-56 showed a decreasing trend in percentage of survival of explants. This could be due to fact that AC is not only adsorb plant growth inhibitory substances (phenolics) in medium produced by explants, but also adsorb the plant growth regulators and organic substances that are very important for explants to grow [22].

Results obtained at 0.4 *gL⁻¹* for C86-12 and 0.3 *gL⁻¹* for C86-56 was similar to the result reported by Huang CM et al. [16]. They reported that AC concentrations from 0.2-0.5 gL⁻¹ resulted in 80% browning free (survived) explants. However, our result contradicts with the result reported by Manchanda P et al. [23] found that MS medium supplemented with 2 *gL⁻¹* AC controlled the effect of phenolic oxidation on the medium and furthermore it increased percent of somatic embryogenesis and percent of regeneration from 80.21 to 84.88% and 75.15 to 81.22% respectively. This difference could be happened due to genotypic difference and the type of regeneration path followed.

Conclusion

In line with the current findings, it is possible to deduce that the two genotypes may have different concentrations of phenolics and polyphenol oxidase enzyme; that is why we found different concentrations of PVP and AC for both genotypes to control browning effects on culture establishment stage of micropropagation. Hence, MS medium supplemented with 0.3 g⁻¹ PVP for C86-12 and 0.2 g⁻¹ PVP for C86-56 and 0.4g⁻¹ AC for C86-12 and 0.3 g⁻¹ AC for C86-56 were obtained to be the optimum concentrations to control effects of phenolic oxidation. These concentrations can be used to have maximum survival percentage of plantlets during culture establishment stage of micropropagation to have enough starter culture for multiplication stage (Figure1).

Acknowledgements

I would like express my deepest gratitude to Ethiopian Sugar Corporation for funding the research budget and Jimma University College of Agriculture and Veterinary Medicine for facilitating the tissue culture laboratory and its facilities. Finally, I do not want to pass without telling my trust for Mami because she is the reason for my existence like now.

| Antioxidant(gL⁻¹) | Genotype | | | |
| | C86-12 | | C86-56 | |
Levels of AC	% Survived ± SD	% Dead (B) ± SD	% Survived ± SD	% Dead(B) ± SD
0.1	20.00ᵇ± 0.00	73.40ᵇᶜ± 0.58	0.00ᶜ± 0.00	86.6ᵃᵇ± 0.58
0.2	20.00ᵇ ± 0.00	60.00ᶜᵈ± 0.00	20.00ᵇ± 0.00	80.0ᵇ± 0.00
0.3	40.00ᵃ ± 0.00	60.00ᶜᵈ± 0.00	40.00ᵃ± 0.00	53.4ᵈᵉ± 0.58
0.4	46.60ᵃ ± 0.58	40.00ᵉ ± 0.00	40.00ᵃ± 0.00	60.0ᶜᵈ± 0.00
0.5	40.00ᵃ± 0.00	60.00ᶜᵈ± 0.00	20.00ᵇ± 0.00	60.0ᶜᵈ± 0.00
CV%	7.53	8.51	7.53	5.81

AC=Activated charcoal; B=Browning (phenolic oxidation).
*Values for percent of explants survived and dead given as mean ± SD.

Table 2: Effects of AC in preventing browning.

A) C86-12 genotype at 0.3 gL⁻¹ of PVP
B) C86-56 genotype at 0.2 gL⁻¹ of PVP
C) C86-12 genotype at 0.4 gL⁻¹ AC
D) C86-56 genotype at 0.3 gL⁻¹ AC

Figure 1: Survived explants.

References

1. Ather A, Khan S, Rehman A, Nazir M (2009) Optimization of the Protocols for Callus induction, Regeneration and Acclimatization of sugarcane cv. thatta-10. Pak J Bot 41: 815-820.

2. Pandey RN, Rastogi J, Sharma ML, Singh K (2011) Technologies for Cost Reduction in Sugarcane Micropropagation. African Journal of Biotechnology 10: 7814-7819.

3. Sengar RS, Sengar K, Garg SK (2011) Role of tissue culture technique in high sugarcane production. Life Sciences 1008-1017.

4. Ali A, Naz S, Siddiqui FA, Iqbal J (2008) an Efficient Protocol for Large Scale Production of Sugarcane through Micropropagation. Pak J Bot 40: 139-149.

5. Ali A, Naz SH, Iqbal J (2007) Effect of Different Explants and Media Compositions for Efficient Somatic Embryogenesis in Sugarcane (saccaharum officinarum). Pak J Bot 39: 1961-1977.

6. Duri M (1969) Private Foreign Investment in Ethiopia. J Eth Studies 7: 53-73.

7. Gallo-Meagher M, English RG, Abouzid A (2000) Thidiazuron Stimulates Shoot Regeneration of Sugarcane Embryogenic Callus. In Vitro Cellular and Developmental Biology Plant 36: 37-40.

8. Anonymous (2010) F.O, licht's International Sugar and Sweetener Report vol.142

9. Manickavasagam M, Ganpati A, Anbazhagan VR, Sudhakar B, Selvaraj N et al. (2004) Agrobacterium mediated genetic transformation and development herbicide resistant sugarcane (Saccharum species hybrids) using auxiliary buds. Plant Cell Rep 23: 134-43.

10. Cox M, Hogarth M, Smith G (2000) Cane breeding and improvement. Manual of cane growing, Bureau of sugar Experimental Stations, Indooroopilly, Australia.

11. Lakshmanan P (2005) Somatic embryogenesis in sugarcane-An addendum to the invited review sugarcane biotechnology: The challenges and opportunities. *In vitro* cellular & developmental biology 42: 201-205.

12. Jalaja NC, Neelamathi D, Sreenivasan TV (2008) Micropropagation for quality seed Production in sugarcane in Asia and the Pacific. Sugarcane pub 13-60.

13. Qin TH, Zhou ZL, Wu CW (1997) Study on the phenol pollution in tissue culture of sugarcane. Sugarcane 4: 12-14.

14. Kumari R, Verma DK (2001) Development of Micropropagation Protocol for sugarcane.

15. Shukla RK, Khan AQ (1994) In vitro clonal propagation of sugarcane: optimization of media and hardening of plants. Sugarcane 4: 21-23.

16. Huang CM, Li YR, Ye YP (2003) Minimizing Phenol Pollution in Sugarcane Stem Apical Meristem Culture. Short Communication 5: 297-300.

17. Bisht SS, Routray Ak, Mishra R (2011) Rapid in vitro propagation techniques for sugarcane variety 018. International journal of pharma and bio science 2: 0975-6299.

18. Murashige T, Skoog F (1962) A Revised Medium for Rapid Growth and Bioassay with Tobacco Cultures. Physiol. Plant. 15: 473–479.

19. Khan SA, Rashid A, Chaudhary MF, Chaudhry Z, Afroz A (2008) Rapid micropropagation of three elite Sugarcane (Saccharum officinarum L.) varieties by shoot tip Culture. African Journal of Biotechnology 7: 2174-2180.

20. Belay T, Mulugeta D, Derbew B (2014) In vitro Aseptic Culture Establishment of Sugarcane (Saccharum officinarum L.) Varieties Using Shoot Tip Explants. Adv Crop Sci Tech 2: 128.

21. Michael PS (2007) Micropropagation of Elite Sugarcane Planting Materials from Callus Culture in Vitro. Journal of the Royal Society 140: 425–426.

22. Liu GL, Liang ZH, Zhu J (2001) Effects of activated charcoal on plant tissue cultures. J Jiangsu Forestry Sci Technol 28: 46-48.

23. Manchanda P, Gosal SS (2012) Effects of Activated charcoal, Carbon Source and Gelling agents on Diret Somatic Embryogenesis and Regeneration in Sugarcane via Leaf roll segments. Sugar Tech 14: 168–173.

Determination of Crop Water Requirements of Sugarcane and Soybean Intercropping at Metahara Sugar Estate

Abera Degefa, Mengistu Bosie, Yohannes Mequanint, Endris Yesuf and Zeleke Teshome*

Sugar Corporation, Research and Development Center, Wonji, Ethiopia

Abstract

Intercropping of short duration crop with sugarcane is a remunerative practices under different irrigation levels. This study was initiated with the objective of determining and evaluating different irrigation depth and intervals under intercropping of sugarcane with soybean on yield and water use efficiency. The experiment was carried out with three depth of (75, 100 and 125% ET) in combination with three irrigation interval of (7, 12 and 16 days) with three replication of randomized complete block design (RCBD). The result revealed that it was noted highly significant difference among treatment on stalk count, tillering, stalk weight and stalk height with a highest value of 116×10^3 ha^{-1}, 126.44×10^3 ha^{-1}, 1.89 kg stalk^{-1} and 2.87 cm, respectively at 100% ET Irrigation depth with interval of 7 days, 75%ET irrigation depth with 12 days interval, 75%ET with 7 days and 75% ET irrigation depth with 12 days irrigation interval. However no significant difference was observed among treatments on cane yield, sugar quality parameters and sugar yield. It was observed highly significant difference among treatments on biomass, plant population, pod per plant and seed per pod of soybean while no significant difference was noted on weight of 1000 seeds and soybean yield. Based on the result obtained it was concluded that intercropping is good practices in realizing and achieving a sustainable advantage of farming at different irrigation levels. It is recommended to use the treatment received 75% ET of irrigation depth and 12 days of irrigation interval especially when shortage of water supply is occurred. It has a net benefit cost ratio of with the advantage of 86.47% and 83.34% over the control. For further recommendation of the treatment it is better to verify T4 (75% ET with 12 days), T6 (125% ET with 12 days) and T10 (Conventional) irrigation depth and irrigation interval.

Keywords: Intercropping; Irrigation depth; Irrigation interval; Water use efficiency; Net benefit cost ratio

Introduction

Sugarcane (*Saccharum officinarum* L.) is a long duration and widely spaced crop in comparison with other field crops. It offers a great scope for using its interspace by growing short duration crops thereby helping to harvest the potential productivity. The concept of intercropping is to obtain optimum plant population of companion crops with the adoption of sustainable planting geometry. Intercropping of short duration crop with sugarcane is a remunerative practices [1]. Intercropping has been a common practice in the tropics and it is often the general assumption that intercropping of field crops with sugarcane increases total productivity per unit area of land that lead to increase food production and income. It is the agricultural practice of cultivating two or more crops in the same space at the same time with the aims to match efficiently crop demands to the available growth resources and labor [2]. It helps in maintaining the soil fertility and making efficient use of nutrients and ensures economic utilization of land, labour and capital resources [3]. In the sugar industry, effective utilization of available resource is one of the means to minimize cost of production and maximize profit. Thus in major sugar producing countries like India, Brazil, Australia, Mauritius and South Africa intercropping is considered as one of the management options, especially for small scale farmers with limited land and inputs [4].

Soya bean is a warm season crop and it is one of the important pulse crops cultivated in Ethiopia. Soya bean is most susceptible to drought damage during flowering and grain filling. It performs well in areas where rainfall is more than 700 mm. While sugarcane being a long duration crop producing huge amounts of biomass is classed among those plants having a high water requirement and yet it is drought tolerant. The critical stages for irrigation of sugarcane are formative or vegetative period (tillering and steam elongation) stage. Due to scarcity of water resources, increasing crops productivity and saving irrigation water are the two interrelated issues raising a lot of concern these days. For achieving higher water use efficiency careful use of water resources is essential. Water use efficiency measures the quantity of water taken up by the crop during its crop life to produce a unit quantity of the output i.e., crop yield. In general, the lower the water resource input requirement per unit of crop yield produced, the higher the efficiency [5]. During intercropping the available water has been used more efficiently and more water has been made available due to the interference of deeper penetrating roots of the intercrop. The water use efficiency for harvested yield of soya bean is 4 to 7 kg/ha-mm [6].

The most important times for soybean plants to have adequate water are during pod development and seed fill. These are the stages when water stress can lead to a significant decrease in yield. Stressful conditions, such as moisture deficiency reduces soybean yield. Since the practice of soya bean intercropping with sugarcane is new no research was done on the determination of appropriate depth and interval of irrigation water on soya bean and sugar cane intercropping. Therefore the objective of this research was to determine appropriated depth and interval of irrigation water for intercropped sugarcane and soya bean.

***Corresponding author:** Zeleke Teshome, Sugar Corporation, Research and Training, PO Box 15, Wonji, Ethiopia, E-mail: zeleketeshome@gmail.com

Objectives

➢ To determine appropriate depth and interval of irrigation water for soybean-sugarcane intercropping at MSF.

➢ To evaluate the effect of different irrigation schedules on the yield of soybean intercropping.

➢ To determine the water use efficiency of the crop under different irrigation depth and interval.

Materials and Methods

Description of the study area

Metahara Sugar Factory is located 200 km from Addis Ababa in the southeast direction. It is situated at 8° 53' N latitude and 39° 52' E longitude at an altitude of 950 meters above sea level (masl). The area has a semiarid climatic condition. Most of soils of the area are Haplic Cambisols and a few are Hypersalic or Haplic Solonchaks because of salinity. The long term climatic condition of the area has a minimum and maximum temperature of 17.36°C and 32.97°C, respectively with the annual rainfall is 533 mm. Whereas the average relative humidity, sunshine hour and wind speed of the area is 77.44%, 8:46 AM, 4.12 respectively.

Experimental design and lay out

The treatments were laid out in randomized complete block design (RBCD) with three replications having three levels of irrigation water depth with three levels of irrigation interval. The experiment was based on daily evapotranspiration of the estates with the value of 75%, 100% and 125% of the irrigation water requirement and an interval of 7, 12 and 16 days respectively. The irrigation water requirement was calculated from the irrigation schedule of Methara and arranging the irrigation interval by a systematic method. The plot size was 14.5 m width (10 furrows) and 10 m furrow length with furrow spacing of 1.45 m. The middle eight furrows were used as net plot furrows for data collection. Space between plots was 2 m while 2 furrows between replications.

Crop establishment and management

Sugarcane was planted on 04 April 2013 whereas soybean was planted after 15 days of the major crop was planted on 19 April 2013. The varieties used were B52-298 for sugarcane and Williams for soya bean, respectively. Soybean was sowing double row at side of ridges and sugarcane was planted in furrows.

Irrigation practice

Equal amount of shallow irrigation was applied to every furrow of each treatment in each plot two days before planting as well as frequently for 15 days to encourage a full and even plant stand at the time of emergence. Treatments application was started the moment soya bean was sown and irrigation was stopped for furrow irrigation systems, two inches parshall flume was installed at the inlet of experimental plots for measuring the amount of water applied for each plot. At each volume of water applied for each plot of a given interval, the height of water in the parshall flume was measured using a ruler. Each depth of water passing through the parshall flume has known amount discharge rate in litter per second.

Water use efficiency (WUE)

Field water use efficiency is a ratio between marketable crop yield and field water supply which includes water used by the plant in metabolic activities, ET and deep percolation losses [5].

$$WUE = \frac{Y}{WR}$$

Where,

- WUE=Field water use efficiency, kg/ha-mm
- Y=Crop yield, kg/ha
- WR=Water used in metabolic activities, ET and deep percolation losses, mm.

Data collection

The major sugarcane and soybean yield and yield components had been collected following the standard procedure of data collection. The data for sugarcane encompasses stalk count, tillering, length, cane yield, sucrose percent, pol (%), Brix (%) and juice purity and sugar yield whereas the data for soybean were plant population count, number of pods/plant, number of seeds/pod, weight of 1000 seeds, aboveground biomass and yield.

Data analysis

The result of analysis was subjected and analyzed by SAS software 9.1 version [7] to saw the effect of among the treatments.

Results and Discussion

Effect of different irrigation depth and interval on sugarcane response under intercropping with soybean

The result depicted in Table 1, stalk count showed highly significant difference (P<0.01) among treatments and it was noted a maximum of 116×000 ha^{-1} the treatment received the 100%ET and an Irrigation interval of 7 days as compared with the controlled T10. The highest tiller count (126.44×10^3 ha^{-1}) was observed at the treatment having a combination of an irrigation depth of 75% ET and 12 days interval followed (125.58×000 ha^{-1}) with irrigation depth of 75%ET and 16 days. To this effect, it was observed no significant relationship among treatments on sugarcane yield though it was noted a numerical difference. These results are in agreement with the findings of Hossain [8]. Whereas the smallest tiller count (111.81×000 ha^{-1}) was noted at the treatment received 125% ET of irrigation depth and 16 days of irrigation interval and this might be due to the presence of waterlogging by the application of high irrigation water which hinders the sprout of the tillering. Similarly it was observed highly significant difference on the stalk weight of sugarcane while comparing with the controlled practice with the highest stalk weight of 1.89 kg per stalk at 75% ET and 7 days of irrigation interval. This might be due to less competition of the tillering in utilization of the applied water. Even though there were no significant relationship observed on stalk height between the treatment having the maximum value and the controlled, it was

Treatment	Irrigation depth, % ET	Irrigation Interval, days
T1	75	7
T2	100	7
T3	125	7
T4	75	12
T5	100	12
T6	125	12
T7	75	16
T8	100	16
T9	125	16

Table 1: Treatment combination.

noted that highest stalk height (T4=2.84 cm) at the treatment with an irrigation depth of 75% ET and 12 irrigation interval.

Brix (%): Irrigation depths in combination with irrigation intervals had no significant effect on brix percent at different treatments which was shown in Table 2. Even though no significant difference among treatments it was noted highest brix per cent (19.11%) at the treatment received an irrigation depth of 75% ET with 16 days of irrigation interval as compared with the controlled. While the lowest brix percentage (17%) was observed at irrigation depths of 75% ET and 7 days of irrigation intervals.

Pol (%): Table 2 showed no significant difference among treatments but numerically it was seen that highest pol percent at (T4=17.29%) with an irrigation depth of 75% ET and 12 days of irrigation intervals while lowest was observed at (T1=15.98) with an irrigation depth of 75% ET and 7 irrigation intervals.

Purity (%): Irrigation depths in combination with irrigation intervals had shown significant effect on purity percent among all treatments which was shown in Table 2. The treatment received an irrigation depth of 100% ET with 7 days of irrigation interval revealed highest purity (92.14%) as compared with the controlled. This might be due to the presence of low soluble solids in the sugar cane. While the lowest purity percentage (89.31%) was observed at irrigation depths of 75% ET and 7 days of irrigation intervals. This result is in consonance with the findings of MS Rahman [8]. Though it was observed no significant variation among treatments, the highest sugar yield was noted at an irrigation depth of 75% ET and 12 days of irrigation intervals.

Effect of soybean as influenced by different irrigation depth and interval under intercropping of sugarcane

Table 3 revealed that highly significant difference in biomass of soybean among treatments of different irrigation depth and interval under intercropping. The highest biomass (20.95 Qt ha⁻¹) was seen the treatment received 125% ET and an irrigation interval of 16 and the lowest (15.34 Qt ha⁻¹) was observed. Similarly, it was observed highly significant difference among treatments on plant population, pod per plant and seed per pod. On the contrary, no significant difference was seen both weight of 1000 seed and yield of soybean at different levels of irrigation depth and irrigation intervals under intercropping with sugarcane.

Overall performance of the water use efficiency of the crops

It was observed a highly significant difference on the water use efficiency of sugarcane while intercropped with soybean as depicted on the Table 1. The highest water use efficiency of sugarcane (906.75 kg/ha-mm) was noted on the treatment received 75% ET of irrigation depth and 12 days of irrigation interval while the smallest was 442.79 Kg/ha-mm on the treatment received at 125% ET irrigation depth and 16 intervals of irrigation. This might be due to less completion of the tillering during the extraction of the water. Similarly, Table 3 revealed highest significant difference with the highest water use efficiency of (1.35 kg/ha-mm) on the treatment received 75%ET irrigation depth and 7 days of irrigation interval but it was noted a smallest water use efficiency (0.69 kg/ha-mm) on the treatment 125% ET irrigation depth and 7 days interval (Tables 4 and 5).

Partial budget analysis

The net benefit cost ratio (NBCR) under intercropping of sugarcane at different irrigation levels was ranged between 2.34 to 2.81. It was observed the highest at the treatment received 75% ET and 12 days of irrigation inter while the lowest was on the treatment of 125% ET of irrigation depth and 7 days interval. This might be due to the reason that the highest return achieved by the good practice of the intercropping of sugarcane with soybean. The higher NBCR was noted at (T4=2.81 and T6=2.72) with an irrigation depth of 75% ET and 12 days of irrigation intervals in 125% ET and 12 days irrigation interval. This might be due to the advantage of higher water use efficiency by both crops and less application cost of the irrigation water. Similarly, it might be due to higher sugar yield 20.71 t ha⁻¹ of the former treatment.

Conclusions and Recommendation

From this study, the highest tiller count (126.44 × 103 ha⁻¹) was observed at the treatment having a combination of an irrigation depth of 75% ET and 12 days interval. It has an advantage of having a tiller with an increasing order of T9>T1>T8>T3>T6>T5>T2>T7 and T4 with 76.54, 79.23, 80.94, 81.91, 82.37, 83.44, 83.72, 85.96, 86.55% as compared with the controlled. Similarly the highest cane yield 170.36 t/ha was observed though there was no significant difference among treatment with same irrigation intervals. The treatment received an irrigation depth of 100% ET with 7 days of irrigation interval revealed highest purity (92.14%) as compared with the controlled. Though no significant difference among treatments it had highest sucrose percent (12.14%) with an irrigation depth of 75% ET with 7 days of irrigation

Treatments	Stalk Count (x'000 ha⁻¹)	Tillering (x'000 ha⁻¹)	Stalk height (cm)	Stalk weight (Kg/stalk)	Cane Yield(t/ha)	Water use efficiency (Kg/ha-mm)
T1	110.26ᶜ	115.75ᵈᵉ	2.76ᵃᵇᶜ	1.89ᵃ	166.99	817.79ᵃᵇ
T2	116.00ᵇᶜ	122.30ᵇᶜᵈ	2.52ᵇᶜ	1.66ᵃᵇᶜ	153.80	625.25ᶜᵈ
T3	115.26ᵇᶜ	119.66ᵇᶜᵈᵉ	2.78ᵃᵇ	1.69ᵃᵇᶜ	156.08	459.91ᵈᵉᶠ
T4	116.87ᵃᵇᶜ	126.44ᵇ	2.87ᵃ	1.82ᵃ	170.04	906.75ᵃ
T5	115.11ᵇᶜ	121.90ᵇᶜᵈ	2.73ᵃᵇᶜ	1.74ᵃᵇᶜ	159.89	598.79ᶜᵈᵉ
T6	117.36ᵃᵇᶜ	120.34ᵇᶜᵈ	2.84ᵃ	1.82ᵃ	170.36	530.20ᶜᵈᵉ
T7	121.15ᵃᵇ	125.58ᵇᶜ	2.48ᶜ	1.48ᶜ	145.25	675.03ᵇᶜ
T8	117.56ᵃᵇᶜ	118.25ᶜᵈᵉ	2.67ᵃᵇᶜ	1.76ᵃᵇᶜ	165.44	540.01ᶜᵈᵉ
T9	116.98ᵃᵇᶜ	111.81ᵉ	2.79ᵃᵇ	1.79ᵃᵇ	168.34	442.79ᵉᶠ
T10	125.98ᵃ	146.09ᵃ	2.80ᵃᵇ	1.49ᵇᶜ	150.78	319.05ᶠ
Mean	117.25	122.81	2.72	1.71	160.70	591.54
CV (%)	4.77	3.81	6.04	10.28	11.76	17.30
Significance level	**	**	**	**	ns	**
LSD (0.05)	11.15	9.32	0.28	0.30	32.42	175.63

Table 2: Effect of Cane yield under intercropping of Sugarcane with soybean as affected by different irrigation depths and intervals.

Treatments	Pol (%)	Brix (%)	Purity (%)	Sugar Yield (t/ha)
T1	15.98	17.88	89.31^{ab}	18.67
T2	17.28	18.76	92.14^{a}	19.04
T3	16.17	18.11	89.28^{ab}	17.50
T4	17.29	19.05	90.76^{ab}	20.71
T5	16.37	18.37	89.12^{ab}	18.23
T6	17.04	19.10	89.18^{ab}	20.19
T7	17.19	19.11	89.97^{ab}	17.62
T8	16.54	18.83	87.87^{b}	18.79
T9	16.28	18.17	89.30^{ab}	19.26
T10	17.46	19.30	90.47^{ab}	18.50
Mean	16.75	18.66	89.73	18.85
CV (%)	6.32	5.26	2.01	16.10
LSD (0.05)	1.82^{ns}	1.68^{ns}	3.10	5.206^{ns}

NB: Treatments in the same column having the same letters are not significantly different but with different letters are significant at P=0.05

Table 3: Response of Sugar yield under intercropping with soybean as affected by different irrigation depths and intervals.

Treatments	Biomass (Qt/ha)	Plant population	Pod/plant	Seed/pod	Wt of 1000 seed	Yield (kg/ha)	Water use efficiency (Kg/ha.mm)
T1	17.54^{ab}	388.77^{b}	31.57^{ab}	2.31^{d}	103.70	309	1.35^{a}
T2	19.95^{ab}	671.67^{a}	34.95^{a}	2.40^{abcd}	107.43	264	0.87^{bc}
T3	16.57^{ab}	664.33^{a}	19.59^{b}	2.36^{bcd}	111.10	266	0.69^{bc}
T4	19.12^{ab}	617.33^{a}	24.46^{ab}	2.56^{a}	106.93	294	1.28^{a}
T5	15.34^{b}	598.00^{a}	21.94^{ab}	2.35^{cd}	110.27	256	0.84^{bc}
T6	15.61^{ab}	588.67^{a}	21.27^{b}	2.51^{abc}	119.53	216	0.56^{c}
T7	18.07^{ab}	635.00^{a}	22.84^{ab}	2.54^{ab}	106.53	266	1.02^{ab}
T8	19.89^{ab}	656.33^{a}	26.44^{ab}	2.33^{cd}	105.07	299	0.86^{bc}
T9	20.95^{ab}	695.33^{a}	25.61^{ab}	2.24^{d}	102.43	334	0.78^{bc}
Mean	18.11	612.82	25.40	2.40	108.11	280	0.92
CV (%)	17.27	17.70	30.64	4.58	9.23	25.99	25.02
Significance level	*	*	*	*	Ns	Ns	*
LSD(0.05)	5.41	13.48	13.48	0.19	17.28	126	0.39

NB: Treatments in the same column having the same letters are not significantly different but with different letters are significant at P=0.05

Table 4: Response of Soybean yield and yield attributing components under intercropping of Sugarcane as affected by different irrigation depths and intervals.

Treatments	Net return, birr	Gross Benefit Cost Ratio (GBCR)	Net Benefit Cost Ratio (NBCR)
T1	212857	4.16	2.52
T2	217377	3.59	2.56
T3	195041	3.37	2.34
T4	242490	3.83	2.81
T5	205745	3.47	2.46
T6	233471	3.74	2.72
T7	197050	3.39	2.37
T8	214733	3.56	2.55
T9	222131	3.64	2.62
T10	204991	3.46	2.43

Table 5: Partial budget analysis of sugarcane with soybean intercropping at different irrigation levels.

interval. Biomass of soybean revealed highly significant difference in among treatments of different irrigation depth and interval under intercropping. The highest biomass (20.95 Qt ha^{-1}) was seen the treatment received 125% ET and an irrigation interval of 16 and the lowest (15.34 Qt ha^{-1}) was observed. While highest soybeans yield (334 Kg ha^{-1}) was noted at 125% ET of irrigation depths and 16 days of irrigation intervals. Based on the result obtained it was concluded that intercropping is good practices in realizing and achieving a sustainable advantage of farming at different irrigation levels. It is recommended to use the treatment received 75% ET of irrigation depth and 12 days

of irrigation interval especially when shortage of water supply is occurred. For further recommendation of the treatment it is better to verify T4 (75% ET with 12 days), T6 (125% ET with 12 days) and T10 (Conventional) irrigation depth and irrigation interval.

References

1. Sharma RK, Bangar KS, Sharma SK, Gwal HB, Ratore OP (1992) Competitive performance of intercropping of pulses in sugarcane under southwestern Madhaya Pradesh. Indian Journal of Agronomy 37: 420-423.

2. Lithourgidis AS, Dordas CA, Damalas CA, Vlachostergios DN (2011) Annual Intercrops: an Alternative Pathway for Sustainable Agriculture. Aus J of Crop Science 5: 396-410.

3. Nazir MS, Jabbar A, Ahmad I, Nawaz S, Bhatti IH (2002) Production Potential and Economics of Intercropping in Autumn Planted Sugarcane. Int J Agric Biol 4: 140-142.

4. Parsons MJ (2003) Successful Intercropping of Sugarcane. Proc S Afr Sug Technol Ass 7: 77-85.

5. Rao VP, Suneetha KB, Hemalatha D (2010) Irrigation water management. Department of Agronomy, College of Agriculture, Rajendranagar, Telangana, India.

6. Doorenbos J, Kassam AH (1986) Yield Response to Water. FAO Irrigation and Drainage Paper, Rome, Italy, pp: 259-279.

7. SAS (Statistical Analysis System) (2002) Institute of Applied Statistics and SAS programming Language. Cary, North Carolina, USA.

8. Hossain SMI, Eusufzai SUK, Rahman MA (2008) Effect of Different Irrigation Levels on Growth and Yield of Sugarcane. Bangladesh J Sugarcane 30: 51-61.

9

Effects of Pre Cutting Nitrogen Application Rate and Time on Seed Cane Quality of Sugarcane (*Saccharum officinarum* L.) Crop at Finchaa Sugar Estate

Mijena Bikila[1]*, Nigussie Dechassa[2] and Yibekal Alemayehu[2]

[1]*Ethiopian Sugar Corporation, Research and Training Division, Sugarcane Production Research Directorate, Agronomy and Protection Research Team, Finchaa Research Center, P.O. Box 5734, Finchaa, Ethiopia*
[2]*Haramaya University, College of Agriculture, Haramaya, Ethiopia*

Abstract

Field experiment was conducted at Finchaa Sugar Estate during the 2010/11 cropping period to assess the effect of rate and time of pre-cutting nitrogen fertilization on seed cane quality of sugar cane (*Saccharum officinarum* L.) crop. The treatments consisted of four levels of N (0, 23, 46, 69 kg N ha^{-1}) and four times of N application (8, 6, 4 and 2 Weeks before Cutting (WBC). The field experiment was laid out as a RCBD in a factorial arrangement and replicated three times. A sugarcane cultivar named N-14 was used as a planting material. The experiment was conducted on a sandy clay luvisol. Analysis of variance of the data revealed that the pre-cutting N application rate had no significant main effect on seed cane sett quality at the time of harvest. In contrast, pre-cutting N application time had significant main effect on seed cane stalk height, girth, reducing sugar, moisture and total N content. Thus, treating seed cane with N 8WBC for commercial planting resulted in improvement of seed cane stalk height, girth, reducing sugar and moisture content. However, significant improvement in total N content occurred in response to treating the crop with N 2WBC. The two main factors interacted to significantly influence only seed cane stalk height, reducing sugar and total N content. Pre-cutting N treatment with the rate of 23 kg ha^{-1} 8WBC resulted in higher seed cane height. However, pre-cutting N treatment with the rate of 69 kg ha^{-1} 8WBC resulted in higher reducing sugar and total N content. On the other hand, treating the seed cane with N at the rate of 23 kg ha^{-1} 2WBC resulted in significantly higher total N content, but this value was in statistical parity with the nitrogen content of the seed cane that was obtained from the treatment of 69 kg N ha^{-1} applied 8 WBC. it was generally observed that early application (8WBC) of N at higher dose (69 kg ha^{-1}) or also late application (2WBC) of N at lower dose (23 kg ha^{-1}) improved seed cane quality through significantly enhancing reducing sugar and total N contents, respectively. Therefore, As the precutting N application rate had no significant main effect, it could be concluded that, treating the seed cane with N at the lower rate of 23 kg ha^{-1} 8WBC had a dual advantage of improving quality (in terms of height, reducing sugar and total N content) of the seed cane crop.

Keywords: Weeks Before Cutting (WBC); Pre-cutting N application; Seed cane; Seed cane quality (Sugar cane)

Introduction

Intensification of sugarcane cultural treatments (fertilization, irrigation, weeding, etc.) and agronomic related production factors play an important role for producing high cane and sugar yield per unit area. However, lack of proper cultural practices specific to fertilizer treatments on seed cane fields are among the major constraints of sugarcane production in Ethiopia in general and at Finchaa Sugar Estate in particular. Usually, fields of light red chromic luvisol have been allocated for the seed cane production and fertilized with Di-ammonium phosphate (DAP) with the rate of 250 kg ha^{-1} at planting and 150 kg urea ha^{-1} at about two and half months after planting in the same way as the commercial cane. However, the purpose of production and quality required to obtain is too different [1].

Seed cane production, which is an integral component of sugar production, often receives less priority than the commercial crop plants in many sugar cane plantations [2,3]. Most of the research works on seed cane has focused also on the mechanics of cutting and fungicidal treatments of setts. Therefore, little effort has been made to improve cultivation of seed cane [3,4]. On the other hand, it is emphasized that improving seed cane production is the basis for satisfactory crop stand establishment. Hebert [5] pointed out that planting sound pieces with high germination capacity is very essential in order to maintain a uniform stand of sugar cane that ultimately produces high cane and sugar yield. Thus, it is unquestionable that seed cane plants should

receive special cultural treatments such as fertilization, irrigation, crop protection measures, and etc.

Apart from other cultural practices, fertilization is one essential input for seed cane production. In general, the importance of balanced fertilizer application to sugar cane has often been emphasized [6-8], since sugar cane by its nature requires nutrients such as nitrogen, phosphorous and potassium in large quantities. According to Russell [9] nitrogen occupies the highest position in the nutrition of sugar cane. Nitrogen fertilization enhances the growth of sugar cane and enables the plants to take up other nutrients [10].

Hebert [5] asserted that increasing the level of nitrogen to the optimum requirements of seed cane plants correspondingly increases

*Corresponding author: Mijena Bikila, Ethiopian Sugar Corporation, Research and Training Division, Sugarcane Research Directorate, Agronomy & crop Protection Research Team, Finchaa Research Center, P.O. Box 5734, Finchaa, Ethiopia, E-mail: selammijena@gmail.com

the quality of seed sett. In the establishment of satisfactory crop stand, attaining good germination or emergence is very important, since germination constitutes a critical stage in the life of the plant [11]. In the physiological process of germination, it is a well-established fact that seed cane is largely dependent upon its own food reserve. Well treated and nourished seed canes have been found to have good germination capacity and vigour of the subsequent crop. According to King [12], plants with high nitrogen content germinate better than setts obtained from a poorly fertilized crop. Kakde [6] Suggested that application of nitrogen on seed cane at 50% greater than the recommended dose applied to the commercial crop leads to higher quality of seed cane sett for the subsequent commercial crop planting.

The intrinsic stalk characteristics are important factors on sprouting and tillering because of the fact that sugarcane is propagated by stalk cuttings and hence sprouting and the early stage of growth depend largely on the reserve food in the sett [13]. The findings of other researchers revealed that the condition of buds, nutrient reserves in setts, size and age of the stalk used affect sprouting and shoot development [6,14]. According to Kakde [6] fresh succulent condition of a bud enhances rapid multiplication of cells and quicker sprouting of buds. Soluble nitrogenous compounds and starch in sett are essential for cell division and elongation while glucose is used for formation of cell walls [15]. Adequate moisture content in setts is important for quicker sprouting as buds remain fresh, plump and healthy [6]. Dellewijin [11] reported that sprouting ability of the buds seems to be positively correlated with the moisture, nitrogen and glucose content of the bud tissue.

Nevertheless, in the Ethiopian Sugar Estates except allotting fertile and good fields for seed cane production, the seed cane plants are fertilized in the same way as the commercial cane fields, despite the difference in the purpose of production. In addition, information on optimum level and time of pre-cutting nitrogen application on seed cane of sugarcane and its effect on sett quality is inadequate. Therefore, the ultimate yield and quality of sugar may still be limited by low quality and poor performance of seed cane production in the country in general and at Finchaa Sugar Estate in particular. This is particularly true with regard to investigating the influence of pre-cutting nitrogen fertilization of sugarcane seed crop on its sett quality. Therefore, this study was initiated with the following objective: To study the effect of pre-cutting nitrogen application rate and time on the quality of seed cane setts.

Material and Method

The study was conducted on sandy clay luvisol soil type of the Estate and the variety Natal 14 (N-14) was used for the study propose. The treatments consisted of four rates of pre-cutting N application (0, 23, 46, and 69 kg N ha^{-1}) and four timings of N application (8, 6, 4, and 2 weeks) prior to cutting for seed cane (i.e. at nine months' crop age) for the subsequent commercial crop planting. At the time of N treatment application (at seven, seven and a half, eight, and eight and a half months crop age after planting of the seed cane nursery), the availability of soil moisture was checked for nutrient dissolution and absorption.

Totally, there were 16 treatment combinations. The experiment was laid out as Randomized Complete Block Design (RCBD) in a factorial arrangement and replicated three times per treatment. The total area of each experimental plot was 4 rows x 5 m x 1.45 m between each rows (29 m^2). The spaces between plots and replications/blocks were kept at 1.5 m and 2.9 m apart, respectively.

Ten months old, healthy and two budded seed cane setts were prepared from initial seed cane field of the Estate. The setts were prepared a day before planting. To protect the sett from various diseases, a Lysol solution (120 ml Lysol in one litre of water) was used for disinfecting cane knives after each cutting of stool and chopping of one stalk. Also to protect the chopped sett from soil borne diseases and pests after planting, the chopped seed material was completely dipped/immersed in a fungicide called Ben-late/Benomyl solution (180 g in 200 litres of water) for 1-2 minutes using a steel wire basket which is made for the free flow of chemical solution in the drum and the solution is drained off into the drum while removing the chopped sett in the steel wire basket after treatment [4].

The treated setts were planted in the furrow with 5 cm overlapping of sett arrangement on the pre-irrigated experimental field on 7th April 2010. Number of sett planted per five meter furrow length were determined based on the average length of the prepared setts before planting by carrying out a simple trial on the spot, i.e. thirty five two bud setts. At planting, the conventional recommended rate of 250 kg DAP (Di-ammonium phosphate, 46% P_2O_5 and 18% N) ha^{-1} (45 kg N + 115 kg P_2O_5) was equally applied for all experimental plots manually in the furrows and immediately covered with a 2-5cm layer of soil. Similarly, Urea (46% N) at the rate of 150 kg ha^{-1} (69 kg N) was equally applied to all experimental plots manually in both side of the furrow after two and a half months of planting the seed cane crop. Other cultural practices like weeding and irrigation were done according to the standard norms of the Estate.

After five months' growth of the crop, the cane fields were dense and had closed canopies that made movement of humans inconvenient for manual operation (fertilizer application). Therefore, to make manual pre-cutting nitrogen treatment application accessible at seven, seven and half, eight and eight and half months after planting of the seed cane nursery, detrashing of dry and lower leaves of all experimental plots was done one day before application of the first treatment (i.e. at seven months' crop age).

Pre-cutting nitrogen treatment in the form of urea (46% N) at the rate of 0, 23, 46, and 69 kg ha^{-1} was applied in both sides of the furrow through band application at 8, 6, 4 and 2 weeks before cutting the seed cane for the subsequent commercial planting, i.e. at seven, seven and a half, eight and eight and a half months after planting of the seed cane nursery, respectively.

Data collection and measurements

Seed canes propagated on experimental field were harvested at nine months crop age for the subsequent commercial crop planting. The following under mentioned seed cane quality parameters were measured, counted and analyzed during the harvest of seed cane:

Stalk length/height: Mean Seed cane stalk length (from zero ground level to the top visible dewlap) was measured from 20 randomly selected representative primary stalk samples of seed cane per plot at harvest by using a tape meter.

Seed cane stalk diameter/girth: Mean seed cane stalk diameter/girth was measured after removal of sheath from three seed cane stalk positions; top, middle and bottom parts by using a standard calliper meter from the above 20 randomly selected representative primary stalk samples of seed cane per plot at harvest.

Number of internodes per seed cane stalk: The number of internodes per seed cane stalk (after removal of sheath and top succulent parts) was counted from the above 20 randomly selected representative primary stalk samples of seed cane per plot at harvest.

Seed cane stalk weight per stalk: The weight of seed cane stalk was taken from the above 20 randomly selected representative primary stalk samples of seed cane per plot at harvest and expressed as weight per stalk. The above samples per specified treatment/plot were subjected to a laboratory analysis. The sample seed cane stalks were crushed with Jeffco cutter grinder having 10HP Model 265BX to disintegrate cane stalk tissue into the required grade. Fiberized cane was sampled for the analysis of different parameters like nitrogen, fibre and moisture content of seed cane stalk sample. To extract the juice, the crushed/fiberized cane samples were pressed with a 110 bar hydraulic press machine to collect the required quantity of juice for analysis of different parameter like reducing sugar, Brix% cane, Pol% cane and total non-sugar content of seed cane stalk sample following the methods and procedures developed for each parameters of cane juice analysis and fiberized cane tissue quality assessment (ICUMSA, 1994).

Total nitrogen content: Fiberized seed cane stalk by Jeffco cutter grinder was sampled and oven-dried at 65°C for 24 hrs and then at 60°C to a constant weight and the dried sample was grind by the standard leaf tissue sample grinding machine. Total nitrogen content of the fiberized seed cane stalk tissue sample was determined by the Micro-Kjeldahl digestion, distillation, and titration method (FAO, 2008).

Reducing sugar content: Reducing sugar content (percentage of other sugars like fructose and glucose) in the planting material/seed cane stalk sample was determined by using the Lane and Eynon's volumetric method outlined by (ICUMSA, 1994).

Cane moisture content (%): Moisture content of seed cane stalk sample was determined from samples taken from crushed/fiberized cane. A test portion of prepared cane was dried under standard conditions using a forced draught drying oven (moisture teller) and the moisture content was determined from the mass loss following the procedure outlined by (ICUMSA, 1994).

Cane moisture content (%) = (M2 - M3)/(M2 - M1)

Where: M1 - weight of empty container; M2 - weight of empty container plus the fresh sample; M3 - weight of empty container plus the dried sample

Fiber content of seed cane stalk (%): The fibre percent of the seed cane stalk sample was calculated using the analytical data of dry substance percent cane and Brix percent extract following the procedure outlined by (ICUMSA, 1994).

Fiber % cane = $(D - 4b_e)/(1 - 0.0125b_e)$

Where: D - Dry substance percent, be - Brix percent extract

Pol% cane, Brix % cane and Total non-sugar content: Pol% cane, Brix % cane and total non-sugar content of the juice were determined according to the methods out lined by ICUMSA (1994).

Brix % cane (Bc) = $b_e (4 - 0.0125 F)$ Where: b_e = Brix extract

Pol % cane (Pc) = $P_e (4 - 0.0125 F)$ p_e = Pol extract

Total non-sugar = 100 – sugar yield   F = Fiber % cane

Statistical analysis

Data collected from the experiments was subjected to analysis of variance (ANOVA) with the GLM procedure of the SAS for windows software version 9.0 [16]. Differences between treatment means were separated using the Least Significant Difference Test (LSD) at 5% level of significance following the methods described by [17].

Results and Discussion

Effect of pre-cutting nitrogen application rate and time on the quality of seed cane stalk/sett

In the study, nitrogen levels had no significant main effect on all seed cane stalk/sett quality parameters. In contrast to this, pre-cutting nitrogen application time had significant ($P \leq 0.05$) main effect only on seed cane stalk height, girth, reducing sugar, total nitrogen and moisture content. The two main factors interacted to significantly ($P \leq 0.05$) influence only seed cane stalk height, reducing sugar and total nitrogen content of seed cane at harvest (Appendix Tables 1 and 2).

Effect of pre-cutting nitrogen application rate and time on seed cane stalk height and girth

The different levels of pre-cutting nitrogen application on seed cane of sugarcane did not significantly ($P \leq 0.05$) affect the stalk height and girth of seed cane at the time of harvest. However, time of pre-cutting nitrogen application on seed cane of sugarcane showed significance main effect ($P \leq 0.05$) on seed cane stalk height and girth/diameter (Appendix Table 1). In addition, the interaction effect of N levels and times of application showed significant differences ($P \leq 0.05$) on seed cane stalk height during the time of harvest for planting (Table 2 and Appendix Table 1).

When the time of pre-cutting nitrogen application was reduced from 8 weeks to 6 weeks before cutting for seed cane, there was no significant change in stem girth (Table 2). However, when the pre-cutting time was shortened to 4 and 2 weeks before harvesting for seed cane, the girth of the plants decreased significantly ($P \leq 0.05$). Thus, plants treated with nitrogen at 6 weeks before cutting for seed cane had about 1.3 and 2.4% wider stem girth than those supplied with nitrogen 4 and 2 weeks before cutting for seed cane, respectively. However, stem girth of plants treated with nitrogen 6 weeks before cutting the seed cane for planting and stem girth of those plants treated with nitrogen 8 weeks before cutting were in statistical parity. Therefore, the highest mean seed cane stalk girth was recorded by treating the seed cane with N at 6 weeks prior to cutting for seed cane but this was statistically the same with that treated 8 WBC (Table 2). Therefore, pre-cutting N treatment 8 and 6 WBC produced stalk girth that was higher by about 2.44% than that produced at 2 WBC. The significant reduction in stem girth with the reduction in the pre-harvest duration of nitrogen application before cutting the seed cane for planting could

Treatment No.	Rate and time of N application prior to cutting for seed cane	Treatment No.	Rate and time of N application prior to cutting for seed cane
T1	0 kg N ha⁻¹@ 8 WBC	T9	0 kg N ha⁻¹@ 4 WBC
T2	23 kg N ha⁻¹@ 8 WBC	T10	23 kg N ha⁻¹@ 4 WBC
T3	46 kg N ha⁻¹@ 8 WBC	T11	46 kg N ha⁻¹@ 4 WBC
T4	69 kg N ha⁻¹@ 8 WBC	T12	69 kg N ha⁻¹@ 4 WBC
T5	0 kg N ha⁻¹@ 6 WBC	T13	0 kg N ha⁻¹@ 2 WBC
T6	23 kg N ha⁻¹@ 6 WBC	T14	23 kg N ha⁻¹@ 2 WBC
T7	46 kg N ha⁻¹@ 6 WBC	T15	46 kg N ha⁻¹@ 2 WBC
T8	69 kg N ha⁻¹@ 6 WBC	T16	69 kg N ha⁻¹@ 2 WBC

Where, WBC = Weeks before cutting for seed cane; N = Nitrogen.

Table 1: Treatment number and the respective amounts of additional nitrogen applied to the seed cane plants at different rates and times prior to cutting for seed cane at Finchaa, 2011.

Treatment	Seed cane stalk girth (mm)
N rate (NR) (kg ha⁻¹)	
0	28.39
23	28.43
46	28.68
69	28.73
SE±	0.17
LSD	NS
Time of N application (NT) (WBC)	
8	28.67ab
6	28.94a
4	28.56b
2	28.25b
SE±	0.17
LSD	0.50
Mean	28.55
CV (%)	2.12

Where, NS = Non-significant; WBC =Weeks before cutting for seed cane; means followed by the same letters are not significantly different at 5% level of significance according to the LSD test.

Table 2: Seed cane stalk girth as influenced by pre-cutting N application rate and time, at Finchaa, 2011.

be attributed to the short time left for the plant to effectively utilize the nitrogen for growth and development. This result is in agreement with that of Lakshimi et al. [18] who also reported that seed cane yield in terms of girth varied significantly with time of nitrogen application.

As a result of the interaction effect of the two factors, significantly tallest seed cane stalk resulted in response to applying nitrogen at the rate of 23 kg N ha⁻¹ to the standing seed cane crop at 8 weeks before cutting for seed cane (Table 3). However, the seed cane stalk height attained at the rate of 23 kg N ha⁻¹ supplied at 8 weeks before cutting for seed cane was in statistical parity with the seed cane stalk height obtained at the treatment combinations of 0 kg N ha⁻¹ + 8 weeks before cutting, 46 kg N ha⁻¹ + 2 weeks before cutting, 69 kg N ha⁻¹ + 6 weeks before cutting, 69 kg N ha⁻¹ + 4 weeks before cutting, and 69 kg N ha⁻¹ + 2 weeks before cutting. The probable reason why higher seed cane stalk height resulted by untreated plot might be due to higher variance in measurement. Pre-cutting N application on standing seed cane crop prior to cutting for planting with a rate of 23 kg N ha⁻¹ 8 WBC resulted in significantly (P ≤ 0.05) higher stalk height than the treatment with 23 kg N ha⁻¹ 2 WBC. In this study, it was observed that pre-cutting application of N with lower dose 8 WBC or early application favored good crop growth/elongation and this helped to increase the seed cane yield in terms of height (Table 3). This result is consistent with that obtained in a previous study by Lakshimi et al. [18]. In line with this, TNAU [19] also reported that applying 23 kg N ha⁻¹ as a top dressing additionally before one month of cutting the seed cane to the nursery crop improved seed cane sett yield.

Bakker [20] illustrates that, growth develops from above to downwards and that it proceeds in the order of leaf blade to sheath and internodes. With respect to the phases of growth, cell elongation is preceded by cell division and cell differentiation. Cell division takes place in the meristem of the growing point, resulting in a continuous formation of new cells. These cells gradually form different kinds of tissues. As a result, stalk elongation and increase in girth comes at a later stage. Therefore, the shorter time gap between fertilizer treatment and cutting of the seed cane stalk for planting may be considered to be one the of the most important factors as the plants need sufficient time to take up the applied nutrient and convert to its stalk elongation and

girth. Therefore, taller and thicker seed cane stalks were obtained from the seed cane plants that were treated with N at the rate of 23 kg ha⁻¹ 8 weeks prior to cutting the seed cane for planting.

Gilbert et al. [21] also illustrates that, the growth of cane plant (increase in size/height or dry weight) is governed by a complex of internal and external factors. The more important external factors affecting growth are soil moisture, temperature, light, soil condition and its nutrient content, of which temperature plays an important role in limiting stalk elongation and girth of sugarcane and the internal factors affecting early establishment and growth of cane are set moisture and nutrient content mainly nitrogen.

Effect of pre-cutting nitrogen application rate and time on seed cane stalk internodes number and weight per stalk

Both the main factor and interaction effects of pre-cutting nitrogen application rate and time did not affect the internodes number and weight per stalk of the seed cane at nine months growth stage (Table 4 and Appendix Table 1).

However, the non-significant main as well as interaction effects of nitrogen level and pre-cutting nitrogen application time on seed cane

Nitrogen rate (NR) kg ha⁻¹	Seed cane stalk height (cm)				
	Time of N application (NT) (WBC)				
	8	6	4	2	Mean
0	189.23ab	177.03cd	181.37bcd	181.50bcd	182.28
23	191.70a	175.03d	182.03bcd	181.77bcd	182.63
46	181.67bcd	182.40bcd	175.90d	187.10ab	181.77
69	178.20cd	186.63ab	185.17abc	187.83ab	184.46
Mean	185.20	180.28	181.12	184.55	182.79
SE±	2.86				
LSD	4.13				
CV (%)	2.70				

Where, WBC = Weeks before cutting for seed cane; Means followed by the same letters are not significantly different at 5% level of significance according to the LSD test.

Table 3: Interaction effect of pre-cutting N application rate and time on seed cane stalk height at Finchaa, 2011.

Treatment	Seed cane stalk internodes (No)	Seed cane stalk weight (kg)
Nitrogen rate (NR) (kg ha⁻¹)		
0	13.42	1.18
23	13.58	1.20
46	13.58	1.23
69	13.83	1.25
SE±	0.16	0.02
LSD	NS	NS
Time of N application (NT) (WBC)		
8	13.58	1.19
6	13.58	1.20
4	13.67	1.23
2	13.58	1.23
SE±	0.16	0.02
LSD	NS	NS
Mean	13.60	1.21
CV (%)	3.99	6.67

Where, NS= Non-significant at 5% level of significance, WBC= Weeks before cutting for seed cane

Table 4: Seed cane stalk internodes number and weight per stalk as influenced by pre-cutting N application rate and time, at Finchaa, 2011.

stalk growth parameters (internodes number and weight per stalk) might be attributed to the shorter time gap that occurred between nitrogen application and cutting of the seed cane for planting the subsequent commercial crop. This suggestion may be substantiated by the fact that, sugarcane has low nitrogen use efficiency, particularly so is variety N-14 [22]. Schumann, et al. [23] also reported that N-14 was less efficient in N use. Supporting this proposition, another experiments also proved that, most of the efficient N uptake takes place within the first six active growth months by the sugarcane crop [20].

Sundara [7] also described that, nitrogen is the key nutrient element influencing sugarcane yield and quality. It is required more for vegetative growth, i.e. tillering, foliage formation, stalk formation, stalk growth (internodes formation, internodes elongation, increase in stalk girth and weight) and root growth. But the author further revealed that to derive maximum benefits from the applied nutrients, it is highly important to apply the fertilizer at optimum time and required dose. In the case of seed cane nurseries, a faster rate of growth is essential in the early stage for maximizing sett yields. Therefore, optimum dosage of nutrients, particularly nitrogen, and their early application would be advantageous since nitrogen requirement of sugarcane is the greatest during the tillering and the early grand growth phase.

Effect of pre-cutting nitrogen application rate and time on seed cane stalk reducing sugar and total nitrogen content

The main effect of time of nitrogen application was significant (P<0.05) on reducing sugar and nitrogen contents of seed cane stalk/sett. However, the main effect of N rate had no influence on reducing sugar and nitrogen contents of the seed cane plant at the time of harvest for planting (Appendix Table 1). The interaction effect of N rate and time of pre-cutting nitrogen application was significant on both reducing sugar and nitrogen contents of seed cane stalk/sett at the time of harvest for commercial planting (Tables 5 and 6, Appendix Table 1).

The interaction effect revealed that treating the standing cane with nitrogen prior to cutting for planting of the subsequent commercial crop at the rate of 69 kg N ha^{-1} 8 weeks before cutting significantly enhanced the reducing sugar (glucose and fructose) content of the seed cane (Table 5). However, the reducing sugar content obtained at the combination of 69 kg N ha^{-1} and applying the N at 8 weeks before cutting the seed cane for planting was in statistical parity with reducing sugar contents obtained at the combination of untreated check plots at 6 as well as 2 weeks before planting (Table 5). Therefore, the result indicated that pre-cutting nitrogen application at the rate of 69 kg N ha^{-1} 8 WBC as well as plots untreated by pre-cutting nitrogen application at 6 and 2 WBC led to the production of stalk/sett having significantly higher contents of reducing sugar. Hence, effect of N applied before treatment application at planting (45 kg N ha^{-1}) and at two and half months after planting (69 kg N ha^{-1}) and/or biological N fixation might be benefited the untreated check plots at 6 as well as at 2 WBC for their significant increase in reducing sugar content of seed cane sett [5].

In line with the result, the overall means of N treated plots reducing sugar content of seed cane stalk/sett in the study was higher than that of matured sugarcane stalk reducing sugar content, which is on average less than 0.1% (35). In addition, Sundara [7] also suggested that pre-fertilizing the nursery crop at about 6 to 8 weeks prior to harvest for planting helps to obtain healthy setts with more moisture, more reducing sugar and with higher nutrient content. Richard and Irvine [24] also reported that young and immature canes have high amount of reducing sugar content.

The results of the interaction effect also showed that treating

standing seed cane crop with 23 kg N ha^{-1} at 2 weeks prior to cutting was highest in total nitrogen content of the seed cane stalk, which was in statistical parity with the mean value produced by untreated/check plots at 2 WBC and plots treated with 46 kg N ha^{-1} 8 WBC, 69 kg N ha^{-1} at 8 WBC, 69 kg N ha^{-1} 4 WBC, 69 kg N ha^{-1} 2 WBC. Hence, effect of N applied before treatment application at planting (45 kg N ha^{-1}) and at two and half months after planting (69 kg N ha^{-1}) and/or biological N fixation might be benefited the untreated check plots at 2 WBC for their significant increase in total nitrogen content of seed cane sett (Table 6).

However, pre-cutting nitrogen application on seed cane of sugarcane with the rate of 23 kg N ha^{-1} 2 WBC had significantly higher nitrogen content in stalk/sett (Table 6). The average means of total nitrogen content of seed cane stalk/sett treated with N prior to cutting for seed cane for all treatment in the study was also higher than that of the untreated conventional nine month crop age stalk of total N content, i.e. 0.16% [20]. In line with this result, Singh et al. [25] reported that pre-cutting nitrogen treatment of seed cane crop during the normal course of growth has a profound effect on the nutritional status of cane stalk/sett which had marked influence on germination of sett for the subsequent commercial crop. Sreewarome et al. [26] also reported that nitrogen is vital for most plant metabolic processes and plays an important role in early growth, tillering and stalk elongation of sugar cane plant.

Kakde [6] reported that the nutritional status of cane stalk/sett had marked influence on germination of sett for the subsequent commercial crop as it is afforded the energy required for sprouting of bud and young shoot till it was established on its own. Therefore, high nitrogen,

Nitrogen rate (NR) kg ha-1	Reducing sugar content (%)				
	Time of N application(NT) (WBC)				
	8	6	4	2	Mean
0	1.83^{bcde}	1.97^{abc}	1.67^{e}	1.99^{ab}	1.87
23	1.95^{bcd}	1.92^{bcd}	1.93^{bcd}	1.82^{bcd}	1.91
46	1.91^{bcd}	1.86^{bcd}	1.79^{de}	1.80^{cde}	1.84
69	2.14^{a}	1.85^{bcd}	1.83^{bcde}	1.93^{bcd}	1.94
Mean	1.96	1.9	1.81	1.89	1.89
SE±	0.06				
LSD	0.09				
CV (%)	5.7				

Where, WBC = Weeks before cutting for seed cane; followed by the same letters are not significantly different at 5% level of significance according to the LSD test.

Table 5: Interaction effect of pre-cutting N applications rate and time on seed cane stalk reducing sugar content at Finchaa, 2011.

Nitrogen rate (NR) kg ha-1	Total nitrogen content (%)				
	Time of N application(NT) (WBC)				
	8	6	4	2	Mean
0	0.21^{d}	0.23^{bcd}	0.21^{cd}	0.24^{abc}	0.22
23	0.22^{bcd}	0.22^{bcd}	0.22^{bcd}	0.26^{a}	0.23
46	0.24^{ab}	0.23^{bcd}	0.23^{bcd}	0.23^{bcd}	0.23
69	0.24^{ab}	0.22^{bcd}	0.24^{ab}	0.24^{abc}	0.24
Mean	0.23	0.22	0.23	0.24	0.23
SE±	0.01				
LSD	0.01				
CV (%)	6.61				

Where, WBC = Weeks before cutting for seed cane; Variable means followed by the same letters are not significantly different at 5% level of significance according to the LSD test.

Table 6: Interaction effect of pre-cutting N application rate and time on seed cane stalk total nitrogen content at Finchaa, 2011.

starch and glucose contents were essential for good seed though their content varied with different varieties. There was also a high positive correlation with amide-N but variable with glucose. However, glucose content of setts was considered as a reliable index of planting material for germination. Thus, the causes of increased germination were attributed to the greater availability of food, water or nutrients from the substrate to the growing bud or an internal metabolic change in the sown sett culminating to a favorable influence on the emergence and after growth of the sugarcane buds [25].

Effect of pre-cutting nitrogen application rate and time on seed cane stalk/sett moisture and fibre content

The main effect of time of pre-cutting nitrogen application significantly ($P \leq 0.05$) influenced the moisture content of the seed cane. However, neither the main effect of rate of nitrogen application nor the interaction effect of the rate and time of nitrogen application had significant ($P \leq 0.05$) influence on moisture content of the seed cane. Fibre content of the seed cane was not affected by the main effect of the rate of nitrogen as well as the time of pre-cutting nitrogen application. It was not also affected by the interaction effects of the two factors (Table 7 and Appendix Table 1).

Sett moisture content increased significantly by about 1% when the time of nitrogen application before cutting was lengthened from 2 to 8 weeks before cutting. However, this increase in moisture content was in statistical parity with the moisture content of the seed cane that occurred when the application of nitrogen was done 4 weeks before cutting the cane for commercial planting. Consistent with this result, Sundara [7] reported that additional fertilizers given to sugarcane crop planted exclusively for seed purpose at about 6 weeks prior to harvest improved seed quality by enhancing sett nutrient status and sett moisture content. Verma [27] also reported that setts with higher moisture content give quicker and higher germination and the seedlings emerging from such setts establish quickly and grow vigorously. Singh et al. [25] also reported that optimum level of sett moisture content for rapid germination was 72 to74%. In addition, Srivastava [28] also described standards of seed cane moisture content and suggested that it should not be less than 65% on weight basis. In general, the average stalk/sett moisture content of all treatments in the study was greater than that of the critical sett moisture content (50.3%) for germination of buds on seed cane sett by 55.35%.

Pre-cutting nitrogen treatment levels and time of application on seed cane had not significantly change the fiber content of the seed cane stalk/sett (Table 7). Corroborating the results of this study, Koochekzadeh et al. [29] revealed that the N application rate had no influence on enhancing the fiber content and trash amounts of sugarcane. This might be more of varietal character. According to SASRI [22] variety N-14 is moderate in fiber content for matured cane (approx. 12.7%). However, the mean result in the study was lower than that of the approximate fiber content of matured cane by 22.71% which is well below the average value for the variety fiber content. This might be due to the fact that the crop supplied with N was harvested earlier for seed cane than the usual time for cane harvesting intended for sugar production. High rate of N application and early harvesting are known to reduce the degree of lignifications of tissues by reducing fiber content of plant organs.

Effect of pre-cutting nitrogen application rate and time on seed cane stalk pol% cane, brix% cane and total non-sugar content

The results of the study indicated that, both the main factors

(N level and time of pre- cutting nitrogen application) and also the interaction effect between them did not significantly ($P \leq 0.05$) affect the Pol% cane, brix% cane and total non-sugar content of the seed cane stalk/sett at the time of harvest for seed cane (Table 8 and Appendix Table 2).

Pol% cane (the apparent sucrose in a juice) is important qualitative parameters used for maturity judgment in sugar cane production [30]. However, the lowest value was recorded for N treated plots and the highest was recorded for untreated check plots. These results are identical with the results obtained by Mohammed [31] who described an inverse relationship between the increasing rate of late fertilizer application and decreasing value of Pol% in juice of seed cane plants.

Treatment	Moisture content (%)	Fiber content (%)
N rate (NR) (kg ha-1)		
0	78.23	10.05
23	78.12	10.49
46	77.80	10.70
69	78.41	10.16
SE±	0.17	0.30
LSD	NS	NS
Time of N application (NT) (WBC)		
8	78.55[a]	9.98
6	78.02[b]	10.62
4	78.22[ab]	10.01
2	77.77[b]	10.80
SE±	0.17	0.30
LSD	0.50	NS
Mean	78.14	10.35
CV (%)	0.77	0.86

Where, NS = Non-significant; WBC; Weeks before cutting for seed cane; means followed by the same letters with in column are not significantly different at 5% level of significance according to the LSD test.

Table 7: Seed cane stalk/sett moisture content and fibre content as influenced by pre-cutting N application rate and time on seed cane of sugarcane at Finchaa, 2011.

Treatment	Pol % cane	Brix% cane	Total non sugar (%)
Nitrogen rate (NR) (kg ha-1)			
0	8.42	11.72	93.77
23	8.20	11.39	93.97
46	8.22	11.50	93.95
69	8.23	11.41	93.86
SE±	0.18	0.17	0.21
LSD	NS	NS	NS
Time of N application (NT) (WBC)			
8	8.09	11.47	94.09
6	8.16	11.36	94.02
4	8.40	11.56	93.79
2	8.43	11.43	93.64
SE±	0.18	0.17	0.21
LSD	NS	NS	NS
Mean	8.27	11.51	93.88
CV (%)	7.38	0.49	0.77

Where, NS = Non-significant at 5% level of significance; WBC = Weeks before cutting for seed cane.

Table 8: Pol, brix and total non-sugar content of seed cane stalk/sett as influenced by pre -cutting N application rate and time on seed cane of sugarcane at Finchaa, 2011.

Brix% cane (apparent/total soluble solids in a juice) is the second qualitative parameters used for maturity judgment in sugar cane production [30]. However, the lowest reading result of brix and pol were recorded for N treated plots and the highest was observed for untreated check plots. These results are in accord with those of Wiedenfeld [32] and who reported that late application of N resulted in poor juice quality (low brix and pol value) in matured cane but in contrary to this, it is a good quality for seed cane plants in maintaining food reserve for the germinating buds [20,33].

Total non-sugar content of seed cane stalk/sett is a constituents of the seed cane stalk/sett other than sucrose, glucose and fructose, i.e. fibre, water and dissolved non sugars (both organic and inorganic). The result obtained in this study is high as compared to the matured cane total non-sugar content. However, the low result of Pol% and Brix% cane and high result of total non-sugar content of the immature seed cane stalk/sett than matured sugarcane stalk might be due to low/minimum rate of reversion of glucose and fructose to sucrose at active growth stage of the plant [20].

Conclusions

Based on the results of this study, it is generally concluded that, early (8 WBC) application of N at higher dose (69 kg ha⁻¹) or also late (2 WBC) application of N at lower dose (23 kg ha⁻¹) improved seed cane quality through significantly enhancing in reducing sugar and total nitrogen contents of the seed cane stalk/sett, respectively. Therefore, As the precutting N application rate had no significant main effect, treating the seed cane with N at 23 kg ha⁻¹ 8WBC the seed cane for commercial planting had evidently the dual advantage of improving seed cane quality (in terms of seed cane height, reducing sugar, and total N content. Moreover, similar studies should be conducted by including more N rates with different times of pre-cutting N application on seed cane for this and other popular varieties under production in the Estate.

Acknowledgements

We would like to thank Ethiopian Sugar Corporation for financing the research and Finchaa Research station for unreserved material and technical support.

References

1. Gutema D, Sime M (2004) Effect of rate and time of N application on growth, yield and quality of seed cane (research report). ESISC.Sh.Co. Co. RTSD. Wonji, Ethiopia

2. Smith D (1978) Cane Sugar World. Palmer Pub., New York.

3. Koeheler PH (1984) Seed Cane production at the Lihue plantation Company. Hawaiian Sugar Technol. Report, 43rd Ann. Conf. pp. 12-13.

4. Anonymous (2004) Finchaa Sugar Factory, Cane Plantation Operation Manual.

5. Hebert LP (1956) Effect of seed pieces and rate of planting on yield of sugar cane, and nitrogen fertilization on yield of seed cane in Louisiana. Proc X Congr ISSCT 1: 301-310.

6. Kakde JR (1985) Sugar Cane Production. Renu Printers, New Delhi.

7. Sundara B (2000) Sugarcane cultivation. Vikas Publishing House Pvt. Ltd. New Delhi.

8. Fauconnier R (1993) Sugar cane. The Tropical Agriculturalist. The Macmillan Press Ltd., London.

9. Russell EW (1988) Soil Condition and plant growth. English Language Book Society. Long man, London.

10. Barnes AC (1974) The sugar cane. 2nd ed. Leonard Hill, Ltd., London.

11. Dellewijin NC (1952) Botany of sugar cane. Chionica Botanica walthan, U.S.A., PP 230.

12. King NJ (1965) Manual of cane growing. Angus and Robertson, Sudmey.

13. Singh K, Ali SH (1973) Germination in Sugar cane. Sugar news 5: 22-29.

14. Clements HF (1980) Sugar cane crop logging and crop control, principle and practices. Hawaii, Honolulu. pp510.

15. Burayu W (1992) The influence of different portions of the stalk cuttings and number of buds per sett on sprouting, tillering and yield of sugar cane, *Saccharum officinarum* L. An M.Sc Thesis presented to School of Graduate Studies of Alemaya University of agriculture. In: Rege RD, Wagle 1939. Problems of sugar cane physiology in the Deccan canal. Indian Journal of Agricultural science.9:423-448.

16. SAS (Statistical Analysis System) (2000) Institute of Applied Statistics and SAS programming Language. Cary, North Carolina.

17. Gomez KA, Gomez AA (1984) Statistical procedures for agricultural research. John Wiley and Sons Inc., New York.

18. Lakshimi MB, Devi TC, Raju DVN (2008) Nitrogen Management in Sugar cane Seed crop. J Sugar Tech 8: 91-94.

19. TNAU (Tamil Nadu Agricultural University) (2008) Nutrient management in sugarcane. TNAU, Agritech. Portal, Coimbatore, India.

20. Bakker H (1999) Sugarcane Cultivation and Management. Kluwer Academic/ Plenum Publishers, New York.

21. Gilbert RA, Shine Jr. JM, Miller JD, Rice RW, Rainbolt CR (2006) The effect of genotype, environment and time of harvest on sugarcane yields in Florida. Field Crops Research 95: 156-170.

22. SASRI (South African Sugarcane Research Institute), 2006. All Variety Information Sheet.

23. Schumann AW, Meyer JH, Nair S (1998) Evidence for different nitrogen use efficiencies of selected sugarcane varieties. Proc S Afr Sug Techno I Ass, South African Sugar Association Experiment Station, Private Bag X02, Mount Edgecombe, 4300.

24. Richard CA, Irvine JE (1993) Associations among production and yield parameters of sugar cane in the United States and other countries. U. S. Dep. Agri., Houma, Louisiana.

25. Singh SB, Rao GP, Solomon S, Gopalasundaram P (2009) Sugarcane Crop Production and Improvement. Studium Press LLC. Texas, USA.

26. Sreewarome A, Saensupo S, Prammannee P, Weerathworn P (2007) Effect of rate and split application of nitrogen on agronomic characteristics, cane yield and juice quality. Prog Int Soc Sugar Cane Technol 26: 465-469.

27. Verma RS (2004) Sugar cane Production technology in India, 1st Eds. International book distributing Co. Lucknow, India.

28. Srivastava AK (2006) Sugarcane at glance. International Book Distributing Co. (IBDC) Lucknow, U.P. India.

29. Koochekzadeh A, Fathi G, Gharineh MH, Siadat SA, Jafari S, et al. (2009) Impacts of Rate and Split Application of N Fertilizer on Sugarcane Quality. International Journal of Agricultural Research 4: 116-123.

30. Sarwar MA, Husain F, Ghaffar A, Nadeem MA (2011) Effect of some newly introduced fertilizers in sugarcane. Shakarganj Sugar Research Institute, Ayub Agri. Res. Instt. Faisalabad, Pakistan.

31. Mohammed BD (1989) Effect of nitrogen fertilizer and harvest time on yield and quality of sugarcane. M.Sc. Agron. Dep. Assuit University, Egypt.

32. Wiedenfeld R (1997) Sugarcane response to N fertilizer application on clay soils. J American Soc of sugarcane tech 17: 14-27.

33. Singh O, Kanwar RS (1986) Sugarcane 2: 7-10.

10

Correlation and Path Coefficient Analysis of Yield and Yield Associated Traits in Barley (*Hordeum vulgare* L.) Germplasm

Azeb Hailu[1], Sentayehu Alamerew[2], Mandefro Nigussie[3] and Ermias Assefa[2,4]*

[1]*Tigray Agricultural Research Institute, Mekelle Agricultural Research Center, P.O.Box: 258, Mekelle, Ethiopia*
[2]*Colleges of Agriculture and Veterinary Medicine, Jimma University, Ethiopia*
[3]*Oxfam America, Horn of Africa Regional Office, Ethiopia*
[4]*Southern Agricultural Research Institute, Bonga Agricultural Research Center, P.O.Box 101, Bonga, Ethiopia*

Abstract

Sixty four barley genotypes were tested in 8 × 8 simple lattice design at Atsbi, Ofla and Quiha environments in Tigray region, in 2009/10. The objective was to estimate the extent of association between pairs of characters in genotypic and phenotypic levels and thereby compare the direct and indirect effects of the characters. Analysis of variance (ANOVA) revealed that there was a significant difference (p<0.001) among the sixty four genotypes for all the characters studied except for 1000-kernel weight at Quiha which was significant (p<0.05) and plant height was non-significant at Atsbi and Ofla. Grain yield had positive and highly significant phenotypic and genotypic correlation with 1000-kernel weight and biological yield in all environments except harvest index at Ofla. Grain yield had positive and highly significant phenotypic and genotypic correlation with 1000-kernel weight and biological yield in all environments except harvest index at Ofla. On the other hand, grain yield had negative and highly significant correlation at genotypic level with days to heading and days to maturity only at Ofla. Path analysis revealed that biological yield exerted maximum positive direct effect on grain yield across location followed by harvest index excluding Ofla. However, days to maturity exhibited highest negative direct effect on grain yield at Ofla.

Keywords: PCV; GCV; Path coefficient analysis; Direct effect

Introduction

The center of origin of cultivated barley has been reported to be the Fertile Crescent of the Middle East [1]. Even though barley was indicated to have been brought to Ethiopia at least 5000 years ago [2,3] new studies supporting the polyphyletic origin of the crop have indicated Ethiopia as one of the centers of origin of barley [4] Furthermore, barley is believed to have originated in Abyssinia (Ethiopia) and Southeast Asia [5]. According to Vavilov [6] declared that nowhere else in nature he has observed such a diversity of forms and genes. Therefore, he proposed Abyssinia (the former Ethiopian Empire) as a center of origin of cultivated barley. The genus *Hordeum* has centers of diversity in central and southwestern Asia, western North America, southern South America, and the Mediterranean [7]. Cultivated barley is adapted to and produced over a wider range of environmental conditions than other cereals. It can grow at a latitude of 70°North in Norway, on the fringe of the Sahara desert in Algeria and below the equator in Ecuador and Kenya. In addition, it was observed at elevations up to 4200 meters on the Altiplano and slopes of the Andes in Bolivia and at 330 meters below sea level near the Dead Sea [8,9].

In genetic studies, it is necessary to distinguish two cause of correlation between characters, genetic and environmental. Two possible causes of correlation are attributed to pleiotropism and/or linkage disequilibrium [10]. Pleiotropy, particularly in a population derived from crosses between divergent strains. The degree of correlation arises from pleiotropy expresses the extent to which two characters are influenced by the same genes. Some genes may increase both characters, while others increase one and reduce the other; the former tends to cause a positive correlation, the latter a negative one.

The association between two characters that can be directly observed is the correlation of phenotypic values, or the phenotypic correlation. This is determined from measurements of the two characters in a number of individuals of the population. The genotypic correlation is the correlation of breeding values, and the environmental correlation, the correlation of environmental deviations together with non-additive genetic deviations [11].

Correlation coefficients measure the absolute value of the correlation between variables in a given body of data. Correlation does not say anything about the cause and effect relationship [12]. Path coefficient analysis is a very important statistical tool that indicates which variables (causes) exert influence on other variables (effects), while recognizing the impacts of multi colinearity [13]. A path coefficient measures the direct influence of one variable upon another and permits the separation of correlation coefficient into components of direct and indirect effects. Path coefficient analysis specifies the cause and measures the relative importance of the characters, while correlation measures only mutual association without considering causation [14].

In any breeding program of complex characters such as yield for which direct selection is not effective, it becomes essential to measure the contribution of each of the component variables to the observed correlation and to partition the correlation into components of direct and indirect effect [15].

Information on the extent and nature of interrelationship among characters helps in formulating efficient scheme of multiple trait selection, as it provides a means of direct and indirect selection of component characters. Therefore, the objective of this study was to estimate the extent of association between pairs of characters in

*****Corresponding author:** Ermias Assefa, Southern Agricultural Research Institute, Bonga Agricultural Research Center, P.O.Box 101, Bonga, Ethiopia, E-mail: ethioerm99@gmil.com

genotypic and phenotypic levels and thereby compare the direct and indirect effects of the characters.

Materials and Methods

Description of the study sites

The experiment was conducted at three locations of the Tigray region, namely Atsbi, Ofla and Quiha where barley grows most with an erratic rainfall where heavy rain alternate with dry periods resulting in alternating floods and dry periods. The region receives the least rainfall compared to other parts of Ethiopia. The average annual rainfall for the period from 1961 to 1987 was 571 mm, which was 38% less than the national average (921 mm) for the same period [16]. The mean annual rainfall ranges from 980 mm on the Central plateau to 450 mm on the northeastern escarpments of the region [17]. The annual rainfall shows a high degree of variation ranging from 20% in the western to 49% in the eastern parts of Tigray [18]. The map of experimental sites is given in Figure 1 and the different characteristics of each location are presented in Table 1.

Experimental materials

A total of 64 barley genotypes from ICARDA and one local check (Saesea) were considered in this study.

Experimental design, management and season

The experiments were conducted in 2009/10 in main cropping season using 8 × 8 Lattice design with two replications at three locations. The varieties were planted in a plot consisted of a four rows with 2 m long and 20 cm apart. The middle two rows were used for data collection. Planting was done by hand drilling using a seed rate of 80 kg/ha for each variety. Nitrogen and phosphorous fertilizers were applied at the rate of 50 kg/ha Urea and 100 kg/ ha DAP at planting. All other management practices were uniformly applied to all plots at planting.

Data collected: Data were collected for the following parameters like plant height, spike length and number of kernels per spike. The data were recorded on plant basis by randomly selecting 10 plants from each plot. Number of productive tillers/m² was recorded by counting the whole second row and then converted into 1 m² area, whereas days for heading, days for maturity, 1000 kernel weight, biological yield, grain yield, and germination test were estimated on plot basis. The germination test was done by soaking 100 seeds of each genotype in water for 12 hours. Then the seeds were planted using top-dressing method on filter paper and two batches of fifty seeds of each genotype were germinated in a box, which were kept under its plastic cover to reduce evaporation. The germination boxes were placed on the laboratory bench at room temperature of 20°C (± 0.5) and were watered every other day.

Evaluation of germination test was done on the seventh day from sowing. A seed was considered to have germinated if the radicle exceeded 2 mm in length [19].

Statistical analysis

Analysis of variance (ANOVA): The data collected for each quantitative trait were subjected to analysis of variance (ANOVA) for simple lattice design. Analysis of variance was done using Proc lattice and Proc GLM procedures of SAS version 9.2, [20] after testing the ANOVA assumptions. Before pooling the data across environments, test of heterogeneity for error of variance was done. The difference between treatment means was compared using DMRT at 5% probability levels. GENRES Version 7.01 [21] was employed for estimation of

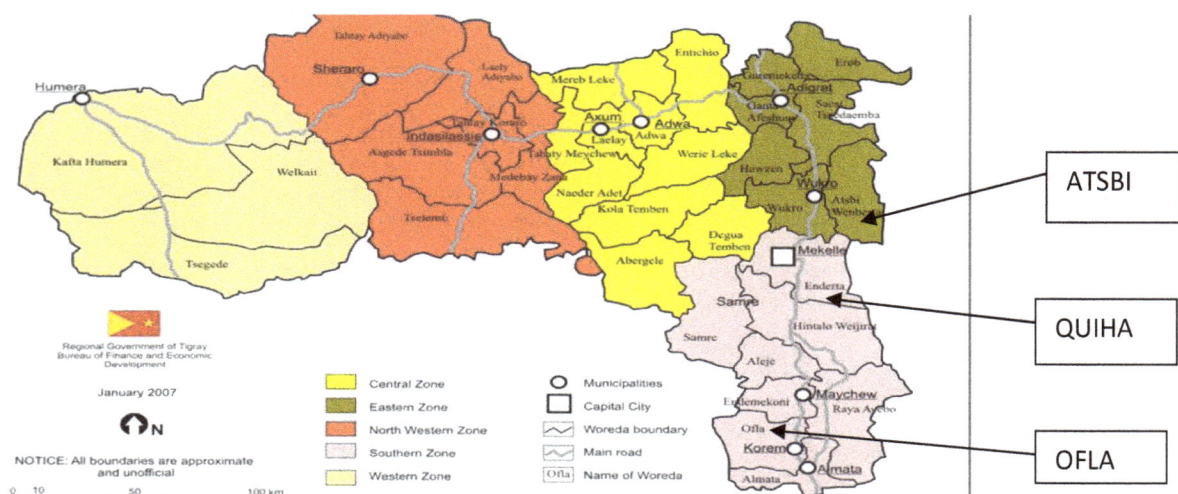

Figure 1: Map of Tigray regional state showing experimental environments.
Source: Regional Government of Tigray Bureau of Finance and Economic development (2007).

Testing location	Agro-ecological zones	Altitude (m.a.s.l)	Location		Annual rainfall (mm)	Annual temperature		Soil type	Soil pH
			Latitude	Longitude		Min.	Max.		
Atsbi	SM2e	2630	13°52'N	39°44'E	500 - 600	15°c	15- 35°c	Sandy loam	6.1
Quiha	Not available	2247	13°30'N	39°29'E	812.4	15.4°c	20.4°c	Clay loam	6.7
Ofla	SM2a	2539	12°30'N	39°31'E	450 - 800	6°c	32°c	Clay loam	5.2

Table 1: Different characteristics of locations.

correlation between traits, phenotypic and genotypic correlation and path coefficient analysis.

The model for lattice design is:

$$Yil(j) = \mu + ti + rj + (b|r)l(j) + eil(j)$$

Where, $Y_{il(j)}$ is the observation of the treatment i (i = 1, ..., v = k^2), in the block l (l = 1, ..., k) of the replication j (j = 1, ..., m); μ is a constant common to all observations; t_i is the effect of the treatment i; r_j is the effect of the replication j; $(b|r)_{l(j)}$ is the effect of the block l of the replication j; $e_{il(j)}$ is the error associated to the observation $Y_{il(j)}$, where $e_{il(j)} \sim N(0, s)$, independent.

Phenotypic and genotypic correlation: Phenotypic correlation (the observation correlation between two variables, which includes both genotypic and environmental components between two variables) and genotypic correlation was computed following the method described in [22].

$$r_p = \frac{p \, cov \, x.y}{\sqrt{\delta^2 px. \delta^2 py}}$$

$$r_g = \frac{g covx.y}{\sqrt{\delta^2 gx. \delta^2 gy}}$$

Where, r_p and r_g are phenotypic and genotypic correlation coefficients, respectively; pcovx.y and g covx.y are phenotypic and genotypic, covariance between variables x and y, respectively; $\delta^2 px$ and $\delta^2 gx$ are phenotypic and genotypic, variances for variable x; and $\delta^2 py$ and $\delta^2 gy$ are phenotypic and genotypic variances for the variable y, respectively. The coefficients of correlation were tested using 'r' tabulated value at n-2 degrees of freedom, at 5% and 1% probability level, where n is the number of treatments (accessions).

Path coefficient analysis: Path coefficient analysis was computed as suggested by Dewey and Lu [14] using phenotypic as well as genotypic correlation coefficients as:

$$rij = Pij + \Sigma rik * Pkj$$

Where, r_{ij} = mutual association between the independent character i (yield-related trait) and dependent character, j (grain yield) as measured by the genotypic correlation coefficients;

P_{ij} = components of direct effects of the independent character (i) on the dependent character (j) as measured by the path coefficients; and $\Sigma r_{ik} P_{kj}$ = summation of components of indirect effects of a given independent character (i) on a given dependent character (j) via all other independent characters (k). Whereas the contribution of the remaining unknown characters is measured as the residual which is calculated as:

$$P_R = \sqrt{(1 - \Sigma P_{ij} \, r_{ij})}$$

Results and Discussion

Analysis of variance

The analysis of variance for different characters at Atsbi, Ofla and Quiha locations is presented in Appendices 1-3, respectively. There were very highly significant differences (P< 0.001) among genotypes for all characters considered in all environments except for 1000 kernel weight in Quiha which was significant (p< 0.05) and plant height was non-significant at both Atsbi and Ofla locations.

The relative efficiency of the two designs showed that for most characters simple lattice design is not more efficient than complete randomized block design (RCBD) (Appendices 1-3). Results obtained from tests of homogeneity for error of variance showed the computed Chi-square test (x^2) value exceeds the corresponding tabular (x^2) value at 5% and 1% level of significance for all traits. Therefore, the hypothesis of homogeneous variance is rejected [23]. Therefore, the analysis of variance and other statistical analysis were run for the three locations separately.

Phenotypic and genotypic correlations

Yield is a very complex character. It is formed by the effect of numerous simple characters that are easily observed and that doesn't change or change a little from one environment to another. If there is no significant correlation between the sample characters, then their effects on yield are direct and determined by applying path analysis method [24]. The analysis of the relationship among these characters and their association with yield is essential to establish selection criteria [25]. Therefore, understanding of the inheritance and interrelationships of grain yield and the magnitudes of genotypic and phenotypic correlations of grain yield and its components among yield related traits is highly crucial to utilize the existing variability through selection.

Phenotypic and genotypic correlations of grain yield with other traits

Phenotypic and genotypic correlations of yield and yield components are presented in Tables 2-4 for Atsbi, Ofla and Quiha environments, respectively. Generally, the estimates of genotypic correlation coefficients were higher than the corresponding phenotypic correlation coefficients for all the character combinations. It was observed that

	DH	DM	PHT	SL	NK	PT	BY	TKW	HI	GEM	GYLD
DH	1.000	0.916**	-0.247	-0.261*	0.467**	-0.433**	-0.341**	-0.445**	0.057	0.021	-0.178
DM	0.9518**	1.000	-0.205	-0.321*	0.496**	-0.471**	-0.314*	-0.444**	0.081	0.020	-0.151
PHT	-0.559**	-0.602**	1.000	1.072	-0.499**	0.658**	0.491**	0.466**	-0.092	-0.190	0.310*
SL	-0.233	-0.297*	0.528**	1.000	-0.444**	0.227	0.110	0.102	-0.371**	-0.118	-0.094
NK	0.403**	0.398**	-0.145	-0.312*	1.000	-0.329**	0.099	-0.266*	0.636**	0.195	0.354**
PT	-0.385**	-0.297*	0.368**	0.375**	-0.482**	1.000	0.485**	0.419**	0.039	-0.322*	0.321*
BY	-0.253*	-0.189	0.485**	0.134	0.123	0.471**	1.000	0.517**	0.607**	-0.001	0.944**
TKW	-0.320*	-0.266*	0.400**	0.167	-0.411**	0.524**	0.588**	1.000	0.330**	-0.120	0.513**
HI	0.067	0.119	0.037	-0.269*	0.605**	0.126	0.516**	0.294*	1.000	0.160	0.836**
GEM	-0.090	-0.092	0.042	-0.013	0.129	-0.065	0.143	-0.109	0.176	1.000	0.064
GYLD	-0.142	-0.070	0.373**	-0.007	0.318*	0.386**	0.944**	0.478**	0.735**	0.168	1.000

* and ** indicate significance at 0.05 and 0.01 probability levels, respectively. PHT=Plant height; DM=Days to maturity; DH=Days to heading; TKW=Thousand kernel weight; PT=Number of productive tillers per meter square area; SL=Spike length; NK=Number of kernels per spike; BY=Biological yield; HI= Harvest index; GEM=Germination test and GYLD=Grain Yield.

Table 2: Genotypic (above diagonal) and phenotypic correlation coefficients at Atsbi.

	DH	DM	PHT	SL	NK	PT	BY	TKW	HI	GEM	GYLD
DH	1.000	0.828**	0.021	-0.141	0.264*	-0.357**	-0.340**	-0.167	-0.234	-0.235	-0.449**
DM	0.972**	1.000	-0.031	-0.203	0.313*	-0.436**	-0.474**	-0.303*	-0.225	-0.337**	-0.601**
PHT	0.310*	-0.107	1.000	0.362**	-0.325**	0.312*	-0.173	0.143	0.119	-0.372**	-0.238
SL	-0.082	-0.168	0.194	1.000	-0.498**	0.239	0.418**	0.227	0.392**	-0.443**	0.074
NK	0.199	0.268*	-0.165	-0.454**	1.000	-0.667**	-0.705**	-0.565**	-0.203	0.470**	-0.352**
PT	-0.182	-0.248*	0.059	0.310*	-0.807**	1.000	0.771**	0.457**	0.249	-0.292*	0.534**
BY	-0.124	-0.158	0.143	0.240*	-0.400**	0.646**	1.000	0.560**	0.345**	-0.112	0.829**
TKW	-0.119	-0.178	0.195	0.334**	-0.655**	0.616**	0.715**	1.000	0.138	0.033	0.649**
HI	-0.187	-0.237	-0.066	-0.416**	0.438**	-0.162	0.030	0.072	1.000	-0.125	0.228
GEM	-0.176	-0.201	0.047	0.341**	-0.184	0.189	0.150	0.081	-0.150	1.000	0.434**
GYLD	-0.188	-0.234	0.119	0.060	-0.200	0.518**	0.925**	0.542**	0.385**	0.084	1.000

* and ** indicate significance at 0.05 and 0.01 probability levels, respectively. PHT=Plant height; DM=Days to maturity; DH=Days to heading; TKW=Thousand kernel weight; PT=Number of productive tillers per meter square area; SL=Spike length; NK= Number of kernels per spike; BY=Biological yield; HI=Harvest index; GEM=Germination test and GYLD=Grain yield

Table 3: Genotypic (above diagonal) and phenotypic correlation coefficients at Ofla.

	DH	DM	PHT	SL	NK	PT	BY	TKW	HI	GEM	GYLD
DH	1.000	0.938**	-0.303*	-0.384**	0.406**	-0.163	-0.083	-0.246	-0.221	-0.224	-0.177
DM	0.966**	1.000	-0.352**	-0.461**	0.498**	-0.314*	-0.212	-0.260*	-0.182	-0.253*	-0.190
PHT	-0.381**	-0.476**	1.000	0.734**	-0.702**	0.740**	0.895**	0.603**	-0.018	0.564**	0.491**
SL	-0.275*	-0.350**	0.598**	1.000	-0.569**	0.328**	0.396**	0.172	-0.152	0.307*	0.133
NK	0.395**	0.454**	-0.453**	-0.388**	1.000	-0.476**	-0.556**	-0.331**	-0.022	-0.234	-0.285*
PT	-0.218	-0.263*	0.538**	0.503**	-0.533**	1.000	0.784**	0.175	0.061	0.365**	0.357**
BY	-0.133	-0.156	0.655**	0.371**	-0.363**	0.500**	1.000	0.353**	0.049	0.365**	0.694**
TKW	-0.131	-0.129	0.300**	0.138	-0.709**	0.487**	0.974**	1.000	0.382**	0.587**	0.819**
HI	-0.090	-0.100	-0.018	-0.144	-0.006	0.047	0.128	0.197	1.000	0.136	0.706**
GEM	-0.235	-0.234	0.439**	0.243	-0.221	0.281*	0.205	0.312*	0.114	1.000	0.252*
GYLD	-0.144	-0.156	0.411**	0.157	-0.221	0.308*	0.727**	0.404**	0.732**	0.228	1.000

* and ** indicate significance at 0.05 and 0.01 probability levels, respectively. PHT=Plant height; DM=Days to maturity; DH=Days to heading; TKW=Thousand kernel weight; PT=Number of productive tillers per meter square area; SL=Spike length; NK= Number of kernels per spike; BY= Biological yield; HI=Harvest index; GEM=Germination test and GYLD=Grain yield.

Table 4: Genotypic (above diagonal) and phenotypic correlation coefficients at Quiha.

grain yield had a positive and significant (P<0.01) phenotypic and genotypic association with biological yield and 1000 kernel weight across location, whereas harvest index except at Ofla. In addition, only at Ofla environment grain yield had positive and significant (P<0.01) correlations with the number of productive tillers/m² at phenotypic and genotypic levels. Therefore, any improvement of these characters would result in a substantial increment on grain yield. In harmony with this, Kole [26] reported that grain yield per plant had positive and significant correlations with tiller number, and 100 grain weight at both genotypic and phenotypic levels. Moreover, Kraljevic-Balalic [27] reported that biological yield was significantly and positively correlated with grain yield.

In addition, at Atsbi environment grain yield had a significant (P<0.01) and positive phenotypic correlation with plant height (r=0.373) and number of productive tillers/m² (0.386) and at genotypic level number of kernels/spike (r=0.354). Similarly, at Quiha a grain yield showed positive and significant (P<0.01) phenotypic and genotypic association with plant height with correlation coefficients of (r=0.411) and (r=0.491) respectively.

On the other hand, at Ofla location grain yield had a negative and significant correlation (P<0.01) at the genotypic level with days to heading (r=-0.449) and days to maturity (r=-0.601). The negative correlations of grain yield with days to heading and maturity at Ofla indicated simultaneous improvement in these two traits and yield seems to be particularly difficult. However, in resource poor environment early maturing genotypes that are able to use existing resources

efficiently and able to complete their life cycle within short period could escape the effect of terminal moisture stress and perform better than late maturing genotypes. Similarly, Bhutta [28] reported that grain yield was negatively correlated with days to heading.

Moreover, grain yield at Ofla and Quiha environments showed a significant and negative association with a number of kernels/spike at the genotypic level. The negative correlation of some important character as the number of kernels/spike with grain yield may lead to some undesirable selection depends on whether the negative association is due to linkage or pleiotropic effect. The negative associations of these character pairs were to impose problem in combining important yield components in one genotype. To improve yield components with negative association with other, suitable recombination may be obtained through bi parental mating, mutation breeding or diallel selective mating by breaking undesirable linkages. Similarly, Khan and Dar [29] reported that seed yield showed a significant negative association with number of seeds/ spikelet at genotypic level in wheat research.

Phenotypic and genotypic correlations among other characters

It was observed that, a positive and highly significant phenotypic and genotypic correlation between biological yield with number of productive tillers/m² and 1000 kernel weight in all environments. Likewise, days to heading and maturity with number of kernels/spike exhibited a positive and highly significant phenotypic and genotypic correlation except at Ofla. However, a negative and highly significant phenotypic and genotypic correlation was found between number of kernels/spike and number of productive tillers/m² and 1000 kernel weigh tin all environments. This indicated that for those traits which were positively associated the improvement for one trait will simultaneously improve the other. Whereas, those traits, which were negatively correlated the improvement for one trait will antagonistically affect the other. Traits like spike length and number of productive tillers/m², plant height with days to heading had a significant phenotypic association but not genotypically correlated except at Quiha. Characters, which are phenotypically correlated but not genotypically, will not produce repeatable estimates of inter-character associations and any selection based on the relationship is likely to be unreliable [30].

Generally, for most characters studied at these three locations, the genotypic correlation coefficients were greater than the phenotypic correlation coefficients. This is similar with the finding of Bhutta [25] where it indicates a greater contribution of the genetic factor in association development. In addition, Ahadu [31] also reported that the magnitudes of genotypic correlation coefficients for most of the characters were higher than their corresponding phenotypic correlation coefficients, except a few cases, which indicate the presence of inherent or genetic association among various characters. From this study, it was suggested that high yielding population in barley may be selected by concentrating upon 1000-kernel weight, biological yield, and number of productive tillers/m². Since the three traits are correlated among themselves, selection in one of the traits will implicitly result in the improvement of the other traits.

Path coefficient analysis

Correlation between yield and its components simply measures mutual relationships without presumption of causation [32] but the result of path coefficient analysis for grain yield and yield components can describe genotypic correlations to direct and indirect effects. Therefore, in this study, it is assumed that grain yield per plot was the end product of days to heading, days to maturity, plant height, 1000-kernel weight, number of productive tillers per m², spike length and number of kernels/spike. Moreover, biological yield, harvest index and germination test were considered as yield components. The residual that represents other factors affecting grain yield but not included in this study was also considered.

Path coefficient analysis in Atsbi revealed that biological yield had the highest positive direct effect on grain yield (0.749) followed by harvest index (0.508). Genotypic correlations were also positive

and significant (r=0.944 and 0.836, respectively. Similarly, at Quiha environment biological yield had the highest positive direct effect on grain yield (0.693) with a positive genotypic correlation coefficient of 0.694. In addition, harvest index had a positive direct effect on grain yield with a positive genotypic correlation coefficient of 0.706. Moreover, at Ofla location the highest positive direct effect was resulted from biological yield (0.846) with a positive genotypic correlation coefficient of 0.649. However, a negative direct effect was resulted from days to maturity (-0.407) with a negative (-0.601) genotypic correlation coefficient.

From this result, it was indicated that there is a true relationship between grain yield and biological yield across locations, and grain yield with harvest index at Atsbi and Quiha sites. In addition, a true association was observed between days to maturity and grain yield only in Ofla condition. Similarly, Milomirka [33] reported that a positive direct effect of harvest index of winter barley on grain yield (0.38).

On the other hand, number of productive tillers/m² at Atsbi had a negative direct effect on grain yield (-0.042) with positive genotypic correlation coefficient of (0.321). Similarly, at Ofla, number of productive tillers/m² had a negative direct effect on grain yield (-0.024) with positive genotypic correlation coefficient of (0.534). Moreover, at Quiha, number of productive tillers/m² had a negative direct effect on grain yield (-0.323) with positive genotypic correlation coefficient of (0.357). The positive genotypic correlation was due to the indirect effect of biological yield with number of productive tillers/m² across locations. This showed there was no true relationship between number of productive tillers/m² and grain yield in all environments. This is in agreement with the findings of Khan and Dar, [29] where path coefficient analysis revealed that number of effective tillers exhibited negative direct effect (-0.170) on seed yield in wheat research. Maximum and positive indirect effect was exhibited by 1000 Kernel Weight through Biological yield (0.675) followed by Plant Height through Biological yield (0.621) and Number of Productive Tillers/m² through Biological yield (0.543) at Quiha.

Residual effect in the present study was 0.03, 0.167 and0.06 at Atsbi, Ofla and Quiha respectively (Tables 5-7). Which means the characters in the path analysis expressed the variability in grain yield by 96.4%, 83.3% and 93.1% and the remaining 3.6%, 16.7% and 6.9% needs additional characterization for the future breeding program at Atsbi, Ofla and Quiha, respectively. Similar result reported by Ali [34], Mollasadeghi [35].

	DH	DM	PHT	SL	NK	PT	BY	TKW	HI	GEM	r_g
DH	**-0.199**	0.222	0.041	-0.023	-0.052	0.018	-0.255	0.041	0.029	-0.001	-0.178
DM	-0.199	**0.222**	0.044	-0.029	-0.055	0.020	-0.235	0.041	0.041	-0.001	-0.151
PHT	0.111	-0.134	**-0.073**	0.096	0.055	0.011	0.368	-0.043	-0.047	0.004	0.310*
SL	0.052	-0.071	-0.079	**0.089**	0.049	-0.016	0.082	-0.015	-0.188	0.003	-0.094
NK	-0.093	0.110	0.037	-0.040	**-0.111**	0.020	0.074	0.038	0.323	-0.005	0.354**
PT	0.086	-0.104	-0.048	0.033	0.053	**-0.042**	0.364	-0.048	0.020	0.008	0.321*
BY	0.068	-0.070	-0.036	0.010	-0.011	-0.020	**0.749**	-0.054	0.308	0.001	0.944**
TKW	0.088	-0.098	-0.034	0.015	0.046	-0.022	0.440	**-0.092**	0.168	0.003	0.513**
HI	-0.011	0.018	0.007	-0.033	-0.071	-0.002	0.454	-0.030	**0.508**	-0.004	0.836**
GEM	-0.004	0.005	0.014	-0.011	-0.022	0.013	-0.001	0.011	0.081	**-0.023**	0.064

Residual effect=0.0366

*, ** indicate significance at the 0.05 and 0.01 probability levels, respectively. PHT=Plant height; DM=Days to maturity; DH=Days to heading; TKW=1000-Kernel weight; PT=Number of productive tillers/m²; SL=Spike length; NK=Number of kernels/ spike; BY=Biological yield; HI=Harvest index and GEM=Germination test; r_g=Genotypic correlation coefficients

Table 5: Path coefficients at genotypic level of direct (diagonal) and indirect effects of the characters studied at Atsbi.

	DH	DM	PHT	SL	NK	PT	BY	TKW	HI	GEM	r_g
DH	**0.283**	-0.395	0.006	0.022	0.029	0.008	0.288	0.015	0.009	-0.077	-0.449**
DM	0.275	**-0.407**	0.002	0.032	0.034	0.010	0.401	-0.028	0.009	0.110	-0.601**
PHT	0.088	0.044	**-0.020**	0.056	0.035	-0.007	0.146	0.013	0.005	0.121	-0.238
SL	0.040	0.083	0.007	**0.155**	0.054	-0.007	0.354	0.031	0.015	-0.144	0.074
NK	0.075	-0.127	0.007	0.077	**0.109**	0.019	-0.597	0.060	0.008	0.153	-0.352**
PT	0.101	0.177	0.006	0.048	0.088	**-0.024**	0.653	0.057	0.009	-0.095	0.534**
BY	0.096	0.193	0.004	0.065	0.077	-0.018	**0.846**	0.066	0.013	0.036	0.829**
TKW	0.047	0.123	0.003	0.052	0.071	-0.015	0.605	**0.092**	0.005	0.011	0.649**
HI	0.066	0.092	0.002	0.061	0.022	-0.006	0.292	0.013	**0.038**	-0.049	0.228
GEM	0.066	0.137	0.008	0.069	0.051	0.007	0.095	0.003	0.006	**0.326**	0.434**

Residual effect=0.1686

*, ** Indicate significance at the 0.05 and 0.01 probability levels, respectively. PHT=Plant height; DM=Days to maturity; DH=Days to heading; TKW=1000-Kernel weight; PT=Number of productive tillers/m²; SL=Spike length; NK=Number of kernels/ spike; BY=Biological yield; HI=Harvest index and GEM=Germination test; r_g=Genotypic correlation coefficients

Table 6: Path coefficients of direct (main diagonal) and indirect effects of the characters studied at Ofla.

	DH	DM	PHT	SL	NK	PT	BY	TKW	HI	GEM	r_g
DH	**0.004**	0.001	-0.023	-0.103	0.128	0.053	-0.057	-0.077	-0.142	0.041	-0.177
DM	0.004	**0.003**	-0.029	-0.124	0.157	0.101	-0.147	-0.081	-0.117	0.046	-0.190
PHT	-0.002	-0.002	**0.061**	0.197	-0.221	-0.239	0.621	0.188	-0.011	-0.102	0.491**
SL	-0.002	-0.001	0.045	**0.268**	-0.179	-0.163	0.274	0.043	-0.098	-0.056	0.133
NK	0.002	0.002	-0.043	-0.153	**0.315**	0.172	-0.385	-0.221	-0.014	0.042	-0.285*
PT	-0.001	-0.001	0.045	0.135	-0.168	**-0.323**	0.543	0.152	0.039	-0.066	0.357**
BY	-0.001	-0.001	0.054	0.106	-0.175	-0.253	**0.693**	0.304	0.031	-0.066	0.694**
TKW	-0.001	-0.001	0.037	0.037	-0.223	-0.157	0.675	**0.312**	0.246	-0.106	0.819**
HI	-0.001	-0.001	-0.001	-0.041	-0.007	-0.020	0.034	0.119	**0.643**	-0.021	0.706**
GEM	-0.001	-0.001	0.034	0.082	-0.074	-0.118	0.253	0.183	0.073	**-0.181**	0.252*

Residual effect=0.0691

*, ** indicate significance at the 0.05 and 0.01 probability levels, respectively. PHT=Plant height; DM=Days to maturity; DH=Days to heading; TKW=1000-Kernel weight; PT=Number of productive tillers/m²; SL=Spike length; NK=Number of kernels/ spike; BY=Biological yield; HI=Harvest index and GEM=Germination test; r_g=Genotypic correlation coefficients

Table 7: Path coefficients of direct (main diagonal) and indirect effects of the characters studied at Quiha.

Conclusion

It was observed that grain yield had a positive and significant (P<0.01) phenotypic and genotypic association with biological yield and 1000-kernel weight across location, whereas harvest index at Atsbi and Quiha and number of productive tillers/m² at Ofla environment. On the other hand, at Ofla location grain yield had a negative and significant correlation (P<0.01) at the genotypic level with days to heading (r=-0.449) and days to maturity (r=-0.601). Generally, for most characters studied at these three locations, the genotypic correlation coefficients were greater than the phenotypic correlation coefficients. From this study, it was suggested that high yielding population might be selected by concentrating upon 1000-kernel weight, biological yield, harvest index and number of productive tillers/m².

Genotypic correlation coefficients of various characters with grain yield were partitioned in to direct and indirect effects. Path coefficient analysis in Atsbi revealed that biological yield and harvest index show high and positive direct effect on grain yield at Atsbi and Quiha, while at Ofla location the highest positive direct effect was resulted from biological yield and days to maturity. On the other hand, number of productive tillers/m² had a negative direct effect on grain yield across location.

Acknowledgement

The Authors would like to thank Ethiopian Agricultural Research Institute for funding the research and we also thank to all crop case team of Mekelle Agricultural Research Center.

References

1. Zohary D, Hopf M (2000) Domestication of Plants in the Old World: The Origin and Spread of Cultivated Plants in West Asia, Europe, and the Nile Valley. 3rd edition. Oxford University Press: 59–69.

2. Harlan JR (1969) Ethiopia: a center of diversity. Economic Botany 23: 309-314.

3. Frost S (1974) Three chemical races in barley. Barley Genetics Newsletter 4: 25-28.

4. Molina CJL, Russell JR, Moralejo MA, Escacena JL, Arias G, et al. (2005) Chloroplast DNA microsatellite analysis supports a polyphyletic origin for barley. Theoretical and Applied Genetics 110: 613-619.

5. Reddy SR (2009) Agronomy of field crops. 3rd edition, Kalyani publishers. India, 169.

6. Vavilov NI (1951) The origin, variation, immunity and breeding of cultivated plants. Chron Bot 13: 1-366.

7. Bothmer, Von R, Seberg O, Jacobsen N (1992) Genetic resources in the Triticeae. Hereditas 116: 141-150.

8. Nilan RA, Ullrich SE (1993) Barley: Taxonomy, origin, distribution, production, genetics, and breeding. In: MacGregor A, Bahatty RS (eds.) Barley: Chemistry and Technology. AACC. St.Paul, MI, pp: 1-29.

9. Harlan JR (1968) On the origin of barley. In Barley: Origin, Botany, Cultivars, Winter Hardiness, Genetics, Utilization and Pests. USDA Agricultural Handbook 338: 9-34.

10. Allard RW (1960) Principles of Plant Breeding. John Wiley and Sons, Inc. New York, pp: 48.

11. Falconer DS, Mackay FCT (1996) Introduction to Quantitative Genetics. (4th eds.) Longman Group Ltd, England, 122-125.

12. Roy D (2000) Plant Breeding Analysis and Exploitation of Variation. Narosa Publishing House. New Delhi, India.

13. Akanda SI, Mundt CC (1996) Path coefficient analysis of the effects of strip rust and cultivar mixtures on yield and yield components of winter wheat. Theory and Applied Genetics 92: 666-672.

14. Dewey DR, Lu KH (1959) A correlation and path coefficient analysis of components of crested wheat grass seed production. Agronomy Journal 51: 515-518.

15. Giriraji K, Vijayakumar S (1974) Path coefficient analysis of yield attributes in mung bean. Ind J Genet 34: 27-30.

16. Webb P, Braun JV (1994) Famine and Food Security in Ethiopia: Lessons for Africa. John Wiley and Sons Ltd. Chichetser, England.

17. Solomon H (1999) Analysis of Irrigation Water Management Practices and Strategies in Betmera - Hiwane, Tigray, Ethiopia. An MSc Thesis presented to the school of Wageningen Agricultural University. Netherlands.

18. CoSAER (1994) Sustainable agriculture and environmental rehabilitation in Tigray. Commission for Sustainable Agricultural and Environmental Rehabilitation in Tigray: Basic Text. Mekelle, Tigray.

19. NSIA (2001) Seed testing laboratory mannual. National Seed Industry Agency. Ethiopia, Addis Ababa, pp: 61-85.

20. SAS Institute Inc (2008) Statistical analysis Software version 9.2, Cary, NC: SAS Institute Inc. USA.

21. Pascal Institute Software solution (1994) GENRES, a statistical package for genetic researchers. Version 7.01.

22. Singh RK, Chaundry BD (1985) Biometrical Methods in Quantitative Genetic Analysis. Kalayani Publishers, New Delhi-Ludhiana78: 318.

23. Gomez KA, Gomez AA (1984) Statistical Procedures for Agricultural Research. 2nd (eds.) John Wiley and sons, Inc. New York, pp: 294-297, 467-469.

24. Dogan R (2010) The correlation and path coefficient analysis for yield and some yield components of durum wheat (triticumturgidumvar. Durum l.)In west anatolia conditions. Pak J Bot 41: 1081-1089.

25. Singh KB, Geletu B, Malhorta RS (1990) Associatition of some characters with seed yield in chick pea collection. Euphytica 49: 83-88.

26. Kole PC (2006) Variability, Correlation and regression analysis in third somaclonal generation of barley. Barley Genetics Newsletter 36: 44-47.

27. Kraljevic BM, Worland AJ, Porceddu E, Kuburovic M (2001) Variability and gene effects in wheat. In: Quarrie SI (eds.) Genetics and breeding of small grains, pp: 9-49.

28. Bhutta WM, Tahira B, Ibrahim M (2005) Path-coefficient analysis of some quantitative characters in husked barley. Caderno de Pesquisa série Biologia 17: 65-70.

29. Khan MH, Dar AN (2010) Correlation and path coefficient analysis of some quantitative traits in wheat. African Crop Science Journal 18: 9 - 14.

30. Ariyo OJ, Aken'ova ME, Fatokun CA (1987) Plant character correlations and path analysis of pod yield in Okra (Abelmoschusesculentus). Euphytica 36: 677–686.

31. Ahadu MA (2008) Genetic variability and association of characters In sesame (SesamumindicumL.) Genotypes. An Msc Thesis presented to the school of Graduate Studies of Haramaya University.

32. Puri YP, Qualset CO, Williams CA (1982) Evaluation of yield components as selection criteria in barley breeding. Crop Science 22: 927–931.

33. Madic M, Paunovic A, Djurovic D, Knezevic D (2005) Correlations and "path" coefficient analysis for yield and yield components in winter barley. Acta Agriculturae Serbica 10: 3-9.

34. Ali Y, Atta BM, Akhter J, Monneveux P, Lateef Z (2008) Genetic variability, association and diversity studies in wheat (Triticumaestivum L.) germplasm. Pak J Bot 40: 2087- 2097.

35. Mollasadeghi V, Shahryari R (2011) Important morphological markers for improvement of yield in bread wheat. Advances Environ. Biol 5: 538–542.

11

Allelopathic Effect of *Parthenium hysterophorus* L. on Germination and Growth of Peanut and Soybean in Ethiopia

Eba Muluneh Sorecha* and Birhanu Bayissa

School of Natural Resources Management and Environmental Sciences, Haramaya University, Dire Dawa, Ethiopia

Abstract

The present study was conducted to investigate the allelopathic effects of *Parthenium hysterophorus* weed on seed germination and early growth stages of peanut and soybean. Leaf, stem, and root aqueous extracts of Parthenium at 0, 2, 4, 6, 8, and 10 g/1000 ml concentrations were applied to determine their effect on both crops seed germination and early growth stages under laboratory conditions. Two runs of laboratory-based experiment with factorial Complete Randomized Design (CRD) with three replications was used to arrange treatments accordingly. The result of study revealed that peanut seed germination only significantly ($P \leq 0.05$) responded to the parthenium stem and root extracts; where on average 2 and 4 seeds were germinated per petridishes, respectively under high concentration treatment of 10 g/1000 ml. whereas, soybean seed germination was significantly responded to all the parthenium plant parts extracts. However, the serious effects have been well observed under the treatments of 10 g/1000 ml stem extracts and 8 g/1000 ml root extracts, where no germination of soybean seed was recorded. Similarly, shoot length was seriously inhibited by the stem extracts for peanut and leaf extracts for soybean, accounting 5.67 and 1 cm, respectively. However, the least average root length of 3.33 cm for peanut and 0.67 cm for soybean has been noticed with the leaf aqueous extracts under 10 g/1000 ml. The study also revealed that the phytotoxicity of parthenium plant parts increase with the increasing concentrations of extracts. Phytotoxicity of parthenium has been more pronounced over soybean germination and early growth stages than peanut.

Keywords: Allelopathic; Germination; *Parthenium hysterophorus*; Shoots and root length

Introduction

Weeds are the most costly category of agricultural pests, causing great yield loss and labor expense. Agricultural weeds can emerge rapidly, resulting in reduction of crop plant growth and quality by competing for nutrients and water provided to crops and producing chemicals that suppress crop growth. *Parthenium hysterophorus* is one of the best known plant invaders in the world linking allelopathy to exotic invasion. The antagonism between weeds and crops in the field of agriculture is a complex interaction which could be allelopathic effect, physical competition, or both. Allelopathy is an interference mechanism in which live or dead plant materials release chemical substances, either inhibit or stimulate the associated plant growth [1]. Several studies [2-5] have been indicating that a number of weeds have an allellophatic effects on seed germination and growth of economically important crops.

The study of allelopathy is a difficult, as there is a difficulty in separating those of competition, because growth and yield may be influenced by each [6]. For example, adverse effect of plant residues on seed germination and plant growth could be the result of immobilization of large amounts of nutrients by micro-organisms involved in decomposition, by allelochemicals, or both [7]. Aqueous extracts of Parthenium leaf and flower inhibited seed germination and caused complete failure of seed germination of crops when the leaf extract concentration of Parthenium weed was increased. Again, yield decline in agricultural crops and reduction in forage production has been reported due to allelopathic effect of Parthenium.

Among those weeds *Parthenium hystherophorus* is a major one posing greater challenges to the economic, food security and sustainable development of many developing countries whose livelihood is of totally or partially depend on agriculture [8-10]. *Parthenium hysterophorus* L., a noxious weed, occurring widely in tropics and sub-tropics, is a major problem in Ethiopia [11], India and Australia [12]. Allelopathic activity of different parts of this cosmopolitan weed has been well documented

[13-16]. Discharge of allelochemicals into the environment occurs by exudation of volatile chemicals from living plant parts, by leaching of water soluble toxins from aboveground parts in response of action of rain, by exudation of water soluble toxins from below ground parts, by release of toxins from non-living plant parts through leaching of litter decomposition.

In Ethiopia despite some stray observations, there are no such sufficient scientific evidences on the extent that allelopathic effect of weeds on growth and productivity of economically important crops like peanut and soybean. Therefore, this study was made to investigate the effects of allelopatic chemicals of *Parthenium hysterophorus* plant part (leaf, Stem, root) aqueous extracts on germination and early growth stages of peanut and soybean crops under a laboratory condition.

Materials and Methods

Test place

Two runs of laboratory-based experiment were conducted under room temperature in Environmental Science Program Laboratory, Haramaya University during February and March, 2017. A factorial Complete Randomized Design (CRD) with three replications was used to arrange treatments accordingly. The experiment has: Factor A: Two crops variety: Peanut and soybean; Factor B: Extracts of the three main plant parts: leaf, stem and root parts, and distilled water as control treatment. Five treatments, solution of parthenium extracted aqueous

---

*Corresponding author: Eba Muluneh Sorecha, School of Natural Resources Management and Environmental Sciences, Haramaya University, PO Box 138, Dire Dawa, Ethiopia, E-mail: ebamule1@gmail.com

(2,4,6,8, and 10 g/1000 ml) and the control, distilled water were prepared and used for the experiment.

Plant material

A vegetative growing fresh tissue of the parthenium weed was collected from fields in Haramaya University. The collected weed (parthenium) was separated into leaf, stem and root parts, then crushed to the size of <2 mm and grinded, with grinding machine particularly used for this purpose. The grinded plant material was mixed in distilled water at 1 g/ml ratio, soaked and blended with blender as of [17,18] for 24 hours. Then after, the mixtures were extracted by using 100 × 100 rpm centrifuge for twenty minutes. The filtered aqueous extracts were poured into long necked and flat bottomed 250 ml volumetric flasks, well covered and preserved in refrigerator set to -5°C for use in a test experiment. Seven healthy, pure-line, viable [tested following ISTA [19] rules] seeds of both crops were procured from Haramaya University, Oil and Pulse Crop Research Team, Ethiopia. Both seeds were sown sparsely in a filter paper covered glass petri dish having 9.5 cm diameter. A 5 ml aqueous extract of parthenium were applied to each petridishes; on the other hand, 5 ml distilled water were applied in the case of control treatment. The treated petridishes were placed at room temperature. Moistening seeds with equal amount of water and germination and early growth stages data for both crops were collected on daily basis after planting.

The formula used to calculate germination percentage and phytotoxicity is indicated hereafter:

$$\% \ Germination = \frac{Number \ of \ seeds \ germinated}{Total \ number \ of \ seeds \ planted} \times 100$$

$$Phytotoxicity = \frac{Radicle \ length \ of \ control \ - \ radicle \ length \ of \ treated \ sample}{Radicle \ length \ of \ control} \times 100$$

Data analysis

The collected data were subjected to analysis of variance procedure with SAS Version 9.1 and the means were compared by using LSD at the 5% level of probability.

Results and Discussion

Germination test

Seed germination of peanut was not significantly (P ≤ 0.05) responded to the parthenium aqueous leaf extracts. However, it significantly responded to the parthenium aqueous stem extracts. Of the 7 seeds sown, only 2 seeds were germinated under high concentration treatment of (10 g/1000 ml) of stem extracts. Under the rest of the treatments (0, 2, 4, 6 and 8 g/1000 ml), germination of the peanut seed were not statistically different (Table 1). Moreover, germination of peanut seed was responded significantly to the aqueous extracts of root having the lowest germinated seed of about 4 per petridishes. Whereas, the highest mean values of seed germination were noticed under 8 g/1000 ml and control about 6 seeds for both treatments (Table 2). Rajendiran [20] reported that allelochemicals present in the parthenium plant parts could prevent the embryonic development and embryo growth and caused death. The extract of *Parthenium hysterophorus* induced a variety of chromosomal aberrations in dividing cells, which increased significantly with increasing concentrations and durations of exposure.

The study also revealed that increase in the concentration of parthenium plant extracts confidentially inhibits the germination of peanut seeds under laboratory condition. Comparing the parthenium

Treatments (g)	Leaf	Stem	Root
0 Control	6.00[a]	6.00[a]	6.00[a]
2	6.00[a]	6.00[a]	5.33[ab]
4	5.67[a]	5.67[a]	6.00[a]
6	6.00[a]	5.33[a]	5.67[ab]
8	5.00[a]	5.67[a]	6.00[a]
10	5.00[a]	2.00[b]	4.67[b]
CV%	15.15	11.30	10.29
LSD	1.512	1.027	1.027

Note: means with the same letter in the same column are not statistically significant different at p<0.05 level according to LSD test.

Table 1: Mean comparison for germination of peanut seed influenced by leaf, stem and root extracts per petridishes.

Plant parts	0 Control	2%	4%	6%	8%	10%
Leaf	85.71	85.71	80.95	85.71	71.43	71.43
Stem	85.71	85.71	80.95	76.19	80.95	28.57
Root	85.71	76.19	85.71	80.95	85.71	76.67

Table 2: Effects of parthenium plant parts extract on the germination of peanut.

plant aqueous extracts, it was stem extract that highly reduce the germination of peanut seeds accounting about 28.57% under high concentration treatment, 10 g/1000 ml, followed by leaf extracts in which about 71.43% seeds were germinated under 8 and 10 g/1000 ml (Table 3).

Germination of soybean seed was significantly responded to all the parthenium plant parts extracts considered in this study. Compared to the control treatment, the least mean value of germinated seeds were recognized under high concentration treatments of 10 g/1000 ml of leaf extracts, about 0.33, followed by 6 g/1000 ml treatment about 1.33, though not statistically significantly different. However, it has been well notified that under 10 g/1000 ml of stem extracts and under 8 g/1000 ml of root extracts, there was no germination of soybean. This might be due to the allelophatic effects of parthenium weeds on the seeds of soybean. According to Tefera [21] aqueous extracts of parthenium leaf and flower inhibited seed germination and caused complete failure of seed germination of teff (*Eragrostis teff*) under high concentration of parthenium exactas 10%.

The higher the concentration of parthenium plant extracts, the higher the influences on the germination of soybean under laboratory condition. Under 2 and 4 g/1000 ml treatments, leaf aqueous extracts had higher germination inhibition than stem and root extracts. Whereas, under 6 g/1000 ml treatment, stem aqueous extract showed the highest germination of soybean seed, about 52.38 compared with leaf and root extracts (Table 4).

Shoot length

Shoot length of peanut significantly responded to the leaf, stem and root aqueous extracts. However, the response depends on the concentrations considered for this particular study. For instance, the lowest average value of peanut shoot length was observed under 10 g/1000 ml treatment, about 6.67 cm, followed by 12.33 cm under treatment of 8 g/1000 ml. whereas, the shoot length was seriously inhibited by the stem aqueous extracts accounting 5.67 cm (Table 5). In all plant parts extracted, the increment in the concentrations inhibits the shoot length of peanut under the laboratory situation.

Similarly, it could be seen from Table 6 that the aqueous extracts of parthenium plant parts reduced the shoot length of soybean comparing with the control treatments. The intensity of shoot length reduction

Treatments (g)	Leaf	Stem	Root
0 Control	6.67[a]	6.33[a]	6.00[a]
2	3.67[b]	3.00[bc]	2.67[b]
4	3.33[b]	3.00[bc]	1.67[bc]
6	1.33[c]	3.67[b]	1.00[cd]
8	1.67[c]	1.00[cd]	0.00[d]
10	0.33[c]	0.00[d]	0.33[d]
CV%	28.8	47.05	29.69
LSD	1.453	2.372	1.027

Note: means with the same letter in the same column are not statistically significant different at p<0.05 level according to LSD test.

Table 3: Mean comparison for germination of soybean seed influenced by leaf, stem and root extracts per petridishes.

Plant parts	0 Control	2%	4%	6%	8%	10%
Leaf	91.43	52.38	47.62	9.52	23.81	4.76
Stem	88.57	42.86	42.86	52.38	14.29	0.00
Root	88.57	38.10	23.81	14.29	0.00	4.76

Table 4: Effects of parthenium weed extracts on the germination of soybean.

Treatments (g)	Leaf	Stem	Root
0 Control	29.67[a]	27.67[a]	27.67[a]
2	27.33[a]	25.33[a]	26.67[ab]
4	21.00[b]	18.00[b]	22.33[b]
6	15.67[c]	12.33[c]	14.33[c]
8	12.33[c]	10.67[c]	13.33[c]
10	6.67[d]	5.67[d]	10.67[c]
CV%	11.23	15.67	13.66
LSD	3.75	4.63	4.63

Note: means with the same letter are not statistically significant different at alpha level of 0.05.

Table 5: Mean comparison for Shoot length (cm) of peanut as influenced by parthenium plant parts aqueous extracts.

Treatments (g)	Leaf	Stem	Root
0 Control	34.33[a]	30.67[a]	27.00[a]
2	27.00[a]	27.00[a]	29.00[a]
4	17.00[b]	19.00[b]	23.00[ab]
6	1.33[c]	2.33[c]	6.33[c]
8	3.33[c]	3.67[c]	12.67[bc]
10	1.00[c]	1.33[c]	6.33[c]
CV%	31.54	19.63	38.43
LSD	7.856	4.89	11.89

Note: means with the same letter are not statistically significant different at alpha level of 0.05.

Table 6: Mean comparison for shoot length (cm) of soybean as influenced by parthenium plant parts aqueous extracts.

increases with the increasing concentrations of aqueous extracts. Leaf aqueous extracts significantly inhibited the shoot length of soybean than the rest two parthenium plant parts extracts (Table 6). The study indicated that the concentrations beyond 6 g/1000 ml of all plant parts (leaf, stem and root) highly influenced the length of soybean (Table 6). Similar trend was reported by Ref. [22] who indicated that increased aqueous extract concentrations of *Parthenium hysterophorus* have increased effects on the germination of crops. Earlier works [23] have also reported that those chemicals found within *Parthenium hysterophorus* inhibited root and shoot elongation of maize and soybean.

Root length

Table 7 shows the root length of peanut as influenced by the aqueous

extracts of parthenium plant parts. Root length of peanut significantly responds to the leaf, stem and root aqueous extracts. The presents study also indicated that 10 g/1000 ml treatment of all plant parts considered highly influenced the root growth of peanut under the laboratory condition. The least average root length (3.33 cm) of peanut had been observed with the leaf aqueous extracts compared with stem and root extracts (Table 7).

Furthermore, the study of analysis of variance showed that the root length of soybean significantly responded to the effects of aqueous extracts of parthenium plant parts. The least average value (0.67 cm) of root length of soybean had been noticed under higher concentration of solutions prepared, which is leaf aqueous extracts followed by stem extracts, about 2.67 cm (Table 8). The effects of root aqueous extracts had showed the least impact on the root length of soybean comparing with the rest two parthenium plant parts extracts (Table 8). This might be due to higher accumulation of allelochemicals found in the other plant parts than in root parts.

Phytotoxicity

The Phytotoxicity of parthenium plant parts against the two crops selected has been investigated and indicated in Tables 9 and 10. The study shows that the phytotoxicity of parthenium plant parts increase with the increasing the concentrations of the treatments from control 0 to the highest treatment of 10 g/1000 ml (Tables 9 and 10). The Phytotoxicity of leaf has been recognized as the highest impact affecting the germination and early growth of the two crops, peanut and soybean. Comparatively, the Phytotoxicity of parthenium plant parts has been more pronounced over soybean germination and early growth stages (Table 10). Some researchers argue that, the effect of *Parthenium hysterophorus* aqueous extract is species specific like in barley, wheat and soybean [24,25].

Treatments (g)	Leaf	Stem	Root
Control	21.67[a]	26.67[a]	26.33[a]
2	19.00[ab]	19.00[b]	22.33[b]
4	15.67[bc]	16.33[bc]	21.67[b]
6	14.67[c]	16.33[bc]	16.00[c]
8	9.67[d]	14.33[c]	14.67[cd]
10	3.33[e]	6.33[d]	11.33[d]
CV%	16.92	10.4	11.67
LSD	4.21	3.05	3.89

Note: means with the same letter are not statistically significant different at alpha level of 0.05.

Table 7: Mean comparison for root length (cm) of peanut as influenced by parthenium plant parts aqueous extracts.

Treatments (g)	Leaf	Stem	Root
Control	21.00[a]	26.33[a]	27.33[a]
2	16.67[b]	23.67[a]	25.33[a]
4	9.00[c]	16.00[b]	21.67[ab]
6	1.00[d]	2.67[c]	6.00[c]
8	2.33[d]	9.00[c]	12.67[bc]
10	0.67[d]	2.67[c]	6.67[c]
CV%	25.12	26.81	47.97
LSD	3.77	6.39	12.40

Note: means with the same letter are not statistically significant different at alpha level of 0.05.

Table 8: Mean comparison for root length (cm) of soybean as influenced by parthenium plant parts aqueous extracts.

Treatments (g)	Leaf	Stem	Root
Control	0.00	0.00	0.00
2	5.00	20.83	6.94
4	12.96	34.67	13.33
6	26.67	18.33	40.74
8	46.30	20.37	51.11
10	87.65	76.54	59.52

Table 9: Mean comparison for Phytotoxicity of parthenium plant parts aqueous extracts against Peanut seed germination and early growth stages.

Treatments (g)	Leaf	Stem	Root
Control	0.00	0.00	0.00
2	16.67	1.39	2.56
4	50.00	36.00	13.33
6	95.00	90.12	77.78
8	87.04	70.00	57.78
10	97.53	90.48	76.19

Table 10: Mean comparison for Phytotoxicity of parthenium plant parts aqueous extracts against soybean seed germination and early growth stages.

Conclusion

The study attempts to investigate the allelophatic effects of parthenium plant parts, leaf, stem and root on the germination and early growth of peanut and soybean seeds. In line with this, it could generalized that germination of both seeds have been seriously inhibited with the aqueous extracts of parthenium plant parts. The severity of parthenium against seeds, peanut and soybean increases with the increasing concentration of the extracts. This implies that in areas with high infestation of this weed, growing both kinds of crops might be at risk. Therefore, further investigation how to control this weed is critical.

Acknowledgments

It's our pleasure to thanks Mr. Arbo Feyissa, Coordinator of Environmental Sciences Program for his kind response while we requested for laboratory activities. Again, we would like to extend our sincere appreciation to Dr. Lemma Wogi, Head, School of Natural Resources Management and Environmental Sciences for all facilitation required for this particular work.

References

1. May FE, Ash JE (1990) An assessment of the allelopatic potential of ecualyptus. Aust J Bot 38: 245-254.
2. Rice EL (1984) Allelopathy. 2nd edn. New York Academic Press.
3. Shibu J, Andrew RG (1998) Allelopathy in black walnut (Juglans nigra L.) alley cropping. II. Effects of juglone on hydroponically grown corn (Zea mays L.) and soybean (Glycine max L. Merr.) growth and physiology. Plant and Soil 203: 199-206.
4. Delabays Mermilled NG, De Jofferey JP, Bohre C (2004) Demonstration in a cultivated fields, of the reality of the phenomenon of allelopathy. 12th Int Conf Weed Boil, pp: 97-104.
5. Mulatu W, Gezahegn B, Solomon T (2009) Allelopathic effects of an invasive alien weed Parthenium hysterophorus L. compost on lettuce germination and growth. Afr J Agric Res 4: 1325-1330.
6. Batish DR, Singh HP, Kohli RK, Kaur S, Saxena DB, et al. (2007) Assessment of parthenin against some weeds. Zeitschrift für Naturforschung 62: 367-372.
7. Wakjira M, Berecha G, Tulu S (2009) Allelopathic effects of invasive alien weed Parthenium hysterophorus L. compost on lettuce germination and growth. African Journal of Agricultural Research 4: 1325-1330.
8. Maharjan S, Shrestha BB, Jha PK (2007) Allelopathic Effects of Aqueous Extract of Leaves of Parthenium Hysterophorous L. on Seed Germination and Seedling Growth of Some Cultivated and Wild Herbaceous Species. Scientific World 5: 33-39.
9. Mcconnachie AJ, Wstrathie L, Mersie W, Gebrehiwot L, Zewdie K, et al. (2010) Current and potential geographical distribution of the invasive plant Parthenium hysterophorus (Asteraceae) in eastern and southern Africa. Weed Research 51: 71-84.
10. Dinwiddie R (2014) Composting of an invasive weed species Parthenium hysterophorus L. An agroecological perspective in the case of Alamata woreda in Tigray, Ethiopia. Master's Thesis in Agricultural Science/Agroecology. SLU, Swedish University of Agricultural Sciences, p: 79.
11. Tamado T, Ohlander L, Milberg P (2002) Interference by the weed Parthenium hystrophorus L with grain sorghum: Influence of weed density and duration of competition. Int J Pest Manage. 48: 183-188.
12. McFadyan RE (1984) Parthenium weed. Aust Weeds 3: 2-4.
13. Mall LP, Dagar JC (1979) Effect of Parthenium hysterophorus extract on the germination and early seedling growth of three crops. J Ind Bot Soc 58: 40-43.
14. Kanchan SD (1975) Growth inhibitors from Parthenium hysterophorus Linn. Curr Sci 42: 729-730.
15. Sarma KV, Girl GS, Subrahmanyam K (1976) Allelopathic potential of Parthenium hysterophorus L. on seed germination and dry matter production in Arachis hypogaea Willd., Crotalariajuncea L. and Phaseolus mungo L. Tropical Ecology 17: 76-77.
16. Sukhada DK, Jayachandra J (1980) Allelopathic effects of Parthenium hysteorphorus L. Part II. Leaching of inhibitors from aerial vegetative parts. Plant and Soil 55: 61-66.
17. Oudhia PN, Pandey P, Thipati RS (1999) Allelopathic effects of weeds on germination and seedling vigor of hybrid rice. Department of Agronomy, Indra Gandhi Agricultural University (IGAU), Raipur, Madhya Pradish, India.
18. Marwat KB, Khan MA, Nawaz A, Amin A (2008) Parthenium hysterophorus L. - A potential source of Bioherbicide. Pak J Bot 40: 1933-1942.
19. International Seed Testing Association (1976) Standard procedures in the field of seed testing.
20. Rajendiran K (2005) Simple and rapid squash schedule for the root tips of Helianthus annuus to determine the environmental clastogens. Journal of Ecotoxicology and Environmental Monitoring 15: 291-295.
21. Tefera T (2002) Allelopathic effects of Parthenium hysterophorus L. extracts on seed germination and seedling growth of Eragrostis tef. J Agron Crop Sci 188: 306-310.
22. Mersie W, Singh M (1988) Effect of phenolic acids and ragweed Parthenium (Parthenium hysterophorus) extract on tomato (Lycopersicon esculentum) growth and nutrients and chlorophyll content. Weed Science 36: 278-281.
23. Bhatt B, Chauhan DS, Todaria NP (1994) Allelopathy is Effect of weed leachates on germination and expected to be an important mechanism in the plant radicle extension of some food crops. Indian Journal Invasion Process 37: 177-179.
24. Shikha R, Jha AK (2016) Allelopathic activity of Parthenium hysterophorus L. leaf extract on Pisum sativum. International Journal of Recent Scientific Research 7: 9461-9466.
25. Shikha R, Jha AK (2016) Species composition and shoot biomass production in Parthenium dominated abandoned fallowland. IOSR Journal of Pharmacy and Biological Science 11: 41-47.

12

Estimating Effect of Vinasse on Sugarcane through Application of Potassium Chloride at Metahara Sugarcane Plantation

Ambachew Dametie, Abiy Fantaye and Zeleke Teshome*

Sugar Corporation, Research and Training, P.O. Box 15, Wonji, Ethiopia

Abstract

A study was conducted at Metahara Sugar Estate in Ethiopia to assess effect of vinasse disposal on sugarcane fields. The experiment was laid out in a randomized complete block design with five replications. To simulate the amount of vinasse to be disposed, three levels of potassium chloride (0, 340, and 580 kg ha^{-1}) were tested on three different soil types (brown clay loam, black non-vertic clay and black vertic clay). Soil analysis result showed that available K in the soil was ranged from 311-547 ppm. Potassium chloride application on cane fields had non-significant ($P<0.05$) effect for cane and sugar yields. Similarly, non-significant ($p<0.05$) effect was found for juice K2O and leaf nutrient contents (N, P and K%). Therefore, the level of potassium chloride applied at the proposed vinasse disposal rates didn't affect nutrient uptake, yield, and juice quality in the subsequent crop; thus at Metahara Sugar Plantation, sugarcane fields can be used as disposing site for vinasse. However, long-term effect of vinasse to sugarcane crop and dynamics of K in relation to availability and fixation in the soils of the plantation should be further investigated through vinasse application to sugarcane fields.

Keywords: Sugarcane; Ratoon crop; Vinasse; Juice quality; Cane yield; Sugar yield

Introduction

Vinasse contains high levels of organic matter, potassium, calcium and moderate amounts of nitrogen and phosphorus [1]; particularly it is rich in potassium [2]. Vinasse produced at Metahara ethanol distillery is acidic in reaction, and contained 0.69 g/l nitrogen and 0.025 g/l phosphorus. In many countries, vinasse is applied to sugarcane fields as a substitute for conventional potassium fertilizer [1,3]. It is also a common practice to dispose vinasse on agricultural fields due to its technical simplicity and economic advantages [3]. Vinasse can modify temporarily some soil chemical characteristics such as pH, organic carbon, and exchangeable acidity [4]. However, in the long term, intensive application of vinasse is reported to increase soil salinity [3].

N, P and K are essential for healthy growth and development of sugarcane; K is required for synthesis of carbohydrate in the leaves and the subsequent translocation of sucrose to parenchyma of the stem. Potassium occurs in highest amount in sugarcane, thus sugarcane is included in the group of high potassium demanding plants [5,6]. Uptake of K varies from 1.0-2.5 kg K$_2$O to produce one tone of cane [7]. Moreover, there is a positive interaction between N and K. The reduction in sugar content caused by high rates of N is ameliorated by an adequate supply of K. Potassium deficiency impairs sucrose transport from the leaf to the stalk. Excessive dosages of K may exert negative effect on apparent sucrose percent in cane (pol % cane) and may promote an increase in the ash content of juice. Increased ash content in cane juice has a negative influence on sugar processing.

The soil potassium unlike to other dominant soil cations is fixed by some clay minerals in the form which renders at least temporarily unavailable to crops [8]. In Ethiopia, fixation of potassium was reported in many soils studied. In spite of the fixation of K in the Awash Valley soils, the soil K is naturally very high and there is no requirement for potassium fertilizer application [9].

High amount of K in the soil may result in higher content of K in sugarcane and cane juice [10]. However, these effects are assumed and could not be confirmed until field trials are established to measure the real effect of added potassium on soil and plant potassium levels.

Metahara Sugar Factory ethanol distillery is generating large amount of vinasse as by-product. It was proposed to apply vinasse produced from the distillery to cane fields as one means of disposal, and initiated a field study by simulating vinasse application with KCl. The objective of the study was to estimate effect of vinasse on nutrient uptake, cane and sugar yields, and juice quality through application of potassium chloride.

Materials and Methods

The study was conducted for plantcane and ratoons (2nd and 4th ratoons) on three soil types specifically brown clay loam, black non-vertic clay and black vertic clay at Metahara sugarcane plantation. The plantcane experiment was established on brown clay loam while the 2nd and 4th ratoon experiments were established on black non vertic clay and black vertic clay soils, respectively. The test varieties were NCo334 and B52-298 for plantcane and ratoons, respectively. To simulate the proposed amount of vinasse to be disposed, the plots were treated with KCl at 0, 340 and 580 kg ha^{-1} assuming that the second and the third rates represent K supply by 50,000 t of vinasse disposal to 75% suitable area and K applied to achieve 1:1 salt ratio balance, respectively.

The treatments were arranged in randomized complete block design with five replications. The plot size was 104.4 m^2 i.e., 6 rows of cane with 1.45 m wide and 12 m long each in which the central 4 rows were used for parameter measurements. Initial composite soil samples were taken from the surface layer (0-30 cm) and analyzed for pH (1:2.5), EC (1:2.5), organic carbon, total nitrogen, Olsen P, and available K.

***Corresponding author:** Zeleke Teshome, Sugar Corporation, Research and Training, P.O. Box 15, Wonji, Ethiopia
E-mail: zeleketesh@yahoo.com

Organic carbon (OC) and total nitrogen (Total N) were determined using Walkley-Black method and Kjeldhal method, respectively.

Leaf sample from each plot was taken at 5 months cane age (2.5 months post KCl application), and analyzed for nitrogen, phosphorus, and potassium [11]. At harvest, stalk length, stalk girth, stalk weight and S% C were taken. Cane yield was calculated from stalk number and stalk weight while sugar yield were calculated from cane yield and S%C. Juice quality parameters such as brix, pol percent cane, and purity were also determined by crushing the cane using Jeffco grinder within two hours after harvest using standard analytical methods [12]. The purity of a juice sample was determined by the percentage ratio of pol in juice to brix in juice [2]. Juice P_2O_5 and K_2O at 10th and 20th month for plantcane were also analyzed using recommended analytical methods [11]. Finally, all the data were subjected to statistical analysis using MSTATC computer software.

Results and Discussion

Physico-chemical properties of the soils

The mean values of pH and EC indicated that the soils were alkaline in reaction and salt free, respectively. Organic carbon and total nitrogen contents of the soil ranged from 0.10-0.12% and 1.29 to 1.91%, respectively (Table 1), and the contents were found to be low.

Nitrogen content is rated as low if the values are between 0.091 and 0.18% [13] while OC of soil is rated as very low if the values less than 2% [14]. Available K was ranged from 311-547 ppm and rated as high.

Effect of potassium on yield attributes and Yield of sugarcane

All measured parameters including millable stalk length, stalk weight, stalk girth, cane and sugar yields, sucrose percent cane, and juice quality (brix, pol and purity) obtained at harvest for plantcane and ratoon crops (2nd and 4th ratoons) were statistically non-significant ($p<0.05$).

For plantcane, on black non-vertic clay soil, application of KCl was not significant ($p<0.05$) for juice quality parameters (brix, Pol % cane, purity) but the trend was decreasing as the KCl rate increased (Table 2).

Juice P_2O_5 and K_2O levels were not significant ($p<0.05$) due to the application of KCl. However, at 10 months cane age juice P_2O_5 was significantly reduced towards higher rate of KCl (Table 2). Critical value of 300 ppm P_2O_5 in juice is required for proper juice clarification [15]. The present finding indicated that the juice P_2O_5 levels, harvested at 20 months cane age, were below the critical value for proper juice clarification. However, as KCl level increased, even if the difference was statistically non-significant, improvement in juice phosphate and potassium was observed (Table 2).

Even if the difference was statistically non-significant ($p<0.05$), juice K_2O showed increasing trend with increasing rate of KCl. Moreover, unlike juice P_2O_5, juice K_2O showed a dramatic increase when the cane gets matured (Table 2). Thus, juice K_2O levels at the age of 20 months were about 29, 31, and 31 times higher than at the age of 10 months for 0, 340 and 580 kg ha^{-1} KCl rates, respectively.

The juice K_2O levels of the study ranged from 2245-2631 ppm for NCo334 which is in agreement with mean juice K_2O levels of many sugarcane varieties ranging from 1265-2867 ppm [16]. There is great variation in juice K_2O levels in different part of the world, thus there is no acceptable juice K_2O level established for all regions. The normal range of K_2O in mixed juice showed, 800-1500 ppm in Java [17]. This study indicated that even at the highest KCl rate the juice K_2O level at

Soil types	KCl (Kg ha^{-1})	pH (1:2.5)	EC (1:2.5) (dS m^{-1})	OC (%)	Total N (%)	Av P (ppm)	Av K (ppm)
Brown clay loam	0	7.47	0.09	1.26	0.11	7.10	319
	340	7.53	0.09	1.28	0.10	8.40	322
	580	7.75	0.09	1.32	0.10	6.10	392
	Mean	**7.58**	**0.09**	**1.29**	**0.10**	**7.20**	**344**
Black non-vertic clay	0	8.10	0.16	1.51	0.09	6.10	373
	340	8.21	0.14	1.58	0.10	7.50	374
	580	8.19	0.15	1.35	0.10	3.60	562
	Mean	**8.17**	**0.15**	**1.48**	**0.09**	**5.73**	**436**
Black vertic clay	0	8.07	0.22	2.01	0.12	5.60	311
	340	8.14	0.22	1.79	0.13	10.20	372
	580	8.10	0.24	1.93	0.12	7.70	547
	Mean	**8.10**	**0.23**	**1.91**	**0.12**	**7.83**	**410.0**

Table 1: Effect of KCl on soil chemical properties.

Soil type	KCl (kg ha^{-1})	Yield (t ha^{-1})		Sucrose percent cane (%)	Juice quality							Length (m)	Girth (cm)	Weight (kg)
		Cane	Sugar		P_2O_5 (ppm)	K_2O (ppm)	P_2O_5 (ppm)	K_2O (ppm)	Brix (%)	Pol (%)	Purity (%)			
Brown clay loam	0	202.90	28.98	14.30	317a	77.8	160.0	2245	22.99	20.01	91.82	3.35	2.55	1.64
	340	185.88	25.84	13.89	315b	80.5	238.8	2467	21.24	19.50	91.76	3.53	2.54	1.73
	580	204.18	27.66	13.56	215c	83.8	289.6	2631	20.80	18.94	91.04	3.38	2.42	1.53
	Mean	197.65	27.49	13.92	282	80.7	229.5	2448	21.68	19.48	91.54	3.42	2.50	1.63
	C.V (%)	8.03	6.80	3.70	19.98	11.44	16.15	-	7.02	3.01	1.80	4.85	3.23	10.44
	Sig	NS	NS	NS	*	NS	NS	NS	NS	NS	NS	NS	NS	NS

NB: NS= non-significant; * = significant at 5%

Table 2: Effect of KCl on yield attributes, yield and juice quality.

Metahara did not exceed the acceptable range reported from most of the sugarcane growing countries.

For 2nd ratoon crop, on black non-vertic clay soil, the trend showed improvement of sucrose percent cane whereas a reduction in cane yield, sugar yield, and stalk girth were observed as the KCl rate increased. Pol % cane, purity (%) and stalk length revealed better improvement for 340 kg ha[-1] than 580 kg ha[-1] KCl (Table 3). For 4th ratoon crop, black vertic clay soil, no definite trend was observed for cane and sugar yields due to KCl application; however, brix, stalk length, and stalk weight showed declining trend as the rate increased (Table 3). Absence of significant differences among the treatments could be attributed to the adequate amount of soil potassium. The result of this study is in agreement with the report indicated that KCl application had non-significant effect on cane and sugar yields when soil available K was in the sufficient range for sugarcane [18].

Effect of KCl on leaf nutrient contents

Leaf tissue analyses indicated that application of potassium chloride was not significant ($p<0.05$) for leaf nutrient concentrations (N, P, and K%) on the three soil types.

Leaf N% and P% showed increasing trend by KCl application when compared with untreated plots, but KCl application had no definite trend on leaf K% on plantcane. Even though there is adequate amount of available K in the soil, it was not expressed in ratoon crops as it was observed in plantcane. This might be due to reduced nutrient utilization efficiency of ratoon crops. The analyses had also indicated that with the exception of leaf P in 2nd and 4th ratoon crops, leaf N and K% were above the critical values. Critical values and optimum ranges were 1.8 and 2.0-2.6%, 0.19 and 0.22-0.30%, and 0.9 and 1.0-1.6% for N, P and K, respectively [19]. The foliar K% of the study ranged from 1.38-1.65% (Table 4) which is found in the optimum range. This is also in agreement with a study that reported sugarcane grown on soil with sufficient potassium had 1.5% of foliar K [20]. It is rated as medium for growth and development of sugarcane [21]. This indicated that though there was high amount of available potassium in the soil, it couldn't be absorbed by the crop. This might be due to K fixation or antagonistic effect of Ca and Mg.

Conclusion

Soil analysis revealed that available K was ranged from 311-547 ppm which is rated as high. The leaf K content was ranged from 1.38-1.61% which is rated as medium. Even though, there was high amount of available K in the soil, it couldn't be absorbed by the crop.

Soil types	KCl (kg ha[-1])	Yield (t ha[-1])		Sucrose percent cane	Brix	Juice quality (%)			Length (m)	Girth (cm)	Weight (kg)
		Cane	Sugar			Pol	Purity				
Black non-vertic clay	0	147.66	20.28	13.78	22.36	20.20	90.32		2.034	2.524	1.42
	340	143.60	19.86	13.84	22.23	20.80	90.88		2.168	2.534	1.47
	580	141.66	19.60	13.89	22.41	20.31	90.64		2.112	2.514	1.42
	Mean	144.31	19.91	13.84	22.33	20.44	90.61		2.10	2.52	1.44
	C.V (%)	9.48	7.99	4.06	1.92	2.19	2.46		3.75	5.16	5.86
	Sig	NS	NS	NS	NS	NS	NS		NS	NS	NS
Black vertic clay	0	97.48	14.46	14.85	22.53	20.83	92.44		1.990	2.540	1.32
	340	112.42	16.70	14.89	22.47	20.85	92.78		1.976	2.506	1.29
	580	110.74	16.30	14.75	22.45	20.72	90.32		1.944	2.508	1.25
	Mean	106.88	15.82	14.83	22.48	20.80	91.85		1.97	2.518	1.29
	C.V (%)	12.42	11.86	1.82	1.68	1.28	2.51		7.05	5.55	8.59
	Sig	NS	NS	NS	NS	NS	NS		NS	NS	NS

NB: NS = non-significant

Soil types	KCl (kg ha[-1])	Leaf nutrient contents, %		
		N	P	K
Brown clay loam	0	2.24	0.22	1.59
	340	2.26	0.24	1.65
	580	2.24	0.23	1.58
	Mean	2.25	0.23	1.61
	C.V (%)	4.97	6.42	5.43
	Sig	NS	NS	NS
Black non-vertic clay	0	1.81	0.15	1.40
	340	1.87	0.16	1.39
	580	1.87	0.16	1.40
	Mean	1.85	0.16	1.40
	C.V (%)	5.29	4.99	4.92
	Sig	NS	NS	NS
Black vertic clay	0	1.58	0.16	1.35
	340	1.65	0.14	1.42
	580	1.67	0.14	1.38
	Mean	1.63	0.15	1.38
	C.V (%)	8.13	20.21	4.71
	Sig	NS	NS	NS

Table 3: Effect of KCl on yield attributes, yield and juice quality.

Soil types	KCl (kg ha⁻¹)	Leaf nutrient contents, %		
		N	P	K
Brown clay loam	0	2.24	0.22	1.59
	340	2.26	0.24	1.65
	580	2.24	0.23	1.58
	Mean	2.25	0.23	1.61
	C.V (%)	4.97	6.42	5.43
	Sig	NS	NS	NS
Black non-vertic clay	0	1.81	0.15	1.40
	340	1.87	0.16	1.39
	580	1.87	0.16	1.40
	Mean	1.85	0.16	1.40
	C.V (%)	5.29	4.99	4.92
	Sig	NS	NS	NS
Black vertic clay	0	1.58	0.16	1.35
	340	1.65	0.14	1.42
	580	1.67	0.14	1.38
	Mean	1.63	0.15	1.38
	C.V (%)	8.13	20.21	4.71
	Sig	NS	NS	NS

NB: NS = non significant

Table 4: Effect of KCl on leaf nutrient contents.

This might be due to K fixation or antagonistic effect of Ca and Mg in the soil. Application of vinasse as simulated by KCl on sugarcane grown on brown clay loam, black non-vertic clay, and black vertic clay soils showed non-significant ($p<0.05$) effect for cane and sugar yields, juice quality parameters and leaf nutrient contents. Juice K_2O levels from both treated and untreated were within the acceptable range for efficient juice clarification. Moreover, the levels of KCl applied at the proposed vinasse disposal rates didn't affect nutrient uptake, yield and juice quality in the subsequent crops. Thus, at Metahara sugarcane plantations, cane fields can be used as disposing site for vinasse. However, the long term effect of vinasse to sugarcane crop and dynamics of K in relation to availability and fixation in the soils of the plantation should be further investigated using vinasse application.

Acknowledgments

The Authors are grateful to the former Ethiopia Sugar Development Agency the then Sugar Corporation, for financing the study and would like to thank those who participated in data collection and summarization. Authors are also indebted to Metahara Sugarcane Plantation Staff for their collaboration while conducting the study.

References

1. Gomez J, Rodriguez O (2000) Effects of vinasse on sugarcane (Saccharum officinarum) productivity. Revista de la Facultad de Agronomia LUZ 17: 318-326.

2. SASTA (South Africa Sugar Technologists' Association (2005) Laboratory Manual for South African Sugar Factories. South Africa. 2-36.

3. Mariano AP, Crivelaro HR, Angelis DF, Bonotto DM (2009) The use of vinasse as an amendment to ex-situ bioremediation of soil and groundwater contaminated with diesel oil. Brazilian Archives of Biology and Technology 52: 1043-1055.

4. Camargo OA, Valadares JMAS, Berton RS, Sobrinho JT, Menk JRF (1987) Alteraçao de caracteristicas químicas de um latossolo vermelho-escuro distrofico pela aplicaçao de vinhaça. Boletim Científico do Instituto Agronomico de Campinas 9: 23.

5. Clements HF (1980) Sugarcane crop logging and crop control principles and practices. University Press of Hawaii, Honolulu. USA.

6. Ing-jye F, Chwan-chou W (1980) The significance of soil non-exchangeable potassium in relation to sugarcane growth. In: Lopez MB, Madrazo CM. General agronomy, plant physiology, and agricultural engineering: International Society of Sugarcane Technologists. Philippines.

7. Rao PN (1990) Recent advances in sugarcane (1st ed.) KCP, India.

8. Ahmad N, Davis CE (1970) Forms of potassium fertilizers and soil moisture content on potassium status of Trinidad soil. Soil Science 109: 121-126.

9. Tekalign M, Haque I (1988) Potassium status of some Ethiopian soils. East African Agriculture Journal 53: 123-130.

10. Blackburn F (1984) Sugarcane. Tropical Agricultural Series. Longman. London.

11. Gomez J, Rodriguez O (2000) Effects of vinasse on sugarcane (Saccharum officinarum) productivity. Revista de la Facultad de Agronomia LUZ 17: 318-326.

12. Kassa H (2010) Handbook of laboratory methods and chemical control for Ethiopia Sugar Factories: Ethiopia Sugar Development Agency, Research Directorate, Wonji, Ethiopia.

13. Bangladesh Agricultural Research Council (BARC) (2005) Fertilizer Recommendation Guide. The Bangladesh Agricultural Research Council, Farmgate, Dhaka-1215.

14. Landon JR (1984) Booker Tropical Soil Manual: A handbook for soil survey and agricultural land evaluation in the tropics. Longman, New York.

15. Honig P (1963) Principles of sugarcane technology. Vol. III, Elsevier, Amsterdam. The Netherlands.

16. Thangavelu S, Rao KC (1997) Potassium content in juice at certain sugarcane genetic stock and its relationship with other traits. Indian Sugar 46: 793-796.

17. Gupta PK (2009) Soil, plant, water and fertilizers (2nd ed.) Agrobios, India.

18. Khadr MS, Negm AY, Khalil FA, Antoun LW (2004) Effect of potassium chloride in comparison with potassium sulphate on sugarcane production and some soil chemical properties under Egyptian conditions. IPI regional workshop on potassium and fertigation development in West Asia and North Africa, Morocco.

19. Anderson DL, Bowen JE (1990) Sugarcane Nutrition. Potash and Phosphate Institute, Alanta, USA.

20. Honig P (1963) Principles of sugarcane technology. Vol. III, Elsevier, Amsterdam. The Netherlands.

21. Humbert RP (1968) The growing of sugarcane. Elsevier Publishing Company (2nd ed.) Amsterdam, The Netherlands.

13

Correlation and Path Coefficient Analysis of Hot Pepper (*Capsicum annuum* L.) Genotypes for Yield and its Components in Ethiopia

Abrham Shumbulo[1*], Mandefro Nigussie[2] and Sentayehu Alamerew[3]

[1]*Department of Horticulture, College of Agriculture, Wolaita Sodo University, Ethiopia*
[2]*Oxfam America, Horn of Africa Regional Office, Addis Ababa, Ethiopia*
[3]*Jimma University College of Agriculture and Veterinary Medicine, Jimma, Ethiopia*

Abstract

Character correlation and path coefficient analysis study was conducted using 55 hot pepper genotypes with the objectives to assess the nature of character correlation at phenotypic and genotypic levels and direct and indirect effects of traits on yield and yield components. The experiment was conducted during 2015-20116 at six environments in Southern Ethiopia using RCBD with three replications. The result revealed that, in most cases, the genotypic correlation coefficients were higher than their respective phenotypic correlation coefficients indicating their inherent association of traits and hence more advantageous for breeding purposes. Phenotypic and genotypic correlation further confirmed that branch number per plant, fruit number per plant, fruit length, fruit diameter and fruit weight were the most important traits for improving the genotypes for higher fruit yield and may be applied for selection in hot pepper productivity. Path analysis revealed that the maximum direct effect on fresh fruit yield was exerted by dry weight (0.6686), average fruit length (0.2185), fruit diameter (0.2085) and average fruit number per plant (0.1444), Thus, on the basis of current result, fruit length, diameter and fruit number per plant could be the most important yield component characters which might be selected for yield improvement while the converse was true with plant height and stem girth (diameter) at phenotypic level.

Keywords: Hot pepper; Correlation; Fruit yield; Path coefficient

Introduction

Hot pepper (*Capsicum annuum* L.), in the family *Solonaceae* (2n=24), is an important spice and vegetable crop [1] which covers 67.98% of all the area under vegetables produced in Ethiopia [2]. The country have been producing paprika and *capsicum* oleoresins for export market. Because of its wide use in Ethiopian diet, the hot pepper is an important traditional crop mainly valued for its pungency and color. The crop serves as the source of income particularly for smallholder producers and also contributes significantly to house hold food security in many parts of rural Ethiopia [3]. When breeders attempt to improve plants, they are generally interested in upgrading several attributes of the phenotype simultaneously. The extent to which these characters are correlated will, therefore, influence the breeder's success [4]. Moreover, development of high yielding cultivars requires knowledge of the existing genetic variation and the extent of association among yield contributing characters (Table 1). Since yield is a complex trait governed by a large number of component traits it is imperative to know the interrelationship between yield and its component traits to arrive at an optimal selection index for improvement of yield [5]. Therefore, selection should be done based on these component characters after assessing their association with the yield.

The correlation between two variables indicates only that the variables are associated; it does not imply a cause and effect relationship [6]. To describe the phenotypic correlation values further, path coefficient analysis was done to identify characters having significant direct and indirect effects on fruit yield [7]. In such situation, the correlation coefficient may be confounded with indirect effect due to common association inherent in trait interrelationships. Path coefficient analysis has proven useful in providing additional information that describes a priori cause-and-effect relationships, such as yield and yield components [8]. Further, path analysis permits the separation of direct effect from indirect effect through other related characters by portioning the correlation coefficients [4,9]. Some researcher argue that Ethiopia is believed to be one of the center of diversity of hot pepper due to

diversity of the existing germplasm in diverse growing agroecological zones in the country. Moreover, the crop is becoming high value cash crop since its demand is extremely growing locally and internationally. Despite its potential, existing variability for improvement works and current demand, the research conducted under Ethiopian condition is almost nil regarding traits association and yield component traits (Table 2). Therefore, the objectives of this study was to assess the nature of character correlation at phenotypic and genotypic level and direct and indirect effects of traits on yield and yield components.

S No.	Genotype	Origin	Code
1	Melka awaze	Ethiopia	G_1
2	Marako fana	Ethiopia	G_2
3	Melka shote	Ethiopia	G_3
4	Melka zala	Ethiopia	G_4
5	AVPP9813	Asian	G_5
6	AVPP0206	Asian	G_6
7	AVPP0514	Asian	G_7
8	AVPP0512	Asian	G_8
9	AVPP0105	Asian	G_9
10	AVPP59328	Asian	G_{10}
11-55	F_1-Hybrids	Cross	G_{11}-G_{55}

Table 1: Hot pepper genotypes used in the study.

*Corresponding author: Abrham Shumbulo, Department of Horticulture, College of Agriculture, Wolaita Sodo University, Ethiopia
E-mail: shumbuloabrham@yahoo.com

Source of variation	Degree of freedom	Mean square of cross product	Expected mean square of product
Location	L-1	$MSCP_{Lxy}$	$\sigma_{exy}+r\,\sigma_{gLxy}+g\,\sigma_{rxy}+rg\,\sigma_{Lxy}$
Replication /Loc	L(r-1)	$MSCP_{rxy}$	$\sigma_{exy}+g\,\sigma_{rxy}+r\,\sigma_{gLxy}$
Genotypes	g-1	$MSCP_{gxy}$	$\sigma_{exy}+r\,\sigma_{gLxy}+rL\,\sigma_{gxy}$
Geno X Loc	(g-1) (L-1)	$MSCP_{gLxy}$	$\sigma_{exy}+r\sigma_{gLxy}$
Error	(r-1)(g-1)	$MSCP_{exy}$	σ_{exy}

Table 2: Analysis of covariance.

Materials and Methods

Description of the study areas

The field experiment was conducted at six different environments (Wolaita Soddo, Alaba, and Humbo locations) that represent major pepper growing areas in the South Ethiopia for two cropping seasons in 2015 and 2016.

Treatments, experimental design and field management

The experiment consisted of 10 parents (six introduced from AVRDC (Asia) and four Ethiopian released varieties obtained from Melkasa Agricultural Research Center). These parents were crossed in half diallel mating design to give 45 F_1 hybrids. Thus, parents and hybrids with a total of 55 genotypes were used in the study. The experiment was laid out using RCBD with three replications. Field planting was done using plant spacing of 70 × 30 cm between rows and plants, respectively. Each plot had 2 rows and 10 plants per row. The total plot area was 1.4 m × 3.0 m=4.2 m². All other recommended agronomic practices were employed during field management as recommended by Melkasa Agricultural Research Center (MARC).

Data collected

Data were collected from randomly taken ten plants from each plot for yield, quality and other related traits. Plant height [cm], Plant canopy width [cm], Stem diameter [cm], Branch number per plant, Number of fruits per plant, Fruit length [cm], Fruit width [cm], Fruit weight [g], Fruit wall thickness [mm], Number of seeds per fruit, Total Fruit yield [kg/ha], Total fruit dry weight [kg/ha], and Oleoresin content [w/w%].

Statistical analysis

Correlation among traits: Correlation was calculated to investigate the degree of relationship between phenotypic and genotypic variances and also to test the degree of character association between parameters or traits studied.

$$\sigma_{gLxy}=\frac{MSCP_{gLxy}-MSCP_{Lxy}}{r}$$

$$\sigma_{gxy}=\frac{MSCP_{gxy}-MSCP_{gLxy}}{rL}$$

Where, r=Number of replications, g=Number of genotypes,

$MSCP_{rxy}$=replication mean square of cross product for traits x and y,

$MSCP_{gxy}$=genotypic mean square of cross product for traits x and y,

$MSCP_{exy}$=environmental mean square of cross product for traits x and y,

$\sigma_{exy}=MSCP_{exy}$;

According to Hartl and Jones, to test significance of phenotypic correlation coefficient, a quantity t can be calculated: $t=\frac{r}{\sqrt{(1-r^2)/(g-2)}}$

Where 'r' is the absolute value of the correlation coefficient and 'g' is number of genotypes.

If t is greater than the value given on the table using g-2 degrees of freedom, r can be considered significantly different from zero. The significance of genotypic correlation coefficient (r_g), can be tested by t-value calculated

as: $t=\frac{r_g}{SEg}$ where, SE_g-Standard error of r_g

$$SE=\sqrt{\frac{1-r_g^2}{2h_x^2h_y^2}}$$

Where, h²x and h²y-heritability of traits X and Y, respectively.

The calculated t value for each genotypic correlation coefficient was tested against tabulated t-value at (g-2) degrees of freedom.

Path coefficient analysis: The advantage of path analysis is that it provides information on the direct and indirect contribution of causal factors to the effect if the cause and effect relationship is well defined. Path coefficient can be defined as the ratio of standard deviation of the effect due to a given cause to the total standard deviation of the effects. If Y is the effect and X_1 is the cause, the path coefficient for the path from cause X_1 to effect Y is σ_x/σ_y [10].

It is computed by the following general formula:

$r_{ij}=p_{ij}+\Sigma\,r_{ij}\,p_{kj}$

Where, r_{ij} is the mutual association between the independent variable (i) and the dependent variable (j) as measured by correlation coefficient.

p_{ij} is component of direct effect of the independent variable(i) on the dependent variable (j) as measured by correlation coefficient.

$\Sigma\,r_{ij}\,p_{kj}$-is the summation of components of indirect effects of a given independent variable(i) on the dependent variable (j) via all other independent variables (k).

The residual effect (U) implies the unexplained variation of the dependent variable that is not accounted by path coefficient, was calculated using the formula:

$$U=\sqrt{1-R^2}$$

Where, $R^2=\Sigma r_{ik}p_{kj}$

Results and Discussion

Phenotypic and genotypic correlation coefficients

The phenotypic and genotypic correlation coefficients of traits studied for 55 hot pepper genotypes at 6 different environments were presented in Table 3.1. The result revealed that in most cases the genotypic correlation coefficients were higher than their respective phenotypic correlation coefficients indicating their inherent association of traits and hence more advantageous for breeding purposes. Similarly, refs. [5,11,12] reported that magnitude of genotypic correlation

Traits	Ph	CW	BN	SD	FN	FL	FD	Tic	SN	yld	Dw
Ph	**1.00**	0.735**	0.369**	0.216ns	0.045ns	0.027ns	0.195ns	0.089ns	0.321**	0.146ns	0.171ns
CW	0.672ns	**1.00**	0.479**	0.187ns	0.086ns	0.214ns	0.327**	0.330**	0.281**	0.250*	0.147ns
BN	0.353**	0.476**	**1.00**	0.638**	0.468**	0.140ns	0.125ns	0.120ns	0.116ns	0.354**	0.405**
SD	0.229ns	0.200ns	0.599ns	**1.00**	0.434**	0.358**	0.052ns	0.217ns	0.019ns	0.374**	0.451**
FN	0.113ns	0.118ns	0.445**	0.424**	**1.00**	0.040ns	-0.211ns	-0.15ns	-0.011ns	0.482**	0.649**
FL	0.036ns	0.203ns	0.133ns	0.338**	0.045ns	**1.00**	0.322**	0.417**	-0.099ns	0.544ns	0.325**
FD	0.121ns	0.277*	0.095ns	0.048ns	-0.184ns	0.311*	**1.00**	0.706**	0.545ns	0.566**	0.294*
Tic	0.012ns	0.239*	0.087ns	0.195ns	-0.119ns	0.388**	0.654ns	**1.00**	0.287*	0.525ns	0.243ns
SN	0.219ns	0.207ns	0.083ns	0.002ns	-0.005ns	-0.090ns	0.504ns	0.238*	**1.00**	0.239ns	0.081ns
yld	0.139ns	0.249*	0.335**	0.369**	0.497**	0.496**	0.512ns	0.438**	0.180ns	**1.00**	0.860**
DW	0.146ns	0.145ns	0.372**	0.430**	0.641ns	0.288*	0.247*	0.183ns	0.033ns	0.863**	**1.00**

**=statistically highly significant at 1%; *= statistically significant at 5% Probability; Ph=plant height(cm); Cw=Canopy width (cm); Bn=Branch number per plant; SD=stem diameter(cm); FN=Fruit number per plant; FL=average fruit length(mm); FD= average fruit diameter(mm); Tic=fruit flesh thickness (mm); SN=Seed number per fruit; yld=fresh fruit yield(kg/ha); and Dw=Fruit/pod dry weight(kg/ha)

Table 3.1: Genotypic (above) and phenotypic (below diagonal) correlation coefficient of selected traits for 55 hot pepper genotypes tested at 6 environments, 2015 to 2016.

coefficients in general was higher than the phenotypic correlation coefficients for traits studied on Chilli genotypes. Again ref. [13] noted the higher genotypic correlation coefficient than the phenotypic ones, which showed the inherent associations between various characters in Ethiopian Capsicums. The current result illustrated correlation coefficient range of -0.211 to 0.86 and -0.119 to 0.863 at genotypic and phenotypic levels, respectively. The result further illustrated that plant height was non-significantly correlated with most of the traits at phenotypic level except branch number (0.353) while at genotypic level it was positively and significantly correlated with canopy width (0.735), branch number per plant (0.369) and seed number per pod (0.321). However, plant height had non significant correlation with both fresh and dry pod yield at genotypic and phenotypic levels (Table 3.1).

The study confirmed the association of branch number and canopy width was significant at phenotypic (0.476) level. Furthermore, branch number had positively significant association with fruit yield (0.335) and dry weight (0.372) at phenotypic level and 0.354 and 0.405 at genotypic, respectively. Again branch number was significantly correlated with fruit number at both phenotypically (0.445) and genotypically (0.468). Fruit length depicted positive significant correlation at both phenotypic and genotypic levels with fruit width but it had positively significant association only at phenotypic level with fresh fruit yield. Some of earlier reports also show the same findings. Singh [5] reported highest phenotypic correlation between fruit length and fruit girth. Zhani [14] observed positive correlation between fruit weight and length.

Although non significant, fruit number showed negative association with fruit diameter and fruit thickness at both levels that might indicate antagonistic effects of gene actions which could not be bred simultaneously. This agrees with finding reported [7], where Fruit length had significant negative correlation with fruit width. Fruit number and fruit yield were associated significantly and positively at phenotypic (0.497) and at genotypic (0.482) level. Fruit number again genotypically positively associated with dry fruit weight. Generally, the current result exhibited that fresh pod yield had significant positive genotypic and phenotypic correlations with canopy width, branch number, stem diameter and fruit number while only at genotypic level with fruit diameter and phenotypically with fruit length. Dry weight also had almost similar association. Hence, these traits were found

to be yield contributing characters towards increased fruit yield and dry weight. This also might indicate complementary gene actions for the traits which could be selected simultaneously. Therefore, branch number per plant, number of fruit per plant, fruit length, fruit diameter and fruit weight were the most important traits for improving the genotypes for higher fruit yield and may be applied for selection in hot pepper productivity. More of the same result was reported in ref. [13] who found high positive genotypic correlation of fruit yield with the number of fruits per plant and pericarp thickness. Usman [11] reported positive and highly significant phenotypic and genotypic associations of fruit length, fruit weight and number of fruits on pepper. Rohini and Lakshmanan [15] found positive and significant correlation of fresh fruit yield per plant with number of branches per plant, fruit length, fruit girth, individual fruit weight and number of fruits per plant. Earlier workers [16-18] also reported more or less same conclusion. Lavinia [19] confirmed the existence of strong correlation between fruit weight to fruit length and diameter and also number and weight of fruits per plant. They further concluded selection made towards increasing length and diameter can be used as indirect selection to obtain higher values of fruit weight.

Path coefficient analysis

The results of phenotypic and genotypic path coefficient analysis was presented in Table 3.2. and 3.3, respectively. Path analysis revealed that the maximum direct effect on fresh fruit yield was exerted by dry weight (0.6686), average fruit length (0.2185), fruit diameter (0.2085) and average fruit number per plant (0.1444) whereas plant height (-0.0173) and stem diameter (-0.0932) depicted negative direct effect and also negative indirect effects though the magnitude is relatively low (Table 3.2). Thus, on the basis of current result, fruit length, diameter and fruit number per plant could be the most important yield component characters which might be selected for yield improvement while the converse was true with plant height and stem girth (diameter) at phenotypic level. This result was consolidated in ref. [15] who reported the direct effect of number of fruits per plant and number of branches per plant on yield and, which could be considered as major yield components and selection indices for improvement. Kumar et al. [18,20,21] indicated that more fruits per plant were highly reliable component on fruit yield. Yatung et al. [22] found number of fruit per plant, fruit weight and number of seed per fruit were the most important traits affecting fruit yield per plant. This investigation also illustrated seed number per fruit and fruit pericarp thickness had positive direct effect but their indirect effect was more magnified through fruit diameter phenotypically indicating fruit diameter was important character both directly and indirectly for improvement and selection of pod yield in hot pepper. Sarkar et al. [7] noticed characters like seeds/fruit, fruit length and fruit width showed direct positive effect on fruit yield with low magnitudes but plant canopy width had negative direct effect on yield. Shimelis et al. [13] also confirmed similar finding.

The results of genotypic path analysis substantiated more of similar effects to that of phenotypic path analysis (Table 3.3). The analysis revealed the existence of positive direct effect of dry weight (0.5685), fruit length (0.2697), fruit number per plant (0.2272) and fruit diameter (0.2141) on fresh fruit yield. Moreover, the magnitude of genotypic direct effects exerted by these yield component characters were relatively higher than their respective phenotypic effects further substantiated the importance and close association of characters to improve yield or to use as selection indices. The result also illustrated that canopy width and stem diameter had exerted negative effects directly and indirectly through other characters that might lead to conclude in such a way that these traits could not be used for yield improvement in hot pepper production. The earlier worker [23] reported the high direct genotypic effect of fruit number per plant on fruit yield whereas plant spread had negative direct effect.

Fruit pericarp thickness showed positive direct effect on fruit yield but its indirect is exerted more by dry weight, fruit length and width. Hence, fruit length and diameter could be considered as characters of indirect selection genotypically to improve yield on hot pepper. In another scenario, both fruit diameter and length exerted their indirect effect via fruit thickness on fruit yield indicating the importance of fruit thickness for indirect selection. Patil [23] found at both phenotypic and genotypic level, the number of fruits per plant recorded positive direct effects. High direct and positive effect of number of fruit per plant [24] again number of fruit per plant and fruit diameter [25,26] on fruit yield had been reported. Kadwey [27] also investigated number of fruits per plant, number of primary branches per plant had positive direct Whereas, negative direct effect was recorded by plant height on fruit yield at genotypic level and fruit width positive indirect effect on green fruit yield via number of fruits per plant.

Conclusion

The phenotypic and genotypic correlation coefficients of traits revealed that, in most, cases the genotypic correlation coefficients were higher than their respective phenotypic correlation coefficients indicating their inherent association of traits and hence more advantageous for breeding purposes. Fruit number and fruit yield were associated significantly and positively at phenotypic (0.497) and genotypic (0.482) level. Fruit number again genotypically positively associated with dry fruit weight. Generally, the current result exhibited

	PH	Cw	BN	SD	FN	FL	FD	TIC	SN	DW	r_p(yld)
PH	**-0.0173**	0.0148	0.0038	-0.0213	0.0164	0.0079	0.0252	0.0014	0.0099	0.0977	0.1385
CW	-0.0116	**0.022**	0.0052	-0.0186	0.017	0.0443	0.0578	0.0271	0.0094	0.0967	0.2493
BN	-0.0061	0.0105	**0.0109**	-0.0558	0.0643	0.0291	0.0199	0.0098	0.0038	0.2486	0.335
SD	-0.004	0.0044	0.0065	**-0.0932**	0.0613	0.0738	0.0101	0.0221	0.0001	0.2875	0.3686
FN	-0.002	0.0026	0.0048	-0.0395	**0.1444**	0.0098	-0.0384	-0.0135	-0.0002	0.4288	0.4968
FL	-0.0006	0.0044	0.0014	-0.0314	0.0064	**0.2185**	0.0648	0.0439	-0.0041	0.1929	0.4962
FD	-0.0021	0.0061	0.001	-0.0045	-0.0266	0.0679	**0.2085**	0.0742	0.0228	0.165	0.5123
TIC	-0.0002	0.0052	0.0009	-0.0182	-0.0172	0.0847	0.1364	**0.1133**	0.0108	0.1223	0.438
SN	-0.0038	0.0045	0.0009	-0.0002	-0.0006	-0.0198	0.1051	0.027	**0.0453**	0.0218	0.1802
DW	-0.0025	0.0032	0.004	-0.0401	0.0926	0.063	0.0515	0.0207	0.0015	**0.6686**	0.8625

Residual=0.325

Ph=plant height(cm); Cw=Canopy width (cm); Bn=Branch number per plant; SD=stem diameter(cm); FN=Fruit number per plant; FL=average fruit length(mm); FD= average fruit diameter(mm); Tic=fruit flesh thickness (mm); SN=Seed number per fruit; yld=fresh fruit yield(kg/ha); Dw=Fruit/pod dry weight(kg/ha); rP(yld)=yield phenotypic correlation coefficient

Table 3.2: Estimates of phenotypic direct effects (bold and diagonal) and indirect effects (off-diagonal) of traits via other independent traits on fresh pod yield of 55 hot pepper genotypes grown at six environments, 2015 to 2016.

	PH	Cw	BN	SD	FN	FL	FD	TIC	SN	DW	r_g(yld)
PH	**0.0186**	-0.0488	0.0157	-0.0323	0.0103	0.0072	0.0418	0.0168	0.0198	0.0974	0.1465
CW	0.0136	**-0.0664**	0.0204	-0.028	0.0194	0.0576	0.0701	0.0622	0.0174	0.0837	0.2500
BN	0.0068	-0.0318	**0.0425**	-0.0954	0.1064	0.0379	0.0269	0.0226	0.0072	0.2305	0.3536
SD	0.004	-0.0124	0.0271	**-0.1494**	0.0986	0.0965	0.0114	0.0408	0.0012	0.2566	0.3744
FN	0.0008	-0.0057	0.0199	-0.0649	**0.2272**	0.0107	-0.0451	-0.0289	-0.0007	0.3687	0.4820
FL	0.0005	-0.0142	0.0059	-0.0535	0.009	**0.2697**	0.0689	0.0786	-0.0061	0.1846	0.5434
FD	0.0036	-0.0217	0.0053	-0.0078	-0.0479	0.0868	**0.2141**	0.1331	0.0336	0.167	0.5661
TIC	0.0017	-0.0219	0.0051	-0.0324	-0.0349	0.1125	0.1513	**0.1884**	0.0177	0.138	0.5255
SN	0.0059	-0.0187	0.0049	-0.0029	-0.0026	-0.0267	0.1167	0.054	**0.0617**	0.0463	0.2386
DW	0.0031	-0.0098	0.0172	-0.0674	0.1474	0.0875	0.0629	0.0457	0.005	**0.5685**	0.8601

Residual = 0.273

Ph=plant height(cm); Cw=Canopy width (cm); Bn=Branch number per plant; SD=stem diameter(cm); FN=Fruit number per plant; FL=average fruit length(mm); FD= average fruit diameter(mm); Tic=fruit flesh thickness (mm); SN=Seed number per fruit; yld=fresh fruit yield(kg/ha); Dw=Fruit/pod dry weight(kg/ha); r_g(yld)=yield genotypic correlation coefficient

Table 3.3: Estimates of genotypic direct effects (bold and diagonal) and indirect effects (off-diagonal) of traits via other independent traits on fresh pod yield of 55 hot pepper genotypes grown at six environments, 2015 to 2016.

that fresh pod yield had significant positive genotypic and phenotypic correlations with canopy width, branch number, stem diameter and fruit number while only at genotypic level with fruit diameter and phenotypically with fruit length. Dry weight also had almost similar association. Therefore, branch number per plant, number of fruit per plant, fruit length, fruit diameter and fruit weight were the most important traits for improving the genotypes for higher fruit yield and may be applied for selection in hot pepper productivity.

Path analysis revealed that the maximum direct effect on fresh fruit yield was exerted by dry weight (0.6686), average fruit length (0.2185), fruit diameter (0.2085) and average fruit number per plant (0.1444). Thus, on the basis of current result, fruit length, diameter and fruit number per plant could be the most important yield component characters which might be selected for yield improvement while the converse was true with plant height and stem girth (diameter) at phenotypic level. The results of genotypic path analysis substantiated more of similar effects to that of phenotypic path analysis.

Acknowledgements

The authors like to thank Wolaita Sodo University and Jimma University College of Agriculture and veterinary medicine for their valuable contribution in both facilitation and financial support. Our recognition also extends to Melkasa Agricultural research center specially vegetable program staff for their cooperation in providing planting materials and AVRDAC for the same. Moreover, our special thanks goes to Prof. Derbew Belew for his unreserved contribution for material transfer from AVRDC.

References

1. Gogula KR (2015) Development of Hybrids and their Stability in Chilli (Capsicum Annuum L.). PhD Dissertation, Department of Vegetable Science, Horticultural College and Research Institute, Dr. YSR Horticultural University.

2. CSA (Central Statistical Authority) (2011/2012) Agricultural Sample Survey, Ethiopia.

3. Shiferaw M, Alemayehu CH (2014) Assessment of Hot Pepper (Capsicum species) Diseases in Southern Ethiopia. International Journal of Science and Research (IJSR) 3: 91-95.

4. Vijaya HM, Gowda APM, Nehru SD (2014) Genetic variability, correlation coefficient and path analysis in chilli (Capsicum annuum L.) genotypes. Research in Environment and Life Science 7: 175-178.

5. Singh PK, Kumar A, Ahad I (2014) Correlation and path coefficient analysis in yield contributing characters in Chilli, Capsicum annum L. International Journal of Farm Sciences 4: 104-111.

6. Pierce BA (2003) Genetics: A Conceptual Approach. WH Freeman and Company, New York, USA.

7. Sarkar S, Murmu D, Chattopadhyay A, Hazra P (2009) Genetic variability, correlation and path analysis of some morphological characters in Chilli. Journal of Crop and Weed 5: 157-161.

8. Gravois KA, Helms RS (1992) Path Analysis of Rice Yield and Yield Components as Affected by Seeding Rate. Agronomy Journal 84: 1-4.

9. Sharon T (1982) Path analysis in Pigeon pea. Indian Journal of Genetics 42: 319-332.

10. Singh RK, Chaudhary BD (1979) Biometrical Methods in Quantitative Genetic Analysis. Kalyani, New Delhi, India, p: 304.

11. Usman MG, Rafii MY, Martini MY, Oladosua Y, Kashianic P (2017) Genotypic character relationship and phenotypic path coefficient analysis in chili pepper genotypes grown under tropical condition. Journal of Science Food and Agriculture 97: 1164-1171.

12. Pandit MK, Adhikary S (2014) Variability and Heritability Estimates in Some Reproductive Characters and Yield in Chilli (Capsicum annuum L.). International Journal of Plant & Soil Science 3: 845-853.

13. Shimelis A, Bekele A, Dagne W, Adeferis TW (2016) Genetic Variability and Association of Characters in Ethiopian Hot Pepper (Capsicum annum L.) Landraces. Journal of Agricultural Sciences 61: 19-36.

14. Zhani K, Hamdi W, Sedraoui S, Fendri R, Lajimi O, et al. (2015) Agronomic evaluation of Tunisian accessions of chili pepper (Capsicum frutescens L.). International Research Journal of Engineering and Technology (IRJET) 2: 28-34.

15. Rohini N, Lakshmanan V (2015) Correlation and Path Coefficient Analysis in Chilli for Yield and Yield Attributing Traits. Journal of Applied and Natural Sciences 4: 25-32.

16. Munshi AD, Behra TK, Gyanedra S (2002) Correlation and path coefficient analysis in chilli. Indian Journal of Horticulture 57: 157-159.

17. Leaya J, Khader KMA (2002) Correlation and path coefficient analysis in chilli (Capsicum annuum L.). Capsicum and Eggplant Newsletter 21: 56-59.

18. Ullah MZ, Hasan MJ, Saki AI, Rahman AHMA, Biswas PL (2011) Association of correlation and cause-effect analysis among morphological traits in chilli (Capsicum frutescens L.). International Journal of Biological Research 10: 19-24.

19. Lavinia S, Madoşă E, Giancarla V, Ciulca S, Avădanei C, et al. (2013) Studies regarding correlations between the main morphological traits in a collection of bell pepper (Capsicum annuum var, grossum) local landraces. Journal of Horticulture, Forestry and Biotechnology 17: 285-289.

20. Kumar BK, Munshi AD, Joshi S, Kaur C (2003) Correlation and path coefficient analysis for yield and biochemical characters in chilli (Capsicum annuum L.). Capsicum and Eggplant Newsletter 22: 67-70.

21. Yadwad A (2005) Genetic studies in chilli (Capsicum annuum L.) with particular reference to leaf curl complex. MSc Thesis, University of Agricultural Sciences, Dharwad, India, p: 144.

22. Yatung T, Dubey R, Singh V, Upadhyay G, Pandey AK (2014) Selection Parameters for Fruit Yield and Related Traits in Chilli (Capsicum annuum L.). Bangladesh J Bot 43: 283-291.

23. Patil CA (2007) Genetic Studies in Capsicum (Capsicum annuum L.). MSc Thesis, University of Agricultural Sciences, Dharwad.

24. Mishra AC, Singh RV, Ram HH (2002) Path coefficient Analysis in Sweet Peppers (Capsicum annuum L.) genotypes under mid hills of Uttaranchal. Veg Sci 20: 71-74.

25. Vani SK, Sridevi O, Salimath PM (2007) Studies on Genetic Variability, Correlation and Path Analysis in Chilli (Capsicum annum L.). Annu Biol 23: 117-121.

26. Sharma VK, Semwal CS, Uniyal SP (2010) Genetic variability and character association analysis in bell pepper (Capsicum annuum L.). Journal of Horticulture and Forestry 2: 58-65.

27. Kadwey S (2014) Genetic Variability, Correlation and Path Coefficient Analysis in chilli (Capsicum annuum L.). MSc Thesis. Department of Horticulture, College of Agriculture, Jawaharlal Nehru Krishi Vishwa Vidyalaya, Jabalpur, Madhya Pradesh, India.

14

Assessment of Genetic Improvement in Grain Yield Potential and Related Traits of Kabuli Type Chickpea (*Cicer arietinum* L.) Varieties in Ethiopia (1974-2009)

Tibebu Belete[1*], Firew Mekbib[2] and Million Eshete[3]
[1]Department of Plant Sciences and Horticulture, College of Dry Land Agriculture, Samara University, Samara, Ethiopia
[2]Department of Plant Science, College of Agricultural Sciences, Haramaya University, Ethiopia
[3]Debre Zeit Agricultural Research Center, Debre Zeit, Ethiopia

Abstract

Kabuli type chickpea is the most important commercial crop in Ethiopia and worldwide. A set of experiment was conducted to estimate the progress made in improving grain yield potential of Kabuli type chickpea varieties and changes in agromorphological traits associated with genetic yield potential. The varieties were laid down in a Randomized Complete Block Design with three replications. The annual rate of increase in yield potential of Kabuli type chickpea was estimated from linear regression of mean grain yields of varieties on year of release was 8.42 kg ha^{-1}yr^{-1} but this increment was not significantly different from zero. This revealed that chickpea breeders have made little/small efforts over the last 35 years to improve the yield of Kabuli type chickpea in Ethiopia. From the linear regression of hundred seed weight (HSW) against the years of release indicated that the annual rate of genetic gain was 1.00 g HSW^{-1} (8.96%) yr^{-1}, reflected that a significant increase was recorded for this trait for the last 35 years of Kabuli type chickpea improvement program in Ethiopia. Hence, better genetic improvement was obtained from breeding for HSW than it was from breeding for grain yield in Kabuli type. In contrast, significant negative trend was observed in number of pods plant^{-1}, seeds per pod^{-1} and seeds plant^{-1}. The correlation coefficients showed that grain yield was significantly and positively correlated with primary branches plant^{-1}, biomass yield and with all productivity traits. However, HSW which is the economical trait in Kabuli type chickpea showed significant negative association with secondary branches plant^{-1}, pods plant^{-1}, seeds pod^{-1} and seeds plant^{-1}. Stepwise regression analysis revealed that most of the variation in grain yield was caused by biomass yield and harvest index.

Keywords: Kabuli type chickpea; Genetic improvement; Harvest index; HSW; Grain yield; Yield components

Introduction

Chickpea (*Cicer arietinum* L.) is one of the principal food legumes in Ethiopia and it covers about 213,187 hectares of land and 2,846,398 quintals of chickpea is produced per annum with average productivity of 1.34 tons per hectare [1]. It, therefore, ranks third in production next to faba bean and haricot bean, but it ranks second in productivity per unit of area next to haricot bean. This clearly indicates the importance of chickpea in Ethiopian agriculture. Ethiopia is the largest producer of chickpea in Africa, accounting for about 46% of the continent's production during 1994 to 2006. It is also the seventh largest producer worldwide and contributes about 2% of the total world chickpea production [2].

According to Bekele [3], Kabuli type chick pea varieties are the most important crop in terms of local and export markets due to their large-seeded type. Therefore, there is a higher economic incentive for farmers to shift from Desi to Kabuli production due to its high price in world market. In Ethiopia, seeds are consumed raw, roasted or in 'wot'. Sometimes, the flour is mixed with other crops for preparing injera and also unleavened bread. Green pods and tender shoots are used as a vegetable. The roasted and salted chickpea is used as snack. It can also be mixed with cereals and root crops as a protein supplement in preparing "fafa" [4]. It is also an important legume crop used in rotation with several cereals like tef or wheat on heavy soils and maintains soil fertility through nitrogen fixation [5,6]. However, both productivity and quality of Ethiopian chickpeas have so far remained threateningly suboptimal due mainly to traditional and inadequate agronomic management practices, low yield potentials of the types under widespread cultivation and ravages of various biotic and abiotic stresses.

More than nine Desi type improved chickpea varieties along with their management practices have been developed and released through the national agricultural research systems in Ethiopia since the inception of chickpea improvement program at Debre Zeit Agricultural Research Center (DZARC) about four decades ago [7]. As can been seen from the annual production statistics above, the national average yield of chickpea is very low (about one tone per hectare) [1]. On the contrary, in areas where improved chickpea technologies were adopted and used, yield levels of up to five tons per hectare have been achieved [7]. This huge productivity gap warrants wider dissemination of the improved chickpea technologies in order substantially boost up the overall productivity and production in the country.

Information on genetic progress achieved over time from a breeding program is absolutely essential to develop effective and efficient breeding strategies by assessing the efficiency of past improvement works in genetic yield potential and suggest on future selection direction to facilitate further improvement [8-11]. Progress made in genetic yield potential and associated traits produced by different crops improvement program and the benefits obtained have been evaluated and documented

*Corresponding author: Tibebu Belete, Department of Plant Sciences and Horticulture, College of Dry Land Agriculture, Samara University, PO Box 132, Samara, Ethiopia, E-mail: tibebelete@gamil.com

in different countries concluded that genetic improvement in those crops have produced modern cultivars with improved yield potential [11-17]. This is also true for some crops in Ethiopia [18-24].

However; despite considerable effort and devotion of resources to Kabuli type chickpea improvement, there has been no work conducted in Ethiopia and worldwide to evaluate and document the progress made in improving the genetic yield potential and associated traits of Kabuli type chickpea varieties from different years in a common environment. Therefore, there is a need to quantify genetic progress in Kabuli type chickpea to design effective and efficient breeding strategy for the future. Hence, this research was initiated with the following objectives:

- To estimate the progress made in improving genetic yield potential of Kabuli type chickpea varieties.

- To assess changes in agro-morphological characters and thereby to identify their association with genetic improvement of Kabuli type chickpea varieties.

Materials and Methods

The experiment was conducted during the main cropping season of 2010 under rain fed condition in the experimental fields of Debre Zeit Agricultural Research Center (DZARC) and Akaki substation. DZARC is located at 08°44'N, 38°58'E and an altitude of 1900 masl. It's mean annual rainfall of 851 mm and mean maximum and minimum temperature of 28.3°C and 8.9°C respectively. Akaki is also situated at 08°52'N and 38°47'E with an altitude of 2200 masl and characterized by long term average annual rainfall of 1025 mm and mean maximum and minimum temperature of 26.5°C and 7.0°C respectively.

The study consisted of nine Kabuli type chickpea varieties released since 1974. The varieties were planted in a Randomized Complete Block Design (RCBD) with three replications at each experimental location. The experimental plot area was 4.8 m^2 having 4 rows each 4 m long and 1.2 m width. Spacing of 0.30 m between rows and 0.10 m between plants were used; the two middle rows with an area of 2.4 m^2 used for data collection. The spacing between plots and blocks were 0.40 m and 1.0 m respectively. Field management and protection practices were applied based on research recommendation for each respective location.

Data on yield and yield related traits were collected on plot and plant basis, such as phenological traits [days to 50% flowering (DF), days to 90% physiological maturity (DM), grain filling period (GFP)], grain yield, biomass yield, harvest index, yield attributes (plant height, number of primary branches per plant, number of secondary branches per plant, number of pods per plant, number of seeds per pod, number of seeds per plant, grain yield per plant, hundred seed weight and productivity traits (biomass production rate, seed growth rate and, grain yield per day).

All measured parameters were subjected to analysis of variance (ANOVA) using PROC ANOVA of SAS software version 9.0 [25] to assess the differences among the tested varieties. The homogeneity of error mean squares between the two locations were tested by F test on variance ratio and combined analyses of variance were performed for the traits whose error mean squares were homogenous using PROC GLM procedure of SAS. Number of pods plant^{-1}, number of seeds plant^{-1}, grain yield plant^{-1}, biomass yield hectare^{-1}, biomass production rate, seed growth rate and grain yield day^{-1} were transformed and their error variances were homogenized by log transformation according to Gomez [26]. Mean separation was carried out using Duncan's Multiple Range Test (DMRT).

The breeding effect was estimated as a genetic gain for grain yield and associated traits in chickpea improvement by regressing mean of each character for each variety against the year of release of that variety using PROC REG procedure. The coefficient of linear regression gives the estimate of genetic gain in kg ha^{-1}yr^{-1} or in % per year [27]. For this study, the year of release was expressed as the number of years since 1974; the year when the first Kabuli type chickpea variety was released. The relative annual gain achieved over the last 35 years (1974-2009) was determined as a ratio of genetic gain to the corresponding mean value of oldest variety and expressed as percentage.

To compute Pearson product moment correlation coefficients among all characters using means of each variety, PROC CORR in SAS was used. Stepwise regression analysis was carried out on the varietal mean using PROC STEPWISE in SAS to determine those traits that contributed much for yield variation among varieties.

Results and Discussion

Grain yield potential

Combined analysis of variance across the two locations showed highly significant (p ≤ 0.01) difference in grain yield among varieties while the effect of location on grain yield was non-significant (Table 1). The location × variety interaction effect was also non-significant for this trait. The grain yield performance of all Kabuli type chickpea varieties averaged over locations was 2018.25 kg ha^{-1}, which ranged from 1451.4 kg ha^{-1} for the variety Monino (recently released variety) to 2789.6 kg ha^{-1} for the variety Arerti (Table 2). The most recently released variety Monino, showed lower grain yield than all varieties represented in the current study. It showed lower grain yield than the first old variety (DZ-10-4) by 76.00 kg ha^{-1} (5%) (Table 3). This clearly indicated that grain yield of Kabuli type chickpea was not improved consistently as per the year of release.

The mean grain yield of varieties released in 1970s, 1990s and 2000s were 1527.40, 2398.75 and 1973.23 kg ha^{-1} respectively. These showed that an increase of 871.35 (57.05%) and 445.83 kg ha^{-1} (29.19%) over the first released variety, respectively. The average grain yield of those varieties which were released in 2000s exceeded that of the first variety but it was smaller than the yield of the variety released in 1990s by 425.52 kg ha^{-1} (17.74%) (Table 4). This clearly indicated that, in Ethiopia, the variety which was released in 1999 (Arerti) was highly productive because of its high yielding potential and is still under cultivation and not yet substituted by other Kabuli type chickpea varieties, but the criteria for releasing other variety were seed size, seed color and other quality parameters [28]. That is why the recently low yielding variety Monino was released. As indicated in Table 5, variety Monino was by far higher in seed size than the first older variety (DZ-10-4). It exceeded the older variety by 51.10 g (456.25%) in hundred seed weight and by 36.60 g (142.41%) in hundred seed weight over the higher yielder variety (Arerti). To this effect, it seems to strategically be advisable that hybridization efforts in the future should give attention to building on the short coming of low yielding modern varieties like Monino with high yielding varieties like Arerti for simultaneous improvement of grain yield and hundred seed weight. Similarly, Pereira [29] reported that there was lack of increase in yield potential during the period 1930-1970 in sunflower. According to these authors, the importance of selection for disease tolerance and grain quality plus a reduced genetic base may have restrained selection for yield potential in sunflower. Another study by Demissew [23] on soybean indicated that the average grain yield of the genotypes in the pipeline exceeded that of the first released varieties by 458.67 kg ha^{-1} or 43.91% but it was smaller than the mean yield of the

S No	Variety/Acc. No	Year of release	Breeder/maintainer€	Source	Seed color
1.	DZ-10-4	1974	DZARC/EIAR	Ethiopia	White
2.	Arerti (FLIP 89-84C)	1999	DZARC/EIAR	ICARDA	White
3.	Shasho (ICCV-93512)	1999	DZARC/EIAR	ICRISAT	White
4.	Chefe (ICCV-92318)	2004	DZARC/EIAR	ICRISAT	White
5.	Habru (FLIP 88-42C)	2004	DZARC/EIAR	ICARDA/ICRISAT	White
6.	Ejeri (FLIP-97-263c)	2005	DZARC/EIAR	ICARDA	White
7.	Teji (FLIP-97-266c)	2005	DZARC/EIAR	ICARDA	White
8.	Yelibey (ICCV-14808)	2006	SRARC/ARARI	ICRISAT	Yellowish
9.	ACOS DUIBIE (Monino)	2009	ACOS and DZARC/EIAR	Mexico	White cream

Source: [2,7,28]; €=Abbreviations: DZARC=Debre Zeit Agricultural Research Center; EIAR=Ethiopian Agricultural Research Institute; SRARC=Sirinka Regional Agricultural Research Center; ARARI=Amhara Regional Agricultural Research Institute.

Table 1: Description of Kabuli type chickpea varieties used in the experiment.

Trait€	Location (1)$^{\Psi}$	Varieties (8)	Location × Varieties (8)	Error (32)	Mean	CV (%)	R²
DF	64.46**	228.27**	62.05**	1.85	46.35	2.94	0.98
DM	9600.00**	127.32**	24.00**	4.05	112.70	1.79	0.99
NPBPP	0.13^{ns}	0.40**	0.13**	0.04	2.40	7.92	0.82
NSBPP	261.36**	13.24**	3.01**	0.88	5.38	17.46	0.93
PH	993.82**	19.32^{ns}	11.18^{ns}	7.18	36.07	7.43	0.85
NPoPP$^{\Psi}$	2038.73(0.28**)	1070.27(0.23**)	166.06(0.01^{ns})	72.36(0.01)	33.87(1.48)	25.12(6.40)	0.84(0.88)
NSPPo	0.01^{ns}	0.18**	0.0038^{ns}	0.006	1.16	6.43	0.90
NSPP$^{\Psi}$	3700.17(0.33**)	2851.92(0.32**)	363.36(0.01^{ns})	92.92 (0.01)	40.52(1.54)	23.79(5.73)	0.91(0.92)
GYPP$^{\Psi}$	313.16(0.46**)	23.90(0.03**)	13.61(0.01^{ns})	7.55 (0.01)	10.96(1.02)	25.07(9.34)	0.72(0.74)
GYPha	127647.92^{ns}	1297018.58**	169690.26^{ns}	161195.78	2018.25	19.89	0.70
HSW	24.81**	1080.43**	2.92^{ns}	1.83	33.51	4.03	0.99
BYPha$^{\Psi}$	1164420.64(0.01^{ns})	3674654.17(0.05**)	336003.28(0.01^{ns})	393922.96(0.01)	3510.56(3.53)	17.88(2.24)	0.73(0.72)
GFP	11237.80**	25.14**	40.00**	8.00	66.35	4.26	0.98
HI	0.008**	0.004**	0.005**	0.001	0.57	5.18	0.75
BPR$^{\Psi}$	1435.72(0.24**)	273.60(0.05**)	52.12(0.01^{ns})	35.54(0.01)	31.73(1.48)	18.79(5.41)	0.79(0.78)
SGR$^{\Psi}$	3227.43(0.54**)	392.19(0.07**)	103.39(0.02^{ns})	50.67(0.01)	32.32(1.48)	22.03(6.38)	0.82(0.82)
GYPD$^{\Psi}$	373.51(0.16**)	99.62(0.06**)	22.92(0.01^{ns})	14.67(0.01)	18.20(1.24)	21.04(7.49)	0.75(0.74)

Table 2: Mean squares from combined analysis of variance for seed yield and other traits in Kabuli type chickpea varieties evaluated over two test locations (Debre Zeit and Akaki).

newest released variety. In contrast, [18] on wheat, [19] on tef, [20] on haricot bean and [22] on barley reported respective increases in grain yield potentials of varieties over the period studied.

Generally, the varieties developed through introduction yielded an average grain yield of 2079.61 kg ha^{-1} and exceeded the variety which was developed through local collection by 552.21 kg ha^{-1} (36.15%) (Table 6). This clearly indicated that varieties developed from introduced material contributed the genetic improvement obtained in grain yield of Kabuli type chickpea over the last 35 years. In line with this study, Kebere [20] also indicated that introduced materials contributed a lot for the improvement of the genetic yield potential of haricot bean varieties in Ethiopia (Table 6). On the contrary, Tamene [21] showed that the local collections and hybridization materials were the most important sources

of genetic variability contributing to the genetic gain of faba bean over the last 30 years period.

The annual rate of increase in yield potential was estimated from linear regression of mean grain yields of varieties on year of release was 8.42 kg ha^{-1}yr^{-1} (Figure 1A). This clearly indicates that chickpea breeders have made efforts over the last 35 years to improve the yield of Kabuli type chickpea in Ethiopia, but this increase was not significantly different from zero (Table 7), rather they get substantial improvement in hundred seed weight. Likewise, Ersullo [24] noticed that an average rate of increase in grain yield potential per year of release since pre-1984 was non- significant (4.329 kg ha^{-1}yr^{-1}) when tested under the four locations for linseed. Similarly, Koemel [30] indicated the more recent entries failed to improve grain yield of hard winter wheat over that of the long-term check cultivars.

Varieties	Locations		Mean
	Debre Zeit	Akaki	
DZ-10-4	1773.90[bc]	1280.80[e]	1527.4[cd]
Arerti	2920.00[a]	2659.20[a]	2789.6[a]
Shasho	1854.20[bc]	2161.70[abc]	2007.9[bc]
Chefe	2661.90[ab]	2278.20[ab]	2470.1[ab]
Habru	2663.10[ab]	2271.80[ab]	2467.4[ab]
Ejeri	2018.90[abc]	1792.80[bcde]	1905.8[cd]
Teji	1670.30[c]	1619.40[ed]	1644.9[cd]
Yelibey	1811.30[bc]	1988.30[bcd]	1899.8[cd]
Monino	1228.30[c]	1674.40[cde]	1451.4[d]
Mean	2066.87	1969.63	2018.25
CV (%)	24.03	13.98	19.89
R²	0.67	0.78	0.70

Means followed by the same letter with in a column are not significantly different from each other at P ≤ 0.05 according to Duncan's Multiple Range Test.

Table 3: Mean grain yield (kg ha⁻¹) of Kabuli type chickpea varieties at Debre Zeit and Akaki and averaged across locations.

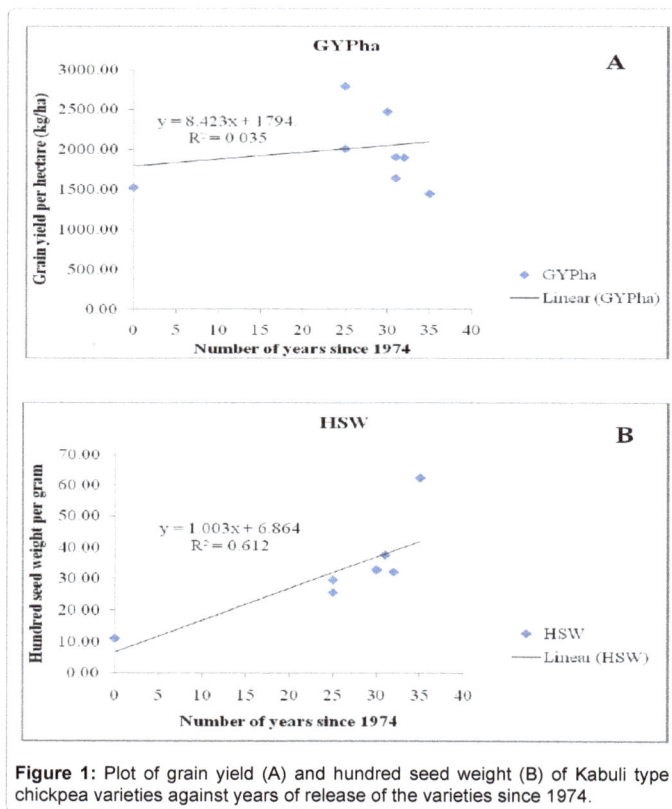

Figure 1: Plot of grain yield (A) and hundred seed weight (B) of Kabuli type chickpea varieties against years of release of the varieties since 1974.

According to Yifru [19] in tef, the genetic gain of some traits was non-significant from 1960 to 1995. Tamene [21] also reported that the genetic gain obtained in faba bean breeding since 1970s was very minimal, that is only 82 kg ha⁻¹ in 30 years period or close to 3 kg ha⁻¹ yr⁻¹. Similarly, Wondimu [22] in grain yield of malt barley reported that slope of regression since 1973 was not significantly different from zero at each location as well as across locations. In addition, Mackey [31] observed no increase in grain yield of oat varieties for the past 50 years in Sweden.

Hundred seed weight

Like grain yield, the combined analysis of variance revealed that there was no location × variety interaction for hundred seed weight of Kabuli type chickpea, but highly significant (p ≤ 0.01) differences were observed for locations and varieties (Table 8). Mean hundred seed weight ranged from 11.20 g (DZ-10-4) to 62.30 g (Monino) with across location average of 33.52 g (Table 9). The most recently released variety, Monino, showed significantly higher hundred seed weight (seed size) than all the varieties represented in the current study. It exceeded the first older variety (DZ-10-4) by 51.10 g (456.25%) in hundred seed weight. The average hundred seed weight of varieties released in 1974, 1999, 2004, 2005, 2006 and 2009 were 11.20, 27.65, 32.80, 37.60, 32.10 and 62.30 g, respectively. This showed that an increase of 16.45 (146.88%), 21.60 (192.86%), 26.40 (235.71%), 20.90 (186.62%) and 51.10 g (456.25%) in hundred seed weight respectively, over the older variety (Table 4). Therefore, hundred seed weight increase was almost consistent and parallel over the year of release of improved Kabuli type chickpea varieties.

Linear regression hundred seed weight against the years of release indicated annual rate of genetic gain of 1.00 g yr⁻¹ (Figure 1B) with a relative annual genetic gain of 8.96% (Table 8), reflecting a significant increase in the trait over the last 35 years of Kabuli type chickpea improvement in Ethiopia (Table 7). Generally speaking, better genetic improvement was obtained from breeding for hundred seed weight than it was from breeding for grain yield as far as Kabuli type chickpea is concerned. Likewise, Amsal [18] in durum wheat, Ortiz [32] in two-row Nordic spring barley, Tamene [21] in faba bean and Ersullo [24] in linseed found that thousand seed weight of modern varieties were heavier than the older ones. Contrary to the present study, [19] in tef, [20] in haricot bean, [22] in food barely noticed non-significant change in seed weight. Highly significant decrease in thousand seed weight with a relative annual reduction of 0.96% was reported in soybean by Demissew [23].

Similar to the grain yield, introduction derived varieties gave an advantage of 25.11 g (224.20%) for HSW over the local collection derived (Table 6) which is contrary to finding of Tamene [21] in faba bean. This indicates that varieties developed from introduced germplasm are the most important sources of genetic material contributing to the genetic improvement in hundred seed weight of Kabuli type chickpea varieties over the last 35 years and the possibility of further improvement in hundred seed weight using this breeding method.

Biomass yield, harvest index and plant height of kabuli type chickpea

Combined analysis of variance for biomass yield indicated non-significant location and location × variety interaction effects. On the other hand, highly significant (p ≤ 0.01) differences were observed among varieties tested for the trait (Table 2). The mean biomass yield of all varieties across the two locations was 3510.56 kg ha⁻¹. The highest mean biomass yield (4948.8 kg ha⁻¹) across locations (Table 9) was recorded from the variety with the highest grain yield, Arerti (Table 3). This variety showed significantly higher biomass yield than all varieties except Chefe and Habru (Table 9). Like that of grain yield, biomass yield also showed inconsistent trend over years of release. Mean biomass yield of varieties developed through introduction was 969.60 kg ha⁻¹ (36.61%) higher than mean biomass yield of variety developed through direct selection from landraces (Table 6). Hence, much of the increase in biomass yield was obtained from introduced materials. Similarly, Kebere [20] reported similar finding in haricot bean in Ethiopia. Yifru and Hailu found that varieties developed through intraspecific hybridization gave higher biomass than the varieties developed through landrace selection in tef breeding program.

Variety	Year of release	Mean grain yield (kg ha⁻¹)	Increment over DZ-10-4		Mean HSW (g)	Increment over DZ-10-4	
			kg	%		g/HSW	%
DZ-10-4	1974	1527.40	---	---	11.20	---	---
Arerti	1999	2398.75	871.35	57.05	27.65	16.45	146.88
Shasho	1999						
Chefe	2004	2468.75	941.35	61.63	32.80	21.60	192.86
Habru	2004						
Ejeri	2005	1775.35	247.95	16.23	37.60	26.40	235.71
Teji	2005						
Yelibey	2006	1899.80	372.40	24.38	32.10	20.90	186.61
Monino	2009	1451.40	-76.00	-5.00	62.30	51.10	456.25

Table 4: Trends in genetic progress in grain yield and hundred seed weight (HSW) for Kabuli type chickpea varieties released in 1999, 2004, 2005, 2006 and 2009 over the older variety (DZ-10-4) released in 1974.

Varieties	Year of release	Mean grain yield (kg ha⁻¹)	Increment over the older variety (DZ-10-4)		Mean HSW (g)	Increment over the older variety (DZ-10-4)		Mean biomass yield (kg ha⁻¹)	Increment over the older variety (DZ-10-4)	
			kg ha⁻¹	%		g/HSW	%		kg ha⁻¹	%
DZ-10-4	1974	1527.40	-	-	11.2	-	-	2648.70	-	-
Arerti										
Shasho	1990s	2398.75	871.35	57.05	27.65	16.45	146.88	4252.25	1603.55	60.54
Chefe										
Habru										
Ejeri										
Teji	2000s	1973.23	445.83	29.19	39.20	28.00	250.00	3406.98	758.28	28.63
Yelibey										
Monino										

Table 5: Trends in genetic progress in grain yield, hundred seed weight (HSW) and biomass yield for Kabuli type chickpea varieties released in 1990s and 2000s over the older variety (DZ-10-4) released in 1974.

Variety	Grain yield (kg ha⁻¹)	Grain yield increment over local collection		Biomass yield (kg ha⁻¹)	Biomass yield increment over local collection		Mean HSW (g)	HSW increment over local collection	
		kg ha⁻¹	%		kg ha⁻¹	%		g/HSW	%
Local collection derived	1527.40	-	-	2648.70	-	-	11.20	-	-
Introduction derived	2079.61	552.21	36.15	3618.30	969.60	36.61	36.31	25.11	224.20

Table 6: Average increments in grain, biomass yield and hundred seed weight (HSW) for Kabuli type chickpea varieties derived from introduction over variety derived from local collection.

Traits	Mean	R²	b	Intercept
Date of flowering	46.35	0.03	-0.10	49.04
Date of maturity	112.70	0.01	-0.04	113.89
Number of primary branches per plant	2.40	0.09	0.01	2.18
Number of secondary branches per plant	5.38	0.26	-0.07	7.32
Plant height	36.07	0.02	-0.02	36.70
Number of pods per plant	33.87	0.65	-1.03**	61.14
Number of seeds per pod	1.16	0.84	-0.02**	1.57
Number of seeds per plant	40.52	0.90	-1.98**	93.14
Grain yield per plant	10.96	0.00	0.00	10.96
Grain yield per hectare	2018.25	0.04	8.42	1794.57
Hundred seed weight	33.51	0.61	1.00*	6.86
Biomass yield per hectare	3510.56	0.04	14.88	3115.40
Grain filling period	66.35	0.08	0.06	64.85
Harvest index	0.57	-0.0002	0.00	0.58
Biomass production rate	31.73	0.05	0.15	27.81
Seed growth rate	32.32	0.01	0.08	30.06
Grain yield per day	18.20	0.04	0.08	16.14

*, **=Significant at $P \leq 0.05$ and $P \leq 0.01$, respectively.
Table 7: Estimates of mean values, coefficient of determination (R^2), regression coefficient (b) and intercept for various traits from linear regression of the mean value of each trait for each Kabuli type chickpea variety against the year of variety release since 1974.

Traits	Mean of the older variety	RGG (% per year)	Correlation coefficients (R)			
			R_{GYPha}	R_{YOR}	R_{HSW}	R_{BYPha}
Date of flowering	44.7	-0.23	0.45	-0.17	-0.36	0.49
Date of maturity	110.5	-0.04	0.41	-0.10	-0.35	0.45
Number of primary branches per plant	2.0	0.40	0.81**	0.31	-0.09	0.85**
Number of secondary branches per plant	6.5	-1.13	0.49	-0.51	-0.89**	0.44
Plant height	36.3	-0.07	0.60	-0.14	-0.35	0.56
Number of pods per plant	55.8	-1.84	0.15	-0.80**	-0.91**	0.12
Number of seeds per pod	1.61	-0.95	0.31	-0.91**	-0.71*	-0.36
Number of seeds per plant	90.1	-2.20	-0.07	-0.95**	-0.89**	-0.11
Grain yield per plant	9.8	0.00	0.40	0.00	-0.46	0.34
Grain yield per hectare	1527.4	0.55	---	0.19	-0.28	0.99**
Hundred seed weight	11.2	8.96	-0.28	0.78*	---	-0.21
Biomass yield per hectare	2648.7	0.56	0.99**	0.2	-0.21	---
Grain filling period	65.8	0.09	-0.42	0.29	0.30	-0.48
Harvest index	0.573	-0.03	0.35	-0.06	-0.52	0.19
Biomass production rate	24.5	0.60	0.97**	0.23	-0.17	0.98**
Seed growth rate	25.4	0.33	0.99**	0.11	-0.33	0.99**
Grain yield per day	14.2	0.55	0.98**	0.2	-0.26	0.96**

*, **=Significant at $P \leq 0.05$ and $P \leq 0.01$, respectively.

Table 8: Estimates of the mean annual relative genetic gain (RGG); and correlation coefficient of all traits with grain yield (R_{GYPha}), year of release of the variety (R_{YOR}), hundred seed weight (R_{HSW}) and biomass yield (R_{BYPha}).

Varieties		Locations				Mean	
		Debre Zeit			Akaki		
	HSW	BYPha	HSW	BYPha	HSW	BYPha$^{\psi}$
DZ-10-4	12.2^f	2967.4^c	10.2^f	2330.1^d	11.2^f	2648.7^e
Arerti	26.7^e	5225.0^a	24.6^e	4672.5^a	25.7^e	4948.8^a
Shasho	30.9^d	3299.9^c	28.2^d	3811.5^b	29.6^d	3555.7^{bc}
Chefe	32.7^{cd}	4513.8^{ab}	32.5^c	3795.3^b	32.6^c	4154.5^{ab}
Habru	34.4^c	4510.8^{ab}	31.6^c	3832.4^b	33.0^c	4171.6^{ab}
Ejeri	38.1^b	3804.7^{ab}c	36.9^b	3087.6^{bcd}	37.5^b	3446.2^{bcd}
Teji	37.3^b	2923.1^c	38.1^b	2731.4^{cd}	37.7^b	2827.2^{de}
Yelibey	33.5^c	2986.8^c	30.8^{cd}	3226.8^{bc}	32.1^c	3106.8^{cde}
Monino	62.0^a	2685.3^c	62.6^a	2785.8cd	62.3^a	2735.6^e
Mean	34.19	3657.41	32.83	3363.72	33.51	3510.56
CV (%)	3.41	21.33	4.61	12.59	4.03	2.24
R²	0.99	0.68	0.99	0.82	0.99	0.73

Means followed by the same letter with in a column are not significantly different from each other at $P \leq 0.05$ according to Duncan's Multiple Range Test; $^{\psi}$=Mean separation and CV based on transformed data.

Table 9: Mean hundred seed weight (HSW) and biomass yield (BYPha) of Kabuli type chickpea varieties evaluated at Debre Zeit and Akaki and combined across the two locations.

Linear regression coefficient revealed that biomass yield did not change significantly during the past three decades of Kabuli type chickpea breeding programs (14.88 kg ha^{-1}yr^{-1}) (Table 7) with a small relative genetic gain of 0.56% yr^{-1}. The present result was in agreement with the findings of Amsal [18] on bread and durum wheat, Hailu [15] on soybean and Wondimu [22] on food barley who observed that non-significant improvement in biomass yield. Ortiz [32] also found non-significant trend observed in biomass yield of spring barely. Ersullo [24] reported non-significant biomass yield in linseed varieties released since 1984. Conversely, Yifru, Kebere and Tamene [19-21] and Demissew [23] reported that biomass yield was linearly related to variety age and positively and significantly associated to grain yield.

As per combined analysis of variance, harvest index showed highly significant (p ≤ 0.01) differences between locations and among varieties. There was also highly significant location × variety interaction for this trait (Table 2). The mean harvest index of Kabuli type chickpea varieties represented in this study was 0.56 at Debre Zeit and 0.59 at Akaki and

0.57 averages over the two locations (Table 9). This is in agreement with the harvest index reported for haricot bean Kebere [20]. Similarly, higher harvest index value of 0.59 for chickpea was reported by Saxena [33].

Linear regression coefficient for harvest index showed a non-significant annual decrease trend (-0.0002) (Table 7), which was almost zero during the 35 years of Kabuli type chickpea improvement with a relative annual genetic reduction of -0.03% (Table 8). Likewise, Yifru [19] and Kebere [20] found that no change in harvest index of tef and haricot bean respectively. Demissew [23] also noticed that non-significant annual reduction in harvest index of soybean. In contrast to this, Hailu [15] and Jin [16] revealed that harvest index increased significantly with year of release of soybean varieties. The varieties in the present study showed a small decrease in harvest index may be the higher non-significant increment of biomass yield than grain yield.

There was a non-significant difference among varieties while the effect of location was highly significant for plant height (Table 2).

However, the annual genetic gain of plant height over the past 35 years of breeding was -0.02 cm and was not significantly different from zero (Table 7) with relative genetic gain of -0.07% yr[-1] (Table 8). Similarly, a non-significant reduction in plant height was reported by Kebere [20] on haricot bean.

Yield components of kabuli type chickpea

Except number of primary branches plant[-1] and number of seeds pod[-1], mean squares of locations from combined analysis of variance were highly significant (p ≤ 0.01) for number of secondary branches plant[-1], number of pods plant[-1], number of seeds plant[-1] and grain yield plant[-1]. Combined analysis of variance across the two locations indicated non-significant location × variety interaction for all yield components except number of primary branches plant[-1] and number of secondary branches plant[-1] which showed highly significant interaction effects. Furthermore, all the above mentioned yield components showed highly significant difference among varieties (Table 2).

Mean number of primary branches plant[-1] and secondary branches plant[-1] from combined analysis was found to be 2.40 and 5.38 respectively (Table 10). In both locations (Debre Zeit and Akaki) the highest yielding variety Arerti, had the highest number of primary branches plant[-1] and secondary branches plant[-1]. Estimated annual gains of both number of primary branches plant[-1] and secondary branches plant[-1] of Kabuli type chickpea varieties over the last 35 years was 0.01 and -0.07 branches plant[-1] yr[-1], which were not significantly different from zero (Table 7); and relative genetic gain of 0.40 and -1.13% yr[-1] (Table 8), respectively.

The average number of pods plant[-1], number of seeds pod[-1] and number of seeds plant[-1] of Kabuli type chickpea varieties, average over locations were 33.87, 1.16 and 40.52 respectively (Table 10). Most recently released varieties which had heavier seed weight (larger seed size) and low yield have low number of pods plant[-1], seeds pod[-1] and seeds plant[-1]. Generally, there was a decreasing trend in number of pods plant[-1], number of seeds pod[-1] and number of seeds plant[-1] over the 35 years period of Kabuli type chickpea improvement as it can be seen from highly significant negative linear regression coefficients (Table 7) and relative genetic gain of -1.84, -0.95 and -2.20% yr[-1] for the three traits (Table 8), respectively. Similarly, Tamene [21] observed that number of seeds plant[-1], number of pods plant[-1], number of podding nods plant[-1]

and number of pods node[-1] followed a decreasing trend against time in faba bean breeding program, that is, the older the variety the higher the value for the component traits and vice versa. However, [20] in haricot bean reported that number of pods plant[-1] and number of seeds pod[-1] showed a non-significant increasing trend for the period studied.

The negative breeding progress in secondary branches plant[-1], number of pods plant[-1], number of seeds pod[-1] and number of seeds plant[-1] may be considered as the result of a negative compensatory response to the radical increment in hundred seed weight (seed size) during the same period. Therefore, for simultaneous improving seed size, number of pods plant[-1] and number of seeds plant[-1] a compromise between selection progresses for both traits must be made, or the breeder must set minimum standards for one trait while selecting for the other.

Phenological and productivity traits

Phenological traits: The combined analysis of variance showed that days to flowering, days tomaturity and grain filling period had highly significant (P ≤ 0.01) differences between locations and among varieties. The location x variety interaction effect also showed highly significant differences for these traits (Table 2). The average values of phenological traits represented in this study was 47.44, 99.37 and 51.93 days at Debre Zeit and 45.26, 126.04 and 80.78 days at Akaki, respectively. In this study, most of recently released varieties relatively took intermediate to short days to reach flowering and maturity. Early maturity is advantageous in chickpea to avoid terminal moisture stress and make adequate use of available soil moisture during growth period, as chickpea is usually grown on conserved soil moisture, where soil moisture reduces towards maturity [34].

From the slope of regression line, there was a negative trend for days to flowering and days to maturity but not significantly different from zero (Table 7). However, grain filling period showed increasing trend, still it was not significantly different from zero. Likewise, Wondimu [22] reported that all of the three phenological traits showed a non-significant decreasing tend in food barely. Hailu [15] observed insignificant increment with delayed flowering and maturity in soybean varieties. However, Kebere [20] and Tamene [21] found a non-significant increase in days to maturity in haricot bean and faba bean

							Trait							
Varieties	DF	DM	NPBPP	NSBPP	PH	NPoPP[Ψ]	NSPPo	NSPP[Ψ]	GYPP[Ψ]	GFP	HI	BPR[Ψ]	SGR[Ψ]	GYPD[Ψ]
DZ-10-4	44.7[cd]	110.5[cd]	2.0[d]	6.5[ab]	36.3	55.8[a]	1.61[a]	90.1[a]	9.8[bc]	65.8[bc]	0.573[abc]	24.5[d]	25.4[cd]	14.2[bc]
Arerti	55.5[b]	118.3[b]	3.0[a]	7.2[a]	35.9	41.7[ab]	1.07[bc]	44.7[bc]	11.3[ab]	62.8[c]	0.56[bc]	43.1[a]	46.4[a]	24.3[a]
Shasho	58.3[a]	122.5[a]	2.5[b]	6.4[ab]	36.6	49.2[a]	1.06[bc]	52.3[b]	14.6[a]	64.2[bc]	0.565[bc]	29.2[cd]	32.0[b]	16.5[bc]
Chefe	43.8[cde]	111.5[c]	2.4[bc]	4.9[c]	37.9	29.6[cd]	1.13[b]	32.7[de]	10.7[ab]	67.7[ab]	0.60[ab]	38.2[ab]	39.2[ab]	22.7[a]
Habru	43.7[def]	110.2[cd]	2.4[bc]	5.6[bc]	38.8	30.5[cd]	1.15[b]	35.0[cde]	11.5[ab]	66.5[abc]	0.59[ab]	38.8[ab]	40.3[ab]	22.9[a]
Ejeri	42.2[efg]	110.0[cd]	2.3[bc]	4.6[c]	36.4	24.6[d]	1.12[b]	27.5[e]	10.1[b]	67.8[ab]	0.55[bc]	32.4[bc]	30.7[bc]	17.8[b]
Teji	41.5[g]	111.3[c]	2.3[bc]	5.0[c]	32.7	26.8[cd]	1.10[bc]	29.4[e]	10.2[b]	69.8[a]	0.58[abc]	25.9[cd]	24.8[cd]	15.0[bc]
Yelibey	45.5[c]	111.7[c]	2.4[bc]	6.1[ab]	35.5	34.9[bc]	1.16[b]	40.8[bcd]	13.0[ab]	66.2[abc]	0.61[a]	28.1[cd]	29.7[bc]	17.2[b]
Monino	42.0[fg]	108.3[d]	2.2[cd]	2.1[d]	34.5	11.9[e]	1.02[c]	12.2[f]	7.5[c]	66.3[abc]	0.52[d]	25.5[cd]	22.3[d]	13.3[c]
Mean	46.35	112.70	2.40	5.38	36.07	33.87	1.16	40.52	10.96	66.35	0.57	31.73	32.32	18.20
CV (%)	2.94	1.79	7.92	17.46	7.43	6.40	6.43	5.73	9.34	4.26	5.18	5.41	6.38	7.49
R[2]	0.98	0.99	0.82	0.93	0.85	0.84	0.90	0.91	0.72	0.98	0.75	0.79	0.82	0.75

Means followed by the same letter with in a column are not significantly different from each other at P ≤ 0.05 according to Duncan's Multiple Range Test;

Ψ=Mean separation and CV based on transformed data.

Table 10: Mean values of phenological traits, yield components and productivity traits of Kabuli type chickpea varieties combined over locations (Debre Zeit and Akaki).

breeding, respectively. In contrast, Demissew [23] noticed that both days to flowering and maturity showed significant increment over years of soybean improvement. This author also indicated that grain filling period showed a non-significant increasing trend in the period studied.

Productivity traits: Combined analysis of variance showed highly significant differences between locations and among varieties tested for all productivity traits (biomass production rate, seed growth rate and grain yield day[-1]) while the location × variety interaction effect was non-significant for all the traits (Table 2). The mean biomass production rate, seed growth rate and grain yield day[-1] of Kabuli type chickpea varieties recorded from the combined analysis averaged over locations were 31.73, 32.32 and 18.20 kg ha[-1]day[-1] (Table 10), respectively.

Linear regression showed a non-significant increasing trend for biomass production rate, seed growth rate and grain yield day[-1] for the past 35 years of Kabuli type chickpea breeding program (Table 7) with relative genetic gains of 0.60, 0.33 and 0.55% yr[-1] (Table 8), respectively. This indicated that breeding did not markedly affect these traits for the last three decades. Similarly, Yifru [19] observed a non-significant increase in both total grain sink filling rate and biomass production rate of tef varieties over the 35 years of variety release. The non-significant increasing trend grain yield day[-1] observed in the present study was in agreement with the finding of Demissew [23] in soybean improvement. Amsal [18] and Wondimu [22] also found that biomass production rate on year of release of the varieties has showed no indication of improvement in the study period. In the same study, however, a significant increasing trend in biomass production rate, seed growth rate and grain yield day[-1] was reported as opposed to the present study in haricot bean for the 26 years period Kebere [20].

Yield related traits associated with grain yield potential improvement

The correlation coefficients of grain yield, hundred seed weight and biomass yield with all the traits studied are presented in Table 8. The correlation coefficients indicated that grain yield showed a highly significant ($p \leq 0.01$) association with biomass yield, while it had non-significant and positive association with harvest index. Hence, the results herein demonstrated that increasing the biomass yield would be a more efficient way to boost up Kabuli type chickpea grain yield than would harvest index. Moreover, biomass yield showed significant positive relation with number of primary branches plant[-1], biomass production rate, seed growth rate and grain yield day[-1], but non-significant association with all other traits. In agreement with the present study, [19] on tef, [20] on haricot bean, [21] on faba bean, [15] and [23] on soybean found that highly significant positive correlations between grain yield and biomass yield, but no significant correlation between grain yield and harvest index. Similarly, Bicer [35] on chickpea reported that biological yield is positively correlated with seed yield, which is an important character for determining seed yield. The reverse is true in the finding of Khan [36] who reported that grain yield positively and highly significantly association with harvest index but non-significantly with biomass yield. As a result, variation in harvest index had a possibility of improving and boosting up grain yield in chickpea. Singh [37] on chickpea found that biological yield and harvest index had significant positive association with seed yield and therefore selection for these traits both together would lead to high seed yield. Conversely, Amsal [18] on bread wheat and Wondimu [22] on food barley reported significant and positive relation between harvest index and grain yield and non-significant association between biomass and grain yield.

The association between grain yield and plant height was also positive and statistically non-significant (Table 8). Different authors also found non-significant correlation between grain yield and plant height [38-40]. Similarly, Yifru, Kebere, Tamene [19-21] and Hailu [15] observed no relation between grain yield and plant height respectively in tef, haricot bean, faba bean, and soybean. However, Wondimu [22] on food barley and Jin [16] on soybean observed negative correlation between plant height and grain yield.

In general, grain yield in the modern varieties appears to be associated more with the production of a higher biomass than with a higher partitioning efficiency to the grain sink. This indicated that biomass yield may serve as an index for identifying chickpea varieties with higher seed yield. Hence, it is of vital importance to give due attention to biomass yield while selecting Kabuli type chickpea varieties for production and commercial cultivation.

Highly significant positive correlation was observed between grain yield and number of primary branches plant[-1], while the association of grain yield with number of secondary branches plant[-1], number of pods plant[-1], number of seeds pod[-1] and grain yield plant[-1] was positive and non- significant. This indicates that number of primary branches plant[-1] is still an important trait used for selection criteria in breeding for further improvement in grain yield of both chickpea types. Among yield components number of seeds plant[-1] and hundred seed weight showed negative and no-significant association with grain yield (Table 8). Similarly, [40] indicated that number of primary branches plant[-1] showed highly significant positive correlation with grain yield whereas number of secondary branches plant[-1] showed non-significant association with grain yield on chickpea. Sharma [41] also reported primary branches plant[-1] showed highly correlation with grain yield. In contrast, Saleem [38] on chickpea found that there was significant and negative relation between grain yield and number of secondary branches plant[-1], but the association of primary branches plant[-1] with grain yield was non-significant. According to Toker, Sharma and Ali [42-44] grain yield was significantly and negatively correlated with hundred seed weight. Similarly, there have been few cases of negative association between seed size and grain yield, apparently as a result of few seeds pod[-1] and few pods plant[-1], characteristics of larger seeded type [45]. Another report which was almost similar to the present study reported by Tamene [21], who reported non-significant association between grain yield and seed weight plant[-1], thousand seed weight, number of seeds pod[-1], number of seeds plant[-1] and number of pods plant[-1].

One of the economical traits in Kabuli type chickpea, hundred seed weight, showed significantly negative associations with number of secondary branches plant[-1], number of pods plant[-1], number of seeds pod[-1] and number of seeds plant[-1] (Table 7). Similarly, Kumar, Naseem and Temesgen [46-48] on Kabuli type chickpea observed that hundred seed weight had significant and negative phenotypic correlation with number of pods plant[-1], number of seeds pod[-1] and number of seeds plant[-1]. Another study by Sharma [41] showed that number of branches plant[-1] and number of pods plant[-1] showed highly negative association with hundred seed weight. Negative association between hundred seed weight and number of seeds per pod indicates a compensatory relationship between them. More seeds per pod could result in the reduction of the average seed size because of competition among seeds for limited food reserves [49].

Positive and non-significant association of grain yields with days to flowering and maturity was observed in the current study (Table 8). The correlation between grain yield and grain filling period was negative but not significant. Hasan [39] also reported positive and non-significant correlation between grain yield and days to flowering as well as days to maturity. Conversely, Arshad [50] showed non-significant negative associations of grain yield with days to flowering and maturity. Similarly, Temesgen [48] reported grain filling period to be positively non-significantly associated with grain yield.

The correlation coefficients for grain yield day[-1], seed growth rate and biomass production rate was highly significant ($P \leq 0.01$) and positive with grain yield (Table 8). This clearly showed that improvement in these traits was markedly concurrent to the yield improvement in the past and can further be exploited in future breeding. In a similar study on haricot bean, Kebere and Tamene [20,21] on faba bean found positive and significant correlation of grain yield with each of these traits.

Stepwise regression analysis, using grain yield as dependent variable (Table 11) indicated that, biomass yield and harvest index were the most important traits which greatly contributed most of the variation in grain yield. Hence, 97% of the total variations in grain yield of Kabuli type chickpea varieties were explained by biomass yield alone and 99.9% by biomass yield and harvest index altogether. In previous study on tef and haricot bean, Yifru and Kebere [19,20] reported that biomass yield was the single most important trait that contributed 56.7 and 82.7% of the variation in grain yield among varieties respectively. About 96% of the variation in faba bean grain yield was explained by economic growth rate, whereas economic growth rate, number of pod plant[-1], harvest index and biomass together accounted for 99% of the variation in grain yield Tamene [21]. Similarly, Wondimu [22] reported that harvest index, biomass yield and biomass production rate were traits which contributed to gain in grain yield of food barley varieties. Demissew [23] also found that biomass yield, harvest index and number of branches plant[-1] were traits which contributed most to the variation in grain yield. Accordingly 93.0% of the variation in grain yield was contributed by biomass yield, 99.5% by biomass yield and harvest index, and 99.8% by biomass yield, harvest index and number of branches plant[-1] together. Therefore, it can be considered that changes in the above trait had probably contributed to the changes in grain yield during the last 35 years of breeding Kabuli type chickpea in Ethiopia.

The stepwise regression analysis also showed that, for hundred seed weight (seed size), which is also another economic trait in Kabuli type chickpea: number of pods plant[-1] had a decreasing effect, contributed to the variation among the varieties in seed size. About 83% of the variation in hundred seed weight was accounted for by number of pods plant[-1]. Similarly, Tamene [21] was reported 88.48% of the variability in thousand seed weight was accounted for by number of pods plant[-1] alone and 92.56% when both number of pods and grain filling period together.

Grain yield				
Independent variables	Intercept	Regression coefficient (b)	R[2]	VIF
Biomass yield hectare[-1]	-1608.12475	0.56669**	0.9717	1.037
Harvest index		2861.87258**	0.9990	1.037
Hundred seed weight				
Number of pods plant[-1]	64.533	-0.915**	0.8298	1.00

** =Significant difference at p ≤ 0.01, VIF=Variance Inflation Factor.

Table 11: Summary of selection from stepwise regression analysis of mean grain yield and hundred seed weight of Kabuli type chickpea as dependent variable on the other traits as independent variables.

Conclusions

Yield potential improvement of Kabuli type chickpea breeding was relatively less marked probably owing to stringent seed size requirements. Therefore, better genetic progress was obtained from breeding for hundred seed weight/seed size within a short period of time than it was from breeding for grain yield for the last three decades for this chickpea type. The improvement of hundred seed weight in Kabuli type chickpea was significantly and negatively correlated with number of secondary branches plant[-1], number of pods plant[-1], number of seeds pod[-1] and number of seeds plant[-1]. Therefore, the negative association between hundred seed weight and with these traits indicates that a compensatory relationship between them.

Finally, about 80% of the varieties were derived from introduction which is crossing materials at ICRISAT and ICARDA whereas the remaining varieties were developed through local selection/collection. Varieties developed from crossing and introduced germplasm was the most important sources of genetic material contributing to the genetic improvement of grain yield, biomass yield and hundred seed weight/ seed size for the last three decades which revealed chickpea breeding effort should focus on crossing works than landrace selection.

Conflicts of Interest

The authors declare that they have no conflicts of interest.

Acknowledgements

It is our pleasure to acknowledge Samara University and Ministry of Education for financial support. Our sincerely appreciation is also extended to Debre Zeit Agricultural Research Center in general and High Land Pulses Improvement Program staff members in particular that provided us with all available resources.

References

1. CSA (Central Statistical Authority) (2010) Agricultural sample survey 2009/10: Report on area and production of crops (Private peasant holdings, Meher season). VI. Statistical Bulletin 446. Addis Ababa, Ethiopia.

2. Menale K, Bekele S, Solomon A, Tsedeke A, Geoffrey M, et al. (2009) Current situation and Future Outlooks of the chickpea sub-sector in Ethiopia. International Crop Research Institute for the Semi-Arid Tropics (ICRISAT), Nairobi Ethiopia Institute of Agricultural Research (EIAR), Debre Zeit Agricultural Research Center, Debre Zeit Ethiopia.

3. Bekele S, Hailemariam T (2007) Structure and functioning of chickpea markets in Ethiopia: Evidence based on analyses of value chains linking smallholders and markets. Improving Productivity and Market Success (IPMS) of Ethiopian Farmers Project Working Paper 6, ILRI (International Livestock Research Institute), Nairobi, Kenya.

4. Senait Y (1990) Uses of chickpea, lentil and Faba bean in Ethiopia. In: Jambunthan R, Hall SD, Sudhir P, Rajan V, Sadhano V (eds.) Uses of Tropical Grains: Proceedings of Consultants Meeting, ICRISAT Center, India, pp: 47- 54.

5. Geletu B, Million E, Yadeta A (1996) Improved cultivars and production technology of chickpea in Ethiopia. Research Bulletin No 2, Debre Zeit Agricultural Research Center, Alemaya University of Agriculture, Debre Zeit, Ethiopia.

6. Kantar F, Hafeez FY, Shivakumar BG, Sundaram SP, Tejera NA, et al. (2007) Chickpea: Rhizobium management and nitrogen fixation. In: Yadav SS, Redden R, Chen W, Sharma B (eds.). Chickpea Breeding and Management. CAB, Wallingford, UK, pp: 101-142.

7. Tabikew D, Asnake F, Kibebew A, Lijalem K (2009) Improved Chickpea (Cicer arietinum L.) Technologies and seed production in Ethiopia. Ethiopia Agricultural Research Center Institute (EIARC), Research bulletin, Debre Zeit, Ethiopia.

8. Waddington SR, Ransom JK, Osmanzai M, Saunders DA (1986) Improvement in the yield potential of bread wheat adapted to North West Mexico. Crop Science 26: 698-703.

9. Cox TS, Shroyer JP, Ben-Hui L, Sears RG, Martin TJ (1988) Genetic improvement in agronomic traits of hard red winter wheat cultivars from 1919 to 1987. Crop Science 28: 756-760.

10. Donmez E, Sears RG, Shroyer JP, Paulsen GM (2001) Genetic gain in yield attributes of winter wheat in the Great Plains. Crop Science 41: 1412-1419.

11. Abeledo LG, Calderini DF, Slafer GA (2003) Genetic improvement of barley yield Potential and its physiological determinants in Argentina (1944-1998). Euphytica 130: 325-334.

12. Ustun A, Allen FL, English BC (2001) Genetic progress in soybean of the US Midsouth. Crop Science 41: 993-998.

13. Zhang JF, Lu Y, Adragna H, Hughs E (2005) Genetic improvement of New Mexico acala cotton germplasm and their genetic diversity. Crop Science 45: 2363-2373.

14. Zhou Y, He ZH, Sui XX, Xia XC, Zhang XK, et al. (2007) Genetic improvement of grain yield and associated traits in the Northern China winter wheat region from 1960 to 2000. Crop Science 47: 245-253.

15. Hailu T, Kamara A, Asafo-Adjei B, Dashiell K (2009) Improvement in grain and fodder yields of early-maturing promiscuous soybean genotypes in the guinea savanna of Nigeria. Crop Science 49: 2037-2042.

16. Jin J, Liu X, Wang G, Mi L, Shen Z, et al. (2010) Agronomic and physiological contributions to the yield improvement of soybean cultivars released from 1950 to 2006 in Northeast China. Field Crops Research 115: 116-123.

17. Khodarahmi M, Nabipour A, Zargari K (2010) Genetic improvement of agronomic and quality traits of wheat cultivars introduced to temperate regions of Iran during 1942-2007. African Journal of Agricultural Research 5: 947-954.

18. Amsal T (1994) Yield potential of rain fed wheat in the central high lands of Ethiopia. MSc Thesis presented to the School of Graduate Studies of Alemaya University.

19. Yifru T, Hailu T (2005) Genetic improvement in grain yield potential and associated agronomic traits of tef (Eragrostis tef). Euphytica 141: 247-254.

20. Kebere B, Ketema B, Prapa S (2006) Genetic gain in grain yield potential and associated agronomic traits in haricot bean (Phaseolu vulgaris L.). Kasetsart Journal (Natural Science) 40: 835-847.

21. Tamene T (2008) Genetic gain and morpho-agronomic basis of genetic improvement in grain yield potential achieved by faba bean breeding in Ethiopia. MSc Thesis Presented to the School of Graduate Studies of Hawasa University.

22. Wondimu F (2010) Assessment of genetic improvement in grain yield potential, malting quality and associated traits of barley (Hordeum vulgare L.) in Ethiopia. MSc Thesis Presented to the School of Graduate Studies of Haramaya University.

23. Demissew T (2010) Genetic gain in grain yield and associated traits of early and medium maturing varieties of soybean [Glycine max (L.) Merrill]. MSc Thesis Presented to the School of Graduate Studies of Haramaya University.

24. Ersullo L (2010) Genetic gain and G x E interaction in grain yield, oilcontent and associated traits of linseed (Linum usitatissimum L.) in Ethiopia. MSc Thesis Presented to the School of Graduate Studies of Haramaya University.

25. Anonymous (2002) SAS Guide for personal computers. Version 9.1 (edn.). SAS Institute Inc., Cary, NC, USA.

26. Gomez K, Gomez A (1984) Statistical Procedure for Agricultural Research. 2nd edn. International Rice Research Institute. John Wiley and Sons Inc., New York.

27. Evans LT, Fischer RA (1999) Yield Potential: Its Definition, Measurement, and Significance. Crop Science 39: 1544-1551.

28. MoARD (2008) Ministry of Agriculture and Rural Development Crop Variety Development Department: Crop Variety Register. Issue number 11. Addis Ababa, Ethiopia.

29. Pereira ML, Sadras VO, Trapani N (1999) Genetic improvement of sun flower in Argentina between 1930 and 1995. I. Yield and its components. Field Crops Research 62: 157-166.

30. Koemel JR, Guenzi AC, Carver BF, Payton ME, Morgan GH, et al. (2004) Hybrid and pureline hard winter wheat yield and stability. Crop Science 44: 107-113.

31. Mackey J (1979) Genetic potentials for improved yield. Proceeding of the workshop on agricultural potentiality directed by nutritional needs. Academia, Kiado, Budapest, pp: 121-143.

32. Ortiz R, Nurminiemi M, Madsen S, Rognli OA, Bjornstad A (200) Genetic gains in Nordic spring barley breeding over sixty years. Euphytia 126: 283-289.

33. Saxena N, Natarajan M, Reddy RS (1983) Chickpea, pigeon pea and groundnut. In: Smith WH, Banta Stephen J (eds.). Potential Productivity of Field Crops under Different Environments. IRRI, Los Banos, pp: 218-305.

34. Upadhyaya HD, Coyne CJ, Singh S, Gowda CL, Lalitha N, et al. (2006) Identification of Large-Seeded High-Yielding Diverse Kabuli Accessions in Newly Assembled Chickpea Germplasm. International Chickpea and Pigeon pea Newsletter 13: 2-5.

35. Bicer BT (2005) Evaluation of chickpea landraces. Pakistan Journal of Biological Sciences 8: 510-511.

36. Khan A, Rahim M, Ahmad F, Ali A (2004) Performance of chickpea genotypes under swat valley conditions. Journal of Research (Science) 15: 91-95.

37. Singh KB, Geletu B, Malhotra RS (1990) Association of some characters with seed yield in chickpea collections. Euphytica 49: 83-88.

38. Saleem M, Tahir MHN, Kabir R, Kabir R, Shahzad K (2002) Interrelationships and path analysis of yield attributes in chickpea (Cicer arietinum L.). International Journal of Agriculture and Biology 4: 404-406.

39. Hasan E, Arshad M, Ahsan M, Saleem M (2008) Genetic variability and interrelationship for grain yield and its various components in chickpea (Cicer arietinum L.). Journal of Agricultural Research 46: 109-116.

40. Ali MA, Nawab NN, Abbas A, Zulkiffal M, Sajjad M (2009) Evaluation of selection criteria in Cicer arietium L. using correlation coefficients and path analysis. Australian Journal of Crop Science 3: 65-70.

41. Sharma LK, Saini DP (2010) Variability and Association Studies for Seed Yield and Yield Components in Chickpea (Cicer arietinum L.). Research Journal of Agricultural Sciences 1: 209-211.

42. Toker C, Cagirgan MI (2004) The use of phenotypic correlations and factor analysis in determining characters for grain yield selection in chickpea (Cicer arietinum L.). Hereditas 140: 226-228.

43. Sharma LK, Saini DP, Kaushik SK, Vaid V (2005) Genetic variability and correlation studies in chickpea (Cicer arietinum L.). Journal of Arid Legumes 2: 415-416.

44. Ali Q, Tahir MHN, Sadaqat HA, Arshad S, Farooq J, et al. (2011) Genetic variability and correlation analysis for quantitative traits in chickpea genotypes (Cicer arietinum L.). Journal of Bacteriology Research 3: 6-9.

45. Muehlbauer FJ, Singh KB (1987) Genetics of Chickpea. In: Saxena MC, Singh KB (eds.). The Chickpea. CAB International, Walling Ford, UK, pp: 11-18.

46. Kumar L, Arora PP (1991) Basis of selection in chickpea. International Chickpea Newsletter 24: 14-15.

47. Naseem BA, Rehman A, Iqbal T (1995) Evaluation of Kabuli chickpea Germplasm. International Chickpea and Pigeon pea. Newsletter 2: 13-14.

48. Temesgen A (2007) Genetic variability and association among seed yield and yield related traits in Kabuli chickpea (Cicer arietinum L.) genotypes. MSc Thesis Presented to the School of Graduate Studies of Haramaya University.

49. Dewey DR, Lu KH (1959) A correlation and path coefficient analysis of components of crested wheat grass seed production. Agronomy Journal 51: 515-518.

50. Arshad M, Bakhsh A, Ghafoor A (2004) Path coefficient analysis in chickpea (Cicer arietinum L.) under rainfed conditions. Pakistan Journal of Botany 36: 75-81.

15

Dry Matter Partitioning and Harvest Index of Maize Crop as Influenced by Integration of Sheep Manure and Urea Fertilizer

Sajid Khan[1], Ahmad Khan[1], Fazal Jalal[3*], Maaz Khan[2], Haris Khan[1] and Said Badshah[1]

[1]Department of Agronomy, The University of Agriculture, Peshawar, Pakistan
[2]Department of Soil and Environmental Sciences, The University of Agriculture, Peshawar, Pakistan
[3]Department of Agriculture, Abdul Wali Khan University, Mardan, Pakistan

Abstract

The experiment was designed to evaluate the effect of sheep manure (SM), its application timing (AT) and N fertilizer (urea) on dry matter partitioning and harvest index. The study was conducted on RCBD split plots arrangement at Agronomic research farm during 2015. Sheep manure (SM1=3 t ha^{-1}, SM2=4 t ha^{-1}, SM3=5 t ha^{-1}) and application timing (AT1=15 days before sowing, AT2=At sowing time) were allotted to main plots however, fertilizer N (N1=0 kg ha^{-1}, N2= 90 kg ha^{-1}, N3=120 kg ha^{-1}) were applied to sub-plots. Application of 5 t ha^{-1} of sheep manure at 15 days before sowing significantly enhanced pre-tasseling (stem and leaves) and physiological maturity (stem, leaves, cobs and grains) dry matter partitioning and harvest index. Pre-tassel and physiological maturity dry matter accumulation were higher with application of 120 kg N ha^{-1} however, Application of 5 t sheep manure ha^{-1} at 15 days before sowing and 120 kg N ha^{-1} was recommended for higher dry matter accumulation of maize crop.

Keywords: Sheep manure; Nitrogen; Application timing; Dry matter; Harvest index

Introduction

Maize (*Zea mays* L.) is the third most important crop in the world after wheat and rice. It is a tropical crop but can be grown profitably in subtropical and temperate climatic zones of the world during spring and summer season [1]. In Pakistan after wheat and rice, maize holds the central position and is extensively cultivated (1.11 million hectares) in Punjab and Khyber Pakhtunkhwa [2]. In Pakistan satisfactory potential yield has not been achieved due to several limitations. The poor soil organic matter and imbalance fertilizer application are the important limitations, which limits plants growth, carbohydrate production and dry matter accumulation [3,4].

Dry matter production is basically a measure of plant photosynthetic efficiency [5], which is influenced by balance nutrient availability [6], and environmental factors [7]. Leaves are the major source of dry matter production through photosynthesis, and then accumulated into various plant parts through different physiological processes [5]. The dry matter production in cereals is highly depended on plant photosynthetic efficiency and the sink capacity to accumulate the photosynthates from the leaf [8]. The balance nutrient provision increase the dry matter production into various plant parts through its impact on more leaf area production and high photosynthetic rates [9]. The dry matter production highly influenced the plant biomass production and grain yield of the crop [10,11].

Nitrogen is very important for crop plants especially for cereals. Uptake of nitrogen by plant roots occurs in the form of ammonium (NH_4^+), and NO_3^- which becomes available through mineralization and nitrification respectively [12]. Nitrogen fertilizer enhances the vegetative growth, maize biomass and dry matter production [13], which ultimately resulted in higher crop productivity [14]. Nitrogen is very important for optimum plant return in term of dry matter and crop yield, but it had also adverse effects on soil fertility, environment and because of its high cost it increase the cost benefit ratio of crop [15]. Combined use of chemical and natural fertilizers is advantageous in enhancing the availability of nitrogen, phosphorus and potassium in plants [16], improve soil fertility and productivity on sustainable basis [17]. These improved soil properties and nutrients availability enhance the fresh and dry biomass production and improve the crop growth [18].

Sheep manure contain high amount of primary macro nutrients and other essential nutrients for plants [19]. The use of organic matter positively influences vegetative and reproductive growth of plants [20] and dry matter production [21]. Naturally available animal manure and plant residues can be used as an alternative cheaper source of synthetic fertilizers [22]. It provides nutrients to plants and adsorb essential nutrients such as Fe^{2+}, Mg^{2+} and NH_4^+ cations, which are essential for enzymes activation and chlorophyll formation, and assimilates production [23]. To get optimum returns from manure application it should be incorporated at proper time, to synchronize the nutrient release from manure with supreme crop growth stage. The better synchronization of nutrients with crop occurs because of rapid mineralization and decomposition at initial stage [24].

Fertilizer is added to soil either in inorganic or organic form for improvement of chemical properties, improving soil fertility, and increase maize production. However, information are limited on the use of sheep manure as a source of organic fertilizers, it application time and also in combination with urea fertilizer. Therefore, this experiment was designed to find out the response maize dry matter and their respective accumulation into various parts by the integrated use of sheep manure and fertilizer N (urea).

Materials and Methods

An experiment was designed at Agronomy Research Farm, The University of Agriculture Peshawar, to find out the impact of combined

---

*Corresponding author: Fazal Jalal, Department of Agriculture, Abdul Wali Khan University, Mardan, Pakistan, E-mail: jalal_146@yahoo.com

application of sheep manure and urea fertilizer. Randomized complete block design with split plot arrangement was used with four replications. Sheep manure at the rate of 3, 4 and 5 t ha^{-1} and application timing at two levels (15 days before sowing, at sowing time) was allotted to main plots and fertilizer N at 3 levels (0, 90 and 120 kg ha^{-1}) was allotted to sub plots of size 5 m × 3.5 m. Sheep manure was incorporated in their respective plots 15 days before sowing and second treatment of SM was applied at sowing time, along with 90 kg N ha^{-1} (from SSP) and half of fertilizer N (urea). The second dose of N was applied at knee height of plant. Azam variety was sown on 19th June 2015 with the help of seed drill. Plant to Plant distance of 20 cm was maintained after proper thinning. Weed was removed from field with the help of hoeing. The crop was harvested at proper maturity and then sundried and threshed. Dry matter was calculated at two stages of crop growth that is given below.

Dry matter partitioning was recorded at pre-tasseling stage by randomly harvesting 1 m row from each sub plot. Harvested plants were divided into stem and leaves (leaf sheath+leaf blade), and these materials were oven dried at 70°C for 24 hours and their weights were recorded with the help of sensitive balance. At physiological maturity, dry matter partitioning was computed by harvesting 1 m long row selected randomly from each sub plot. The harvested plants were separated into stem, leaf, husk, grains, cobs and tassel. These materials were oven dried at 70°C for 24 hours and their weights were recorded.

The data was analyzed statistically with analysis of variance technique appropriate for RCBD split plots arrangement. Means was compared by LSD technique at 0.05 level of probability.

Results

Dry matter partitioning at pre-tasseling (g plant^{-1})

Data regarding dry matter partitioning in stem, leaves and total dry matter (TDM) at pre-tasseling is demonstrated in Table 1. Data analysis clarified that stem dry matter was significantly influenced by different sheep manure rates, its application time and N levels. Among various interactions only SM × N was found to have significant effects for pre-tasseling dry matter partitioning.

Treatments	DM in plant components (g plant^{-1})		Total
	Stem	Leaf	
Sheep manure (t ha^{-1})			
3	8.9 b	21.8 b	30.8 b
4	9.9 a	22.8 ab	32.7 ab
5	10.0 a	25.5 a	35.5 a
LSD$_{0.05}$	0.4	2.7	2.7
Application timing (Days before sowing)			
15	9.9 a	24.6 a	34.4 a
0	9.3 b	22.2 b	31.5 b
LSD$_{0.05}$	0.3	2.2	2.2
Nitrogen (kg ha^{-1})			
0	8.7 c	21.7 b	30.4 b
90	9.2 b	22.2 b	31.5 b
120	10.8 a	26.3 a	37.1 a
LSD$_{0.05}$	0.4	1.6	1.6
Interactions	P-value	P-value	P-value
SM × AT	0.098	0.140	0.230
SM × N	0.000	0.027	0.008
AT × SM	0.832	0.386	0.356
SM × AT × N	0.410	0.842	0.951

Table 1: Stem, leaf, and total dry matter (g plant^{-1}) of maize at pre-tasseling as affected by integrated management of sheep manure and urea fertilizer.

Sheep manure application of 5 t ha^{-1} amassed maximum stem, leaf and total pre-tasseling dry matter (10, 25.5, and 35.5 g plant^{-1}), respectively followed by 4 t ha^{-1} (9.9, 22.8, 32.7 g plant^{-1}), while lowest dry matter (8.9, 21.8, 30.8 g plant^{-1}) was noticed from plots receiving 3 t ha^{-1} sheep manure. Dry matter production is a function of nutrients availability and uptake by the plant [6], environmental and genetic factors [7]. At this stage higher dry matter were produced in leaves than stem. The higher dry matter accumulation in leaves and stem with addition of sheep manure might possibly since manure provides macro and micro nutrients to soil, improves soil properties and water use efficiency, better soil fertility might have increased the photosynthetic efficiency and partitioning of photo assimilates [25]. These findings are fully supported by Ayeni and Adetunji [26] who concluded that manure incorporation into the soil supply essential nutrients (N, P, K, Ca, Mg, Fe, Cu, Mn and Zn) to the maize crop, that resulted in maximum dry matter production with higher rates of manure. Greater dry matter (9.9, 24.6, 34.4 g plant^{-1}) was resulted from plots receiving sheep manure 15 days before sowing while its incorporation at planting time accumulated lowest dry matter (9.3, 22.2, 31.5 g plant^{-1}) respectively in stem, leaves and whole plant. Maximum dry matter accumulation recorded from plots receiving sheep manure prior to planting could be due to better mineralization of sheep manure and synchronization of nutrient release with crop growth stage, this timely nutrient availability might have increased the crop growth rate, leaf area, and dry matter partitioning in stem and leaves [27].

At pre-tasseling highest dry matter in stem, leaves and whole plant (10.8, 26.3, 37.1 g plant^{-1}) respectively was accumulated by applying 120 kg N ha^{-1}, that was higher than dry matter (9.2, 22.2, 31.5 g plant^{-1}) accumulated by 90 kg N ha^{-1} treatment, while lowest stem dry matter (8.7, 21.7, 30.4 g plant^{-1}) was observed in stem, leaves, and whole plant respectively in plots with no nitrogen application. Maximum dry matter accumulation with higher N level could be explained by the fact that N fertilization is indispensable for photosynthesis and protein synthesis, and higher photosynthetic efficiency ultimately resulted in maximum dry matter production. These findings are supported and confirmed by Desta [28] who concluded increase in dry matter production with increasing N. Leaf, stem and ultimately total dry matter production in plots receiving 120 kg N ha^{-1} was probably due to essentiality of nitrogen for protein synthesis, amino acids, enzymes, coenzymes, nucleic acids, phytochromes, and chlorophyll formation, that have the effects on dry matter production [29], high chlorophyll content, photosynthesis and photo assimilates [25].

Interaction SM × N (Figure 1) reveals that increasing sheep manure rates from 3 to 4 t ha^{-1}, stem dry matter production was not significantly affected with enhancing N levels, from 0 and 90 kg ha^{-1}, however it increased significantly with 120 kg N ha^{-1}, but with application of 5 t SM ha^{-1} it increased linearly across all the three levels of N. The SM × N interaction (Figure 2) indicated that in plots having 3 and 4 t SM ha^{-1}, the leaf dry matter production increased slightly with nitrogen rates from 0 and 90 kg ha^{-1}, however leaf dry matter greatly decreased with 5 t SM ha^{-1}, and was further increased with increasing N from 90 to 120 kg ha^{-1}. The relative increase in leaf dry matter was highly pronounced in 5 t SM ha^{-1} treated plots. SM × N interaction (Figure 3) reveals a mild increase in TDM from 3 and 4 t SM ha^{-1} with increasing N levels from 0 to 90 kg ha^{-1} resulted in non-significant differences. In 5 t ha^{-1} sheep manure treated plots TDM accumulation first decrease with nitrogen levels from 0 to 90 kg ha^{-1}, whereas more increase in N enhanced total dry matter accumulation in maize.

Highest dry matter at pre-tasseling with combined application of SM

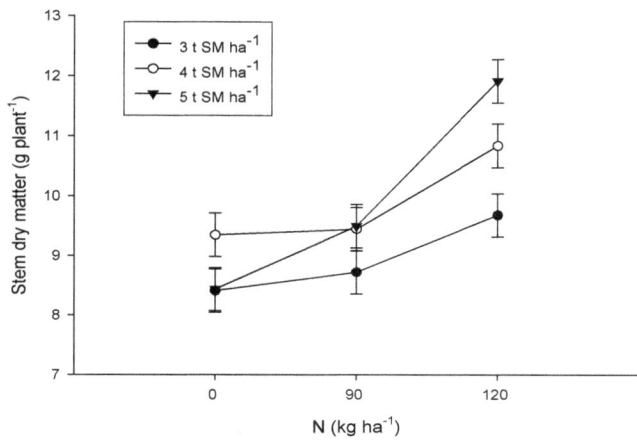

Figure 1: Interactive response of sheep manure (SM) and nitrogen (N) for stem dry matter at pre-tasseling.

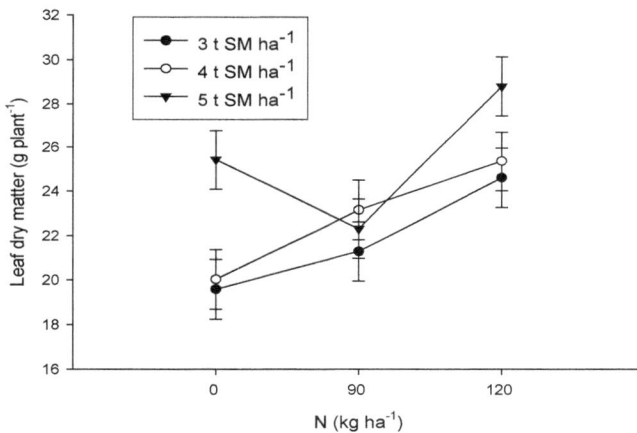

Figure 2: Interactive response of sheep manure (SM) and nitrogen (N) for leaf dry matter at pre-tasseling.

and urea fertilizer could be related to the balanced nutrients provision, improved soil physical and chemical properties, and higher N that might have improved plant metabolism, photosynthetic efficiency, leaf expansion and plant growth and might have increased the dry matter production [6,29] observed maximum DM accumulation from combined use of manure and synthetic fertilizer which was in conformity to our finding. The maximum dry matter production by leaves compared to stem was in negation [30] who observed maximum dry matter accumulation in stem compared to leaves after 8 weeks of sowing the crop. Dry matter accumulated in leaves up to leaf area index of 5, thereafter further increases in dry matter translocated in to stem [7].

Dry matter portioning at physiological maturity (g plant⁻¹)

Perusal of the data (Table 2) showed that dry matter (DM) assimilation at physiological maturity stages into different plant components (stem, leaf, cob, grains) was significantly affected by sheep manure, its application timing, nitrogen levels, and M × N interaction, while the remaining interactions except AT × M for grains and total dry matter was found non-significant. Regarding remaining plant (husk

and tassel) parts only dry matter accumulation in husk was significantly affected by N levels, while dry matter partitioning to shank, silk and tassel was found non-significant at all levels of sheep manure, its application time and nitrogen rates.

Mean comparison indicated that maximum dry matter (40.3, 33.6, 11.4, and 60.2, 176.9 g plant⁻¹) were partitioned in to stem, leaf, cob, ear, and whole plant respectively, in plot having 5 t SM ha⁻¹, which was followed by 4 t ha⁻¹ (36.3, 31.6, 10.1, 53.5, 168.0 g DM plant⁻¹), while lowest dry matter accumulation (34.5, 30.7, 9.4, 49.9, 151.0 g plant⁻¹), respectively was reckoned from 3 t ha⁻¹ sheep manure incorporation. Incorporation of 3 and 4 t ha⁻¹ sheep manure to plants accumulated statistically comparable dry matter at physiological maturity stage across stem, cobs, grains and TDM. In stem more DM accumulation was observed than leaves at physiological maturity. Increased dry matter

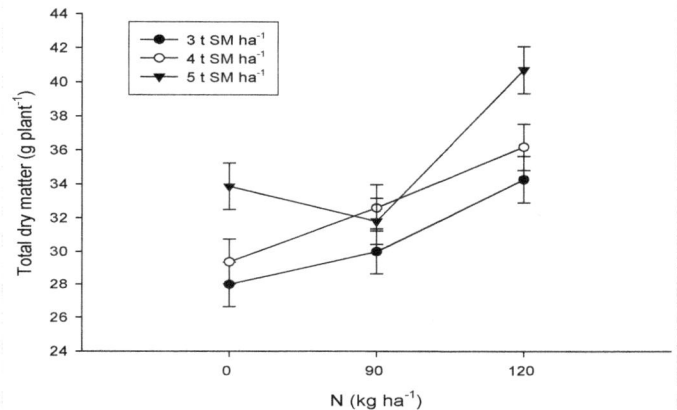

Figure 3: Interactive response of sheep manure (SM) and nitrogen (N) for total dry matter at pre-tasseling.

Treatments	DM in plant components (g plant⁻¹)						
	Stem	Leaf	Husk	Cob	Grains	Tassel	Total
SM (t ha⁻¹)							
3	34.5 b	30.7 b	12.0	9.4 b	49.9 b	3.7	151.0 c
4	36.3 b	31.6 ab	11.3	10.1 b	53.5 b	3.7	168.0 b
5	40.3 a	33.6 a	11.5	11.4 a	60.2 a	4.2	176.9 a
LSD₀.₀₅	2.9	2.11	N.S	1.22	5.13	N.S	8.81
AT (days before sowing)							
15	38.5 a	32.9 a	11.8	10.8 a	56.8 a	4.0	169.6 a
0	35.5 b	31.1 b	11.4	9.8 b	52.3 b	3.8	161.0 b
LSD₀.₀₅	2.3	1.72	N.S	0.99	4.19	N.S	7.19
N (kg ha⁻¹)							
0	33.2 c	29.7 c	9.9 c	8.7 c	48.1 c	3.7	150.6 c
90	37.1 b	32.0 b	11.7 b	10.2 b	54.0 b	4.0	161.6 b
120	40.8 a	34.3 a	13.2 a	12.0 a	61.6 a	4.0	183.7 a
LSD₀.₀₅	3.0	1.63	1.22	1.14	4.80	N.S	6.63
Interactions				P value			
SM × AT	0.488	0.172	0.225	0.972	0.00	0.257	0.02
SM × N	0.036	0.006	0.820	0.018	0.02	0.539	0.01
AT × N	0.072	0.580	0.152	0.060	0.95	0.800	0.22
SM × AT × N	0.064	0.219	0.722	0.063	0.71	0.987	0.57

SM=Sheep manure, AT=Application time, N=Nitrogen; Means bearing similar letter(s) are statistically comparable within the same category using LSD test at P ≤ 0.05

Table 2: Dry matter partitioning (g plant⁻¹) at physiological maturity of maize as affected by integrated management of sheep manure and urea fertilizer.

production from higher levels of sheep manure might be associated with more nutrients availability with higher amounts of sheep manure [19], improved soil aeration, water holding capacity, improved adsorption of calcium, magnesium, and potassium by forming clay humic complexes, and activities of soil microorganism [31], that might have improved plant growth [32], and photosynthetic production [25] and accumulated maximum dry matter. These finding are supported by Buriro [33] who observed higher dry matter production with increasing levels of manure. Likewise, with sheep manure incorporation 15 days before sowing higher dry matter was accumulated in stem, leaves, cobs, grains, and in whole plant (38.5, 32.9, 10.8, 56.8, 169.6 g plant[-1]) than its application at sowing time (35.5, 31.1, 9.8, 52.3, 161.0 g plant[-1]), respectively. This greater production of dry matter with sheep manure application before planting might be due to optimum mineralization of manure, and with more nutrients availability which might have helped in increasing root growth, water use efficiency and better soil fertility [34].

Nitrogen application of 120 kg ha[-1] reckoned optimum dry matter at physiological maturity in stem, leaves, cobs, grains and whole plant (40.8, 34.3, 12, 61.6, 183.7 g plant[-1]), which was higher than 90 kg ha[-1] N (37.1, 32.0, 10.2, 54, 161.6 g plant[-1]), while lowest dry matter (32.2, 29.7, 8.7, 48.1, 150.6 g plant[-1]) was observed from control plots, respectively. Regarding husk optimum dry matter (13.2 g plant[-1]) at physiological maturity was accumulated in plots receiving 120 kg N ha[-1], which was followed by 90 kg N ha[-1] (11.7 g plant[-1]), while lowest (9.9 g plant[-1]) was observed from control. Maximum stem, leaves, cobs, grains, husk and total dry matter accumulation with higher rates of nitrogen might have the fact of its influence on vegetative growth of the plant, photosynthesis. These results are in accordance with the findings of Quansah, Ammanullah, and Nasim [34-37], who observed maximum dry matter production with increasing levels of inorganic fertilizer.

The data in Figure 4 (SM × N) showed no significant increase in stem dry matter accumulation with increasing N from 0 to 120 kg N ha[-1], at 3 t SM ha[-1]. With changing N from 0 to 90 kg ha[-1], plots having 4 t SM ha[-1] showed non-significant effect on dry matter accumulation, but higher sheep manure showed a marked increase with increase in N to 120 kg ha[-1]. A strong increase was shown by 4 t ha[-1] sheep manure with increasing N up to 120 over 5 tons SM ha[-1]. From SM × N interaction (Figure 5) it is evident that with increase in N from 0 to 120 kg ha[-1]. No apparent differences for leaf dry matter accumulation in plots having 5 t SM ha[-1] were observed. However, with 3 and 4 t SM ha[-1] the response was opposite, mean with increase in N increase in leaf dry matter was observed up to 90 kg N ha[-1], and with further increased the leaf dry matter accumulation was decreased. The interaction SM × N (Figure 6) showed that nitrogen increase from 0 to 90 kg ha[-1] dry matter accumulation in cob increased across all the three levels of sheep manure, however with 3 t ha[-1] sheep manure the increase was highest, and slighter with other two levels of sheep manure. The AT × M interaction (Figure 7) demonstrated that increasing SM incorporation from 3 to 4 t ha[-1] no differences for grains dry matter partitioning were observed with its incorporation in plots at the time of sowing compared to increased grain DM when SM was applied 15 days before sowing. The further increasing sheep manure to 5 t ha[-1] non-significantly differences for grains dry matter were observed whether it was used at the time of sowing or before sowing. The SM × N interaction (Figure 8) illustrated a linear increase in grains dry matter accumulation in 4 t ha[-1] sheep manure treatment from 0 through 90 to 120 kg N ha[-1]. While in plots received 3 t ha[-1] grains dry matter increased but slightly decreased in 5 t ha[-1] at nitrogen levels of 0 and 90 kg ha[-1], beyond 90 kg ha[-1] increase in N to 120 kg ha[-1] it significantly enhanced in 5 t ha[-1] sheep manure but remain constant in 3 t ha[-1]. The AT × SM interaction

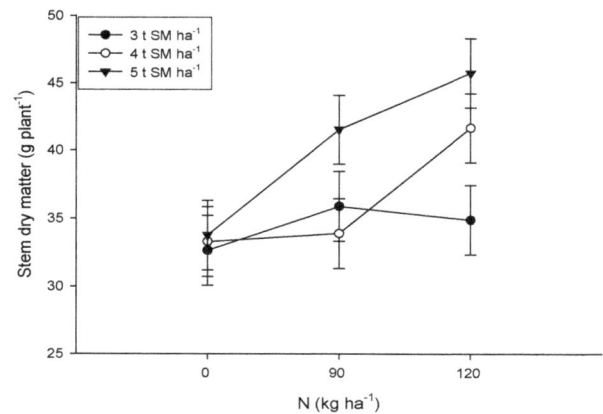

Figure 4: Sheep manure (SM) and nitrogen (N) interactive response for stem dry matter (g plant[-1]) at physiological maturity.

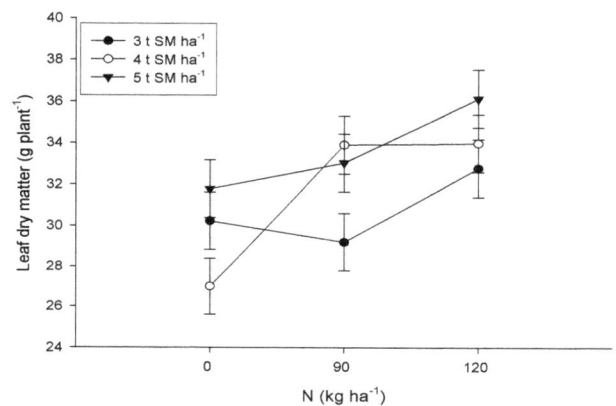

Figure 5: Sheep manure (SM) and nitrogen (N) interaction for leaf dry matter at physiological maturity.

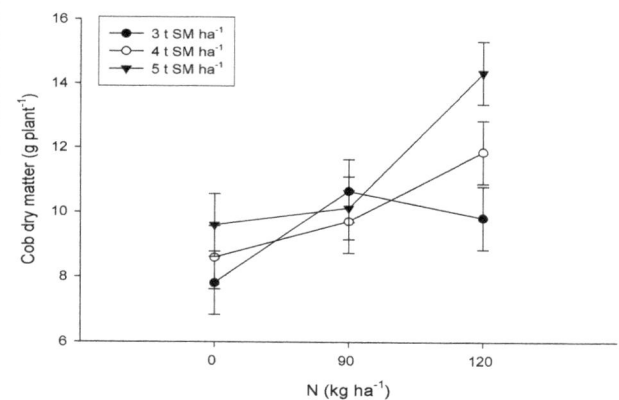

Figure 6: Interactive response of sheep manure (SM) and nitrogen (N) for cob dry matter (g plant[-1]).

(Figure 9) clearly pointed out that sheep manure application 15 days before planting had significantly increased total dry matter production

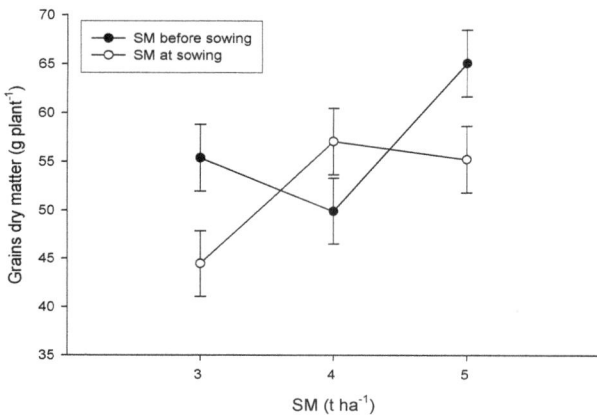

Figure 7: Interactive response of sheep manure (SM) and nitrogen(N) for grains dry matter (g plant⁻¹).

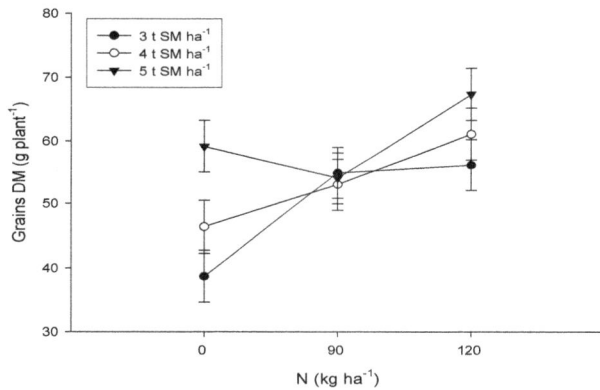

Figure 8: Sheep manure (SM) and nitrogen (N) interactive response for grain dry matter (g plant⁻¹).

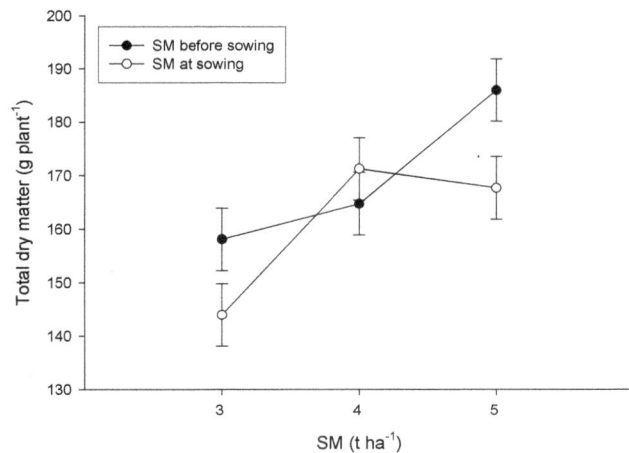

Figure 9: Interactive response of sheep manure (SM) and application time (AT) for total dry matter (g plant⁻¹).

dry matter accumulation of plants in 5 t SM ha⁻¹ over sheep manure levels of 3 and 4 t ha⁻¹. The SM × N interaction (Figure 10) revealed that total dry matter accumulation increased with N incorporation from 0 to 120 kg ha⁻¹ across 4 and 5 t SM ha⁻¹, however this increase was small from 0 to 90 kg ha⁻¹, and from onward this point the increase was higher. In case of 3 t ha⁻¹ sheep manure total dry matter increased with N application from 0 to 90, and with no differences thereafter up to 120 kg N ha⁻¹. Higher dry matter production with combined use of sheep manure and urea fertilizer might be due to more leaf area and LAI by these plants, optimum LAI is an indication of better photosynthesis and more assimilates production [26]. These results are in line with Ibeawuchi, Quansah, Iqbal and Burio [5,6,33,34].

Data analysis showed that (Table 3) harvest index was significantly affected by SM, nitrogen levels, AT × SM and SM × N interaction, while application time of SM, and the interactions AT × N and M × AT × N were found non-significant for Harvest index. SM applied at the rate of 5 t ha⁻¹ resulted maximum harvest index (35.8%), followed by 4 t ha⁻¹ (33.6%) when compared to the minimum harvest index (31.9%) recorded from 3 t SM ha⁻¹. The physiological efficiency of crop plants in converting photosynthetic products into grain yield is termed as harvest index [38]. Higher harvest index were recorded from sheep manure incremental levels. This increased harvest index with higher sheep manure levels could be associated with enhanced soil cation exchange capacity, increased C, N, and P content, and lowered hydraulic conductivity of soil [39], That might had improved photosynthetic efficiency (Liu et al.) [40] and enhanced assimilates translocation to economic portion [41], which all have direct effects on grain yield, that might have resulted in higher harvest index. This conclusion was supported by Farhad [2]. Maximum harvest index (36.1%) was recorded from 120 kg ha⁻¹ nitrogen application, followed by 90 kg N ha⁻¹ (33.6%) over the minimum (31.6%) observed from control. The increased harvest index with higher levels of N might be due to efficient portioning of assimilates towards the economic portion. These results are in line with Hokmalipour and Darbandi [42] who recorded optimum harvest index from higher N levels. The harvest index further boost up with the integrated use of sheep manure and urea fertilizer. Report of Mohsin [43] supported our finding. The interaction AT × M (Figure 11) indicated that harvest index was greater in plots having low SM (3 t SM ha⁻¹) when applied before sowing, however, it was greater in plots having greater SM (5 t SM ha⁻¹) when soil incorporated at the time

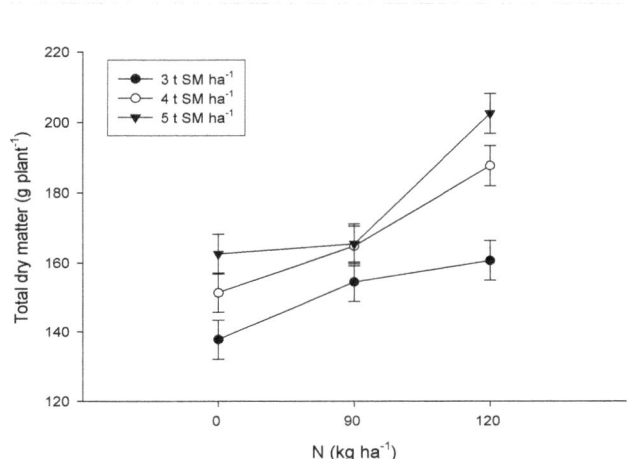

Figure 10: Interactive response of sheep manure (SM) and nitrogen (N) for total DM (g plant⁻¹).

across all the three sheep manure levels, however the increase with 5 t ha⁻¹ sheep manure was higher than both 3 and 4 t SM ha⁻¹. Likewise, sheep manure incorporation at the sowing time highly enhanced total

of sowing. The SM × N interaction (Figure 12) revealed that increasing N from 0 to 120 kg ha^{-1} had significantly increased harvest index (%) in plots having 3 t SM ha^{-1}. However, no significant increases in harvest index were observed with increasing N in plots having higher SM incorporation (4 to 5 t ha^{-1}). Higher harvest index recorded from SM application prior to planting, might be due to more nutrient uptake and higher dry matter portioning toward grain.

Conclusion

From the results, it was concluded that application of 5 t sheep manure ha^{-1} at 15 days before sowing and 120 kg N ha^{-1} produce more dry matter accumulation in terms of leaves, stem, cobs and grains.

Sheep manure (SM, t ha^{-1})	Application time of SM (AT, days)[†]	Nitrogen rates (N, kg ha^{-1})			SM × AT
		0	90	120	
3	15	26.2	36.3	39.3	33.9
4		27.9	33.4	35.2	32.2
5		38.6	34.5	38.9	37.3
3	0	27.3	33.2	29.3	29.9
4		35.7	31.4	37.8	34.9
5		33.7	32.6	36.4	34.2
	15	30.9	34.7	37.8	34.5
	0	32.2	32.4	34.5	33.0
3		26.7	34.8	34.3	31.9b
4		31.8	32.4	36.5	33.6ab
5		36.2	33.5	37.7	35.8a
Mean		31.6	33.6 ab	36.1 a	

[†]SM application was made 15 or 0 days before sowing

Interactions	P-values	Interactions	P-values
SM × AT	0.053	AT × N	0.191
SM × N	0.033	SM × AT × N	0.131
LSD$_{0.05}$ for SM and AT	2.93, NS	LSD$_{0.05}$ for N	2.68

Means bearing identical letter(s) in the same column are comparable using LSD test at P ≤ 0.05.

Table 3: Harvest index (%) of maize as influenced by sheep manure and urea fertilizer.

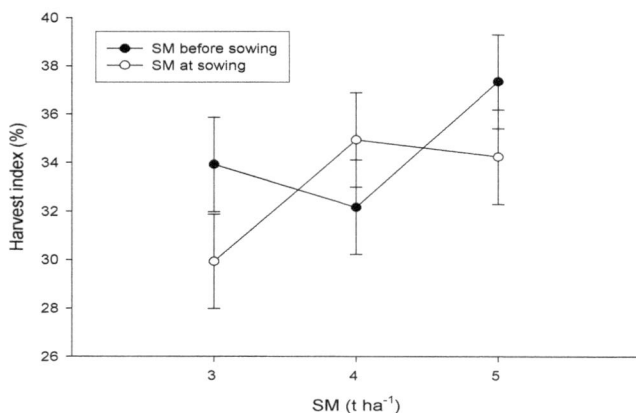

Figure 11: Interactive response of sheep manure (SM) and application time (AT) for harvest index (%).

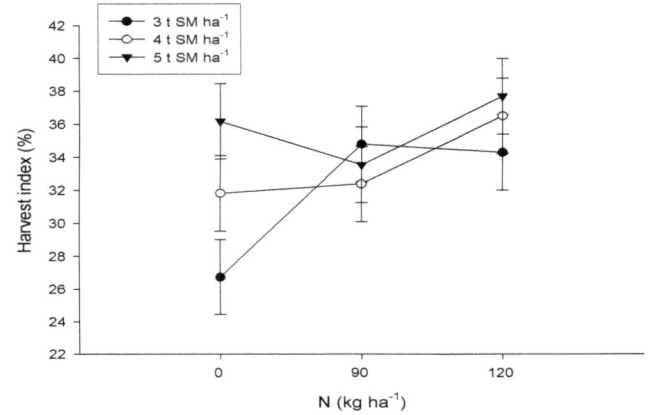

Figure 12: Interactive response of sheep manure (SM) and nitrogen (N) for harvest index (%).

Hence, it was recommended for higher dry matter returns in maize crop.

References

1. Tagne A, Feujio TP, Sonna C (2008) Essential oil and plant extracts as potential substitutes to synthetic fungicides in the control of fungi. International Conference Diversifying crop protection, 12-15 October La Grande-Motte, France.

2. Farhad W, Saleem MF, Cheema MA, Hammad HM (2009) Effect of poultry manure levels on the productivity of spring maize. J Animal Plant Sci 19: 122-125.

3. Oad FC, Buriro UA, Agha SK (2004) Effect of organic and inorganic fertilizer application on maize fodder production. Asian J Plant Sci 3: 375-377.

4. Tanaka A, Yamaguchi J (1972) Dry matter production, Yield components and grain of the maize plant: Hokkaido University.

5. Iqbal A, Iqbal MA, Raza A, Akbar N, Abbas RN, et al. (2014) Integrated nitrogen management studies in forage maize. Ameri Eura J Agric & Envi Sci 14: 744-747.

6. Ibeawuchi II, Opara FA, Tom CT, Obiefuna JC (2007) Graded replacement of inorganic fertilizer with organic manure for sustainable maize production in Owerri Imo State, Nigeria. Life Sci J 4: 82-87.

7. Amin MEMH (2011) Effect of different nitrogen sources on growth, yield and quality of fodder maize (Zea mays L.). J Saudi Soc Agri Sci 10: 17-23.

8. Warriach EA, Ahmad N, Basra SM, Afzal L (2002) Effect of nitrogen on source sink relationship in wheat. Int J Agri Bio 4: 300-302.

9. Gasim S (2001) Effect of nitrogen, phosphorus and seed rate on growth, yield and quality of forage maize (Zea mays L). MSc Thesis, Faculty of Agric, Univ of Khartoum.

10. Hocking P, Kirkegaard J, Angus J, Gibson A, E Koetz (1997) Comparison of canola, Indian mustard and linola in two contrasting environments. I. Effects of nitrogen fertilizer on dry matter production, seed yield and seed quality. Field crops Research 49: 107-125.

11. Plaut Z, Butow B, Blumenthal C, Wrigley C (2004) Transport of dry matter into developing wheat kernels and its contribution to grain yield under post anthesis water deficit and elevated temperature. Field Crop Res 86: 185-198.

12. Havlin JL, Tisdale SL, Nelson WL, JD Beaton (2005) Soil fertility and fertilizers: An introduction to nutrient management. Pearson Education, incorporation upper Saddle River, New Jersey.

13. Ogola JBO, Wheeler TR, Harris PM (2002) Effects of nitrogen and irrigation on water use of maize crops. Field Crop Res 78: 105-117.

14. Habtegebrial K, Singh BR, Haile M (2007) Impact of tillage and nitrogen fertilization on yield, nitrogen use efficiency of eragrostis, trotter and soil properties. Soil Till Res 94: 55- 63.

15. Ali S, Uddin S, Ullah O, Shah S, Din SU, et al. (2012) Yield and yield components of maize response to compost and fertilizer-nitrogen. Food Sci Qual Manag 38: 39-44.

16. Rautaray SK, Ghosh BC, Mittra BN (2003) Effect of fly ash, organic wastes and chemical fertilizers on yield, nutrient uptake, heavy metal content and residual fertility in a rice-mustard cropping sequence under acid lateritic soils. Bioresource Technol 90: 275-283.

17. Satyajeet RK, Nanwal N, Yadav VK (2007) Effect of integrated nutrient management in nitrogen, phosphorus and potassium concentration, uptake and productivity in pearl millet. J Maharastra Agri Universities 32: 186-188.

18. Ghosh P, Bandyopadhyay K, Manna M, Mandal K, Misra A (2004) Comparative effectiveness of cattle manure, poultry manure, phosphocompost and fertilizer npk on three cropping system in vertisols of semi-arid tropics II. Dry matter yield, nodulation, chlorophyll content and enzyme activity. Bioresour Technol 95: 85-93.

19. Dekisissa T, Short I, Allen J (2008) Effect of soil amendment with compost on growth and water use efficiency of amaranath. In: Proc. of UCOWR/NIWR Annual Conf. Int'l. Water Resources: Challenges for the 21st Century and Water Resources Education: 22-24, Durham, NC.

20. Shadanpour F, Mohammadi TA, MK Hashemi (2011) The effect of cow manure vermicompost as the planting medium on the growth of marigold. Annals Biol Res 2: 109-115.

21. Xie R, Mackenzie A (1986) Urea and manure effects on soil nitrogen and corn dry matter yield. Soil Science Society of America J 50: 1504-1509.

22. Khan M, Abid AM, Hussain N, MU Masood (2005) Effect of phosphorous levels on growth and yield of maize cultivars under saline conditions. Int J Agric Biol 3: 511-514.

23. Elhindi K (2012) Evaluation of composted green waste fertigation through surface and sub surface drip irrigation systems on pot marigold plants (Calendula officinalis L.) grown on sandy soil. Australian J Crop Sci 6: 1249-1259.

24. Thulasizwe SM, Simeon AM (2013) Influence of kraal manure application time on emergence, growth and grain yield of maize grown in two soils with contrasting textures. J Food Agri Env 11: 422-427.

25. Baiyeri KP, Tenkouano A (2008) Manure placement effects on root and shoot growth and nutrient uptake of 'PITA 14' Plantain hybrid (Musa sp. Aaab). Africa J Agric Res 3: 13-21.

26. Ayeni LS, Adetunji MT (2010) Integrated application of poultry manure and mineral fertilizer on soil chemical properties, nutrient uptake, yield and growth components of maize. Nature Sci J 8: 60-67.

27. Amanullah MM, Yasin MM, Somasundaram E, Vaiypapuri K, Sathyamoorthi K, et al. (2006) N availability in fresh and composted poultry manure. Res J Agric Bio Sci 2: 406-412.

28. Desta HA (2015) Response of maize (Zea mays L.) to different levels of nitrogen and sulfur fertilizers in Chilga District, Amhara National Regional State, Ethiopia. Basic Res J Soil Env Sci 3: 38-49.

29. Schroder JJ, Neeteson JJ, Oenema O, Struik PC (2000) Does the crop or the soil indicate how to save nitrogen in maize production. Reviewing the state of the art. Field Crops Res 66: 151-164.

30. Akongwubel AO, Ewa UB, Prince A, Jude O, Martins A, et al. (2012) Evaluation of agronomic performance of maize (Zea mays L.) under different rates of poultry manure application in an Ultisol of Obubra, cross river state, Nigeria. Int J Agric Forest 2: 138-144.

31. Nyle C, Brady R (2003) Nature and properties of soil. 13th edn, New York, USA, p: 960.

32. Dauda SN, Ajayi FA, Ndor E (2008) Growth and yield of water melon (Citrullus lanatus) as affected by poultry manure application. J Agric Soc Sci 4: 121-124.

33. Buriro M, Oad A, Nangraj T, Gandahi AW (2014) Maize fodder yield and nitrogen uptake as influenced by farm yard manure and nitrogen rates. European Acad Res 2: 11624-11637.

34. Quansah GW (2010) Effect of organic and inorganic fertilizers and their combinations on the growth and yield of maize in the semi-deciduous forest zone of Ghana. Bsc (hons) thesis: department of crop and soil sciences, university of science and technology, kumasi, Ghana.

35. Amanullah, Khalid S (2015) Phenology, growth and biomass yield response of maize (Zea mays L.) to integrated use of animal manures and phosphorus application with and without phosphate solubilizing bacteria. J Microb Biochem Tech 7: 439- 442.

36. Amanullah, Bashir SF, Qahar A, Shah S, Ahmad B, et al. (2015) Interactive response of nitrogen and sulfur on growth, dry matter partitioning and yield of maize. Pure Appl Biol 4: 164-170.

37. Nasim W, Ahmad A, Khaliq T, Wajid A, MFH Munis, et al. (2012) Effect of organic and inorganic fertilizer on maize hybrids under agro-environmental conditions of Faisalabad-Pakistan. African J Agri Res 7: 2713-2719.

38. Khaliq T, Mahmood T, Kamal J, Masood A (2004) Effectiveness of farmyard manure, poultry manure and nitrogen for corn (Zea mays L.) productivity. Int J Agri Biol 6: 60-63.

39. Uzoma KC, Inoue M, Andry H, Fujimaki H, Zahoor A, et al. (2011) Effect of cow manure biochar on maize productivity under sandy soil condition. Soil Use & Mgt 27: 205-212.

40. Liu X, Herbert SJ, J Jin, Q Zhang, G Wang (2004) Responses of photosynthetic rates and yield/quality of main crops to irrigation and manure application in the black soil area of Northeast Chin. Plant and Soil 261: 55-60.

41. Smaling EMA, Nandwa SM, Prestle H, Roetter H, Muchena FN (2002) Yield response of maize to fertilizers and manure under different agro-ecological conditions in Kenya, Elsevier Dordrecht, Netherlands. 41: 241-252.

42. Hokmalipour S, Darbandi MH (2011) Investigation of nitrogen fertilizer levels on dry matter remobilization of some varieties of corn (Zea mays L). World Appl Sci J 12: 862-870.

43. Mohsin AU, Ahmad J, Ahmad AUH, Ikram RM, Mubeen K (2012) Effect of nitrogen application through different combinations of urea and farm yard manure on the performance of spring maize (Zea mays L.). J Animal Plant Sci 22: 195-198.

16

Bioinformatic Characterization of Desaturase Partial Gene in *Sesamum indicum*

Spandana B[1]*, DSRS Prakash[2] and Nagalakshmi S[1]

[1]*Institute of Biotechnology (IBT), Acharya NG Ranga Agricultural University (ANGRAU), Guntur, Andhra Pradesh, India*
[2]*Department of Human Genetics, Andhra University, Visakhapatnam, Andhra Pradesh, India*

Abstract

By using Bioinformatic tools three sets of codehop, webprimer and manually, gene specific degenerate primers were designed for Sesamum desaturase. Through Gradient PCR approach maximum homology among desaturase sequences was observed to be with *Arachis hypogaea* than the other oil seed crops compared. Primers designed through Codehop and webprimershowed amplification of the gene sequence. The aim of this present study was to utilize bioinformatic tools for sesame improvement programs towards diversifying the fatty acid composition of the seed oil.

Keywords: Degenerate primers; Desaturase; Sesamum; Fatty acids; Conserved regions

Introduction

Fatty acid desaturases are the enzymes that are responsible for the insertion of cis-double bonds into pre-formed fatty acid chains in reactions that require oxygen and reducing equivalents. In the last few decades considerable work has been carried out in several plant species to alter the fatty acid composition of the oil by classical plant breeding methods and were able to generate low erucic acid. Brassica (canola oil) and sunflower oil containing high oleic acid. Currently Canola oil isthe lowest saturated fat vegetable oil. Fatty acids are synthesized from acetyl-Co A by a series of reactions that are localized in the plastids. Desaturases play a key role in polyunsaturated fatty acid homeostasis i.e., mostly the unsaturation which is the initial step in the essential fatty acid metabolism. Depending upon the intermediate product, Desaturases are of different types i.e., Stearoyl-ACP desaturases, oleate desaturases, linoleate desaturases etc. By understanding the enzymes involved in the synthesis and accumulation of seed- specific fatty acids, it may be possible to manipulate the composition of seed oil by generating transgenic plants by molecular biology techniques. It is necessary first to identify plant species which synthesize economically valuable oils and characterize the proteins/enzymes involved in the triacylglycerol biosynthesis.

Material and Methods

Genomic DNA Isolation: Total Genomic DNA of sesamum was extracted using a modified Cetyl trimethyl ammonium bromide (CTAB) method and quantified. All the reagents required for CTAB buffer preparation were obtained from Sigma-Aldrich (USA)

Designing of degenerate primers

The oil plant species containing the desaturase genes which have already been reported were collected from NCBI and EBI data banks. Desaturase nucleotide sequences of *B. juncea, G.hirsutum, H.anus, S.indicum, A. hypogeae, R. communis, G. max.* were taken from NCBI database and they are converted into protein sequences by web tool http://www.expasy.ch/tools/dna.html (Figures S1 and S2) The consensus amino acid and nucleotide sequences of desaturases (Fad D) were selected by using the Protein Domain database, Pro Dom (http://prodes. toulouse. inra. Fr /prodom /doc/ prodom.html), NCBI (http://ncbi.nlm.nih.gov) and EBI (www.ebi.ac.uk) tools and desaturase as the query. Clustal W software is used for multiple sequence alignment to identify the most conserved regions (Figures S3 and S4). The similarity matrix and the phylogenetic trees were produced using DNA Star (version 2004). The use of software in biological applications has given a new dimension to the field of bioinformatics. Many different programs for the design of primers are now available [1].

Degenerate primers for conserved regions of desaturases were designed manually and by using few software programmes like Web Primer and Codehop [2,3]. The consensus of both these protein and DNA sequence alignments gave rise to 3 sets of primer pairs which are listed in Table 1. Primer sequence degeneracy was also taken into consideration. In the present study Apart from the above mentioned softwares, the universal site for most of the bioinformatics tools were also used. (www.expasy.org). All the sequences were given for multiple sequence analysis by Clustal W. The resulted aligned sequences were given to BLOCKMAKER (http://bioinformatics.weizmann.Ac.il/blocks/blockmkr/www/make_blocks.html) with each block at a time to BLOCKMAKER. The blocks with maximum similarity and minimum degeneracy were selected from the results. These blocks were given to web programme Codehop (http://bioinformatics.weizmann.ac.il/blocks/codehop.html) for designing of primers. Primers were selected based on minimum degeneracy minimum degeneracy, length and percentage of GC. With one DNA sample three sets of gene specific degenerate primers (Manual, Web primer, Codehop) synthesized by Biotech Desk, India Pvt. Ltd.) were tested for PCR amplification of the gene PCR reactions were carried out in a DNA thermocycler (Eppendorf, USA) with a heated lid. Each 20 µl reaction volume contained about 100 ng of template DNA, 1X PCR buffer (10 mM Tris-HCl pH 5.3, 50 mM KCl), 2 mM $MgCl_2$ (Invitrogen, USA), 2 mM dNTP, (Sigma -Aldrich., USA), 5 pmol each of Forward and Reverse primer and 1U of *Taq* DNA polymerase (Stratagene, Germany) and

*Corresponding author: Spandana B, Institute of Biotechnology (IBT), Acharya NG Ranga Agricultural University (ANGRAU), Guntur, Andhra Pradesh, India
E-mail: spandanabandila@gmail.com

programmed for an initial denaturation step of 5 min at 94°C, followed by 35 cycles of 1 min at 94°C, 1 min at 50°C, extension was carried out at 72°C for 1 min and a final extension at 72°C for 10 min and a hold temperature of 14°C at the end. Negative controls were also run without template DNA to avoid non-specific amplification.

On 1.4% agarose gels PCR were electrophoresed, in 1X TAE Buffer at 50 V for 3 hrs and then stained with ethidium bromide (0.5 ug/ul). To ascertain reproducibility of the raeaction PCR reactions were repeated three times reaction.

Gel elution

Fragments ranging from 400-700 bp were eluted by using Column method (Qiagen Kit)

Ligation

A 10 µl Ligation reaction volume was set up containing 1 µl of pGEM easy vector (Promega, USA); 5 µl of PCR enriched (gel eluted) product; 1 µl 10 X T₄ DNA ligation buffer (New England Biolabs, USA), 2.4 U/µl of T₄ DNA ligase (New England Biolabs, USA) and incubated at 14°C overnight. The samples were stored at - 20°C until used.

Transformation

Ligated product was transformed. The plasmid isolation was done by Alkaline lysis method. After PCR amplified plasmids amplification range (400 bp desaturase) was selected and was given for sequencing. Using Vecscreen (NCBI) the vector sequences were removed form the sequenced clones using and homology search was carried.

Sequence alignment and phylogenetic analysis

The nucleotide sequences and deduced aminoacid sequenceswere analysed using comparitive and bioinformatic analysis that are available online at the websites (http://www.ncbi.nlm.nih.gov/) and (http://cn.expasy.org/). The nucleotide sequence, deduced aminoacid sequence, and open reading frame were analyzed and the sequence comparision was conducted through a database search using BLAST programs (Figure S5). The phylogenetic analysis of desaturase gene from other oilcrop species was aligned with Clustal W program using default parameters [4].

From the edible oil yielding species *Glycine max, Brassica rapa, Helianthus annuus, Arachis hypogaea* the sequences were selected for characterization and converted to amino acid sequences to identify the conserved regions which are given Figure S1.

Desaturase

The Degenerate primers were designed by manually, webprimer and codehop methods which are given in Tables 1 and 2.

Results

Three sets of gene specific degenerate primers (manual, Web primer, Codehop) synthesized by Biotech Desk, India Pvt. Ltd.) were tested for PCR amplification by gradient PCR with one sesame sample Figure 1 The amplified products of each primer is eluted and transformed and isolated plasmids from selected colonies are sequenced. A total of 10 positive clones of which 6 clones were sequenced. The sequences obtained was subjected to BLAST search.

Search results of desaturase in the NCBI for homologs

The Six sequenced desaturase sequences were submitted to the

Primer pair	Primer sequence (5'-3')	bp	Product size(bp)
Des M	F -ACYGGNRTNTGGGTNMTNGCNCAYG	25	400
	R- NGCYTGYTAYTGNGGCATNGTNGAC	25	
Des W	F -GGGCGTGTCACTAAGATTGAA	21	400
	R- TGTACCAGAGCACACCTTTGT	21	
Des C1	F- TGGACATCATGCTTTCTCTGAT-3'	24	400
	R- GGTACGATACCTCCGATGATTCC-5'	24	
Des 2	F -AGTGCTTTCACGCGATTTCT	20	400
	R -TCTCTCGAAGCAGTGTGGTG5'	20	

Degenerate bases information

N=A/T/G/C R=A/G M=A/C W=A/T S=G/C Y=C/T K=G/T V=A/G/C H=A/C/T D=A/G/T B=C/G/T

Table 1: PCR primers designed and targeted fragments of the Desaturase gene in *Sesamum indicum*.

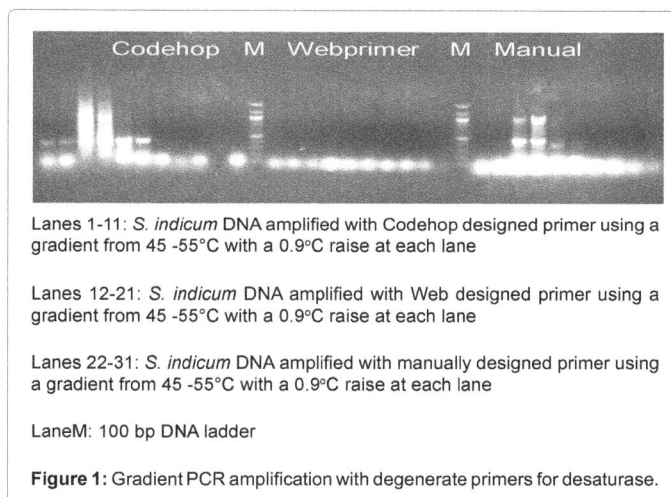

Lanes 1-11: *S. indicum* DNA amplified with Codehop designed primer using a gradient from 45 -55°C with a 0.9°C raise at each lane

Lanes 12-21: *S. indicum* DNA amplified with Web designed primer using a gradient from 45 -55°C with a 0.9°C raise at each lane

Lanes 22-31: *S. indicum* DNA amplified with manually designed primer using a gradient from 45 -55°C with a 0.9°C raise at each lane

LaneM: 100 bp DNA ladder

Figure 1: Gradient PCR amplification with degenerate primers for desaturase.

NCBI (National Centre for Biotechnology Information) website for BLAST-N (http://www. ncbi.nlm. nih.gov/ BLAST/) which showed high level of sequence homology both at the nucleotide and amino acid level with those from *P. crysosporium, L. longiflorum, P. farinose* and *S. bicolor* as shown in Figure 2.

Most of the sequences showed high homology with *Sesamum indicum* omega-6 fatty acid desaturase (FAD2) mRNA, complete cds and also with *Phanerochaete chrysosporium* Pc-ole1 gene for delta9-fatty acid desaturase, complete cds, *Lilium longiflorum* phytoene desaturase mRNA, partial cds, *Primula farinosa* sphingolipid delta-8 desaturase mRNA, complete cds *Mortierella* sp. strain M10 mRNA for delta-9 fatty acid desaturase, partial *Sorghum bicolor* fatty acid desaturase mRNA, complete cds, *Chlamydomonas reinhardtii* fatty acid desaturase (CHLREDRAFT_32523) mRNA, partial cds.

A comparison of Fatty acid desaturase sequences showing phylogenetic relationship at the deduced amino acid level with those of other oilseed crops is shown in the dendrogram (Figure 3). The homology was based on small portion of entire coding sequence, as the sequence length of the obtained sequence was 181 to 538 bp, where as the length of the selected sequences was 1149 to 1095 bp.The expected DNA fragment range was 400 bp and the primers were designed excluding N terminal region with C terminal region only as N terminal

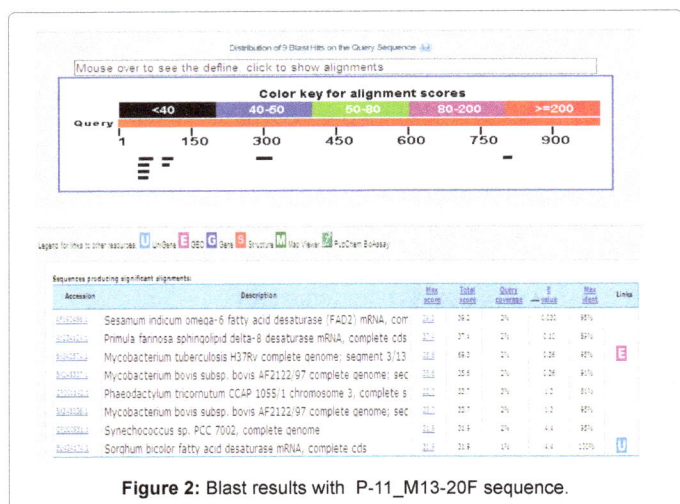

Figure 2: Blast results with P-11_M13-20F sequence.

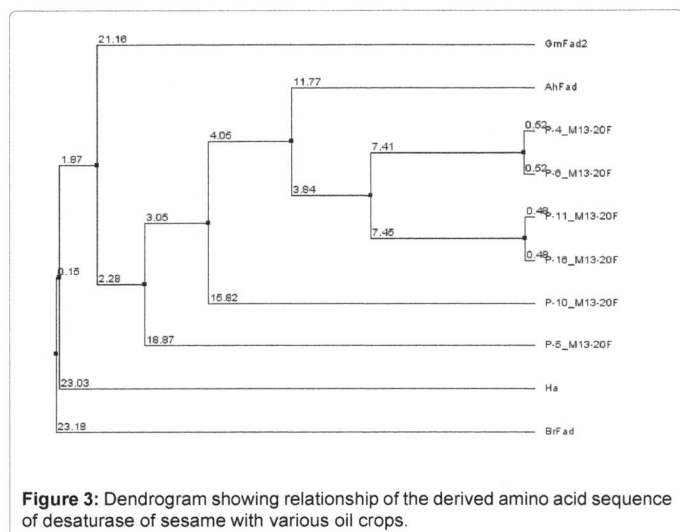

Figure 3: Dendrogram showing relationship of the derived amino acid sequence of desaturase of sesame with various oil crops.

Accession	Organism	Max score
AB 262185.1	*Phanerochaete chrysosporium* Pc-ole1 gene for delta9-fatty acid desaturase, complete cds	87.8
AY500378.1	*Lilium longiflorum* phytoene desaturase mRNA, partial cds	87.8
AY 234124.1	*Primula farinosa* sphingolipid delta-8 desaturase mRNA, complete cds	41.0
EU 424174.1	*Sorghum bicolor* fatty acid desaturase mRNA, complete cds	31.9

Table 2: Homology results of Desaturase.

region was having more INDELS (Insertions and deletions).

A sequence homology search revealed that the putative protein had high homology with the desaturase sequences of *Arachis hypogaea* indicating that the desaturase isolated from *S. indicum* was closer to *A. hypogaea* than the other oil crops compared.

Discussion

Degenerate primers based on the amino acid sequence of conserved regions were also used to search for members of a gene family [2]. Computer programs have also been developed specifically

for degenerate primer design [3].

Bioinformatics approaches to lipid research have recently begun using large amounts of mass spectrometry and microarray data. Phylogenetic analysis of protein families related to fatty acids has also been performed. However, to our knowledge, there is no report that describes the investigation of fatty acid structures based on the comprehensive analysis of the gene contents in the genomes. The fatty acid structures is hard to predict from genomic information due to the the functional diversity of the key enzymes. Many studies in plant systems were also conducted, which eventually led to the purification of a soluble stearoyl-acyl carrier protein (ACP) 9-desaturase from safflower [5]. Later studies allowed for the full nucleotide sequence of stearoyl-ACP 9-desaturases from castor and cucumber to be reported [6]. Various genes encoding enzymes involved in fatty acid synthesis have been isolated from the species and characterized. Carotene desaturase gene was cloned and characterized from chlorella protothecoides cs-41 to elucidate lutein biosynthesis pathway [7]. Cloning of Psy1, *Pds* and *Zds* cDNAs encoding the enzymes responsible for lycopene biosynthesis, namely phytoene synthase 1 (PSY1), phytoene desaturase (PDS) and ζ -carotene desaturase (ZDS), respectively, from high lycopene tomato cultivar, *Solanum lycopersicum* KKU-T34003 [8]. Although the pathway for sesame is not documented, the fatty acid profile suggests synthesis via the known route common to most major oil crops Desaturase, This enzyme in fatty acid biosynthesis in sesame appear to be showing homology to different oilseed *helianthus* and *arachis*. The reasons are unknown but a more detailed study of these enzymes across the different oilcrops will through some light on the fatty acid biosynthesis pathway.

In the present study after analysis of available sequences of desaturase related proteins, retrieved desaturase subgroup clusters and selected six representative sequences of most members of desaturase families which specifically belonged to dicots as *Sesamum* is a eudicot species. In the present study codehop and Webprimer designed primers showed amplification for desaturase whereas manual designed primer gave no amplification. Desaturase derived sequences showed high homology with desaturase genes of *Phanerochaete, lillium, primula sp.*

References

1. Kamel A, Abd-Elsalam A (2003) Bioinformatic tools and guideline for PCR primer design. African Journal of Biotechnology 2: 91-95.

2. Wilks AF, Kurban RR, Hovens CM, Ralph SJ (1989) The application of the polymerase chain reaction to cloning members of the protein tyrosine kinase family. Gene 85: 67-74.

3. Chen H, Zhu G (1997) Computer program for calculating the melting temperature of degenerate oligonucleotides used in PCR or hybridization. Biotechniques 22: 1158-1160.

4. Thompson JD, GibsonTJ, Plewniak F, Jeanmougin F, Higgins DG (1997) The CLUSTAL X windows interface: flexible strategies for multiple sequence alignment aided by quality analysis tools. Nucleic Acids Research 25: 4876-4882.

5. McKeon TA, Stumpf PK (1982) Purification and characterization of the stearoyl-acyl carrier protein desaturase and the acyl-acyl carrier protein thioesterase from maturing seeds of safflower. Journal of Biological Chemistry 257: 1214-12147.

6. Shanklin J, Sommerville CH (1991) Stearoyl- acyl-carrier-protein desaturase from higher plants is structurally unrelated to the animal and fungal homologs. Proc Natl Acad Sci 88: 2510-2514.

7. Meiya Li, Zhibing Gan, Yan Cui, Chunlei Shi, Xianming Shi (2011) Cloning and Characterization of the ζ-Carotene Desaturase Gene from Chlorella protothecoides CS-41. Journal of Biomedicine and Biotechnology 2011: 731542.

8. Krittaya S, Preekamol K (2013) cDNA cloning and expression analyses of phytoene synthase 1, phytoene desaturase and ζ -carotene desaturase genes from Solanum lycopersicum KKU-T34003 Songklanakarin. Journal of Science and Technology 35: 517-527.

17

Genetic Diversity Studies on Selected Rice (*Oryza Sativa*) Genotypes Based on Amylose Content and Gelatinization Temperature

Mawia A. Musyoki[1]*, Wambua F. Kioko[1], Ngugi P. Mathew[1], Agyirifo Daniel[1], Karau G. Muriira[2], Nyamai D. Wavinya[1], Matheri Felix[1], Lagat R. Chemutai[1], Njagi S. Mwenda[1], Mworia J. Kiambi[1] and Ngari L. Ngithi[1]

[1]Department of Biochemistry and Biotechnology, School of Pure andApplied Sciences, Kenyatta University, P.O. Box 43844-00100, Nairobi, Kenya
[2]Molecular Laboratory, Kenya Bureau of Standards, Kenya

Abstract

Improving cooking and eating quality of rice is one of the important objectives of many breeding programs. The aim of the study was to carry out genetic diversity studies on selected rice (*Oryza sativa* L.) genotypes from Kenya and Tanzania based on amylose content and gelatinization temperature using microsatellite markers. Power marker version 3.25 and GenALEx version 6.5 softwares were used to analyze the data. The number of alleles per locus ranged from 2 to 4 alleles with an average of 2.75 alleles across 8 loci obtained in this study. The polymorphic information content (PIC) values ranged from 0.2920 (RM 202) to 0.6841 (RM 141) in all 8 loci with an average of 0.4697. Pair-wise genetic dissimilarity coefficients ranged from 0.9003 to 0.2251 with an average of 0.5627. Maximum genetic similarity was observed between *R 2793* and *BS 17*, *Supa* and *IR 64*, *R 2793* and *ITA 310*, *Saro 5* and *ITA 310*, *Saro 5* and *R 2794*. Minimum similarity of was observed between *Wahiwahi* and *BW 196*, *IR 64* and *BW 196*. The dendogram based on cluster analysis by microsatellite polymorphism grouped the rice genotypes into 2 clusters effectively differentiating Kenyan and Tanzanian rice genotypes based on amylose content and gelatinization temperature. The results obtained suggested that use of microsatellite markers linked to Quantitative Trait Loci (QTLs) controlling these two traits could effectively be utilized for diversity analysis among diverse rice genotypes.

Keywords: *Oryza sativa; Oryza glaberrima;* Heterozygosity; Germplasm; Genotypes

Introduction

Rice (*Oryza spices*) is a monocotyledonous plant belonging to the family Granineae and subfamily Oryzoidea. It is cultivated under diverse eco-geographical conditions in various tropical and subtropical countries [1]. Due to its importance as a food crop, rice is being planted on approximately 11% of the Earth's cultivated land area [2]. It is the grain with the third highest production globally after sugarcane and maize (FAOSTAT, 2012). *Oryza sativa* and *Oryza glaberrima* are the only two cultivated species of rice while the other species are wild. *Oryza sativa* is commonly grown in Asia, North and South America, Europe and Africa. *Oryza glaberrima* is highly grown in West African but due to higher yields of *O.sativa* and *O. glaberrima-sativa* varieties; it is being replaced in most parts of Africa [3].

There are two classes of rice based on starch content, that is, waxy and non waxy rice. Glutinous or waxy rice in which endosperm starch lacks or has very little amylose content consists mainly of amylopectin starch [4]. The ratio of amylose to amylopectin has a major effect on the physical properties of starch. When cooked, the semi-crystalline structure of rice starch is disrupted thus transforming the starch into a softer, edible, and gel-like material [5]. Generally, the amylose content of milled rice is classified into five classes: waxy (0-2%), very low amylose (3-9%), low amylose (10-19%), intermediate amylose (20-24%) and high amylose (above 24%) [6].The cooking temperature at which water is absorbed and the endosperm starch granule swell irreversibly with subsequent loss of crystalline structure is referred to as gelatinization temperature (GT) [7].Gelatinization temperature is an important component of rice cooking quality. Rice grain with low gelatinization temperature takes shorter cooking times leading to significant potential savings in fuel costs [8]. Three classes of GT are recognized in rice breeding programs: high (>74°C), intermediate (70-74°C), and low (<70°C) [9,10]. Besides waxy and alk genes that controls amylose content and gelatinization temperature in rice, there are other several QTLs within the rice genome that are linked to these two genes.

There is a wide range of rice varieties grown both in Kenya and Tanzania. These rice cultivars are either local landraces or improved varieties and they express different levels of amylose and amylopectin that influences amylose content and gelatinization temperature in rice respectively. Since these two traits are key determinant in cooking and eating qualities of rice, unscrupulous traders often blend rice grains which have good cooking and eating quality traits with grains which have poor cooking and eating quality traits based on amylose content and gelatinization temperature to make more profit from their trade. This causes a negative impact on rice trade and consumption. Accurate evaluation of these two traits is difficult and has hindered development of better varieties with good eating and cooking qualities by rice breeders both in Kenya and Tanzania. The various physicochemical methods commonly used to determine amylose content and gelatinization temperature in rice are often inaccurate and time consuming. However, genetic diversity analysis on these selected rice genotypes from Kenya and Tanzania based on amylose content and gelatinization temperature using microsatellite markers has not yet been studied.

Molecular markers can have a number of applications in agriculture, and their application in rice improvement has been reviewed [11]. Simple Sequence Repeat (SSR) markers are easily available for any

*Corresponding author: Mawia A. Musyoki, Department of Biochemistry and Biotechnology, School of Pure and Applied Sciences, Kenyatta University, P.O. Box 43844-00100, Nairobi, Kenya, E-mail: amosyokis@gmail.com

region of the genome. In rice, SSR markers have been effectively utilized for many purposes such as genetic diversity and relatedness [12], QTL mapping [13], mutation analysis [14] and maker assisted selection [15]. This study was carried out to estimate the pattern and level of genetic diversity and relatedness among selected rice genotypes from Kenya and Tanzania along with other germplasm from Philippine based on amylose content and gelatinization temperature.

Materials and Methods

Plant materials

A total of 13 rice genotypes comprising of local landraces and improved rice genotypes were collected from Mwea Irrigation Agricultural Development Centre (MIAD) in Mwea, Kenya and Kilimanjaro Agricultural Research Institute in Moshi, Tanzania. The name, country of origin and category of the rice genotype chosen for the study are given in (Table 1). Genotype IR 64 was used as the check variety since it is known to have both intermediate amylose content and gelatinization temperature [16].

DNA extraction and SSR marker analysis

Genomic DNA was extracted from seed samples using a modified CTAB method [17]. The quality of the genomic DNA was determined in a 1% agarose gel in 100ml TBE electrophoresis by running 10 μl of genomic DNA at a voltage of 75 for 45 minutes. DNA purity for each sample solution was evaluated using a spectrophotometer which employed the Thermo Scientific Nano drop 2000 system (Wilmington, USA). A set of 8 microsatellite markers shown in (Table 3) covering different genomic regions of rice were selected from published data-based search for rice Simple Sequence Repeats (SSR) markers [18,19].

The PCR reactions were carried out in Thermal cycler (Bio Rad Inc. USA) with the total reaction volume of 25μl containing, 5 μl of genomic DNA, 1X assay buffer, 200 μM of dNTPs, 2 μM $MgCl_2$, 0.2μM of forward and reverse primer and 1 unit of Taq DNA polymerase (Fermentas Life Sciences). The PCR cycles were programmed as 95°C for 2 min, 94°C for 1 min, 55°C 72°C for 2 min for 35 cycles and an additional temperature of 72°C for 10 min for final extension. The amplified products were separated on 1.0 percent agarose gel prepared in 0.5X TBE buffer pre-stained with 10 μl of ethidium bromide then electrophorized at 100V for 1 hour. The gel was then visualized under UV trans-illuminator and photographs were taken using gel documentation instrument. The PCR products were sized against l00bp DNA ladder (Life sciences-USA). Clearly resolved, unambiguous bands were scored visually for their presence or absence with each primer. The scores were obtained in the

form of matrix with '1' and '0', indicating the presence and absence of bands in each genotype respectively.

Data management and analysis

Using the Power Marker version 3.25 software package [20], the diversity of each accession was analysed on the basis of three statistical parameters: allele number, gene diversity and polymorphism information content (PIC), which measures the genic diversity [21]. Genetic distance was calculated using "C.S Cord 1967" distance [22] followed by phylogeny reconstruction using rooted UPGMA as implemented in Power Marker with the tree viewed using Treeview. To visualize the relationship between the sample genotypes among the 13 rice varieties, principle coordinate analysis (PCoA) was conducted using GenALEx 6.5 software. It was chosen to complement the UPGMA cluster analysis. Furthermore, to reveal the partition and variation within and among the genotypes, analysis of molecular variance (AMOVA) was carried out using GenALEx 6.5 statistical software [23].

Results

Assessment of polymorphism from SSR profiles

Out of 12 SSR markers used, only 8 markers were polymorphic and showed consistent banding patterns and amplification of each genotype and were ultimately chosen for assessing genetic diversity among the rice genotypes studied. A total of 22 alleles were detected at the loci of 8 microsatellite markers across 13 rice genotypes. The allelic richness per locus varied from 2 to 4, with an average of 2.75 alleles. RM 141 produced the highest number of polymorphic alleles while RM 125, RM 202, and RM 253 produced the least number of polymorphic alleles as shown in (Table 2).

Occurrence of null allele was also observed among some genotypes whenever an amplification product could not be detected in their combination. Experiments detecting null alleles were all repeated at least once to ensure that the absence of an amplified product was not as a result of experimental error. Nineteen SSR loci showed null alleles in two to eight of the 13 rice genotypes. The genotypes having the largest proportion of null alleles were *Kahogo* and *IR 54* (null alleles at four loci) and *IR 64* and *Supa* (null alleles at two loci). The frequency of null allele was not included in the genetic diversity calculations for each SSR locus since they might decrease the apparent heterozygosity in a population leading to deviation of genotypes in a sample from Hardy-Weinberg expectations.

Since gene diversity is the measure of expected heterozygosity, the average gene diversity among the 13 rice populations was 0.5503. The gene diversity values ranged from 0.3550 (RM 202) to 0.7337 (RM 141) as shown in Table 3 Polymorphism information content (PIC) value is a measure of polymorphism among varieties for a marker locus used in linkage analysis. The eight SSR markers used in this study produced polymorphic bands as shown in (Figure 1). The PIC value of each marker, which can be evaluated on the basis of its alleles, varied greatly for all tested SSR loci. The values ranged from 0.2920 (RM 202) to 0.6841 (RM 141) with an average of 0.4697 per locus as shown in (Table 2).

Pairwise genetic dissimilarity

A dissimilarity matrix based on the "C.S Cord 1967" shared SSR alleles was used to determine the level of relatedness among the rice genotypes based on amylose content and gelatinization temperature. The pairwise genetic dissimilarity values as shown in table 3 ranged

Genotype	Origin	Category of rice
R 2793	Kenya	Improved variety
BS 217	Kenya	Improved variety
BS 370	Kenya	Improved variety
BW 196	Kenya	Improved variety
ITA 310	Kenya	Improved variety
SARO 5	Tanzania	Improved variety
IR 64	Philippine	Improved variety
KILOMBERO	Tanzania	Local land race
RED AFAA	Tanzania	Local land race
KAHOGO	Tanzania	Local land race
SUPA	Tanzania	Local land race
IR 54	Philippine	Improved variety
WAHIWAHI	Tanzania	Local land race

Table 1: Names, origin and category of the rice genotypes used in this study.

from 0.9003 to 0.2251 with an average of 0.5627. *Wahiwahi* and *BW 196*; and *IR 64* and *BW 196* exhibited the highest genetic dissimilarity of 90.03% followed by *IR 64* and *BS 370*; *Wahiwah*i and *BS 217*; and *Kahogo* and *ITA 310* with dissimilarity of78.78%. On the other hand, *R 2793* and *BS 217*; *Supa* and *IR 64*; *R 2793* and *ITA 310*; *Saro 5* and *ITA 310*; and *Saro 5* and *R 2794* depicted the lowest genetic dissimilarity of 22.51%.

Clustering of rice genotypes

A rooted UPGMA tree presented in (Figure 2) was constructed using the "C.S Chord 1967" distance values (Cavalli-Sforza and Edwards, 1967) in Power Marker with tree viewed using Treeview software. It revealed genetic relatedness among the 13 rice genotypes based on amylose content and gelatinization temperature using the 8 microsatellite markers. Genotypes that are derivatives of genetically similar types clustered together. The rice genotypes were clustered into two major groups; that is: group I and group II as shown in Figure 2. Group I consisted of the 5 improved genotypes form Kenya and one improved genotype from Tanzania. Group I was further divided into two subgroups; IA and IB. IA was further divided into two small groups; IA1 and IA2. Group IA1 comprised of *R 2793* and *BS 217* while group IA2 comprised of *Saro 5* and *ITA 310*. Group IB comprised only of *BW 196* and *BS 370*.

On the other hand, Group II was comparatively diverse since it consisted of 7 genotypes; 5 local landraces from Tanzania and 2 improved genotypes from Philippine. Group II was divided into two major subgroups; IIA and IIB. Major subgroup IIA was further divided into minor subgroups; IIA1 and IIA2. Group IIA1 consisted only of *Kahogo* while subgroup IIA2 consisted of *Supa* and *IR 54*. The other major subgroup IIB was further divided into two minor subgroups;

IIB1 and IIB2. Subgroup IIB1 consisted only of *Red Afaa* while subgroup IIB2 comprised of *Kilombero, IR 64* and *Wahiwahi*.

Principle coordinate analysis (PCoA)

Principle coordinate analysis (PCoA) based on Nei's genetic distance (Nei 1972) was used to visualize the genetic relationship among the accessions as shown in (Figure 3). The PCoA supports the results obtained from UPGMA cluster analysis. The first two principle axes accounted for 28.42% and 25.58% variation respectively. The first principle axis comprised of *Saro 5, R 2793, ITA 310,* and *BS 217*. The second quadrant comprised of *Kilobero, Wahiwahi* and *IR 64*. The third quadrant comprised of *Red Afaa, Kahogo* and Supa while the forth quadrant comprised of *IR 54, BS 370* and *BW 196*.

Analysis of molecular variance (AMOVA)

Analysis of molecular variance (AMOVA) was used to determine the proportion of genetic variation partitioned among and within the 13 rice genotypes. Twenty four percent (24%) (P<0.001) of genetic variation partitioned among genotypes and 76% (P<0.001) within the genotypes as shown in (Table 4).

Discussion

Assessment of genetic diversity is a key factor for germplasm conservation, characterization and breeding. Classical breeding affects genetic diversity by selection of combination of outcomes from diverse allele frequencies and leads to favorable effects and loss of diversity [24]. Little was known about genetic diversity among Kenyan and Tanzanian

Marker	Motiff	Chr*	T**	Allele No	Gene Diversity	PIC
RM 125	(GCT)8	7	55	2.0000	0.4970	0.3735
RM 141	(CT)12	6	55	4.0000	0.7337	
RM 488	(GA)17	1	55	3.0000	0.5444	0.4836
RM 225	(CT)18		55	3.0000	0.6627	0.5888
RM 202	(CT)30	11	55	2.0000	0.3550	0.2920
RM 434	(TC)12	9	55	3.0000	0.5917	0.5186
RM 55	(GA)17	3	55	3.0000	0.5207	0.4434
RM 253	(GA)25	6	55	2.0000	0.4970	0.3735
Mean				2.7500	0.5503	0.4697

Chr*-chromosome on which marker is located; T**-annealing temperature

Table 2: List of the 8 polymorphic SSR markers used in this study, their motif, allele number, gene diversity and PIC values.

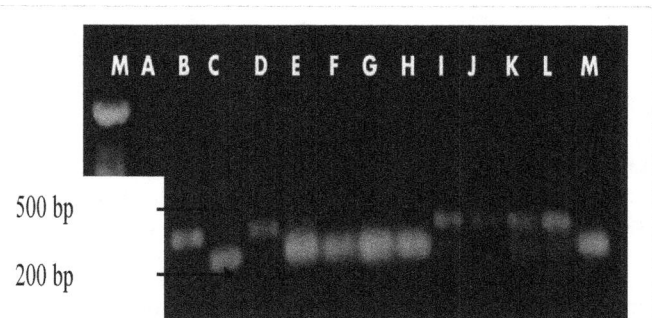

Figure 1: SSR RM 141 showing polymorphic pattern of 13 rice genotypes. The letters represent: M-100bp molecular ladder.

A-R 2793; B-BS 217; C-BS 370; D-BW 196; E-ITA 310; F-Saro 5; G-IR 64; H-Kilomber;, I-Red Afaa; J-Kahogo; K-Supa; L-IR54; M-Wahiwahi.

	BS 217	BS 370	BW 196	IR 54	IR 64	ITA 310	KAHOGO	KILOBERO	R 2794	RED AFAA	SARO 5	SUPA	WAHIWAHI
BS 217	0.0000												
BS 370	0.5627	0.0000											
BW 196	0.4502	0.3376	0.0000										
IR 54	0.4502	0.4502	0.4502	0.0000									
IR 64	0.6752	0.7878	0.9003	0.6752	0.0000								
ITA 310	0.4502	0.4502	0.3376	0.6752	0.6752	0.0000							
KAHOGO	0.6752	0.6752	0.6752	0.3376	0.4502	0.7878	0.0000						
KILOBERO	0.5627	0.6752	0.6752	0.6752	0.3376	0.3376	0.5627	0.0000					
R 2794	0.2251	0.5627	0.4502	0.5627	0.5627	0.2251	0.6752	0.4502	0.0000				
RED AFAA	0.5627	0.5627	0.6752	0.6752	0.3376	0.6752	0.5627	0.4502	0.5627	0.0000			
SARO 5	0.4502	0.5627	0.4502	0.6752	0.4502	0.2251	0.5627	0.3376	0.2251	0.4502	0.0000		
SUPA	0.4502	0.4502	0.5627	0.2251	0.6752	0.7878	0.3376	0.6752	0.6752	0.4502	0.5627	0.0000	
WAHIWAHI	0.7878	0.6752	0.9003	0.5627	0.3376	0.5627	0.5627	0.3376	0.6752	0.5627	0.5627	0.5627	0.0000

Table 3: SC. Cord coefficients of dissimilarity among pairs of 13 rice genotypes.

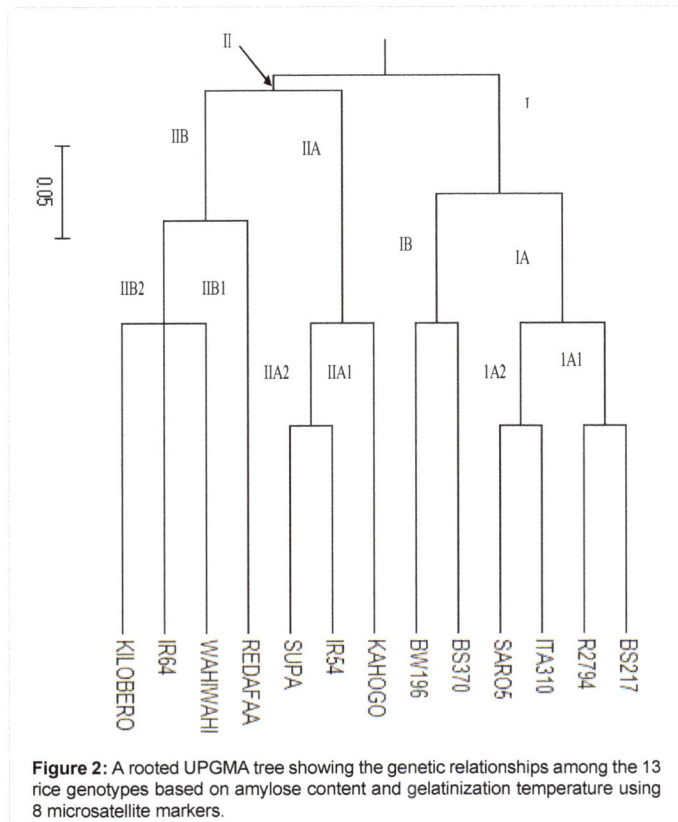

Figure 2: A rooted UPGMA tree showing the genetic relationships among the 13 rice genotypes based on amylose content and gelatinization temperature using 8 microsatellite markers.

rice varieties compared to those from Philippine. In the present study, a total of 8 SSR markers were used to assess genetic diversity of 13 rice genotypes. The results indicated a considerable level of genetic diversity among the genotypes used. Different numbers of alleles per locus were detected in all the 8 SSR markers used. This could have been attributed by remnant heterozygosity in some varieties and expected varietal heterogeneity where landrace varieties consist of mixtures of pure lines that contribute to their broad adaptation in traditional farming systems. These results were similar to those reported by Shahid MS et al. [25] using a different group of rice genotypes. In contrast, the average number of alleles per locus obtained in the present study was smaller than that reported in previous studies [26,27]. This difference in average allele per locus might be due to diverse nature of genotypes used by these authors and selection of SSR markers with different scorable alleles.

A null allele is a mutant copy of a gene at a locus that completely lacks that gene's normal function. Occurrence of null alleles in some rice genotypes could have been attributed by chromosomal mutation within a particular gene locus. As a result, it was difficult to amplify the targeted gene sequence using the SSR primer pairs. Similar null allele occurrence was observed by Lapitan VC et al. [28].

Gene diversity is the measure of expected heterozygosity. The genetic diversity mean value of 0.5503 depicted relative heterezygosity based on amylose content and gelatinization temperature among the 13 rice genotypes studied. The average gene diversity value was lower than what was reported by Lapitan VC et al. [28] whom obtained a value of 0.71 but slightly higher than the one obtained by Shahid MS et al. [25]. The high mean gene diversity value reported by Lapitan VC et al. [28]. Could be as a result of high rate of exchange of genetic

materials among the rice genotypes studied mostly during their genetic improvement. The relative heterozygozity reported in our study could have been attributed by mixing or exchange of genetic materials from different parental lines especially during rice improvement strategies. Conventional breeding of rice by rice breeders is one of major contributory factor to the exchange of genetic material between two lines resulting into a hybrid variety.

The level of polymorphism was determined by calculating polymorphism information content (PIC). Polymorphism information content is a measure of allele diversity at a locus. Loci which are highly informative have PIC value > 0.5), reasonably informative loci have PIC values between 0.25 and 0.5, and slightly informative loci have PIC value <0.25. Out of the 8 SSR markers used in this study, only 3 markers (RM 141, RM 225, and RM 434) had a PIC value greater than 0.5. These markers appeared to be highly informative and could therefore be utilized in marker assisted selection of rice genotypes because they are capable of distinguishing between genotypes. The mean PIC value observed in this study was higher than the PIC value of 0.31 recorded by Sivaranjani AKP et al. [29].This indicated that the genotypes used in the present study were more diverse due to differences in origin.

Maximum genetic distance values observed in our study between *Wahiwahi* and *BW 196*; *IR 64* and *BW 196* indicated high genetic dissimilarity between them and they showed more divergence. Chromosomal mutation and diverse geographical origin could be the contributory factor for the observed genetic dissimilarity between these rice genotypes. The minimum genetic distance values observed between *R 2793* and *BS 217*; *Supa* and *IR 64*; *R 2793* and *ITA 310*; *Saro 5* and *ITA 310*; *Saro 5* and *R 2794* portrayed less genetic dissimilarity and having a very close relationship. The observed genetic dissimilarity is an indication of a common ancestral origin, or perhaps high rate interbreeding which results to sharing of similar alleles in their genome. This result was comparable with what was reported by Islam MM et al. [30] But slightly higher than what was observed by Shahid MS et al. [25]. A different value of average genetic similarity of 0.79 between 40 rice genotypes was reported by Ravi M et al. [31]. this genetic difference

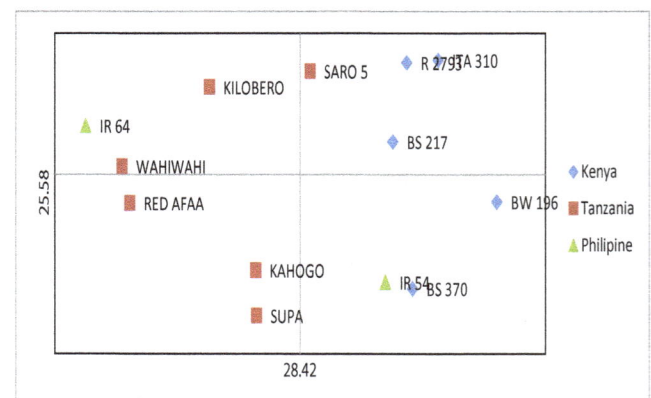

Figure 3: Principle coordinate analysis of the 13 rice genotypes based on amylose content and gelatinization temperature using 8 SSR markers.

Source of variation	Df	SS	MSD	% variation	P-value
Among genotypes	1	131.1	65.6	24%	
Within genotypes	12	1113.8	48.4	76%	<0.001

Table 4: Summary of analysis of molecular variance (AMOVA). Degrees of freedom (df), sum of squares (SS), mean of square deviation (MSD), % variation and P-values.

could be due to different group of rice genotypes used. The high level of similarity recorded by this author could be due to the intra specific variation in the germplasm used.

On comparing Kenyan and Tanzanian rice genotypes, it was found that Kenyan genotypes were closely related compared to Tanzanian genotypes, with a mean genetic dissimilarity coefficient of 0.3939 against that of 0.5064 for the Tanzanian genotypes. The genetic closeness of Kenyan rice varieties could be as a result of high intra specific variation, evolution from a common ancestry and introgression of similar traits during the time of genetic improvement. On the other hand, the relatively high genetic dissimilarity witnessed among the Tanzanian varieties could be as a result of having diverse ancestral origins, high gene flow caused by cross pollination among these varieties and chromosomal mutations in their genome.

Clustering of these genotypes together, for example, *BW 196* and *BS 370* could be as a result of sharing common ancestry or similar genes were introgressed into their genome during their breeding. Surprisingly, group I genotypes which were all improved varieties were genetically distinct when compared with the *IR 64* that was used as a check variety. Therefore, based on these results, it is evident that the six improved varieties studied have different levels of amylose content and gelatinization temperature. Thus further breeding on these genotypes should be carried out so as to introgress favorable genes conferring intermediate amylose content and gelatinization temperature in their genome so as to make them highly competitive in rice market.

Supa, a local landrace from Tanzania and *IR 54*, an improved cultivar with low amylose content from Philippine clustered together. Based on these results, these two genotypes share common alleles for *waxy* gene responsible for high amylose content and *alk* alleles associated with low gelatinization temperature. Therefore, *Supa* genotype does not have good cooking and eating quality characteristics. Genes expressing good quality traits should be introgressed into genome of this genotype. *Kilombero*, *Wahiwahi* and *IR 64* were clustered together. Factors such as sharing of common ancestry and gene flow caused by interspecific gene transfer could be reason behind clustering together. Clustering of *Kilombero* and *Wahiwahi* together with *IR 64* is an indication that these two Tanzanian local landraces have good cooking and eating qualities like those of *IR 64*.

Principle coordinate analysis is a method that visualizes similar and dissimilar data. It assigns similar or dissimilar matrix a location in a three dimensional space. It was chosen to complement the UPGMA cluster analysis by visualizing the relationship between the sample genotypes using genetic distances. The Kenyan improved genotypes were clearly separated from Tanzanian local landraces. Genotypes that grouped together were interpreted to have similar characteristics (closely related) while those apart interpreted to be different or distantly related.

Analysis of molecular variance revealed percentage variation between and among the Kenyan and Tanzanian rice genotypes used in this study. The high genetic variation within the sample populations could be due to increased gene flow or mutations of a number of repeats of a given genotype for a given SSR. In addition, natural selection mechanism could be another source of this high genetic variation within the rice genotypes studied. On the other hand, the relatively low genetic variation among these rice genotypes could be attributed by sharing of same SSR profiles among themselves. The low genetic variation among these genotypes could explain the probability of sharing a common ancestry despite the fact that they are grown in different countries. Similar huge differences in percentage variation between and among a group of rice genotypes studied using SSR markers.

Conclusion

The present study provided an overview of genetic diversity based on amylose content and gelatinization temperature among the rice genotypes studied. The UPGMA cluster analysis showed that all 13 rice genotypes could be easily distinguished based on the information generated by the 8 polymorphic SSR markers. The PIC values revealed that RM 141, RM 225, and RM 434 might be the best markers for identification and diversity estimation of rice genotypes. The genetic distance revealed that the Kenyan genotypes had relatively narrow genetic base compared to the Tanzanian genotypes. Therefore, it is highly important not only to conserve these genotypes, but also to reveal their gene pool and unlock other valuable genes for breeding purposes.

References

1. Causse MA1, Fulton TM, Cho YG, Ahn SN, Chunwongse J, et al. (1994) Saturated molecular map of the rice genome based on an interspecific backcross population. Genetics 138: 1251-1274.

2. Khush GS1 (2005) What it will take to feed 5.0 billion rice consumers in 2030. Plant Mol Biol 59: 1-6.

3. Linares OF (2002). African rice (Oryza glaberrima): History and the future potential. Proceedings of the National Academy of Science of United States of America 99: 16360-16365.

4. Bertoft E, Koch K (2000) Composition of chains in waxy-rice starch and its structural units. Carbohydrate Polymers 41: 121-132.

5. Ramesh M, Ali SZ, Bhattacharya KR (1999) Structure of rice starch and its relation to cooked-rice texture. Carbohydrate Polymers 38: 337-347.

6. Suwannaporn P, Pitiphunpong S, Champangern S (2007. Classification of rice amylose content by discriminant analysis of physicochemical properties. Starch-Stärke 59: 171-177.

7. Cuevas RP, Daygon VD, Corpuz HM, Reinke RF, Waters DL, et al. (2010) Melting the secrets of gelatinization temperature in rice. Functional Plant Biology 37: 439-447.

8. Fitzgerald MA1, McCouch SR, Hall RD (2009) Not just a grain of rice: the quest for quality. Trends Plant Sci 14: 133-139.

9. Juliano BO (2003) Rice chemistry and quality. Philippine Rice Research Institute 8-21.

10. Waters DL1, Henry RJ, Reinke RF, Fitzgerald MA (2006) Gelatinization temperature of rice explained by polymorphisms in starch synthase. Plant Biotechnol J 4: 115-122.

11. Collard BC1, Mackill DJ (2008) Marker-assisted selection: an approach for precision plant breeding in the twenty-first century. Philos Trans R Soc Lond B Biol Sci 363: 557-572.

12. Pervaiz ZH, Rabbani MA, Khaliq I, Pearce SR, Malik SA (2010) Genetic diversity associated with agronomic traits using microsatellite markers in Pakistani rice landraces. Electronic Journal of Biotechnology 13: 4-5.

13. Guo Y, Cheng B, Hong D (2010) Construction of SSR linkage map and analysis of QTLs for rolled leaf in Japonica Rice. Rice Science 17: 28-34.

14. Wang Q1, Sang X, Ling Y, Zhao F, Yang Z, et al. (2009) Genetic analysis and molecular mapping of a novel gene for zebra mutation in rice (Oryza sativa L.). J Genet Genomics 36: 679-684.

15. Thomson MJ (2009) Marker-assisted breeding for abiotic stress tolerance in rice: Progress and future perspectives. Metro: 43-55.

16. Khush GS, Virk PS (2005) International Rice varieties and their impact. International Rice Research Institute, Manilla 115-135.

17. Doyle JJ, Doyle JL (1990) Isolation of plant DNA from fresh tissue. Focus 12: 13-15.

18. Panaud O1, Chen X, McCouch SR (1996) Development of microsatellite markers and characterization of simple sequence length polymorphism (SSLP) in rice (Oryza sativa L.). Mol Gen Genet 252: 597-607.

19. McCouch SR, Teytelman L, Xu Y, Lobos KB, Clare K et al. (2002). Development and Mapping of 2240 New SSR Markers for Rice (Oryza sativa L.). DNA Research 9: 199-207.

20. Liu K1, Muse SV (2005) PowerMarker: an integrated analysis environment for genetic marker analysis. Bioinformatics 21: 2128-2129.

21. Botstein D, White RL, Skolnick M, Davis RW (1980) Construction of a genetic linkage map in man using restriction fragment length polymorphisms. Am J Hum Genet 32: 314-331.

22. Cavalli-Sforza LL, Edwards AW (1967) Phylogenetic analysis. Models and estimation procedures. Am J Hum Genet 19: 233-257.

23. Peakall R, Smouse P (2009) GenALEx6.5. Gene analyses in excel. Population genetic software for teaching and research. Molecular Ecology Notes 28: 2537-2539.

24. Singh RK, Sharma RK, Singh AK, Singh VP, Singh NK et al. (2004). Suitability of mapped sequence tagged microsatellite site markers for establishing distinctness, uniformity and stability in aromatic rice. Euphytica, 135: 135-143.

25. Shahid MS, Shahzad AN, Muhammad A (2013) Genetic diversity in basmati and non-basmati rice varieties based on microsatellite markers. Pakistan Journal of Botany 45: 423-431.

26. Kuroda Y, Sato YI, Bounphanousay C, Kono Y, Tanaka K (2007) Genetic structure of three Oryza AA genome species (O. rufipogon, O. nivara and O. sativa) as assessed by SSR analysis on the Vientiane Plain of Laos. Conservation Genetics 8: 149-158.

27. Shefatur RM, Rezwan MM, Samsul AM, Rahman L (2009) DNA fingerprinting of rice (Oryza sativa L.) cultivars using microsatellite markers. Australian Journal of Crop Science 3: 122-128.

28. Lapitan VC, Brar DS, Abe TE, Redoña ED (2007) Assessment of genetic diversity of Philippines rice cultivars carrying good quality traits using SSR markers. Breeding Science 57: 263-270.

29. Sivaranjani AKP, Manish K, Pandey I, Sudharshan G, Kumar GR, et al. (2010) Assessment of genetic diversity among basmati and non-basmati aromatic rice of India using SSR markers. Research communications 99: 2-25.

30. Islam MM, Ali MS, Prodhan SH (2012) SSR marker-based molecular characterization and genetic diversity analysis of aromatic landreces of rice (Oryza sativa L.). Bioscience and Biotechnology 1:107-116.

31. Ravi M, Geethanjali S, Sameeyafarheen F, Maheswaran M (2003). Molecular marker based genetic diversity analysis in rice (Oryza sativa L.) using RAPD and SSR markers. Euphytica 133: 243-252.

18

Response of Improved Durum Wheat (*Triticum durum* L.) Varieties to Wheat Stem Rust in Central Ethiopia

Abiyot Lemma[1], Getaneh Woldeab[2] and Selvaraj T[3]

[1]EIAR, Worer Agricultural Research Center, P.O. Box 2003, Addis Ababa, Ethiopia
[2]EIAR, Ambo Plant Protection Research Center, P.O. Box 37, Ambo, Ethiopia
[3]Ambo University, P.O. Box 19, Ambo, Ethiopia

Abstract

Wheat stem rust caused by *Puccinia graminis* f.sp. *tritici* is amongst the biotic factors which causes up to 100% yield loss during epidemic years. Therefore, the present study was carried out to observe the reaction of improved durum wheat varieties to virulent stem rust isolates at seedling growth stage and to stem rust population at adult growth stages. The finding of this experimental study was based on isolation and multiplication of virulent races of *P. graminis,* then ten durum wheat varieties including the susceptible check morocco were tested for the virulent isolates at seedling growth stage; and the reaction of these varieties under natural field conditions were evaluated in the Eastern and Western Showa of Central Ethiopia at adult growth stages. At seedling growth stage, 60% of varieties tested with the virulent stem rust isolate from Debrezeit exhibited resistance, while all varieties were resistant to Ambo isolate. Ten durum wheat varieties evaluated at adult plant growth stage to stem rust population of Debrezeit (Eastern Showa) showed severities of 20S in the variety Geredo to 70S in the variety Foka, while at Ambo (Western Showa) only the susceptible check, Morocco was infected up to 20S. The result indicated that stem rust from Debrezeit was virulent to durum wheat varieties as compared to the Ambo population.

Keywords: Durum wheat; *P. graminis*; Reaction; Ambo; Debrezeit

Introduction

Wheat is the most important cereal crop in the world and widely grown occupying 17% of the world cultivated land [1,2]. Ethiopia is the second largest producer of wheat in sub-Saharan Africa [3] and it is represented as hexaploid (bread wheat) and tetraploid (durum and emmer wheat types) [4]. The crop has considered as the main staple food of Ethiopian population particularly in highlands of the country [5] where it has produced in a large volume and 95% of the total production is produced by small scale farmers. Wheat accounts for 17.5% of major crops produced in Ethiopia [6]. Currently it ranks second both in terms of volume of production and productivity after Maize and third in terms of area coverage after Maize and Tef [7].

Durum wheat which was believed to be originated in the Abyssinian and Mediterranean [8] is traditionally grown on heavy black clay soil (vertisol) and it has very narrow adaptation and lower yield potential as compared to bread wheat [9]. The grain is differentiated by its big size and weight, mainly suitable for pasta, macaroni, pastini and other manufacturing products. Besides, its stalk as other cereal crops residue is a good source of animal feed and serves as mulch for different agronomic practices in Agriculture.

The production and productivity of wheat crop in Ethiopia has increased in the last decades, though the national average yield has not exceeded 1.7 tones/ha [7]. This is by far below the world's average yield/ha which is about 3.3 tones/ha [10]. This low yield is attributed to multi-faced abiotic and biotic factors such as cultivation of unimproved low yielding varieties, low and uneven distribution of rainfall, poor agronomic practices, insect pests and serious disease like rusts [11].

Rust fungal pathogens are among the major stresses that cause high yield losses in wheat crop. Over 30 fungal wheat diseases are identified in Ethiopia, stem rust caused by *Puccinia graminis* f.sp. *tritici* is one of the major production constraints in most wheat growing areas of the country; causing yield losses of up to 100% during epidemic years [12].

The high lands of Ethiopia are considered to be hot spot for stem rust

development where an estimated loss due to this disease ranges from 40% in endemic areas to 100% where epidemics occur on susceptible varieties [13]. Epidemics of stem rust occurred in 1993 resulted in a tremendous loss of grain in the variety Enkoy which ranged from 67 to 100% in Arsi and Bale zones of Ethiopia and which causes for the demise of this variety [14].

Recent studies in the country showed that most previously identified races were virulent on most of varieties grown in the country [12,15,16]. This may show that there is potential danger of resistance breakdown in Ethiopian released bread, durum and emmer wheat varieties. Resistance could be expressed as reduction in number of lesions, a reduction in the size of sprouting area, an increase in the length of the latent period and a reduction in length of sporulating period [17]. Selection of Indigenous germplasm, selection from introduction, hybridization and evaluation of selected lines are the breeding methods used for screening the targeted disease problem and yield improvement by the national durum wheat program in Ethiopia [18]. Therefore the present study was carried out to determine the reaction of improved durum wheat genotypes to virulent isolates of stem rust at seedling growth stage and to stem rust population under natural field condition at adult growth stage.

Materials and Methods

Adult plants and seedlings of ten commonly grown improved

*Corresponding author: Abiyot Lemma, EIAR, Worer Agricultural Research Center, P.O. Box 2003, Addis Ababa, Ethiopia
E-mail: alemma2009@gmail.com

durum wheat varieties and a susceptible check, Morocco (Table 1) were screened against stem rust populations of Debrezeit and Ambo under natural field conditions, and with one stem rust isolate each from Debrezeit and Ambo in the green house. The varieties were included in the Ethiopian Wheat Rust Trap Nursery (WRTN) and were tested at many locations. For this study, data were collected from Debrezeit Agricultural Research Center (DZARC) and Ambo Plant Protection Research Center (APPRC) experimental fields.

Response of durum wheat varieties to stem rust population under natural field conditions of DZARC and APPRC at adult growth stages

The ten durum wheat varieties were planted at Debrezeit Agricultural Research Center (DZARC) and Ambo Plant Protection Research Center (APPRC) experimental fields in un replicated trials with two rows per plot. Each plot has a size of 1 m long with two rows and 20 cm spacing between rows. Plant population (density) were regulated to favor maximum disease development in which the plant is able to expresses its genetic resistance potential, and agronomic practices which ensures optimum plant growth and development was followed. At APPRC, the test materials were planted on July 14, 2009, while it was planted in the first week of July at DZARC. Artificial fertilizers were applied at the rate of 50 kg Urea (N=46 kg/100 kg) and 100 kg of DAP (N P K=18:18:0). Three-four times hand weeding was carried out during the crop growing season to make the plots were weed free at both locations. Data were scored during October, in which the disease was reached at its maximum severity level ([19] [20] using Peterson et al., [21] severity scoring scale (Table 2), and the crop response classification scale of Stubbs et al. ([22] under field conditions. Based on the size of the pustule, Stubbs et al. (1986) has classified the response as: 0 when no visible infection has observed; Resistant (R), when necrotic areas are with or without small pustules; Moderately Resistant (MR), when small pustules are surrounded by necrotic areas; Intermediate (M), when pustules of variable size & when there is some necrosis or chlorosis; Moderately Susceptible (MS), when there is medium sized pustules and no necrosis but some chlorosis are observed; and Susceptible (S), for large pustules with no chlorosis and necrosis have observed on the crop.

Response of improved durum wheat varieties to stem rust isolates at seedling stages

Stem rust isolate from Debrezeit (TTTTF) and from Ambo

Variety	Pedigree
Cocorit-71	RAE/4* TC 6011 STW 63 \3/AA S, DZ 27617- 18-64-0M
Gerardo	VZ 466/61-130xGII "s", CM 9605
LD357	CI 8188 No. 58-40
Boohai	Cr "s"/21563/61-130xLds candeal II=Coo s/CII,CD 3062-Bs-OGR
Foka	COCRIT 71/CANDEAL-II, CD 3369
Kilinto	ILLUMILLO/INRAT69//BHA/3/HORA/4/CIT 71 JORI, DZ 918
Bichena	Illumillo/ cocorit 71, DZ 393-2
Robe (DZ1640)	Hora/Cit's'//Jo's'/GS's'/3/Some's'/4/Hora/Raspinegro// CM9908/3/
Ude (CD 95294-2y)	chen / ALTAR 84// Ald CD 95294-2y
Yerer(CD 94026-4y)	chen/Tez/3/Guil//cll CD 94026-4y-040m-030y-pAp-0y
Morocco	

Source Kulumsa Agricultural Research Center (KARC)

Table 1: List of durum wheat lines used for screening against stem rust and their pedigree

Actual percentage	Visual percentage
0.37	1
1.87	5
3.7	10
7.4	20
11.1	30
14.8	40
18.5	50
22.2	60
25.9	70
29.6	80
33.3	90
37.0	100

Table 2: The actual percent and visual percentage through considering combination of pustule size and distribution of severity of rust disease [21].

(PTKTK) were multiplied on the universally susceptible variety Morocco. On the other hand, six seeds from each of the ten improved durum wheat varieties and a susceptible check Morocco were sown in plastic pots containing soil, compost and sand in the ratio of 2:1:1 using Complete Randomized Design (CRD) in a greenhouse. When the seedlings were 7 days old (the first leaf is fully expanded and the second leaf is just emerged), they were inoculated with one race from Debrezeit and another from Ambo and maintained in greenhouse. Data on infection types were recorded 14 days after inoculation according to the host response and infection type description given by Stakman et al. [23] (Table 3).

Results and Discussion

In this study, adult plants and seedlings of ten commonly grown improved durum wheat varieties and a susceptible check, Morocco (Table 1) were screened against stem rust populations of Ambo and Debrezeit under natural field conditions, and with two virulent isolates of *P. graminis* f.sp *tritici* in greenhouse test. The results showed that at adult plant growth stages the test varieties reacted differently to stem rust populations of Ambo and Debrezeit. At Ambo experimental field, these varieties were very resistant to the existing stem rust populations except on the universally susceptible variety Morocco, this variety was infected by stem rust up to 20S (Table 4). Different result was observed at Debrezeit experimental field where except the variety LD 357 which showed moderately susceptible/moderately resistant reactions, all the rest were susceptible/moderately susceptible to the stem rust populations. The universally susceptible variety Morocco was infected up to 60S. This data showed that the pathogen populations at the two centers were probably different.

The green house test at the seedling growth stages revealed that none of varieties were immune to both Ambo and Debrezeit isolates/ races. All varieties tested with Ambo stem rust isolate/race "PTKTK" at seedling stage had resistant reactions of "1- 2" in a "0-4" scoring scale. The only variety which showed a susceptible reaction "4" was the susceptible check Morocco. The durum wheat varieties tested were resistant/effective to the stem rust population as well as to the isolate of Ambo at both growth stages. However, the Debrezeit stem rust isolate/race "TTTTF" used in the greenhouse test was virulent when it was compared with Ambo's isolate. In the greenhouse test, 40% of the durum wheat varieties included in the study was not effective, while 60% or six varieties namely LD 357, Bichena, DZ-1640, Foka, Kilinto and Yerer were effective to this isolate. Nevertheless, at adult plant growth stage, the later four varieties were non-effective under natural field conditions of Debrezeit. The reason could probably be the difference in virulence between the stem rust field population and

Class	IT	Description of symptoms
Immune	0	No sign of infection to the naked eye
Very Resistant	0;	No uredia, but distinct flakes of varying sizes, usually a chlorotic yellow but occasionally necrotic
Resistant	1	Small uredia surrounded by yellow chlorotic and necrotic area
Moderately Resistant	2	Small to medium sized uredia, typically in a dark green island surrounded by a chlorotic area
Mesothentic or Heterogeneous	X	A range of infection type from resistant to susceptible scattered randomly on a single leaf caused by a single isolate not a mixture
Moderately Susceptible	3	Medium sized uredia. Usually surrounded by a light green chlorotic
Susceptible	4	Large uredia with a limited amount of chlorosis: may be diamond shaped
Modified characters		
Lower uredinia	=	uredia much smaller than typical and at the lower limit of the infection type
Small uredinia	-	uredia smaller than normal
Larger uredinia	+	uredia larger than normal
Largest uredinia	++	uredia much larger than typical and at the upper limit for the infection type

Table 3: Description of infection types used in classifying the reactions to stem rust on seedling wheat leaves as it was adopted from [23].

Variety	Reactions of durum varieties to stem rust at			
	Adult plant growth stage		Seedling growth stage	
	Ambo (%)	Debrezeit (%)	PTKTK / Ambo isolate (0-4 score)	TTTTF/DZ isolate (0-4 score)
Cocorit 71	0	40MS	1+	3-
Geredo	0	20MS	1	3-
LD 357	0	30MS/MR	1	2
Boohai	0	40S/MS	1	3-
Foka	0	70S	1	1
Kilinto	0	60S	1	2
Bichena	0	20S/MS	1	2
DZ 1640	0	40S	1	2
CD-95294-2y (Ude)	0	40MS/S	2	3
CD-94026-4y (Yerer)	0	20S/MS	1	1
Morocco(Suscp.check)	20S	60S	4	3+

Table 4: Reaction of some durum wheat varieties to stem rust at Adult and seedling growth stage of Ambo and Debrezeit during 2009 growing season.

the isolate used in the greenhouse. According to this study, varieties Foka and Yerer had better resistance to stem rust to both Ambo and Debrezeit stem rust isolates at seedling stage. In addition to those varieties LD 357, Bichena, DZ-1640 and Kilinto were resistant to Debrezeit stem rust isolates and very resistant to Ambo stem rust isolates at seedling stage. Cocorit and Geredo which were found to be resistant at seedling stage for two Debrezeit isolates of *P. graminis* f.sp. *tritici* in 2003 [24] were found to be susceptible at seedling stage for the virulent isolate of Debrezeit this year. Generally, the seedling test with isolates from Debrezeit showed that 60% of the varieties were resistant to the currently available virulent race, and therefore, Ethiopian land races could be a valuable source of resistance to the stem rust virulent race(s), which is in harmony with reports of Mengistu and Yeshi [25]. They mentioned as the large proportion of Ethiopian durum wheat accessions are resistant or moderately resistant to stem rust, and the land races are found to be a good source of resistance to stem rust. At Ambo, all varieties could be grown as long as the population of stem rust is not changed. So, this study indicated the type of varieties we have at hand in regard to infection by stem rust.

Conclusion

All the tested durum wheat varieties were resistant to stem rust of Ambo at adult growth stages under natural field condition, while 20MS in variety Geredo to 70S in variety Foka were recorded in the Debrezeit experimental field. The seedlings of all ten Durum wheat varieties tested with Ambo stem rust isolate showed resistant reactions, while 60% of these varieties were resistant to Debrezeit isolate. Therefore, the development of stem rust under natural field condition was favored by the environmental condition at Debrezeit than at Ambo. Moreover the pathogen populations in the two locations were different, in which the pathogen from Debrezeit is virulent.

References

1. Carter B (2002) The importance of wheat quality. Agricultural horizons contemporary issues for agriculture Washington State University.

2. Roohparvar (2007) Drug transporters of the fungal wheat pathogen Mycosphaerel graminicola dissertation no. 4144. Wageningen Unvieristy.

3. White JW, Tanner DG, Corbett JD (2001) An agroclimatological characterization of bread wheat production area in Ethiopia. CIMMIT, Mexico NRG.

4. Belayneh B, Lind V, Friedt W, Ordon F (2008) Virulence analysis of Pucciniagraminis f.sp. tritici populations in Ethiopia with special consideration of Ug 99.

5. CIMMYT (2005) Sounding the Alarm on Global Stem Rust: an assessment of race Ug99 in Kenya and Ethiopian and the potential for impact in neighboring regions and beyond. MexicoCity, CIMMYT.

6. Ethiopian Commodity Exchange (ECX), 2009. ECX website

7. Central Statistical Agency (CSA) (2007) Agricultural sample survey 2006/2007 volume I. report on area and production of crops (private peasant holdings, Meher season). Statistical bulletin 388. CSA, Addis Ababa, Ethiopia. 13

8. Martin JH, Leonard WH, Stamp DL (1976) Principles of field crop production. Newyork: Mcmillan pub.co

9. Ministry of Agriculture and Rural development (MoARD) (2007) Crop development Department variety registration, Addis Ababa, Ethiopia. 7-22.

10. FAO (2008) Newsroom.

11. Dereje G, Yaynu H (2000) Yield losses of crops due to disease in Ethiopia. Pest Mgt Journal Eth. 5: 55-67.

12. Emebet F, Belayneh A, Kassaye Z (2005) Identification of sources of resistance to stem rust of wheat, P. graminis f.sp. tritici. Report of completed research project from 1999-2004, Volume I Pathology. PPRC, Ambo.

13. Centro International de Mejoramiento de Maize y Trigo (CIMMYT) (1989) The wheat revolution revisited: Recent trends and future challenges. In the1987/88 CIMMYT World Wheat Facts and Trends. Mexico, D. F. CIMMYT. 1-10.

14. Shank R (1994) Wheat stem rust and drought effects on Bale Agricultural production and future prospects. Summarized report. 17-18, February 1994. UNDP, Emergency Unit in Ethiopia, Addis Ababa, UNDP, 5.

15. Naod B (2004) Resistance Sources and Variation Among Ethiopian Cultivated Tetraploid Wheat (Triticum durum and T. dicoccum) to Stem Rust (P. graminis f. sp. tritici) Infection. Plant Scince M.Sc. Thesis presented to School of Graduate Studies of Alemaya University, Ethiopia.

16. Belayneh B, Lind V, Friedt W, Ordon F (2008) Virulence analysis of Pucciniagraminis f.sp. tritici populations in Ethiopia with special consideration of Ug 99.

17. Roelfs AP, Singh RP, Saari EE (1992) Rust Diseases of Wheat: Concepts and methods of disease management. Mexico, DF. CIMMYT. 81.

18. Tesfaye T, Getachew B (1991) Aspects of Ethiopian tetraploid wheats with emphasis On Durum wheat Genetics and Breeding research. In: Hailu G, Tanner DG, Mengistu H (eds.) Wheat Research in Ethiopia: A Historical Perspective. Addis Ababa. IAR /CIMMYT.

19. Mandefro A (2000) Influence of Micro environment in the development of Rust Epidemics, On Durm wheat (Triticum turgidum L. var durum) M.sc. Thesis. The School of Graduate Studies of Alemaya University, Ethiopia.

20. Serbessa N (2003) Wheat Stem Rust (P. graminis f. sp. tritici) Intensity and Pathogenic Variability in Arsi and Bale zones of Ethiopia. Plant Science MSc. Thesis presented to School of Graduate Studies of Alemaya University, Ethiopia.

21. Peterson RF, Campbell AR, Havnah AE (1948) A diagrammatic scale for estimating rust intensity on leaves and stem of cereals. Canad J Res C-26: 496-500 Pretorius, ZA.

22. Stubbs RW, Prestcott JM, Saari EE, Dubin HJ (1986) Cereal disease Methodology Manual CIMMYT, Mexico. 46.

23. Stakman EC, Stewart DM, Loegering WQ (1962) Identification of physiologic races of Puccinia graminis var. tritici. USDA ARS, E716. United States Government Printing Office: Washington.

24. Emebet F, Belayneh A, Kassaye Z (2004) Identification of sources of resistance to stem rust of wheat, P. graminis f.sp. tritici. Report of completed research project from 1999-2004, Volume I Pathology. PPRC, Ambo.

25. Mengistu H, Yeshi A (1992) Stability of the reaction of wheat Differential lines to stem and leaf rust at Debrezeit Ethiopia. In: Tanner DG, Mwangi W (eds.) Seventh Regional wheat workshop for Eastern, Central and South Africa. Nakuru, Kenya CIMMYT.

19

Interaction Effects of 6-Benzylaminopurine and Kinetin on *In vitro* Shoot Multiplication of Two Sugarcane (*Saccharum officinarum L.*) Genotypes

Dereje Shimelis[1]*, Kassahun Bantte[2] and Tilaye Feyissa[3]

[1]Ethiopian Sugar Corporation, Research and Training Division, Variety Development Directorate, Biotechnology Research Team, Wonji Research Center, P. O. box 15, Wonji, Ethiopia
[2]Jimma University College of Agriculture and Veterinary Medicine, Jimma, Ethiopia
[3]Addis Abeba University, Science and Technology faculty, Addis Abeba, Ethiopia

Abstract

In Ethiopia, sugarcane is grown as an important cash and industrial crop. It is not an ideal crop for conventional breeding and it lacks rapid multiplication procedures to commercialize newly released varieties within a short period of time. Hence, the objective of this work was to optimize the optimum concentration of 6-benzylaminopurine (0.0, 0. 5, 1.0, 1.5, 2.0, 2.5, and 3.0 mgL-l) and Kinetin (0.0, 0.1, 0.5, and 1.0 mgL-l) combination for shoot multiplication of C86-12 and C86-56 genotypes in completely randomized design with 5x4x2 factorial treatment combinations arrangements. The analysis of variance showed that the interaction effects of BAP, kinetin and genotypes on the number of shoots per explant, number of leaves per shoot and average shoot length were highly significant ($p < 0.001$). Murashige and Skoog (MS) medium supplemented with 1.5 mgL-l BAP and 0.5 mgL-1 of Kin for B86-12; and 1.5 mgL-1 of BAP and 1.0 mgL-1 of Kin for C86-56 were found to be the optimum media for shoot multiplication. B86-12 showed 33.8 ± 0.837 number of shoots per explant with 13.04 ± 0.089 average number of leaves per shoot and 8.4 ± 0.008 cm shoot length whereas C86-56; 25.6 ± 0.548 number of shoots per explant with average number of leaves per shoot of 9.8 ± 0.447 and shoot length 8.65 ± 0.72 cm was obtained after 30 days of sub culturing. Thus, the optimized protocol can be used for rapid multiplication of the planting materials for commercializing the newly released genotypes within a short period of time.

Keywords: Sugarcane; BAP; Kin; Shoot multiplication

Introduction

Sugarcane (*Saccharium officinarum L.*) belongs to the Saccharium genus of the Andropogoneae tribe of the Poaceae (Gramineae/grass) family with an octaploid 2n=8x=80 number of chromosomes [1]. It is a tall perennial tropical grass that tillers at the base to produce unbranched stems, 3-4 m or more in height with a thickness of approximately 5 cm in diameter. It is one of the most efficient photosynthesizing plant or converter of solar energy to sugar stored in the internode [2]. The commercially cultivated crops of sugarcane have two geographic centers of origin in New Guinea and Northern India [3]. Although the major industries are found in Brazil, China and India, the crop is also commercially produced in many other countries, including South Africa [4]. But there is no well documented reference on how, where and when sugarcane was introduced to Ethiopia, although some records claim its introduction during the early 18th century [5].

Sugarcane accounts approximately 75% of the world's sugar and it is economically important cash crop in tropical and sub-tropical regions of many countries [6]. Its properties such as efficient photosynthesis and biomass production make it an excellent target for industrial processing, valuable alternative for animal feed and for the production of by-product such as ethanol from molasses [7]. In Ethiopia, this crop is grown as an important cash and industrial crop among many crops and it has an immense importance for the development of the economy of the country. It is used for the production of white and brown sugar and by-products like molasses, bagasse and press mud (filter cake) which have been used for different purposes in a daily life and there is no by-product thrown as non- useful matter. Furthermore, production of sugar in Ethiopia has created employment opportunities and foreign currency.

Day by a day increasing use of sugar and its relevant by-products has created a challenging situation for sugar producing countries,

researchers and growers [8]. In Ethiopia the annual yield of sugar from three factories was nearly around 300,000 tons while the annual domestic demand is close to 450,000 tons [9] and the deficit was covered by importing from abroad. In addition, the yield per hectare of this crop is the lowest all over the world [8]. Considering the availability of abundant water resource coupled with a vast fertile land favorable for sugarcane cultivation, suitable agro-ecological conditions, cheap labor and huge domestic and foreign demand for sugar and for its by-products [9], this yield is very minimal in the country. Hence, by considering these opportunities the government has planned to establish ten sugar factories on 370,000 ha of plantation area.

Yield potential of sugarcane varieties is deteriorating day by day due to segregation, susceptibility to diseases, insects, admixture, and changes in edaphic and climatic factors [10]. Improvement of sugarcane varieties is very difficult, because it is not an ideal candidate crop for conventional plant breeding, since its flowering is not-synchronized, it has low sexual seed viability and it is a perennial crop [11]. Hence, its improvement takes up to ten years from initial crosses to final agronomic assessments [12,3]. But also the lack of rapid multiplication

*Corresponding author: Dereje Shimelis, Ethiopian Sugar Corporation, Research and Training Division, Variety Development Directorate, Biotechnology Research Team, Wonji Research Center, P. O. box 15, Wonji, Ethiopia
E-mail: d.shimelis@yahoo.com

procedure has long been a serious problem in sugarcane conventional breeding programs as it takes 10-15 years of work to complete a selection cycle [13,8]. Commercially, sugarcane is propagated from stem cutting with each cutting or set having two or three buds [14,15]. Vegetative propagation by cutting is a very low rate of propagation which is about 1:6 to 1:8 [15] and 1:7 to 1:10[16,17]. In addition to low rate of propagation on an open field, it favors pathogens to keep on accumulating generation after generation which reduces the yield and quality of sugarcane [14,18]. For instance Ratoon stunting disease is a common disease in sugarcane and conventionally it is treated with hot water that could be ineffective or could damage the set [19]. Hence, availability of quality planting material of newly released varieties is a major constraint in their adoption and commercialization within a short period of time. The time spent for conventional multiplication is considered as a serious economic problem, mainly in view of the higher yields that would be obtained by planting the new variety earlier on a large commercial scale, therefore efficient propagation systems are mandatory for mass multiplication.

Tissue culture of sugarcane has got a considerable research attention because of its economic importance as a cash crop. Plant multiplication or regeneration via tissue culture is a viable alternative for improving the quality and quantity of sugarcane [20]. Plant tissue culture (Micro propagation) holds immense potential for mass multiplication, subsequent rejuvenation and quality production of sugarcane [21]. By in vitro propagation, it is possible to produce some 260,000 shoots in four months [22] and $2x10^8$ plantlets within 4-5 weeks [14] from single shoot tip of sugarcane. It is demonstrated that micro propagated system exhibited a potential in vitro production of 75600 shoots from a single shoot apex explant in a period of about 5.5 months [23]. It is reported that around 2500 seedlings could be generated from one bud in a 12 week period on MS medium supplemented with 1.5 mgL^{-1} of BAP and GA$_3$[24]. Rapid micro propagation is also achieved [25] by producing 78408 plantlets in three months on MS media supplemented with BAP (0.2 mgL^{-1}) and Kinetin (0.1 mgL^{-1}) and he conclude that by using tissue culture it would be possible to commercialize a new variety within 1-2 years. In Ethiopia, there is no sugar breeding facility and new varieties have be imported by sett and propagated for commercialization by cutting so far. Hence, this experiment was launched to optimize a protocol for in vitro mass propagation of newly introduced genotypes to supplement the conventional vegetative propagation.

Materials and Methods

The experiment was conducted at Plant Tissue Culture Laboratory of Jimma University College of Agriculture and Veterinary Medicine (JUCAVM), Ethiopia. Two sugarcane genotypes (C86-12 and C86-56) were used for the study. They were imported from Cuba in 2006 and passed through agronomic performance evaluation. They were among the selected ones to be commercialized. The sets of these genotypes were prepared and treated with hot water. The setts were taken to JUCAVM green house and planted. After two to three months of growing, shoot tip explants were taken from the sugarcane plants. The explants were prepared according to [14] procedures. The surrounding leaf sheaths of sugarcane tops were carefully removed one by one until the inner white sheaths were exposed. The explants were sized to about 10 cm length by cutting off at the two ends, locating the growing point somewhere near to top. They were washed under running tap water and liquid detergents. They were socked in fungicide solution (0.3% kocid) for 30 minutes under laminar flow cabinet containing three drops of tween-20. After the kocid was properly washed off from the explants, they were rinsed three times with distilled water and disinfected with

70% ethanol for one minute. The ethanol was poured off and the explants were rinsed again with sterile distilled water. Disinfection of explants was done with 0.1% of HgCl$_2$ for 10 minutes [26] followed by 3-4 washing with sterile distilled water. The required amounts of all stock solutions of MS [27] media, 30 gL^{-1} [28] sucrose and combinations of different concentrations of BAP and Kin were mixed in a beaker and the pH was adjusted to be 5.8. This was followed by addition of 0.8% agar for solidifying the media. Then, it was heated to melt the agar and 30 ml media was dispensed in to culture jars. Finally, it was autoclaved at temperature of 121°C for 20 minutes with 15 psi of pressure.

Initiated explants were cultured under laminar flow hood aseptically and transferred to the growth room at which growth conditions were adjusted to be 16 hours of photoperiod with 25 μmolm^{-2}s^{-1} photosynthetic photon flux intensity and 26 ± 2°C of temperature. The experiment was laid down in factorial treatment combination in complete randomized design with two factor factorial treatment combination arrangements. Each of treatment was replicated three times. Data on number of shoots per explant, number of leaves per shoot and shoot length were collected after 30 days of culturing. Finally data were subjected to two-way analysis of variance (ANOVA) using SAS statistical software version 9.2 (SAS Inc., 2008) and treatments' means were separated by using REGWQ (Ryan-Einot-Gabreil-Welsch Multiple range test) mean separation method.

Results and Discussions

Analysis of variance revealed that the interaction among BAP and Kin combinations and genotypes was highly significant (p=0.001) on number of shoots per explant, number of leaves per shoot and shoot length. MS medium without PGRs did not result in shoot multiplication on both genotypes (Table 1). However, increasing the concentration of kinetin from 0.0 mgL^{-1} to 1 mgL^{-1} without BAP increased the mean number of shoots per explant from 0.0 ± 0.0 to 10.2 ± 0.445 and 0.0 ± 0.0 to 13.6 ± 0.548 for C86-12 and C86-56 respectively. Similarly, increasing the concentration of BAP from 0.0 mgL^{-1} to 2.0 mgL^{-1} for C86-12 and 0.0 mgL^{-1} to 1.0 mgL^{-1} for C86-56 without Kin showed a significant increase in the mean number of shoots per explant from 0.0 ± 0.0 to 17.8 ± 0.447 and 0.00 ± 0.00 to 14.0 ± 0.707 respectively. This showed that addition of exogenous PGRs is a must to have shoot multiplication. Moreover, the increasing trend in shoot number per explant is due to the fact that cytokinin (BAP and Kin) stimulate protein synthesis and participate in cell cycle control in a cell division [29]. If cytokines are used for shoot culture media, they can overcome apical dominance and release lateral buds from dormancy and enhance shoot multiplication [29].

From the two genotypes, C86-12 gave higher mean number of shoots per explant (33.8 ± 0.837) with 13.04 ± 0.089 mean number of leaves per shoot and mean shoot length of 8.4 ± 0.008 cm on MS medium supplemented with 1.5 mgL^{-1} of BAP and 0.5 mgL^{-1} of Kin (Table 1 and Figure 1A). With the same medium composition, C86-56 gave only 17.4 ± 0.548 mean number of shoots per explant with 8.4 ± 0.548 mean number of leaves per shoot and 3.22 ± 0.567 cm mean shoot length. In this genotype, the highest mean number of shoots per explant (25.6 ± 0.548) was obtained with mean number of leaves per shoot of 9.8 ± 0.447 and mean shoot length of 8.65 ± 0.724 cm on MS medium supplemented with 1.5 mgL^{-1} BAP and 1.0 mgL^{-1} Kin(Table 1 Figure 1B). However, the same medium in C86-12 resulted in 20.0 ± 0.707 mean number of shoots per explant; 6.3 ± 0.447 mean number of leaves per shoot and 4.27 ± 0.013 cm mean shoot length. For C86-12, as the concentration of kinetin increased from 0.0 mgL^{-1} to 0.5 mgL^{-1}

Figure 1: Shoot multiplication showing best results of
A: C86-12 genotype on MS medium containing 1.5 mgL⁻¹ BAP and 0.5 mgL⁻¹ Kin
B: C86-56 genotype on MS medium containing 1.5 mgL⁻¹ BAP and 1 mgL⁻¹ Kin.

PGRs(mgl⁻¹)		C86-12			C86-56		
BAP	Kin	Number of shoots per explant ± SD	Number of leaves per shoot ± SD	Shoot length(cm) ± SD	Number of shoot per explant ± SD	Number of leaves per shoot ± SD	Shoot length(cm) ± SD
0	0	$0.0^{s} \pm 0.000$	$0.0^{p} \pm 0.00$	$0.0^{v} \pm 0.00$	$0.00^{s} \pm 0.00$	$0.00^{p} \pm 0.00$	$0.00^{v} \pm 0.00$
0	0.1	$2.2^{rs} \pm 0.447$	$3.9^{o} \pm 0.224$	$3.52^{m-p} \pm 0..013$	$3.00^{qr} \pm 0.00$	$3.98^{o} \pm 0.044$	$3.04^{p-t} \pm 0.089$
0	0.5	$8.2^{p} \pm 0.0.447$	$5.1^{mn} \pm 0.224$	$7.04^{bc} \pm 0.089$	$5.2^{q} \pm 0.447$	$5.08^{mn} \pm 0.179$	$4.04^{h-n} \pm 0.094$
0	1	$10.2^{n-p} \pm 0.445$	$6.06^{i-m} \pm 0.134$	$4.72^{e-g} \pm 0.013$	$13.6^{h-m} \pm 0.548$	$4.7^{no} \pm 0.975$	$5.74^{d} \pm 0.004$
1	0	$12.8^{i-n} \pm 0.433$	$7.8^{f-h} \pm 0.477$	$5.47^{de} \pm 0.241$	$14.0^{h-l} \pm 0.707$	$9.28^{b-e} \pm 0.438$	$4.44^{g-j} \pm 0.458$
1	0.1	$16.2^{f-h} \pm 0.447$	$7.1^{g-j} \pm 0.894$	$5.8^{d} \pm 0.811$	$9.1^{op} \pm 0.224$	$8.24^{e-g} \pm 0.537$	$4.57^{f-i} \pm 0.297$
1	0.5	$15.1^{g-j} \pm 0.548$	$10.4^{b} \pm 0.548$	$6.76^{c} \pm 1.327$	$20.2^{cd} \pm 0.834$	$9.42^{b-e} \pm 0.83$	$5.19^{f-i} \pm 0.495$
1	1	$15.0^{g-j} \pm 1.225$	$9.4^{b-e} \pm 0.548$	$5.22^{d-f} \pm 0.367$	$17.8^{d-f} \pm 1.095$	$8.9^{c-f} \pm 0.549$	$3.77^{j-p} \pm 0.223$
1.5	0	$15.9^{f-h} \pm 0.224$	$8.0^{f-h} \pm 0.000$	$3.82^{i-o} \pm 0.008$	$12.2^{i-n} \pm 0.447$	$8.0^{f-h} \pm 0.000$	$3.28^{n-r} \pm 0.008$
1.5	0.1	$16.9^{fg} \pm 0.224$	$9.0^{c-f} \pm 0.000$	$4.04^{h-n} \pm 0.089$	$14.0^{h-l} \pm 0.00$	$8.8^{d-f} \pm 0.447$	$3.59^{l-p} \pm 0.004$
1.5	0.5	$33.8^{a} \pm 0.837$	$13.04^{a} \pm 0.089$	$8.4^{a} \pm 0.008$	$17.4^{e-g} \pm 0.548$	$8.4^{ef} \pm 0.548$	$3.22^{o-s} \pm 0.567$
1.5	1	$20.0^{cd} \pm 0.707$	$6.3^{i-l} \pm 0.447$	$4.27^{g-m} \pm 0.013$	$25.6^{b} \pm 0.548$	$9.8^{b-d} \pm 0.447$	$8.65^{a} \pm 0.724$
2	0	$17.8^{d-f} \pm 0.447$	$9.2^{b-e} \pm 0.433$	$4.48^{f-j} \pm 0.171$	$12.8^{i-n} \pm 0.447$	$7.00^{h-j} \pm 0.707$	$3.29^{h-q} \pm 0.350$
2	0.1	$16.0^{f-h} \pm 0.00$	$7.88^{f-h} \pm 0.521$	$3.45^{n-p} \pm 0.172$	$12.8^{i-n} \pm 1.095$	$6.4^{i-l} \pm 0.548$	$3.91^{i-o} \pm 0.004$
2	0.5	$12.8^{i-n} \pm 0.447$	$6.62^{ij} \pm 0.567$	$4.31^{g-j} \pm 0.050$	$17.0^{fg} \pm 2.121$	$6.4^{i-l} \pm 0.548$	$4.41^{g-j} \pm 0.004$
2	1	$10.2^{n-p} \pm 0.447$	$9.78^{b-d} \pm 0.491$	$4.84^{e-g} \pm 0.014$	$11.8^{k-n} \pm 0.837$	$9.00^{c-f} \pm 0.00$	$4.89^{e-g} \pm 0.007$
2.5	0	$11.0^{m-o} \pm 0.00$	$6.16^{i-m} \pm 0.447$	$2.57^{q-t} \pm 0.108$	$12.8^{i-n} \pm 0.447$	$6.6^{ij} \pm 0.548$	$2.52^{st} \pm 0.005$
2.5	0.1	$11.2^{m-o} \pm 0.447$	$6.84^{h-j} \pm 1.314$	$2.41^{tu} \pm 0.101$	$14.2^{g-j} \pm 1.923$	$6.4^{i-l} \pm 0.548$	$2.67^{q-t} \pm 0.004$
2.5	0.5	$10.2^{n-p} \pm 0.837$	$5.26^{k-n} \pm 0.581$	$2.42^{tu} \pm 0.121$	$12.4^{i-n} \pm 3.647$	$5.8^{i-n} \pm 0.447$	$2.73^{q-t} \pm 0.039$
2.5	1	$17.2^{e-g} \pm 0.447$	$7.16^{g-j} \pm 0.851$	$4.96^{e-g} \pm 0.604$	$14.8^{h-k} \pm 3.271$	$6.6^{e-l} \pm 0.547$	$4.38^{g-k} \pm 0.000$
3	0	$20.6^{c} \pm 0.548$	$10.02^{bc} \pm 0.447$	$7.46^{b} \pm 0.380$	$13.2^{i-m} \pm 2.588$	$5.2^{l-m} \pm 0.447$	$3.96^{i-o} \pm 0.054$
3	0.1	$12.8^{i-n} \pm 0.00$	$6.34^{i-l} \pm 0421$	$3.64^{k-p} \pm 0.215$	$19.6^{c-e} \pm 3.286$	$6.66^{ij} \pm 0.615$	$3.74^{i-p} \pm 0.004$
3	0.5	12.0 ± 0.447	$6.46^{i-k} \pm 0.639$	$2.55^{r-t} \pm 0.152$	$15.08^{g-j} \pm 0.179$	$7.1^{g-j} \pm 0.224$	$2.71^{q-t} \pm 0.004$
3	1	$11.4^{l-o} \pm 0.548$	$6.6^{ij} \pm 0.616$	$2.38^{tu} \pm 0.225$	$11.4^{l-o} \pm 0.548$	$6.00^{i-m} \pm 0.00$	$1.73^{u} \pm 0.02$
CV	(%)	8.45	7.25	7.89	8.45	7.25	7.89

Table 1: Effects of 6-benzylaminopurine and kinetin on shoot multiplication
PGRs=Plant growth regulators. Values for number of shoots per explant, number of leaves per shoot and shoot length given as mean ± SD. Numbers with in the same column with different letter(s) are significantly different from each other according to REGWQ at p ≤ 0.05.

keeping BAP at 1.5 mgL^{-1}, the mean numbers of shoots per explant, mean number leaves per shoot and mean shoot length showed a significant increase from 15.9 ± 0.224 to 33.8 ± 0.837, 8.0 ± 0.000 to 13.04 ± 0.089 and 3.82 ± 0.008 to 8.4 ± 0.008 respectively. However, for C86-56 only mean number of shoots per explant showed a significant increment from 12.2 ± 0.447 to 17.4 ± 0.548.

The best result obtained in C86-12 is in agreement with the result reported by [30]. They reported that optimum multiplication from HSF-240 genotype exhibited 16.5 cm mean shoot length, 11 number of shoots per explant and 32 leaves per explant on medium supplemented with 1.5 mgL^{-1} BAP, 0.5 mgL^{-1} Kin and 30 gL^{-1} sucrose after 30 days of culturing. Though they found higher number of leaves per shoot and shoot length, the present study is better in terms of mean shoot number per explant. This difference could be due to genotypic difference. The best results in both genotypes of the present study contradict with results reported in [31-33]. Best results were obtained from CO678 genotype on MS medium supplemented with 2 mgL^{-1} BAP+0.5 mgL^{-1} Kin with 9.1 ± 0.1 mean number of shoots, 6.83 ± 0.12 mean shoot length and 5.67 ± 0.04 leaves per shoot. He also obtained 7.87 ± 1.06 mean number of shoots, 5.44 ± 0.19 mean number of leaves and 6.33 ± 0.21 mean shoot length on MS medium supplemented with 2 mgL^{-1} BAP+0.25 mgL^{-1} kin+30 gL^{-1} sucrose from Co449 genotype but from both genotypes he reported much less number of shoot per explant than the result of this study[31]. Tilahun M (2011) [31] reported more number of shoots per explant (34 ± 1.54) than the current result but with less number of leaves per shoot (12 ± 0.17) and shoot length (6.95 ± 0.01 cm) on MS medium supplemented with 1 mgL^{-1} BAP+0.5 mgL^{-1} Kin+30 gL^{-1} sucrose for B41-227 genotype. Comparable mean shoot number per explant (29.7 ± 1.0069) from BL-4 genotype on MS medium supplemented with 0.25 mgL^{-1} BAP and Kin each was reported [8]. These differences happened because it is an established fact that different genotypes may give different results on MS medium supplemented with different concentrations of plant growth regulator and combinations. Sharma M [33] found 20 ± 0.15; 24 ± 0.22 mean number of shoots per explant and 7.0 ± 0.27; 7.4 ± 0.06 mean shoot length for CoJ 83 and CoS 8436 genotypes respectively after 21 days of culturing on MS medium supplemented with 1.0 mgL^{-1} BAP+1.5 mgL^{-1} Kin+30 gL^{-1} sucrose . The difference is not only due to genotypic variation but also due to the number of days taken for culturing.

1. Therefore, the best results obtained on MS medium supplemented with1.5 mgL^{-1} BAP and 0.5 mgL^{-1} kin for B85-12 and 1.5 mgL^{-1} BAP and 1 mgL^{-1} kinetin for C86-56 showed genotypic difference in relation to concentrations of BAP and Kinetin combinations to be used for optimal shoot multiplication. It is because of the fact that different genotypes possess specific receptor proteins and differed in concentration for plant growth regulators [29]

Conclusion

The present results showed that MS medium fortified with 1.5 mgL^{-1} BAP and 0.5 mgL^{-1} of Kin for B86-12; and 1.5 mgL^{-1} of BAP and 1.0 mgL^{-1} of Kin for C86-56 were found to be the optimum media for shoot multiplication. Hence, by using these media combinations (protocol), these genotypes can be commercialized within a short period of time and supplement the conventional propagation which improves both the quality and quantity of the planting materials.

Acknowledgments

We would like to gratitude the Ethiopian sugar corporation for funding the research budget and, Jimma University College of Agriculture and Veterinary Medicine (JUCAVM) for providing the Tissue Culture laboratory.

References

1. Ather A, Khan S, Rehman A, Nazir M (2009) Optimization of the Protocols for Callus induction, Regeneration and Acclimatization of sugarcane cv. Thatta-10. Pak. J. Bot. 41: 815-820.

2. Naturland EV (2000) Organic farming in the tropics and Sub tropics. Kleinhaderner, Germany 10-15.

3. Sengar K, Garg SK (2011) Role of tissue culture technique in high sugarcane production. A review Life Sciences 1008-1017.

4. Snyman SJ, Meyer GM, Koch AC, Banasiak M, Watt MP (2011) Applications of in vitro culture systems for commercial sugarcane production and improvement. In Vitro Cell. Biol. Plant. 47: 234-249

5. Duri M (1969) Private Foreign Investment in Ethiopia. J. Eth Studies 7: 53-73.

6. Pandey RN, Rastogi J, Sharma ML, Singh K (2011) Technologies for Cost Reduction in Sugarcane Micro propagation. African Journal of Biotechnology 10: 7814-7819.

7. Gallo-Meagher M, English RG, Abouzid A (2000) Thidiazuron Stimulates Shoot Regeneration of Sugarcane Embryogenic Callus. In Vitro Cell Dev Biol-Plant 36: 37-40.

8. Ali A, Naz Sh, Siddiqui FA, Iqbal J (2008) An Efficient Protocol for Large Scale Production of Sugarcane through Micropropagation. Pak. J. Bot. 40: 139-149.

9. Anonymous (2010) F.O, licht's International Sugar and Sweetener Report.

10. Malik KB (1990) Proposal for Approval of BF-162. Sugarcane Research Institute, Faisalabad 7-10.

11. Manickavasagam M, Ganpati A, Anbazhagan VR, Sudhakar B, Selvaraj N, et.al. (2004) Agrobacterium mediated genetic transformation and development herbicide resistant sugarcane (Saccharum species hybrids) using auxiliary buds. Plant Cell Rep. 23: 134-43.

12. Cox M, Hogarth M, Smith G (2000) Cane breeding and improvement. Manual of cane growing, Bureau of sugar Experimental Stations, Indooroopilly, Australia.

13. Birader S, Biradar DP, Patil VC, Patil SS, Kambar NS (2009) In vitro plant regeneration using shoot tip culture in commercial cultivar of sugarcane. Karnataka J. Agric. Sci. 22: 21-24.

14. Cheema KL, Hussain M (2004) Micropropagation of Sugarcane through Apical Bud and Axillary Bud. Int J Agric Biol 6: 257-259.

15. Jalaja NC, Neelamathi D, Sreenivasan, TV (2008) Micropropagation for quality seed Production in Sugarcane in Asia and the Pacific. Sugarcane pub 13-60.

16. Dash M, Mishra PK, Mohapatra D (2011) Mass propagation via shoot tip culture and detection of genetic variability of Saccharium Officinarum clones using biochemical markers. Asian Journal of Biotechnology 10:1996-0700.

17. Gosal SS, Thind KS, Dhaliwal HS (1998) Micropropagation of sugarcane an efficient protocol for commercial plant production. Crop Impro. 25: 167-171.

18. Nand L, Singh HN (1994) Rapid clonal multiplication of sugarcane through tissue culture. Plant Tissue Cult 4:1-7.

19. Hoy JW, Bischoff KP, Milligan SB, Gravois KA (2003) Effect of tissue culture explant source on sugarcane yield components. Euphytica 129: 237-240.

20. Baksha R, Alam R, Karim MZ, Paul SK, Hossain MA, et.al. (2002) In vitro Shoot Tip Culture of Sugar-cane (Saccharium officinarum) Variety Isd. Biotechnology 1: 67-72.

21. Heinz DJ, Mee GWP (1969) Plant differentiation from callus tissue of Saccharium species. Crop Sci. 9: 346-348.

22. Hendre RR, Iyer RS, Kotwal M (1983) Rapid multiplication of sugarcane by tissue culture. Sugarcane 1: 58.

23. Lal J, Pande HP, Awasthi SK (1996) A general Micropropagation protocol for sugarcane varieties. New Bot. 23: 13-19.

24. Chattha MA, Imran MI, Abida A, Muhammad I, Akhtar A (2001) Micropropagation of sugarcane (Saccharum sp.). Pakistan Sugar J. 16: 2-6.

25. Lee TSG, (1987) Micropropagation of sugarcane (Saccharium sp.) Plant Cell, Tissue and Organ culture 10: 47-55.

26. Bisht SS, Routray AK, Mishra R (2011) Rapid in vitro propagation techniques for sugarcane variety 018. IJPBS 2: 0975-6299.

27. Murashige T, Skoog F (1962) A Revised Medium for Rapid Growth and Bioassay with Tobacco Cultures. Physiol. Plant. 15: 473-479.

28. Khan SA, Rashid A, Chaudhary, MF, Chaudhry Z, Afroz A (2008) Rapid Micropropagation of three elite Sugarcane (*Saccharium officinarum L.*) varieties by shoot tip Culture. AJB 7: 2174-2180.

29. George EF, Machakova I, Zazimalova E (2008) Plant propagation by tissue culture. 3rd edition 175-205.

30. Khan SA, Rashid H, Chaudhary MF, Chaudhry Z, Fatima Z et.al. (2009) Effect of Cytokinin on Shoot Multiplication in Three Elite Sugarcane Varieties. Pak. J. Bot. 41: 1651-1658.

31. Tilahun M (2011) Protocol Optimization for In Vitro Mass Propagation of Two Sugarcane (Saccharum officinarum L.) Clones Grown in Ethiopia. An MSc Thesis Presented to School of Graduates Studies of Jimma University College of Agriculture and Veterinary Medicine.

32. Belay T, Mulugeta, B and Derbew, B (2014) Effects of 6-Benzyl amino purine and Kinetin on In Vitro Shoot Multiplication of Sugarcane (Saccharum officinarum L.) Varieties. Adv Crop Sci. Tech. 2: 129

33. Sharma M (2005) In vitro regeneration studies of sugarcane. An MSc Thesis, Patiala, India 24-32.

20

Marker Assisted Selection: Biotechnology Tool for Rice Molecular Breeding

Y.M.A.M. Wijerathna*

The Division for International Studies, Robert H. Smith Faculty of Agriculture, Food and Environment, The Hebrew University of Jerusalem. P.O.Box 12, Rehovot 76100, Israel.

Abstract

The biotechnology tool of MAS has irreversibly changed the disciplines of conventional rice breeding. Molecular markers are indispensable tools for measuring the diversity of rice varieties and rice breeding. However, MAS is not always advantageous, so careful analysis of the costs, convenience, ease of assay development and automation are important factors to be considered when choosing a technology relative to the conventional breeding programs. This review focuses on possibilities for the application of marker-assisted selection in the genetic improvement of rice breeding.

Keywords: Rice; Biotechnology; Molecular breeding; Hybridization

Introduction

Rice is a dietary staple food for at least 62.8% of the world population. In Asia it accounts for 29.3% [1]. Global rice consumption is projected to increase from 450 million tons in 2011 to about 490 million tons in 2020 and to about 650 million tons by 2050 [2]. The main challenge encountered by scientists involved in rice research and production in the world is to find appropriate solutions for major issues such as the impacts of climate change viz. temperate, water use efficiency and availability of pollution will play a key role in determining food security in large parts of the world. Also the environmental crisis and plant diseases and pests have been the factors that decrease rice production in many countries all around the world [3]. At this time when the world's population is increasing rapidly and the demand for food is high, these problems have threatened food security and people health worldwide. In order to meet these growing problems in future ahead, it is necessary to use rice varieties with higher yield potential, durable resistance to diseases and insects and tolerance to abiotic stresses.

Yield potential of rice can be improved with the help of various strategies; conventional hybridization and selection procedures, ideotype breeding, heterosis breeding, wide hybridization and molecular breeding [4]. There are two strategies in biotechnological application in molecular rice breeding; one is by Marker-Assisted Selection (MAS), also called marker-assisted breeding (MAB) and the other one is by developing the Genetically Modified crops. A genetic marker is any visible character or otherwise assayable phenotype, for which alleles at individual loci segregate in a Mendelian manner. MAS is a technique that does not replace traditional breeding, but can help to make it more efficient. It does not include the transfer of isolated gene sequences such as genetic engineering, but offers tools for targeted selection of the existing plant material for further breeding.

The genetic markers covered include (1) morphological markers (2) biochemical markers (alloenzymes and other protein markers) and (3) molecular markers (based on DNA-DNA hybridization).

DNA-based molecular markers

DNA marker is a small region of DNA sequence showing polymorphism between different individuals. They arise from different classes of DNA mutations such as substitution mutations (point mutations), rearrangements (insertions or deletions) or errors in replication of tandemly repeated DNA [5]. These markers are selectively neutral because they are usually located in non-coding regions of DNA. DNA markers are the most widely used type of marker predominantly due to their abundance. Unlike morphological and biochemical markers, DNA markers are practically unlimited in number and are not affected by environmental factors and/or the developmental stage of the plant [6].

Properties which desirable for ideal DNA markers include highly polymorphic nature, codominant inheritance (determination of homozygous and heterozygous states of diploid organisms), frequent occurrence in the genome, selective neutral behavior (the DNA sequences of any organism are neutral to environmental conditions or management practices), easy access (availability), easy and fast assay, high reproducibility, and easy exchange of data between laboratories [7]. Also should follow Mendelian inheritance, genetically linked to trait in question and not affected by pleiotropism and epistatic interactions [8].

There are two basic methods to detect the polymorphism: Southern blotting, a nuclear acid hybridization technique and polymerase chain reaction (PCR) technique. Using PCR and/or molecular hybridization followed by electrophoresis (e.g. Polyacrylamide gel electrophoresis, Agarose gel electrophoresis, Capillary electrophoresis), the variation in DNA samples or polymorphism for a specific region of DNA sequence can be identified based on the product features, such as band size and mobility [9].

Among the techniques that have been extensively used on plant breeding, are the Restriction Fragment Length Polymorphism (RFLP), Amplified Fragment Length Polymorphism (AFLP), Random Amplified

*Corresponding author: Y.M.A.M. Wijerathna, the Division for International Studies, Robert H. Smith Faculty of Agriculture, Food and Environment, the Hebrew University of Jerusalem. P.O.Box 12, Rehovot 76100, Israel
E-mail: akila.yapa@mail.huji.ac.il.

Polymorphic DNA (RAPD), Microsatellites Or Simple Sequence Repeat (SSR), Inter Simple Sequence Repeat (ISSR), Expressed Sequence Tag (EST), Cleaved Amplified Polymorphic Sequence (CAPS), Diversity Arrays Technology (DArT), Sequence Characterized Regions (SCARs), Sequence Tag Sites (STSs) and Single Nucleotide Polymorphism (SNP) [10,11].According to a causal similarity of SNPs with some of these marker systems and fundamental difference with several other marker systems, the molecular markers can also be classified into SNPs (due to sequence variation, e.g. RFLP) and non-SNPs (due to length variation, e.g. SSR) [10]. RFLP is the most widely used hybridization-based molecular marker [12]. The various PCR-based techniques are of two types depending on the primers used for amplification: 1) Arbitrary or semi-arbitrary primed PCR techniques that developed without prior sequence information (e.g., AP-PCR, DAF, RAPD, AFLP, ISSR). 2) Site targeted PCR techniques that developed from known DNA sequences (e.g., EST, CAPS, SSR, SCAR, STS, SNP)[13,14].

PCR-based markers are more attractive for MAS, due to the small amount of template required and more efficient handling of large population sizes. AFLP, RAPD and Sequence tagged site (STS) are dominant markers, which limits its application for differentiation of homozygous and heterozygous individuals in segregating progenies. Among the DNA markers, the most widely used markers in major cereal crops are SSRs or microsatellites [14,15].

Marker assisted selection

Marker assisted evaluation of breeding materials involves cultivar identity, assessment of purity and genetic diversity, parental selection, study of heterosis and identification of genomic regions under selection [16]. MAS refer to the use of DNA markers that are tightly-linked to target loci as a substitute for or to assist phenotypic screening. These DNA markers should reliably predict phenotype. By determining the allele of a DNA marker, plants that possess particular genes or quantitative trait loci (QTLs) may be identified based on their genotype rather than their phenotype. A marker can either be located within the gene of interest or be linked to a gene determining a trait of interest, which is the most common case. Thus MAS can be defined as selection for a trait based on genotype using associated markers rather than the phenotype of the trait [17].

Molecular markers can be used in many steps of a rice breeding program, e.g. germplasm characterization, pedigree and evolution studies, parental selection for crossing, test for F1 hybrid confirmation, test for genetic purity of seeds, cultivar protection, breeding strategies establishment, link- age map construction, and mapping of genes and QTLs associated with biological processes.

Application of DNA markers in MAS indicated in five main considerations viz. i.) Reliability: Molecular markers should co-segregate or tightly linked to traits of interest, preferably less than 5 cM genetic distance. The use of flanking markers or intragenic markers will greatly increase the reliability of the markers to predict phenotype. ii.) DNA quantity and quality: Some marker techniques require large amounts and high quality DNA, which may sometimes be difficult to obtain in practice and this, adds to the cost of the procedures. iii.) Technical procedure: Molecular markers should have high reproducibility across laboratories and transferability between researchers. The level of simplicity and time required for the technique are critical considerations. High-throughput simple and quick methods are highly desirable. iv.) Level of polymorphism: Ideally, the marker should be highly polymorphic in breeding material and it should be co-dominant for differentiation of homozygous and heterozygous

individuals in segregating progenies. v.) Cost: Molecular markers should be user-friendly, cheap and easy to use for efficient screening of large populations. The marker assay must be cost-effective in order for MAS to be feasible [18].

Applications of MAS

With respect to important MAS strategies, three main uses of molecular markers in rice breeding can be emphasized: Marker assisted evaluation of breeding material, Marker assisted introgression and Marker-assisted pyramiding.

Marker assisted evaluation of breeding material

To improve early generation selection, markers should decrease the number of plants retained due to their early generation performance, and at the same time they should ensure a high probability of retaining superior lines [19]. Markers are also frequently used to select parents with desirable genes and gene combinations, and Marker-assisted recurrent selection (MARS) involve several successive generations of crossing individuals based on their genotypes. This type of evaluation has the potential to make parental selection more efficient, to expand the gene pool of modern cultivars and to speed up the development of new varieties [20].

Marker assisted introgression

Introgression is the procedure of the transfer of genetic information from one species to another as a result of hybridization between them and repeated backcrossing. The process, where a gene or a QTL from a population A is introduced to a population B by crossing A and B and then repeatedly backcrossing to B, is called introgression [21]. Here, molecular markers can be used to control the presence of the target gene or QTL and to accelerate the return of background genome to recipient type. Marker-assisted introgression is very effective for introgressing genes or QTLs from landraces and related wild species, because is reduces both the time needed to produce commercial cultivars and the risk of undesirable linkage drag with unwanted traits of the landrace or wild [22].

Marker-assisted pyramiding

Marker assisted pyramiding is the process of combining several genes together into one genotype and using DNA markers for selection [23].Gene pyramiding is a useful approach to the durability or level of pest and disease resistances, or to increase the level of abiotic stress tolerance. Genes controlling resistance to different races or biotypes of a pest or pathogen and genes contributing to agronomic or seed quality traits can be pyramided together to maximize the benefit of MAS through simultaneous improvement of several traits in an improved genetic background [22].

Discussion

Adequate genotyping and phenotyping are both important for the success of rice breeding with MAS. MAS can be used in any breeding method (e.g. backcross marker assisted method) for any single gene transfer procedure if reliable markers exist and the indirect selection is more advantageous than the direct selection of the trait.

Many agronomically important traits/genes of rice have been mapped with linked markers [Table 1]. When the selected trait is expressed late in plant development, like seeds and flower features or adult characters in crop with a juvenile period, faster selection process because an individual's phenotype can be predicted at a very early stage since screen can follow at the seedling stage or even as seeds rather

Table 1: Some of the success stories related to application of MAS in rice [28-37].

No.	Target trait Gene(s)/ QTL(s)	Type/name of marker(s) used	Remarks
1.	eating, cooking and sensory quality : Wx gene	microsatellite	MAS applied for Marker-assisted backcross breeding
2.	Os2AP or badh2 gene Wx gene	microsatellite	Marker Assisted Backcrossing
3.	Sub1, Wx gene, Osbadh2, SSIIa loci,	microsatellite	Ideotype breeding
4.	Bacterial blight (BB) resistance And Blast resistance xa5, xa13, Xa21 &Pi25	CAPS, STS	MAS applied for pyramiding multiple genes
5.	Blast resistance Pi-9	STS	Introgressed the broad-spectrum blast resistant gene Pi-9(t)
6.	Genetic Diversity Analysis	RAPD and SSR	SSR markers for the accurate determination of relationships between accessions
7.	Bacterial blight (BB) resistance xa13 and Xa21	CAPS , STS	Improved the two traditional BB-susceptible Basmati varieties
8.	Bacterial blight (BB) resistance And stem borer Xa21, Bt & Chitinase	STS	MAS applied for pyramiding of target traits. Bt gene and Chitinase gene originally
9.	BPH resistance Bph1 Bph2	STS	MAS applied for gene pyramiding
10.	Submergence tolerance Sub1QTL	SSR	MAS applied for backcross breeding

than having to wait for the individual to develop to a stage where the adult phenotype is apparent [24].With the fast and constant advance of molecular technologies, it is plausible to predict that the main constraint in the near future will be the ability of the breeder to make a high quality phetotyping. The total number of lines that need to be tested can be reduced. Since many lines can be discarded after MAS early in a breeding scheme, this permits more efficient use of glasshouse.

MAS using co-dominance markers (e.g. SSR and SNP) can allow effective selection of recessive alleles of desired traits in the heterozygous status. No selfing or test crossing is needed to detect the traits controlled by recessive alleles, thus saving time and accelerating breeding progress[25].

With comparison to transgenesis, MAS can be considering that there are not major issues of biosafety and intellectual property rights. MAS respect species barriers and is accepted by the public. Genetic engineering is not the most effective tool to develop crops with complex traits such as drought and salinity tolerance, nor is it necessary.

Startup expenses and labor costs are higher in many cases. Therefore, as other new methods of rice breeding like transgenic breeding or genetic manipulation do, MAS cannot replace conventional breeding but is and only is a supplementary addition to conventional breeding. High costs and technical or equipment demands of MAS will continue to be a major obstacle for its large-scale use in the near future, especially in the developing countries [26,27]. For marker assisted backcrossing, the initial cost of using markers would be more expensive compared to conventional breeding in the short term however time savings could lead to an accelerated variety release which could translate into greater profits in the medium to long term.

Conclusion

Adoption of a completely new variety by farmers could take considerable time, whereas chances of acceptability of converted popular varieties are relatively higher. Improvement of these traits viz. yield, grain quality, select efficiency, disease and pest resistance, and stress tolerance, illustrates the superiority of using marker assisted selection in crop improvement compared to conventional breeding. The integration of these techniques towards producing improved varieties should solve farmers demands while contributing to the protection of national food and environmental security. Therefore, integration of MAS into conventional breeding programs will be an optimistic strategy for crop improvement in the future.

References

1. Timmer CP (2010) The Changing Role of Rice in Asia's Food Security. Asian Development Bank, Working paper series.

2. Rejesus RM, Mohanty S, Balagtas JV (2012) Forecasting global rice consumption, Department of Agricultural and Resource Economics,North Carolina State University.

3. Tran DV (1997) World rice production: main issues and technical possibilities. International Rice Commission, FAO, Rome 24: 57-69.

4. Khush GS (2005) what it will take to feed 5.0 billion rice consumers in 2030. Plant Mol Biol 59: 1-6.

5. Paterson AH (1996) Making genetic maps. In Genome mapping in plants, ed. A.H. Paterson. Academic press, Austin, Texas 23-29.

6. Winter P, Kahl G (1995) Molecular marker technologies for plant improvement. World journal of Microbiology and Biotechnology 11: 438-448.

7. Joshi S P, Ranjekar P K, Gupta V S (1999) Molecular markers in plant genome analysis. Current. Science 77: 230–240.

8. Bhat Z A, Dhillon WS, Rashid R, Bhat JA, Dar WA, et al. (2010) the Role of Molecular Markers in Improvement of Fruit Crops. Not Sci Biol 2: 22-30.

9. Collard BCY, Jahufer MZZ, Brouwer JB, Pang ECK (2005) an introduction to markers, quantitative trait loci (QTL) mapping and marker assisted selection for crop improvement: The basic concepts. Euphytica 142: 169-196.

10. Gupta PK, Roy JK, Prasad M (2001) Single nucleotide polymorphisms: a new paradigm for molecular marker technology and DNA polymorphism detection with emphasis on their use in plants. Current Science 80: 524-535.

11. Doveri S, Maheswaran M, Powell W (2008) Molecular markers – history, features and applications. In: Kole C, Abbott AG (eds) Principles and Practices of Plant Genomics. Vol 1: Genome Mapping. Science Publishers, Enfield, Jersey, Plymouth.

12. Semagn K, Bjornstad A, Ndjiondjop MN (2006) an overview of molecular marker methods for plants. Afr J Biotechnol 5: 2540–2568.

13. Semagn K, Bjornstad A, Ndjiondjop MN (2006) Progress and prospects of marker assisted backcrossing as a tool in crop breeding programs. Afr J Biotechnol 5: 2588-2603.

14. Kumar P, Gupta VK, Misra AK, Modi DR, Pandey BK (2009) Potential of Molecular Markers in Plant Biotechnology. Plant Omics Journal 2:141-162.

15. Tautz D, Trick M, Dover G (1986) Cryptic simplicity in DNA is a major source of genetic variation. Nature 322: 652–656.

16. Collard BC, Mackill DJ (2008) Marker-assisted selection: an approach for precision plant breeding in the twenty-first century. Philos Trans R Soc Lond B Biol Sci 363: 557-72.

17. Foolad MR, Shama A (2005) Molecular Markers as Selection Tools in Tomato Breeding. Acta Horticulturae 695: 225-240.

18. Mohler V, Singrun C (2004) General considerations: marker-assisted selection in Biotechnology in agriculture and forestry. Molecular marker systems 55: 305– 317.

19. Eathington SR, Dudley JW, Rufener GK (1997) Usefulness of Marker-QTL Associations in Early Generation Selection. Crop Science 37: 1686-1693.

20. Xu Y, Crouch JH (2008) Marker-assisted selection in plant breeding: from publications to practice. Crop Science 48: 391 – 407.

21. Hospital F (2009) Challenges for effective marker-assisted selection in plants. Genetica 136: 303-10.

22. Dwivedi SL, Crouch JH, Mackill DJ, Xu Y, Blair MW et al. (2007) The molecularization of public sector crop breeding: Progress, problems, and prospects. Advances in Agronomy 95: 163–318.

23. Ejeta G (2007a) Breeding for Striga resistance in sorghum: Exploitation of intricate host -parasite biology. Crop Science Society of America 47: 216-S227.

24. Brumlop S, Finckh MR (2010) Applications and potentials of marker assisted selection (MAS) in plant breeding. Final report of the F+E project "Applications and Potentials of Smart Breeding" (FKZ 350 889 0020) On behalf of the Federal Agency for Nature Conservation.

25. Jiang GL (2013) Plant Marker-Assisted Breeding and Conventional Breeding: Challenges and Perspectives. Adv Crop Sci Tech 1: e105.

26. Collard B, Mackill D (2006) Marker assisted breeding for rice improvement.

27. Ribaut JM, de Vicente MC, Delannay X (2010) Molecular breeding in developing countries: challenges and perspectives. Current Opinion in Plant Biology 13: 213-218.

28. 28 Datta K, Baisakh N, Thet KM, Tu J, Datta SK (2002) PyTamiding transgenes for multiple resistance in rice against bacterial blight, yellow stem borer and sheath blight. Theor Appi Genet 106: 1-8.

29. 29 Imam J, Alam S, Variar M, Shukla P (2013) Identification of Rice Blast Resistance Gene Pi9 from Indian Rice Land Races with STS Marker and Its Verification by Virulence Analysis. Proceedings of the National Academy of Sciences 83: 499–504.

30. 30 Jantaboon JM, Siangliw S, Im-mark, Jamboonsri W, Vanavichit A et al.(2011) Ideotype breeding for submergence tolerance and cooking quality by marker-assisted selection in rice. Field crops research 123: 206–213.

31. 31 Jin L, Lu Y, Shao Y, Zhang G, Xiao P et al. (2010) Molecular marker assisted selection for improvement of the eating, cooking and sensory quality of rice (Oryza sativa L.). J Cereal Sci 51: 159-164.

32. Neeraja C, Maghirang-Rodriguez R, Pamplona A, Heuer S, Collard B et al.(2007) A marker-assisted backcross approach for developing submergence tolerant rice cultivars. Theoretical and Applied Genetics 115: 767-776.

33. Pandey MK, Shobha Rani N, Sundaram RM, Laha GS, Madhav MS et al. (2013) Improvement of two traditional Basmati rice varieties for bacterial blight resistance and plant stature through morphological and marker-assisted selection. Molecular Breeding 31: 239–246.

34. Ravi RS, Geethanjali F, Sameeyafarheen M, Maheswaran (2003) Molecular Marker based Genetic Diversity Analysis in Rice (Oryza sativa L.) using RAPD and SSR markers 133: 243-252.

35. Sharma HC, Sharma KK, Crouch JH (2004) Genetic transformation of crops for insect resistance: Potential and limitations. Critical Reviews in Plant Sciences 23: 47–72.

36. Yi M, Nwe KT, Vanavichit A, Chai-arree W, Toojinda T (2009) Marker assisted backcross breeding to improve cooking quality traits in Myanmar rice cultivar Manawthukha. Field Crops Res 113: 178-186.

37. Zhan Xiao-deng, Zhou Hai-peng, Chai Rong-yao, Zhuang Jie-yun, Cheng Shi-hua (2012) Breeding of R8012, a Rice Restorer Line Resistant to Blast and Bacterial Blight through Marker-Assisted Selection. Rice Science 19: 29-35.

21

Plant Regeneration of Kenyan Cassava (*Manihot Esculenta* Crantz) Genotypes

Mathew Piero Ngugi *, Oduor Richard Okoth , Omwoyo Richard Ombori , Njagi Joan Murugi , Mgutu Allan Jalemba and Cheruiyot Richard Chelule[2]

[1]Department of Biochemistry and Biotechnology, Kenyatta University, P.O Box 437844-00100, Nairobi, Kenya
[2]Department of Plant and Microbial Sciences, Kenyatta University, P.O Box 437844-00100, Nairobi, Kenya
[3]Department of Environmental Health, Kenyatta University, P.O Box 437844-00100, Nairobi, Kenya

Abstract

A reproducible regeneration system based on direct somatic embryogenesis is described for Kenya cassava lines. Cassava plants were regenerated at high frequency by inducing shoot primordial on explants derived from cotyledons of cassava somatic embryos. Various parameters were evaluated on their effects on callus induction, somatic embryogenesis, maturation and germination of somatic embryos as well as recovery of regenerated plantlets. Immature leaf lobes were used as explants for somatic embryogenesis. Three Kenyan cassava genotypes viz; *Adhiambo Lera*, *Kibanda Meno* and *Serere* along with a model cultivar, TMS 60444 were used this system. Remarkable regeneration frequencies were observed in all the evaluated genotypes with *Adhiambo Lera* showing the best responses. As a result, a highly efficient plant regeneration protocol via germination of somatic embryos was achieved. This system enriches the scope of *in vitro* regeneration protocols for cassava and is envisaged to be a reliable prerequisite to genetic transformation of African cassava genotypes.

Keywords: Somatic embryogenesis; Immature; leaf lobes; Kenyan cassava genotypes; Regeneration

Introduction

Cassava (*Manihot esculenta* Crantz, Euphorbiaceae, 2n=36) is one of the most important food crops in the world, especially in the tropics [1]. Its starchy tuberous roots provide a valuable source of cheap calories for about 500 million people in the developing world commonly plagued by chronic food deficiency and malnutrition [2]. In addition, the leaves and tender shoots are eaten in many parts of Africa as a source of vitamins, minerals and protein [3]. It is tolerant to low fertility and drought, which makes it popular among small-scale farmers in places with infertile soils and adverse climates [4]. Cassava is also used in the production of ethanol for fuel, for animal feed, and as a raw material for the starch industry.

Despite these pleasant qualities, cassava has its drawbacks. It contains cyanogenic glycosides which when not effectively removed may cause a variety of health problems in nutritionally-compromised peoples [5]. Due to its low protein content (1-2%) and limited amounts of sulphur-containing amino acids, additional food sources are required to ensure a diet balanced in protein, vitamins and minerals [6,7]. During the long cultivation period (up to 18 months) of cassava, repeated attacks by various insect pests and virus diseases can cause 20-50% yield losses worldwide, and locally they can lead to total crop failures [8,9]. Further, cassava suffers from postharvest physiological deterioration during transport, storage and marketing [10]. Although the roots can remain in the ground for many months, once they are harvested they deteriorate rapidly and within 48hr they are unmarketable.

In spite of its high importance to food security in third world countries, cassava has long been neglected in plant breeding programmes [11]. The shortcomings associated with cassava production as well as the undesirable traits are potentially amenable to genetic engineering techniques particularly since genetic improvement through traditional breeding has been problematic due to low seed production and long generation time.

A critical requirement for the generation of transgenic cassava is an efficient and reproducible plant regeneration system that is compatible with the available transformation methods. Somatic embryos have been induced from cassava young leaf lobes and cotyledons leading to plant regeneration [12-17].

The reported cassava regeneration protocols have seldom optimally worked for Kenyan cassava varieties. Apart from a few studies byKonan, et al. [17] and Taylor, et al.[18] little if any efforts have focused on developing regeneration and transformation systems for African cassava varieties [2]. Further, it is reported that African cassava varieties from Africa respond differently in culture from varieties from other parts of the world. Environmental, biological, and human demands for African varieties may be responsible for genetic divergence of African cassava from their South American progenitors potentially accounting for the different responses in culture (Sayre, personal communication). In this regard, it is imperative to develop and optimize a regeneration protocol that can feasibly work for Kenyan and perhaps African cassava varieties, as a pre-requisite for successful genetic transformation of Kenyan cassava varieties. In this paper, we report successful development of an improved regeneration protocol for Kenyan cassava varieties via somatic embryogenesis.

Materials and Methods

Plant materials

Four cassava varieties viz; *Adhiambo Lera, Kibanga Meno, Serere*

***Corresponding author:** Mathew Piero Ngugi, Department of Biochemistry and Biotechnology, Kenyatta University, P.O Box 437844-00100, Nairobi, Kenya
E-mail: matpiero@gmail.com, piero.mathew@ku.ac.ke

and *TMS60444* were sourced from Kenya Agricultural Research Institute (KARI), Kakamega. The plantlets were grown *in vitro* on Murashige and Skoog media [19] supplemented with 20 g/L sucrose, MS Vitamins (Duchefa, Germany) and 8 g/L of noble agar. All media used for *in vitro* propagation of cassava was sterilized through autoclaving. The growth chamber conditions were set at a temperature of 28°C and a 16 hr day/8 night cycle.

Callus induction and somatic embryogenesis

Meristematic leaf lobes (2-6 mm long) from *in vitro*-grown plants were cultured on MS basal medium supplemented with 2% (w/v) sucrose, B5 vitamins, 50 mg/L casein hydrolysate, 0.5 mg/L $CuSO_4$ [20] and 4-16 mg/L 2,4-dichlorophenoxyacetic acid (2,4-D). The same set of meristematic leaf lobes was put in the same media substituted with Picloram. The media pH was adjusted to 5.7 and it was solidified with 0.8% (w/v) noble plant agar. Factors affecting embryogenesis were studied with 0-8% sucrose, 0-1.0 mg/L additional $CuSO_4$ and keeping the embryogenesis cultures in darkness or on a 12 hr day/12 hr night cycle. The cultures were maintained at a temperature of 28°C.The explants were left in the induction medium for 4-6 weeks. The type of calli was observed at each step and the frequency of embryogenic calli formation was recorded four weeks of culture on callus induction medium (CIM).

Maturation of somatic embryos

This entailed the development of globular stage embryos into green cotyledonary embryos with defined shoot and root axes [14]. The globular stage somatic embryos were subcultured on maturation media composed of MS salts [19] supplemented with 2% (w/v) sucrose, 1 mg/L thiamine-HCl, 100 mg/L myo-inositol, 0.01 mg/L 2,4-D, 1.0 mg/L BA, and 0.5 mg/L GA_3; MS medium supplemented with 2% (w/v) sucrose, 0.01 mg/L NAA, 0. 1 mg/L BA and 0.1 mg/L GA_3; or MS basal medium supplemented with 2% (w/v) sucrose. The media pH was solidified with 0.8% (w/v) Difco-Bacto agar. The embryos were maintained in the maturation medium in the dark for 4 weeks.

Germination and plant recovery

Mature somatic embryos were transferred to basal MS medium supplemented with 2% (w/v) sucrose; MS medium supplemented with 2% (w/v) sucrose, 1 mg/L thiamine-HCl, 100 mg/L myo-inositol, 0.01

mg/L 2,4-D, 1.0 mg/L BA and 0.5 mg/L GA_3; MS basal salt supplemented with 2% (w/v) sucrose and 0.8% (w/v) activated charcoal. All the media were solidified with 0.8 % (w/v) noble agar for germination and plant recovery. Germination and conversion rates were recorded after four weeks in culture. The cultures were exposed to a daily photoperiod of 12 hr. All cultures were kept at 28°C. Some regenerated plantlets were maintained in growth room ready for transformation experiments while others were hardened in the glass house.

Data analysis

All data were expressed as mean ± SEM (n=3). The collected was subjected to one-way ANOVA. Statistically significant means were analysed by Tukey's Pairwise comparisons at 95% level of significance.

Results

Effects of 2,4-D and Picloram on callus induction and somatic embryogenesis

It was possible to induce calli in all the cassava varieties at all the concentrations of 2,4-D and picloram. Four concentrations of 2, 4-D and picloram (4, 8, 12 and 16 mg/L) were tested for their ability to induce calli and somatic embryogenesis. The different auxin treatments were used to determine which concentration was best for calli and somatic embryo induction of the cassava cultivars. All the treatments produced calli and somatic embryos.

For both callus induction and somatic embryogenesis, the best auxin concentration was 8 mg/L of both 2,4-D and picloram, although 2,4-D had better response than picloram (Tables 1 and 2). 12 mg/L of both 2,4-D and picloram also induced calli and embryogenesis appreciably compared to 4 mg/L and 16 mg/L, which induced the least responses. *Adhiambo Lera* and *Serere* were the best responding varieties with regards to callus induction and somatic embryogenesis (Tables 1 and 2). There was no significant difference in the frequency of callus induction and somatic embryogenesis at auxin concentrations of 4 mg/L and 16 mg/L (P>0.05). Formation of embryogenic calli was consistent with the frequency of callus induction in all the cassava varieties.

Embryogenesis began as swollen regions at the cut ends and mid-veins on the adaxial surface of the immature leaf lobes. The

	Average Number of Calli per Explant							
	4		8		12		16	
	2,4-D	Picloram	2,4-D	Picloram	2,4-D	Picloram	2,4-D	Picloram
TMS 6044	64.3 ± 0.9[b]	60.3 ± 0.9[b]	85.0 ± 0.6[a]	81.0 ± 0.6[c]	74.3 ± 0.9[b]	70.0 ± 0.6[b]	68.2 ± 2.4[bc]	58.0 ± 1.2[b]
Adhiambo Lera	75.0 ± 1.5[a]	71.7 ± 1.2[a]	93.0 ± 1.8[c]	90.0 ± 1.2[a]	82.7 ± 0.9[a]	78.7 ± 1.3[a]	68.3 ± 0.9[a]	65.0 ± 2.3[a]
Kibanda Meno	58.3 ± 1.2[c]	55.3 ± 0.9[c]	74.7 ± 2.4[b]	71.0 ± 0.6[b]	67.7 ± 1.5[b]	63.0 ± 1.5[b]	57.0 ± 1.7[c]	54.3 ± 2.3[b]
Serere	66.3 ± 0.9[b]	61.3 ± 0.9[b]	78.7 ± 0.9[b]	75.3 ± 1.8[b]	68.7 ± 1.9[bc]	65.7 ± 2.9[b]	63.3 ± 0.9[b]	57.1 ± 2.1[b]

Values are expressed as mean ± SEM. Values followed by different letters are statistically significant by ANOVA followed by Tukey's pairwise comparisons (P>0.05)

Table 1: Effects of the different concentrations 2,4-D and Picloram on callus induction.

	Average Number of Embryonic Calli per Explant							
	4		8		12		16	
	2,4-D	Picloram	2,4-D	Picloram	2,4-D	Picloram	2,4-D	Picloram
TMS 6044	20.7 ± 1.2[a]	20.5 ± 3.4[a]	33.3 ± 2.0[b]	29.7 ± 3.0[a]	26.7 ± 0.9[a]	27.7 ± 1.5[b]	21.0 ± 0.6[a]	19.3 ± 1.8[b]
Adhiambo Lera	27.0 ± 2.7[a]	26.3 ± 2.3[b]	40.7 ± 0.9[a]	39.4 ± 3.2[b]	37.3 ± 1.2[b]	33.8 ± 1.9[a]	26.7 ± 0.9[b]	27.7 ± 1.5[a]
Kibanda Meno	22.0 ± 2.3[a]	23.7 ± 3.6[a]	34.7 ± 0.9[ab]	33.6 ± 2.6[c]	33.0 ± 0.6[c]	29.8 ± 2.5[ab]	22.7 ± 0.3[a]	18.7 ± 1.9[b]
Serere	27.0 ± 0.6[a]	22.2 ± 2.4[a]	37.7 ± 1.5[ab]	42.4 ± 3.1[b]	33.3 ± 1.7[b]	30.5 ± 2.3[ab]	31.3 ± 0.9[c]	19.8 ± 1.6[b]

Values are expressed as mean ± SEM. Values followed by different letters are statistically significant by ANOVA followed by Tukey's pairwise comparisons (P>0.05)

Table 2: Effects of the different concentrations 2,4-D and Picloram on somatic embryogenesis.

swollen segments then gave rise to two kinds of calli viz; a loose non-embryogenic friable white callus and a translucent gelatinous embryogenic callus that formed globular stage embryos (Figure 1).

Effect of light on somatic embryogenesis

The effect of light on somatic embryogenesis was done by exposing the somatic embryo cultures to 16 hr/day photoperiod versus 8 hr/day dark incubation. This was observed to affect the process of somatic embryogenesis. Following four weeks in culture, the light-exposed explants formed more non-embryogenic calli than those kept in the dark (Table 3). Generally, a few embryos were produced when the explants were given 12 hr of light followed by 12 hr of darkness. With dark incubation, *Adhiambo Lera* and *Serere* showed the best responses with 36.6% and 33.8% of light-exposed explants developing somatic embryos respectively compared to 87.7% and 80.1% of dark incubated explants respectively. The two cassava varieties had the highest mean numbers of embryogenic calli per explants compared to TMS 60444 and *Kibanda Meno* (Table 3). TMS 60444 and *Kibanda Meno* had 32.5% and 31.5% of light-exposed explants producing somatic embryos respectively compared to 75.4% and 74.1% of dark incubated explants respectively. Nevertheless, light is required for germination of embryos [21]. *Adhiambo Lera* had the best response with regards to callus induction of explants exposed to darkness compared to other genotypes

(P<0.05) (Table 3). However, there was no significant difference in the number of embryonic calli produced per explants between *Adhiambo Lera* and *Serere* genotypes.

Effects of Copper Sulphate

The effect of additional Copper in the form of $CuSO_4$ was studied with all the cultivars by the inclusion of 0.25, 0.5, 0.75 and 1.0 mg/L of additional $CuSO_4$ in the induction medium. The results were cultivar-dependent and consistent (Table 4). For cultures devoid of additional copper and those with 0.25 mg/L, formation of embryogenic calli was comparable in all varieties (P>0.05). For all varieties except *TMS 60444*, formation of embryogenic calli with inclusion of 0.5, 0.75 and 1 mg/L of additional Copper was also comparable. However, 0.5 mg/L $CUSO_4$ had the best response with regards to the number of embryonic calli per explants (Table 4).

Effect of sucrose on embryogenesis

Callus formation and embryogenesis hardly took place with the exclusion of sucrose from the callus induction medium (Table 5). The explants died after four weeks in culture. The number of embryos produced in 2-6% sucrose media was cultivar dependent.2% Sucrose concentration promoted formation of healthier embryos than the others. Explants grown on 6% sucrose were all cream-colored and most turned brown after 4 weeks in culture. This is a sign of cell death caused by formation of phenolic compounds in the cultures (Ihemere, 2003). *Adhiambo Lera* and *Serere* genotypes produced the highest number of embryonic calli with all the sucrose concentrations (Table 5).However, the response of *Adhiambo Lera* significantly different from the response of *Serere* at 4% and 6% sucrose concentrations (P<0.05). *Kibanda Meno* and the model cultivar TMS 60444 had comparable responses at 2% and 4% sucrose concentrations (Table 5).

Maturation of somatic embryos

The early stage somatic embryos were transferred to the maturation medium after 4 weeks of culture with adjoining callus. All the embryos grew to cotyledonary embryos with distinct root and shoot axes in all media compositions albeit with differences in frequencies. Three different maturation media differing in cytokinins content were used in this experiment. Embryos maintained on MS medium supplemented with 2% sucrose, 1 mg/L thiamine-HCl, 100 mg/L myo-inositol, 0.01 mg/L 2,4-D, 1.0 mg/L BA and 0.5 mg/L GA_3 was superior to

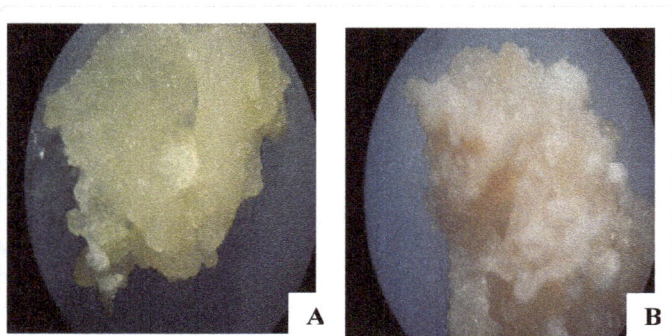

A: A loose non-embryogenic friable white callus
B: Translucent gelatinous embryogenic callus

Figure 1: Tissue culture profile somatic emryogenegenis of cassava lines.

	Total number of explants		No. of explants with developed callus		Mean No. of embryonic calli /explants	
	12	0	12	0	12	0
TMS 60444	124.0 ± 1.7	125.0 ± 1.7	40.3 ± 1.2ᵃ **(32.5%)**	94.3 ± 2.3ᵃ **(75.4%)**	12.7 ± 1.8ᵃ	127.3 ± 2.3ᵃ
Adhiambo Lera	125.7 ± 2.9	121.7 ± 4.9	46.0 ± 2.3ᵇ **(36.6%)**	106.7 ± 1.8ᵇ **(87.7%)**	17.7 ± 2.0ᵃᵇ	161.3 ± 5.8ᵇ
Kibanda Meno	115.3 ± 2.9	123.7 ± 2.7	36.3 ± 1.8ᵃᵇ **(31.5%)**	91.7 ± 1.2ᵃ **(74.1%)**	13.0 ± 1.7ᵃ	130.3 ± 0.9ᵃ
Serere	125.0 ± 2.5	122.0 ± 1.2	42.3 ± 2.7ᵃᵇ**(33.8%)**	97.7 ± 2.6ᵃ**(80.1%)**	25.7 ± 2.2ᵇ	159.3 ± 1.2ᵇ

Values are expressed as mean ± SEM. Values with the same number are not statistically significant by ANOVA followed by Tukey's pairwise comparisons (P>0.05).

Table 3: Effect of light on somatic embryogenesis.

	$CUSO_4$ concentrations (mg/l)				
	0.0	0.25	0.5	0.75	1
	Average Number of Embryonic Calli per Explant				
TMS 6044	23.7 ± 2.0ᵃ	25.3 ± 1.2ᵃ	43.7 ± 1.9ᵃ	34.0 ± 0.6ᵃ	27.3 ± 3.1ᵃ
Adhiambo Lera	31.0 ± 1.2ᵃ	33.0 ± 1.7ᵃ	62.7 ± 1.2ᵇ	53.0 ± 1.7ᵇ	44.3 ± 2.5ᵇ
Kibanga Meno	25.7 ± 2.2ᵃ	32.0 ± 3.2ᵃ	52.3 ± 1.9ᵇ	50.0 ± 0.6ᵇ	39.0 ± 0.6ᵇ
Serere	29.0 ± 1.2ᵃ	32.0 ± 1.2ᵃ	59.0 ± 0.6ᵇ	45.7 ± 5.2ᵃᵇ	40.0 ± 1.2ᵇ

Values are expressed as mean ± SEM. Values with the same letter are not significantly different by ANOVA followed by Tukey's pairwise comparisons (P>0.05)

Table 4: Effect of different CUSO4 concentrations on somatic embryogenesis of cassava.

MS medium supplemented with 2% sucrose, 0.01 mg/L NAA, 0.1 mg/L BA and 0.1 mg/L GA$_3$ and MS medium supplemented with 2% sucrose. As Figure 2 shows, more than 50% percent of embryonic calli produced cotyledonary embryos in the medium containing MS salts supplemented with 2% (w/v) sucrose, 1 mg/L thiamine-HCl, 100 mg/L myo-inositol, 0.01 mg/L 2,4-D, 1.0 mg/L BAP and 0.5 mg/L GA3. In

Cultivar	Sucrose concentration (percent w/v)			
	0	2	4	6
	No. of Embryonic calli			
TMS 6044	0.0 ± 0.0	84.0 ± 2.3^a	77.3 ± 1.7^a	39.3 ± 1.2^a
Adhiambo Lera	0.0 ± 0.0	110.7 ± 5.8^c	108.0 ± 2.3^b	52.0 ± 1.2^c
Kibanga Meno	0.0 ± 0.0	89.3 ± 2.9^{ab}	79.0 ± 1.5^a	32.0 ± 1.2^b
Serere	0.0 ± 0.0	106.3 ± 3.6^{bc}	97.0 ± 1.5^c	37.3 ± 1.8^{ab}

Values are expressed as mean ± SEM. Values with the same letter are not significantly different by one-way ANOVA followed by Tukey's pairwise comparisons (P>0.05)

Table 5: Effect of different sucrose concentrations on somatic embryogenesis.

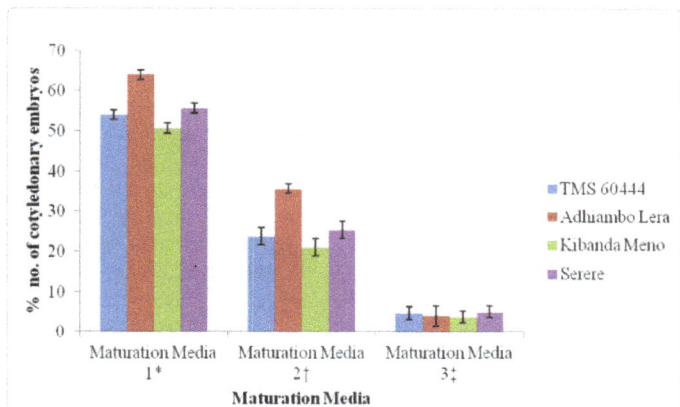

*MS salts supplemented with 2% (w/v) sucrose, 1 mg/L thiamine-HCl, 100 mg/L myo-inositol, 0.01 mg/L 2,4-D, 1.0 mg/L BAP and 0.5 mg/L GA3

†MS salts supplemented with 2% (w/v) sucrose, 0.01 mg/L NAA, 0.1 mg/L BA and 0.1 mg/L GA3

‡MS basal medium supplemented with 2% (w/v) sucrose

Figure 2: Effects of cytokinins on maturation of somatic embryos.

the medium supplemented with 0.01 mg/L NAA, 0.1 mg/L BA and 0.1 mg/L GA$_3$, more than 70% of embryonic calli hardly produced cotyledonary embryos. The medium supplemented with 2% sucrose promoted the least frequency of maturation (Figure 2). Overall the cultivar *Adhiambo Lera* had the highest frequency of maturation of somatic embryos followed by *Serere* and *TMS 60444. Kibanda Meno* had the least response in all the media compositions. As (Figure 3) shows, the embryos first of all turned green and then progressively formed distinct shoot and root apices.

However, there was no significant difference in the frequency of maturation of somatic embryos of *Adhiambo Lera* and *Serere* in the media containing MS salts supplemented with 2% (w/v) sucrose, 1 mg/L thiamine-HCl, 100 mg/L myo-inositol, 0.01 mg/L 2,4-D, 1.0 mg/L BAP and 0.5 mg/L GA$_3$ (P>0.05) (Table 6). In the same media, the responses of TMS 60444 and *Serere* genotypes were comparable.

Germination of somatic embryos and plant recovery

The germination medium comprising MS salts, B5 vitamins, 2% sucrose, and 0.8% activated charcoal was superior to basal MS medium plus 2% sucrose, and MS supplemented with 2% sucrose, 1 mg/L thiamine-HCl, 100 mg/L myo-inositol, 0.01 mg/L 2,4-D, 1.0 mg/L BA and 0.5 mg/L GA$_3$ in terms of germination time and frequency (Table 7). Single cotyledonary embryos did not always produce single plantlets. In some cases up to five plantlets arose from one embryo. The recovery of plantlets from embryos was genotype-dependent and it was consistent. Overall, *Adhiambo Lera* was the best genotype in terms of both germination time and frequency.

As Table 7 shows, there was significant difference between *Adhiambo Lera* and other genotypes in all media compositions in terms of germination frequency (P<0.05). On the other hand, the other genotypes; *TMS 60444, Kibanda Meno* and *Serere* had comparable germination frequencies in the basal MS medium supplemented with B5 vitamins and 2% (w/v) sucrose and in MS medium supplemented with 2% (w/v) sucrose, 1 mg/L thiamine-HCl, 100 mg/L my-oinositol, 0.01 mg/L 2,4-D, 1.0 mg/L BAP and 0.5 mg/L GA$_3$ (P>0.05).

Upon germination, the recovered plants were grown in the growth room for two weeks following which they were hardened in the glass house. Growth of the regenerated plantlets is summarized in (Figure 4).

A: Greening of embryogenic embryos
B: Initiation of shoots
C: Formation of distinct shoots

Figure 3: Tissue culture profile of maturation of embryogenic somatic embryos.

Discussion

In this study, a protocol for somatic embryogenesis and plant recovery for Kenyan cassava varieties has been optimized. Studies on African cassava varieties by Ihemere UE [2] established that 12 mg/L 2, 4-D was better than 12 mg/L picloram for embryogenesis induction for most of the cultivars. Conversely, in this study, it was established that 8 mg/L 2,4-D and Picloram produced the best responses in terms of callus induction and somatic embryogenesis in all cultivars. Moreover, 2,4-D (8 mg/L) was better with regards to frequency of leaf

Cultivar	Mean No. of Cotyledonary Embryos			
	Initial no. of Embryonic Calli	Maturation Media 1*	Maturation Media 2†	Maturation Media 3‡
TMS 60444	98.3 ± 1.5	53.3 ± 0.9ac	23.9 ± 0.9b	4.7 ± 1.2a
Adhiambo Lera	91.3 ± 0.9	58.3 ± 1.2a	33.0 ± 1.2a	6.3 ± 1.8a
Kibanga Meno	90.7 ± 1.2	46.3 ± 1.5b	19.3 ± 2.0b	3.3 ± 1.2a
Serere	90.7 ± 1.8	50.7 ± 1.2bc	23.0 ± 1.7b	4.3 ± 0.9a

Values are expressed as mean ± SEM. Values with the same letters are not significantly different by ANOVA followed by Tukey's pairwise comparisons (P>0.05).
*MS salts supplemented with 2% (w/v) sucrose, 1 mg/L thiamine-HCl, 100 mg/L myo-inositol, 0.01 mg/L 2,4-D, 1.0 mg/L BAP and 0.5 mg/L GA3.
†MS salts supplemented with 2% (w/v) sucrose, 0.01 mg/L NAA, 0.1 mg/L BA and 0.1 mg/L GA3.
‡MS basal medium supplemented with 2% (w/v) sucrose

Table 6: Effects of different cytokinins on maturation of somatic embryos.

Cultivar	Number of Regenerated Plants			
	Initial No. of Cotyledonary Embryos	Germination Media 1*	Germination Media 2†	Germination Media 3‡
TMS 60444	77.7 ± 0.6	22.0 ± 1.5b	23.7 ± 1.8b	63.3 ± 1.5bc
Adhiambo Lera	78.7 ± 1.5	31.7 ± 1.2a	37.0 ± 2.5a	76.3 ± 2.3a
Kibanga Meno	82.3 ± 1.5	18.7 ± 0.9b	23.7 ± 0.3b	58.3 ± 1.2c
Serere	79.3 ± 1.9	23.0 ± 0.6b	28.3 ± 0.9b	67.7 ± 1.5b

Values are expressed as mean ± SEM. Values with the same letters are significantly different by ANOVA followed by Tukey's pairwise comparisons (P>0.05).
*basal MS medium supplemented with B5 vitamins and 2 % (w/v) sucrose; †MS medium supplemented with 2% (w/v) sucrose, 1 mg/L thiamine-HCl, 100 mg/L myo-inositol, 0.01 mg/L 2,4-D, 1.0 mg/L BAP and 0.5 mg/L GA3; ‡MS basal salt supplemented with 2 % (w/v) sucrose and 0.8 % (w/v) activated charcoal

Table 7: Germination and Regeneration Frequencies of the four cassava genotypes.

lobe embryogenesis than Picloram (8 mg/L)in all cultivars (Tables 1 and 2). Nevertheless, embryogenesis was induced at all auxin concentrations in all cultivars. Successful somatic embryogenesis in *Glycine max* on medium supplemented with 40 mg/L 2, 4-D has been accomplished in the past [22]. This is in spite of the studies by Konan NK et al. [17] reported an inhibition of embryogenesis in cassava by the supplementation of the induction-medium with more than 12 mg/L 2, 4-D. The observed disparities in the results of this study can be blamed on use of different explants. This study used young leaf lobes, while Konan's group used the cotyledons of zygotic embryos.

Many plant species regenerated *in vitro* through callus are bedeviled by somaclonal variation [2]. This is postulated to be due to the high rate of cell division during callus formation induced by high auxin levels in the induction medium. This drawback is reduced in plants recovered through somatic embryogenesis. Konan NK, et al. [17] observed that the incidence of somaclonal variation in cassava regenerated by somatic embryogenesis with 4-16 mg/L 2,4-D is negligible. As a convention, embryogenesis in most plants starts with callus induction, followed by transfer to another medium for embryo emergence. In cassava, the explant on the induction medium proliferates to globular stage embryos in about two weeks of culture, hence reducing the length of time the regenerative tissues stay on high auxin (8-12 mg/L) medium [15]. Perhaps this ameliorates the incidence of somaclonal variation further in cassava plants recovered from somatic embryos. Recently, there have been more reports of somaclonal variation in cassava regeneration systems [16,21]. The drive to improve on the efficiency of cassava transformation has led to the development of additional techniques for plant regeneration. A classical example of such efforts is the use of friable embryogenic callus to regenerate somatic embryos. This technique requires the callus to be on the high auxin medium (50 mg/L) for six or more months resulting in high rates of somaclonal variation [21,23].

Copper is thought to enhance growth of embryos during somatic embryogenesis. Somatic embryogenesis experiments performed with media lacking additional copper produced embryos difficult to distinguish from the non-embryogenic callus because they were too small [2]. However, the inclusion of additional copper made the embryos distinct at the early stage of embryogenesis. This was also the case in this study. Presumably, it is due to the fact that Ihemere worked on African cassava varieties.

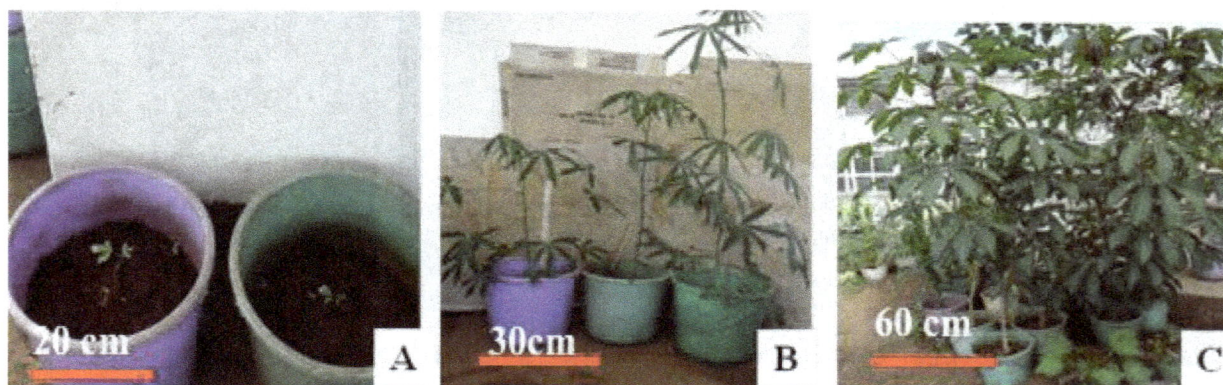

A: Young cassava plantlets immediately transferred to soil in glass house from growth room plantlets
B: Mature cassava plants in glass house
C: Mature cassava plantlets in the field outside glasshouse

Figure 4: Recovery of regenerated cassava plantlets.

Callus formation is often considered to inhibit plant regeneration and in cassava plant regeneration from callus has never been possible, neither via organogenesis nor via embryogenesis. It is therefore assumed that the increased capacity of shoot organogenesis is at least partially related to the inhibition of callus formation [24]. Cassava regeneration via organogenesis involves the induction of shoots from young leaf lobes and embryo cotyledons on MS medium supplemented with cytokinins (23 μM zeatin and 44 μM benzylaminopurine) [25,26]. This method results in the regeneration of multiple shoots but the origin of each shoot is doubtful making it suspect for plant transformation purposes [2].

Embryo development was achieved in medium supplemented with 2% sucrose, 1 mg/L thiamine-HCl, 100 mg/L my-oinositol, 0.01 mg/L 2,4-D, 1.0 mg/L BA and 0.5 mg/L GA_3 was superior to MS medium supplemented with 2% sucrose, 0.01 mg/L NAA, 0.1 mg/L BA and 0.1 mg/L GA3 and MS medium supplemented with 2% sucrose but devoid of phytohormones. This agrees with Mathews H, et al. [15], who cultured globular stage embryos of cassava on medium without plant growth regulators and achieved the highest rate of regeneration yet in cassava. It can therefore be inferred that phytohormones may only be critical in the induction medium for embryogenesis to occur.

Culturing explants in light at the beginning of the induction process led to reduction in the number of embryos formed. This indicates that light-intensity affects embryogenesis. Embryogenesis in cassava is enhanced by lowering light intensity [21].Studies on the effect of light on callus growth and somatic embryogenesis from *Lavandula vera* and *Teucrium chamaedrys* showed that the relative length of the incubation period under illumination significantly affected callus growth and somatic embryo induction and proliferation, although in a species-specific fashion. Lavender callus growth was improved under increased incubation in darkness while somatic embryogenesis was remarkably reduced under the same conditions [27].Further, somatic embryogenesis from some species, such as cucumber, squash, melon and gardenia is promoted under initial culture incubation in darkness [28,29].It is, thus, thought that lowering the light intensity of the cultures could have accounted for the high rates of embryogenesis in all the cultivars studied in this research.

Conclusion

In conclusion, this study has optimized a reproducible *in vitro* regeneration pathway of Kenyan genotypes using three Kenyan genotypes viz; *Adhiambo Lera, Kibanda Meno* and *Serere* along with an exotic model cultivar TMS 60444 using immature leaf lobes as explants for somatic embryogenesis. It avails an opportunity for further studies on cassava with varying research interests to mitigate the various genetic and phenotypic drawbacks associated with cassava. This also arrests the view that African cassava genotypes are recalcitrant to *in vitro* manipulations.

Acknowledgements

This work is part of a continuing project on genetic engineering of Kenyan cassava genotypes funded by the Danish International Development Agency (DANIDA). The cassava cultivars used in this study were kindly provided by the Kenya Agricultural Research Institute (KARI). Duncan Ogweda is hereby acknowledged for maintenance of explants sources and regenerated plants.

References

1. Zhang P (2000) Studies on cassava (ManihotesculentaCrantz) transformation: towards genetic improvement, Swiss Federal Institute of Technology Zürich.

2. Ihemere UE (2003) Somatic embryogenesis and transformation of cassava for enhanced starch production, The Ohio State University.

3. Cock JH (1982) Cassava: a basic energy source in the tropics. Science 218: 755-762.

4. Koch BM, Sibbesen O, Swain E, Kahn RA, Liangcheng D et al. (1994) Possible use of a biotechnological approach to optimize and regulate the content and distribution of cyanogenicglucosides in cassava (ManihotesculentaCrantz) to increase food safety. Acta Horticultura 375: 45-60.

5. Rosling H, Mlingi N, Tylleskar T, Banea M (1993) Causal mechanisms behind human diseases induced by cyanide exposure from cassava.InW Roca A Thro eds proceedings of the First International Scientific Meeting of the Cassava Biotechnology Network, Centro Internacional de Agricultura Tropical Cali Colombia 366-375.

6. Omole TA (1977) Cassava in the nutrition of layers IDRC, Ottawa Canada.

7. Cock JH (1985) Cassava: new potential for a neglected crop. Westview Press Boulder, London.

8. Thresh JM, Fargette D, Otim-Nape GW (1994) Effects of African cassava mosaic geminivirus on the yield of cassava. Trop Sci 34: 26-42.

9. Belloti AC, Smith L, Lapointe SL (1999) Recent advances in cassava pest management. Ann R Entomol 44: 343-370.

10. Wenham J (1995) Post-harvest deterioration of cassava: A biotechnology perspective. FAO Plant Production and Protection Paper 130.

11. Li H-Q, Huang YW, Liang CY, Guo JY, Liu HX et al. (1998) Regeneration of cassava plants via shoot organogenesis. Plant Cell Rep 17: 410-414.

12. Stamp JA, Henshaw GG (1987) Somatic embryogenesis from clonal leaf tissue of cassava. Ann Bot 59: 445-450.

13. Szabados L, Hoyos R, Roca W (1987) In vitro somatic embryogenesis and plant regeneration of cassava. Plant Cell Rep 6: 248-251.

14. Taylor NJ, Clarke M, Henshaw GG (1993) The induction of somatic embryogenesis in fifteen African and one South American cassava cultivars. Food and agriculture organization of the united nations 123: 134-140

15. Mathews H, Schopke C, Carcamo R, Chavarriaga P, Fauquet L et al. (1993) Improvement of somatic embryogenesis and plant recovery in cassava. Plant Cell Rep 12: 328-333.

16. Raemakers CJJM, Bessembinder J, Staritsky G, Jacobsen E, Visser RGF (1993) Inducion, germination and shoot development of somatic embryos in cassava. Plant Cell Tiss Org Cult 33: 151-156.

17. Konan NK, Sangwan RS, Sangwan-Norreel BS (1994) Somatic embryogenesis from cultured mature cotyledons of cassava (ManihotesculentaCrantz). Plant Cell Tiss Org Cult 37: 91-102.

18. Taylor NJ, Clarke M, Henshaw GG (1992) The induction of somatic embryogenesis in 15 African and one South American cassava cultivars. In: Roca WM, Thro AM (eds). Proceedings of the First International Scientific Meeting of the Cassava Biotechnology Network, Cartagena, Colombia August. Working Document 123. Cali, Colombia CIAT.

19. Murashige T, Skoog F (1962) A revised medium for rapid growth and bioassays with tobacco tissue culture. Plant Physiology15: 473-497.

20. Schopke C, Franche C, Bogusz D, Chavarriaga P, Fauquet C, et al. (1992) Transformation in cassava (ManihotesculentaCrantz) In: Biotechnology in Agriculture and Forestry (Ed) Bajaj YPS, Springer Verlag (in press).

21. Raemakers CJJM, Sofiari E, Jacobsen E, Visser RGF (1997) Regeneration and transformation of cassava. Euphytica 96: 153-161.

22. Finer JJ, Nagasawa J (1988) Development of an embryogenic suspension culture of soybean [Glycine max (L.) Merrill]. Plant Cell Tiss Org Cult15: 125-136.

23. Puonti-Kaerlas J, Frey P, Potrykus I (1998) Competence for embryogenesis and organogenesis in cassava. In: Pires de Matos A, Vilarinhos (eds). Proc. IV Int Scientific Meeting of the Cassava Biotechnology Network,Brasil.

24. Zhang P, Phansiri S, Puonti-Kaerlas J (2001) Improvement of cassava shoot organogenesis by the use of silver nitrate in vitro. Plant Cell Tiss Org Cult 67: 47-54.

25. Guohua M (1998) Effects of cytokinins and auxins on cassava shoot organogenesis and somatic embryogenesis from somatic embryo explants. Plant Cell Tiss Org Cult 54: 1-7.

26. Mussio I, Chaput MH, Serraf I, Ducreux G, Sihachakr D (1998) Adventitious

shoot regeneration from leaf explants of an African clone of cassava (ManihotesculentaCrantz) and analysis of the conformity of regenerated plants. Plant Cell Tiss Org Cult 53: 205-211.

27. Spiridon K, Panagiotis T, Charalambos P, Vlassios G, John D (2002) The Effects of Light on Callus Growth and Somatic Embryogenesis fromLavandulaveraand Teucriumchamaedrys: A Preliminary Study. Journal of Herbs, Spices & Medicinal Plants 9: 223-227.

28. Cade RM, Wehner TC, Blazich FA (1988) Embryogenesis from cotyledon-derived callus ofCucumissativus L. Cucurbit. Genet. Coop. Rep 11: 3-4.

29. Kintzios S, Hioureas G, Shortsianitis E, Sereti E, Blouchos P et al. (1998) The effect of light on the induction development and maturation of somatic embryos from various horticultural and ornamental species. ActaHorticulturae 461: 427-432.

22

Genotypic Response to Salt Stress: I – Relative Tolerance of Certain Wheat Cultivars to Salinity

Ravi Sharma*

Eco-physiology Laboratory Department of Post-graduate Studies and Research in Botany K R College Mathura Formerly Head Department of Botany K R College, Mathura and Ex-Principal ESS ESS College, Agra (Dr B R Ambedkar University formerly Agra University, Agra) 281 001 UP India

Abstract

Forty two wheat (*Triticum aestivum* L) cultivars screened for their relative salt resistance raising seedlings in half-Hoagland solution (control) salinized with NaCl and maintained at 4, 8, 12 and 16 dsm^{-1} showed a wide range of salt resistance. The growth response to salinity, judged by the shoot and root lengths, ranged from a stimulation in the case of some cultivars at lower salinity levels (4 and 8 EC) to a severe suppression in most of the cultivars at higher levels (12 and 16 EC). It was further observed that the shoot growth was often suppressed more than the root growth with this a level of 12 EC also found to be critical for most of the cultivars except HD–2160 which showed good stand even at a salinity level of 16 EC. Based on these observations, cultivar IWP–72 of the 42 cultivars tested was found to have the maximum sensitivity to salt stress whereas cultivar HD–2160 showed highest salt tolerance. The remaining 40 cultivars fell between the two extremes and were categorized into *sal–sensitive, moderately salt–tolerant* and *salt – tolerant* groups exhibiting more than 60%, 40 – 60% and less than 40% reduction respectively in shoot length at 12 EC dsm^{-1} over control.

Keywords: Wheat (Triticum aestivum L); Salt stress; Critical level; Salt–tolerant; Moderately salt–tolerant; Salt–sensitive genotypes

Introduction

The complexion of salt tolerance and the multitude of ways in which plants adapt to it have caused much confusion. Sodium (Na$^+$) and chloride (Cl$^-$) are among the most common ions found in excess in saline soils, and some plant species are especially sensitive to one or both of these ions [1-8]. A general suppression of growth is probably the most common plant response to salinity [9]. Crop plants differ greatly in their tolerance to salinity. Differences between species and varieties in regard to salt tolerance have been reported by several workers Bernstein, Hayward, Shannon, Ogra, Sharma , Baijal, Nauhbar, Yadav, Rani , Gautam and Parashar [1-4,6-8,10-20]. In saline soils [2,4,8,21-24] the control of water, the proper techniques of planting and the choice of tolerant crops are essential for their successful use in crop production. The choice of crops is based on: (1) the tolerance to salt; (2) adaptability to climatic or soil characteristics and (3) value of the crop in the individual farm activity. The chances of a crop failure are less if an adequately salt tolerant crop or its variety is selected. The key to improving salt tolerance in plants and studying its inheritance lies in finding sufficient variation within breeding populations and devising a screening procedure capable of identifying resistant or tolerant genotypes.

Further, as the period of seed germination and early seedling stage is the most crucial and important stage in the life cycle of species growing in saline environment [25] the present investigation was, therefore, undertaken to analyze the relative salt tolerance in wheat (*Triticum aestivum* L) at the early seedling stage and to select varieties that could withstand varying concentrations of the salts in their environment.

Materials and Methods

Forty two wheat cultivars (*Triticum aestivum* L) were procured from Wheat Directorate, Cummings Laboratory, Division of Genetics and Plant Breeding, Indian Agricultural Research Institute, New Delhi and Chandra Sekhar Azad University of Agriculture and Technology, Kanpur (UP), India. Screening of wheat cultivars for salt resistance was made by Garrad's Technique (1945) as modified by Sarin and Rao [26]

and Sharma [2] and as per method of Sheoran and Garg [11] wherein shoot and root lengths of seedlings were recorded at definite intervals. Here test tubes of uniform size (30 ml capacity) were fitted with rolls of filter paper folded at the top into a cone to support the seeds. The tubes were filled to one-third part with the test solutions so that the solution might not come in direct contact with the growing roots, the salt solution being supplied to the roots through capillary action of the filter paper. Distilled water (represented the mean loss of water from the blanks) was added to each test tube after every 24 hr of interval in order to maintain salt concentration near the target levels throughout the germination period. The seeds were initially sterilized with 0.1% mercuric chloride (HgCl$_2$) solution and later washed thoroughly with distilled water. Three seeds per tube were then transferred to the edge of the filter paper cone and were allowed to grow between the filter paper roll and the wall of the test tube in dark growth chamber at 25 ± 2°C. Fifteen replicates (five tubes each having three seeds) were maintained for each treatment including the controls (half-strength Hoagland solution grown). Observations on the influence of salinity levels at 4, 8, 12 and 16 EC dsm^{-1} of salt solution and the controls on the total length of coleoptile and root at early seedling stage were recorded at 24 hour intervals from 48 hr after sowing up to the end of 120 hr under green safe light. The relative tolerance of different cultivars was evaluated on the basis of the percentage reduction in shoot growth at 12 EC.

All parameters were analyzed by 'Analysis of Variance' (ANOVA)

Corresponding author: Ravi Sharma, Eco-physiology Laboratory Department of Post-graduate Studies and Research in Botany K R College Mathura , Formerly Head Department of Botany K R College, Mathura and Ex-Principal ESS ESS College, Agra (Dr B R Ambedkar University formerly Agra University, Agra) 281 001 UP India E-mail: drravisharma327@yahoo.com.

method as given by Panse and Sukhatme [27] wherein Critical Differences (CD at 1 and 5% probability were calculated wherever the results were significant.

Results and Discussion

The observations summarized here clearly demonstrate that exposure to salinity during early seedling stage resulted in stunting of growth of the shoot and root at higher salinity levels. This reduction in shoot and root growth is one of the most commonly observed responses to salinity [2-4,6-8,12-20,28,29].

In agreement with Richards [30] it is observed that the changes induced by addition of NaCl to the growth medium became more distinct with increasing salinity and with prolongation of the period of exposure to salinity. This is perhaps due to a higher intake of ions [2,13,16-18,22] which resulted in toxicity [31-33]. Osmotic effects might also have contributed to the low growth rates under saline conditions [34].

Seed lots of 42 wheat cultivars screened for salinity tolerance at the early seedling stage for shoot and root lengths under varying salinity levels (0,4,8,12 and 16 dsm⁻¹) induced by NaCl as indicated (Table 1), all the main effects viz., variety, treatment and seedling age and their interactions (V × D, V × T, D × T and V × D × T) were highly significant at 0.01 probability with significant differences noticed in the shoot and root growth of all the cultivars studied (Figure 1). The highest mean shoot growth (3.091 cm) was recorded in the cultivar Kharchia followed by HD-2009, Sonalika, Sharbati sonora, WL-410, HD-2236,

UP-262, HS-43, IWP-503, HP-1303, HD-2177, HD-2135, WH-246, K-7634, HD-2260, Raj-1556, UP-115, WL-711, Moti, HD-2282, WL-2200, Raj-1482, HD-1980, IWP-72, CC-464, HD-2275, Raj-1409, HD-2160, HD-1593, Raj-1494, HD-2252, WL-908, HD-2267, UP-171, Raj-1493, HD-1977, HD-2204, K-7631, WL-1531, WG-1559, UP-154 and lastly WG-1558 with the lowest shoot length of 0.282 cm (Table 2). Similarly, significant differences were also noticed in the root growth of the cultivars studied. The maximum root length (5.974 cm) was observed in the cultivar Kharchia followed by HD-2009, IWP-503, Sonalika, Sharbati sonora, HS-43, WL-410, CC-464, UP-262, HD-2135, HD-2177, Raj-1556, UP-115, HP-1303, WL-2200, Moti, HD-2275, HD-2160, HD-2252, WL-711, WH-246, HD-1980, IWP-72, Raj-1494, Raj-1482, K-7634, HD-2260, HD-1593, Raj-1409, UP-171, WL-903, HD-2282, Raj-1493, HD-2236, HD-1977, HD-2267, K-7631, HD-2204, WL-1531, UP-154, WG-1558, and minimum (0.658 cm) was observed in WG-1559 (Table 2).

As indicated in the Table 3 only 11 cultivars showed less than 60% reduction in shoot growth while majority of the 31 cultivars had more than 60% reduction at 16 EC. This is in contrast with root growth (Table 3) where almost a reverse trend was noticed, i.e, out of the 42 cultivars only 15 showed more than 60% reduction at 16 EC whereas 27 had less than 60% reduction. This clearly showed that the shoot is more sensitive to salinity than the root growth. This differential response of shoot and root growth is shown in Table 4 and Figure 2 where the mean shoot growth was found to be more adversely affected than the root growth. Thus, it was interesting to find that not all plant parts were equally affected. In spite of the fact that the roots were directly exposed to the saline environment it seemed significant that shoot growth was affected more adversely than the root growth. With this also 12 EC was found to be a critical level for most of the cultivars. Thus, shoot growth seemed to be better criterion for relative salt tolerance of the cultivars of the same species at early seedling stage. Based on these observations all the 42 wheat (Triticum aestivum L) cultivars were categorized into three groups viz., salt–tolerant, moderately salt–tolerant and salt–sensitive, showing <40%, 40–60% and >60% reduction in shoot growth at 12 EC over respective controls (Table 3). Further, the different rates of shoot growth of the three groups (Figure 4) as affected by increasing level of salinity showed a gradual decline in both the salt–tolerant and moderately salt–tolerant cultivars. On the other hand, the salt–sensitive cultivars had a sharp decline in growth with increasing salt concentrations.

Source of Variation	DF	Characters (MSS)	
		Shoot Length	Root Length
Replication (R)	4	0.486375**	0.061000
Varieties (V)	41	45.705478**	161.962530**
Duration (D)	3	1477.620900**	5062.824300**
Treatment (T)	4	298.855950**	884.461750**
V X D	123	10.662409**	12.680032**
V X T	164	3.96484**	7.322207**
D X T	12	70.434100**	87.099666*
V X D X T	492	0.989345**	0.970510**
Error	3356	0.053137**	0.319951**

Shoot Length: G.M. = 1.259 S.Em. ± 0.231 C.V. = 18.310 ** P = 0.01
Root Length: G.M. = 3.093 S.Em. ± 0.566 C.V. = 18.289 ** P = 0.01

Table 1: ANOVA Table (Shoot and Root Growth in 42 Wheat Cultivars).

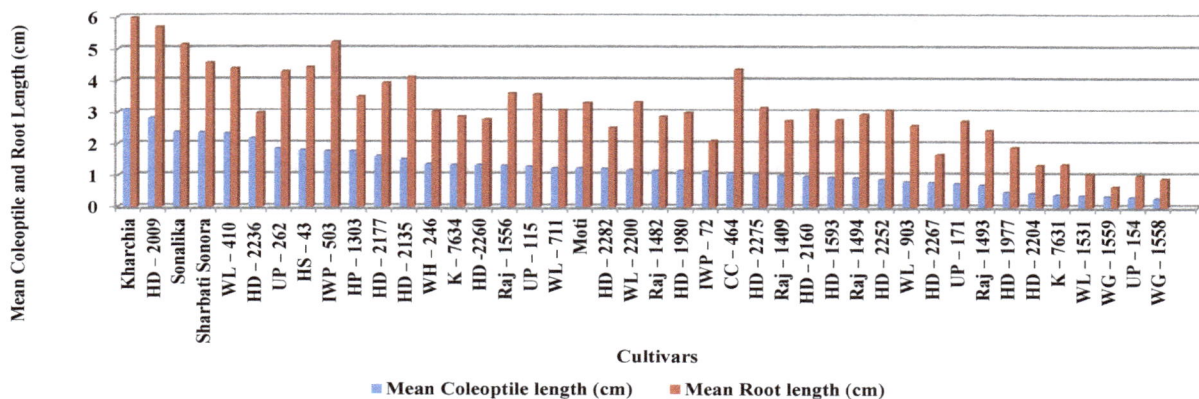

Figure 1: Relative Shoot and Root Growths of Certain Wheat (*Triticum Aestivum* L) Cultivars under Salt Stress at the Early Seedling Stage.

S No	Cultivar	Shoot Growth (cm)						Root Growth (cm)					
		Cont	4EC	8EC	12EC	16EC	Mean	Cont	4EC	8EC	12EC	16EC	Mean
1	HD-2236	3.768	4.279	2.142	0.433	0.211	2.167	4.502	5.004	3.783	1.017	0.581	2.977
2	WL-410	3.326	3.767	2.488	1.455	0.623	2.332	5.122	5.856	4.736	3.849	2.417	4.396
3	Sharbati sonora	3.263	2.956	2.639	1.585	1.315	2.352	5.271	5.190	5.017	4.221	3.158	4.571
4	Moti	2.171	2.063	1.441	0.248	0.217	1.228	5.075	4.463	3.591	1.886	1.501	3.303
5	Sonalika	3.406	2.947	2.409	1.641	1.472	2.375	5.945	5.661	5.319	4.813	3.965	5.140
6	HD-2160	1.069	1.017	0.974	0.911	0.883	0.970	3.627	3.432	3.038	2.814	2.582	3.098
7	HD-2135	2.790	2.052	1.452	0.992	0.276	1.512	6.451	5.287	4.232	2.783	1.843	4.119
8	IWP-503	3.135	2.527	1.869	0.921	0.437	1.778	7.118	6.626	5.745	4.340	2.266	5.219
9	HS-43	2.710	2.509	1.557	1.191	0.793	1.792	6.012	5.568	4.517	3.555	2.454	4.421
10	UP-262	3.374	2.647	2.003	0.977	0.275	1.855	6.630	5.683	4.429	2.969	1.735	4.289
11	HD-2177	2.948	2.329	1.719	0.841	0.198	1.607	5.019	5.482	4.488	2.790	1.853	3.926
12	WG-1559	0.748	0.623	0.193	0.153	0.070	0.357	1.442	1.211	0.293	0.227	0.095	0.653
13	HD-2267	1.516	1.358	0.745	0.180	0.125	0.785	3.658	2.891	1.039	0.619	0.105	1.662
14	IWP-72	2.430	2.015	0.950	0.190	0.125	1.142	4.119	3.496	2.006	0.607	0.240	2.093
15	HD-2282	1.613	1.543	1.445	0.927	0.568	1.219	3.046	2.902	2.844	2.007	1.719	2.503
16	WL-711	1.537	1.429	1.373	1.098	0.709	1.229	3.678	3.586	3.259	2.756	1.999	3.055
17	Raj-1482	1.711	1.654	1.229	0.660	0.554	1.161	3.909	3.762	3.072	1.893	1.729	2.873
18	HD-2260	1.935	1.491	1.406	1.360	0.443	1.327	3.470	3.213	3.096	2.692	1.395	2.773
19	WH-246	2.069	1.909	1.227	0.903	0.702	1.362	3.811	4.353	2.964	2.152	1.904	3.036
20	WL-2200	1.644	1.028	1.850	0.767	0.580	1.174	3.934	3.342	4.148	2.796	2.370	3.318
21	K-7634	1.583	1.533	1.465	1.272	0.826	1.336	3.449	2.941	3.132	2.776	2.046	2.869
22	Raj-1556	1.895	1.542	1.267	1.002	0.835	1.308	4.855	3.957	3.618	2.901	2.704	3.607
23	UP-154	0.420	0.381	0.319	0.255	0.208	0.316	1.196	1.092	1.034	0.898	0.821	1.008
24	HD-1977	0.744	0.620	0.409	0.320	0.301	0.479	2.842	2.506	1.556	1.366	1.178	1.889
25	WG-1558	0.409	0.388	0.292	0.177	0.144	0.282	0.988	1.290	1.191	0.645	0.428	0.908
26	HD-2204	0.681	0.447	0.524	0.292	0.238	0.436	1.716	1.332	1.573	1.068	0.954	1.328
27	WL-1531	0.490	0.445	0.412	0.348	0.162	0.371	1.384	1.265	1.105	0.926	0.566	1.049
28	K-7631	0.560	0.410	0.385	0.315	0.265	0.387	1.834	1.411	1.338	1.164	1.080	1.365
29	Raj-1409	1.824	1.263	0.849	0.711	0.323	1.006	4.377	3.571	2.583	1.952	1.223	2.741
30	Raj-1493	1.104	0.839	0.716	0.561	0.295	0.703	3.599	2.636	2.415	2.189	1.335	2.427
31	Raj-1494	2.095	0.925	0.825	0.594	0.235	0.935	5.685	3.311	2.393	2.013	1.265	2.933
32	WL-903	1.172	0.883	0.797	0.700	0.455	0.801	3.187	2.969	2.711	2.222	1.825	2.583
33	UP-171	1.355	1.116	0.814	0.233	0.198	0.743	4.435	3.592	2.643	1.693	1.245	2.722
34	HD-2275	1.760	1.463	1.127	0.453	0.305	1.021	4.840	3.869	3.138	2.085	1.828	3.152
35	HD-1593	2.148	0.665	0.986	0.596	0.321	0.943	5.365	2.390	3.030	1.870	1.098	2.750
36	HD-2252	1.139	1.216	0.858	0.749	0.403	0.873	3.830	4.189	2.935	2.736	1.637	3.065
37	HP-1303	2.640	2.275	1.504	1.430	1.032	1.776	4.886	3.869	3.366	3.288	2.054	3.493
38	UP-115	1.775	1.523	1.289	1.181	0.713	1.296	4.808	4.312	3.934	2.803	2.050	3.581
39	HD-1980	1.634	1.536	0.987	0.889	0.725	1.154	4.335	4.048	2.534	2.295	1.760	2.994
40	CC-464	1.931	1.103	0.985	0.905	0.465	1.078	6.199	4.515	4.293	3.835	2.968	4.362
41	HD-2009	4.077	3.627	2.583	2.337	1.514	2.824	7.755	6.441	5.383	4.891	3.909	5.675
42	Kharchia	5.291	3.661	2.610	2.277	1.616	3.091	7.838	7.070	5.522	5.110	4.332	5.974
	Means	1.997	1.666	1.267	0.834	0.527	1.259	4.315	3.799	3.167	2.416	1.767	3.092

CD at 5% P = 0.064 S.Em. ± 0.023   CD at 5% P = 0.351 S.Em. ± 0.126

Table 2: Shoot and Root Growth of Forty two Wheat Cultivars at Different Salinity Levels.

A significant reduction in shoot and root growth with increasing salinity levels was observed irrespective of cultivars and seedling age (Table 4 and Figure 2). The reduction was more pronounced after 8 EC salinity level. It was observed that the cultivars showed the first sign of germination at 48 hr after sowing irrespective of salinity level and thereafter shoot growth increased significantly with seedling age till 120 hr (Table 4 and Figure 2). In the significant interaction of varieties with treatment the cultivars showed a decrease in shoot growth with salinity levels; however, the varietal variations were quite evident. All the cultivars except **HD-2160, Sharbati sonora, Sonalika, WL-171, K-7634, Raj-1556, UP-154, HD-1977, K-7631, UP-115,** and **HD-1980** showed more than 60% reduction in shoot growth at 16 EC salinity level (Table 3). Like shoot growth, salinity in general, resulted in a

reduction in root growth irrespective of cultivars and duration. This decline in root growth was significant at all EC levels. On the other hand, root growth increased significantly with the age of the seedling (Table 5 and Figure 3). Further, it was observed that the cultivars differed significantly in their response to increasing salinity levels and all other cultivars except **HD-2160, UP-154, Sonalika,** and **WL-2200** showed less than 60% root growth at 16 EC level (Table 6).

The relative comparisons of seedling growth between different wheat cultivars indicated better performance of HD–2160 at almost all levels of salinity when compared with controls. It showed highest tolerance to salinity (i.e., 82.60 percent shoot growth at 16 EC over control) and IWP–72 showing highest inhibition in shoot growth (i.e.,

S.No.	Cultivar	Shoot Growth				Root Growth			
		4EC	8EC	12EC	16EC	4EC	8EC	12EC	16EC
1	HD-2236	113.561*	56.847	11.491	05.599	111.150*	84.029	22.589	12.905
2	WL-410	113.259*	74.804	43.746	18.731	114.330*	92.463	75.146	47.188
3	Sharbati sonora	90.591	80.876	48.574	40.300	98.463	95.181	80.079	59.912
4	Moti	95.025	66.374	11.423	09.995	87.940	70.758	37.162	29.576
5	Sonalika	86.523	70.728	48.179	43.217	95.222	89.470	80.958	66.694
6	HD-2160	95.135	91.113	85.219	82.600	94.623	83.760	77.584	71.188
7	HD-2135	73.548	52.043	35.555	09.892	81.956	65.602	43.140	28.569
8	IWP-503	80.606	59.617	29.346	13.939	93.087	80.710	60.972	31.834
9	HS-43	92.583	64.833	43.948	29.261	92.614	75.133	59.131	40.818
10	UP-262	78.452	59.844	28.956	08.150	85.716	66.802	44.781	26.168
11	HD-2177	79.002	58.310	28.527	06.716	109.224*	89.420	55.588	36.919
12	WG-1559	83.288	25.802	20.454	09.358	83.980	20.319	15.742	6.588
13	HD-2267	89.577	49.142	11.873	08.245	79.032	28.403	16.921	02.870
14	IWP-72	82.921	39.094	7.818	05.144	84.874	48.701	14.736	05.826
15	HD-2282	95.600	89.584	57.470	35.213	95.272	93.368	65.889	56.434
16	WL-711	92.973	89.329	71.437	46.128	97.498	88.607	74.932	54.350
17	Raj-1482	96.668	71.829	38.573	32.378	96.239	78.587	48.426	44.231
18	HD-2260	77.059	72.661	70.284	22.894	92.593	89.221	77.579	40.201
19	WH-246	92.266	59.304	43.644	33.929	114.221*	77.774	56.468	49.960
20	WL-2200	62.530	112.530	46.654	35.279	84.951	105.439*	71.072	60.244
21	K-7634	96.841	92.545	80.353	52.179	85.271	90.808	80.487	59.321
22	Raj-1556	81.372	66.860	52.875	44.063	81.503	74.521	59.752	55.695
23	UP-154	90.714	75.952	60.714	49.523	91.303	86.454	75.083	68.645
24	HD-1977	83.333	54.973	43.010	40.456	88.177	54.750	48.064	41.449
25	WG-1558	94.865	71.393	43.276	35.207	130.566*	120.546*	65.282	43.319
26	HD-2204	65.638	76.945	42.878	34.948	77.622	91.666	62.237	55.594
27	WL-1531	90.816	84.081	71.020	33.061	91.401	79.841	66.907	40.895
28	K-7631	73.214	68.750	56.250	47.321	76.935	72.955	63.467	58.887
29	Raj-1409	69.243	46.546	38.980	17.708	81.585	59.013	44.596	27.941
30	Raj-1493	75.996	64.855	50.815	26.721	73.242	67.101	60.822	37.093
31	Raj-1494	44.152	39.379	28.353	11.217	58.240	42.093	35.408	22.251
32	WL-903	75.341	68.003	59.726	38.822	93.159	85.064	69.720	57.263
33	UP-171	82.361	60.073	17.195	14.612	80.992	59.594	38.173	28.072
34	HD-2275	83.125	64.034	25.738	17.329	79.938	64.834	43.078	37.768
35	HD-1593	30.959	45.903	27.746	14.944	44.547	56.477	34.855	20.465
36	HD-2252	106.760*	75.329	65.759	35.381	109.373*	76.631	71.436	42.741
37	HP-1303	86.174	56.969	54.166	39.090	79.185	68.890	67.294	42.038
38	UP-115	85.802	72.619	66.535	40.169	89.683	81.821	58.298	42.637
39	HD-1980	94.002	60.403	54.406	44.369	93.379	58.454	52.941	40.599
40	CC-464	57.120	51.009	46.866	24.080	72.834	69.253	61.864	47.878
41	HD-2009	88.962	60.902	57.321	39.097	83.056	69.413	63.068	50.406
42	Kharchia	69.192	49.329	43.035	30.542	90.201	70.451	65.195	55.269

Table 3: Shoot and Root Growth of Forty two Wheat Cultivars at Different Salinity Levels (Data expressed as percent over control).

	Interaction Duration Seedling Age (hours)				Interaction Treatment Salinity Level dsm⁻¹				
	48hrs	72hrs	96hrs	120hrs	Control	4EC	8EC	12EC	16EC
Shoot	0.192	0.562	1.419	2.863	1.997	1.666	1.267	0.835	0.529
	CD at 5% P = 0.048 SEm ± 0.017				CD at 5% P = 0.022 SEm ± 0.008				
Root	0.693	2.122	3.759	5.798	4.314	3.799	3.167	2.147	1.767
	CD at 5% P = 0.020 SEm ± 0.007				CD at 5% P = 0.054 SEm ± 0.020				

Table 4: Relative Shoot and Root Growth (cm) of Certain Wheat Cultivars at Varying Salinity Levels (dsm⁻¹).

only 5.14 percent growth at 16 EC over control). The next cultivars which were relatively lesser tolerant but close to HD–2160 were K-7634, WL-711, WL-1531, HD-2260, UP-115, HD-2252 and UP-154. Based on these growth responses other cultivars of wheat followed a sequence of decrease as shown in Table 3 as far as their resistance to salt stress was concerned.

On the other hand, all the cultivars showed an increase in shoot growth with seedling age. It was evident that the different cultivars exhibited marked differences in their early seedling growth with increasing age of the seedling and that with advancement of seedling age the effect of salt declined and that, in general, tolerance to salinity increased. It was observed that root length increased with age of the seedlings in all the 42 cultivars studied irrespective of the salinity levels. This table also shows that the cultivars differed significantly in their relative root growth. Like shoot, it was observed in the present investigation that irrespective of the cultivars studied the seedlings exhibited increase in salt tolerance with the advancement of age (Tables 4 and 5, Figures 2 and 3).

A stimulation observed in growth of some cultivars as shown in Table 3 marked with asterisk (*) at moderate levels of salinity (4 and 8 dsm-1) confirmed similar observations of Eaton [35] Nieman [9] Ogra and Baijal [36] Sharma [2,7] Nauhbar [16], Yadav [17], Rani [18], Gautam [19] Parashar [20] in certain crop plants. Poljak off-Mayber and Gale [37] reported that Na+ and Cl- ions play important roles in the life of the plant within the range of suitable concentrations. The stimulation in growth might be attributed to the nutritional supplementation at low concentrations of the salt [2,4,13].

Thus, it is clear from the data that the cultivars differed in their ability to grow as seedlings under high salinity levels. That wheat showed fairly large varietal differences to salt stress had also been reported earlier by Bhardwaj [38] Sarin and Narayanan [39]. Varietal

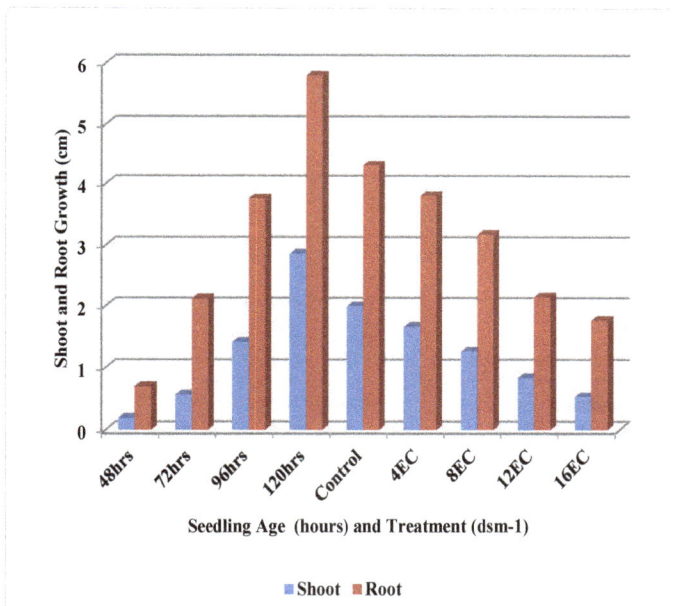

Figure 3: Effect of varying salinity levels on progressive shoot and root growth (cm) of certain wheat (*Triticum aestivum* L) cultivars at the early seedling stage. (Treatment X Duration).

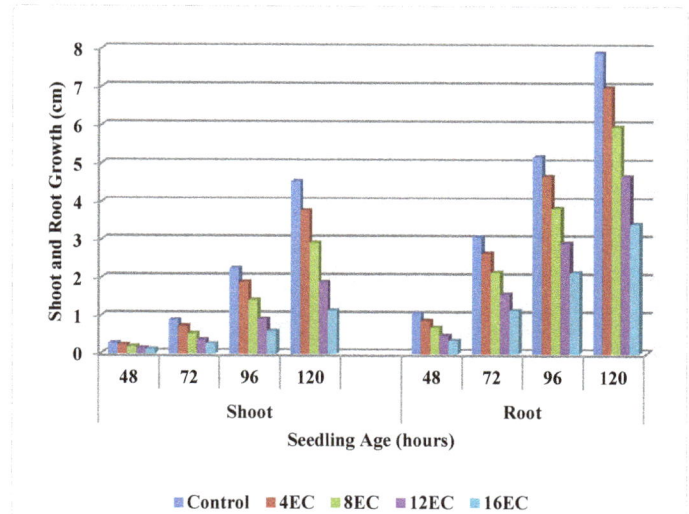

Figure 2: Relative Shoot and Root Growths of Certain Wheat (*Triticum Aestivum* L) Cultivars Under Salt Stress at the Early Seedling Stage

	Group I Salt-tolerant (**Less than 40% reduction**)	Group II Moderately Salt-tolerant (**40 – 60% reduction**)	Group III Salt-sensitive (**More than 60% reduction**)
CULTIVARS	HD-2160 K-7634 WL-711 WL-1531 HD-2260 UP-115 HD-2252 UP-154	WL-903 HD-2282 HD-2009 K-7631 HD-1980 HP-1303 Raj-1556 Raj-1493 Sharbati Sonora Sonalika CC-464 WL-2200 HS-43 WL-410 WH-246 WG-1558 Kharchia HD-1977 HD-2204	Raj-1409 Raj-1482 HD-2135 IWP-503 UP-262 HD-2177 Raj-1494 HD-1593 HD-2275 WG-1559 UP-171 HD-2267 HD-2236 Moti IWP-72

Table 6: Showing Relative Tolerance of Certain Cultivars of Wheat Based on the Percent Reduction in Coleoptile Growth at 12 EC (dsm-1) Salinity Level.

differences to salt stress were also reported in other agricultural crops by several workers Ayers [40] , Sarin [41], Bhumbla and Singh[42], Puntamkar et al.[43] Taylor[44] Epstein [45] Maas and Hoffman[46], Garrard A[47], Sheoran[48].

Conclusion

The observations recorded clearly indicated that the shoot is more sensitive to salt stress than the root and that shoot growth is a better index of relative salt tolerance of different cultivars of the same species at early seedling stage with this also 12 EC salinity level was found to be a critical level for majority of the cultivars. Thus, on the basis of the percent reduction in shoot growth at 12 EC salinity level over respective control all the cultivars were categorized into three groups *viz.*, *salt-tolerant*, *moderately salt-tolerant* and *salt-sensitive*, showing less than 40%, 40–60% and more than 60% reduction respectively.

	Seedling Age (hours)	Salinity Level dsm-1				
		Control	4EC	8EC	12EC	16EC
Shoot	48	0.280	0.236	0.184	0.145	0.114
	72	0.901	0.741	0.531	0.372	0.266
	96	2.263	1.905	1.417	0.922	0.588
	120	4.544	3.783	2.935	1.903	1.148
	CD at 5% P = 0.044 SEm ± 0.016					
Root	48	1.065	0.868	0.697	0.486	0.347
	72	3.083	2.653	2.152	1.571	1.148
	96	5.192	4.674	3.848	2.934	2.148
	120	7.915	7.002	5.972	4.675	3.425
	CD at 5% P = 0.039 SEm ± 0.108					

Table 5: Relative Shoot and Root Growth (cm) of Certain Wheat Cultivars at Varying Salinity Levels (dsm-1) (Treatment X Duration).

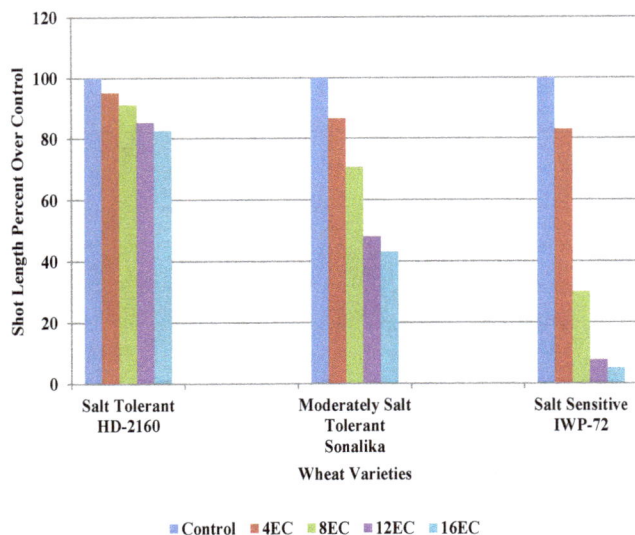

Figure 4: Relative Salt Tolerance of Three Groups (Salt-Tolerant; Moderately Salt-Tolerant and Salt-Sensitiv Wheat (*Triticum Aestivum* L) Cultivars under Salt Stress at the Early Seedling Stage (Data Expressed As Percent Over Control).

Acknowledgements

Author is indebted to Dr B D Baijal (Retd Professor Plant Physiology Department of Botany Agra College, Agra) for expert comments and to the Principal K R College, Mathura for providing necessary facilities.

References

1. Bernstein L, Hayward HE (1958) Physiology of salt tolerance. Ann Rev Plant Physiol 9: 25-46.

2. Sharma, Ravi (1982) Physiology of plant tolerance to salinity at early seedling stage. Ph D Thesis, Agra Univ., Agra.

3. Sharma, Ravi, Baijal BD (1985a) genotypic response to salt stress I: Screening for salt-resistance – selection of salt-tolerant and salt-sensitive wheat varieties. National Seminar on Plant Physiology. Institute of Agricultural Sciences BHU Varanasi 12-23 127:74-75.

4. Sharma, Ravi (1987) towards an understanding of the physiology of salt tolerance in wheat (Triticum aestivum L) at early seedling stage; XIV International Bot Cong Berlin, W Germany, 19th July-1 Aug Sym 22: 32.

5. Rani, Saroj (2007) Investigation on salt tolerance parameters specially growth and biochemical traits for selection of salt tolerant lines in legumes at the early seedling stage. PhD Thesis Dr B R Ambedkar Univ. formerly Agra University, Agra.

6. Rani, Saroj, Sharma SK, Ravi Sharma (2009) Germination and early seedling growth in six leguminous crops under salt stress. Plant Archives 9: 145 – 151.

7. Sharma, Ravi, Nidhi Parashar, Sharma SK, Singh DK, et al. (2011) Toxic effects of city and industrial effluents vis – a – vis effects of salinity and heavy metal stresses on certain crop plants. XXXIV All India Botanical Conference of The Indian Botanical Society, Oct., 10 – 12, 2011, Department of Botany University of Lucknow, Souvenir and Abstracts, Section VII Plant Physiology, Biochemistry and Pharmacology, O. VII. 29: 285.

8. Sharma (2013) Screening for salt tolerance – Selection of salt tolerant and salt sensitive wheat cultivars; Third National Conference on Innovations in Indian Science, Engineering and Technology (Bilingual Hindi & English) Organized by Swadeshi Science Movement of India, Delhi at CSIR National Physical Laboratory and IARI, New Delhi, Feb. 25 – 27.

9. Nieman RH (1962) some effects of NaCl on growth, phtosynthesis and respiration of twelve crop plants. Bot Gaz 123: 279-285.

10. Hayward HE, Bernstein L (1958) Plant growth relationships on salt-affected soils. Bot Rev 24: 584-635.

11. Shannon M C (1978) Testing salt tolerance variability among tall wheat grass populations. Agron J 70: 719-722.

12. Ogra RK (1981) Physiological studies on salt tolerance in Sorghum. Ph D Thesis. Agra University, Agra.

13. Sharma Ravi, Baijal BD (1984) Ion uptake and ATPase activity in certain wheat cultivars under salt stress conditions. VIII All India Bot Conference, Rajasthan University. J Ind Bot Soc Abst 63: 97.

14. Sharma, Ravi, Baijal BD (1984) Carbohydrate metabolism in salt tolerant and salt sensitive wheat cultivars under salt stress conditions. VII All India Botanical Conference, Rajasthan Univ. J Ind Bot Soc 63: 93.

15. Sharma, Ravi, Baijal BD (1985) Genotypic response to salt stress II: Differential physiological and biochemical response of salt-tolerant and salt-sensitive wheat cultivars; National Seminar on Plant Physiology, Institute of Agricultural Sciences, BHU Varanasi 12-23, Feb. 128: 75-76.

16. Nauhbar, Suman (2005) Relative Tolerance of Crop Plants to Salt Stress at the Early Seedling Stage, PhD Thesis, Dr B R Ambedkar Univ. formerly Agra University, Agra.

17. Yadav, Neetu (2006) Physiology of Salt Tolerance for Effective Biological Control of Salinity, Ph D Thesis, Dr B R Ambedkar Univ. formerly Agra University, Agra.

18. Rani, Saroj, Sharma SK, Ravi Sharma (2007) Effect of salinity on germination and early seedling growth in six leguminous pulse crops. XXX Annual Conference Indian Botanical Society, Jiwaji University, Gwalior (MP) India, 28 – 30 Nov. Abst & Souvenir S7. 27: 156.

19. Gautam, Aruna (2009) the Problem of Saline Wastelands and their Management – A Biological Approach with Special Reference to Mathura. PhD Thesis, Dr B R Ambedkar Univ. formerly Agra University, Agra.

20. Parashar, Nidhi (2011) Planning and Investigation for City and Industrial Effluent Utilization in Abating Pollution of River Yamuna and Improving Agricultural Production. Ph D Thesis, Dr B R Ambedkar Univ. formerly Agra University, Agra.

21. Sharma, Ravi, Baijal BD, Goyal AK(1981a) Salinity – an increasing environmental hazard; First International Conference on Environmental Education, Vigyan Bhavan, New Delhi 16th – 20th Dec. Proc Indian Environmental Society 115: 116 – 117.

22. Sharma, Ravi, Baijal BD (1987) The problem of saline wastelands and their management – a biological approach; National Seminar on Pollution, Conservation awareness and Wastelands development, K.R. College, Mathura 14 – 15 Feb.

23. Babu, Dinesh, Singh DK, Sharma SK, Arti et al. (2011) The problem of saline wastelands and their management – a biological approach. XXXIV All India Botanical Conference of The Indian Botanical Society, Oct., 10 – 12, 2011, Department of Botany University of Lucknow, Lucknow, Souvenir and Abstracts, Section VII Plant Physiology, Biochemistry and Pharmacology 6: 274.

24. Sharma (2012) Strategy for sustainable improvement of selected crops-production under saline conditions emphasizing biocontrol of salinity. National Conference on Emerging Trends in Biotechnology and Pharmaceutical Research Feb 18 – 19, Mangalayatan University, Aligarh, Souvenir and Abstract Book, Technical Session II, OP-013: 60.

25. Ranganathan R, Rajalakshmi AK (2006) the effect of NaCl and Na2SO4 salinity on germination behavior of Avicennia officinalis L. Plant Archives 6: 585 – 588.

26. Sarin MN, Rao IM (1956) Effect of sodium sulphate on early seedling growth of gram and wheat. Agra Univ J Res Sci 5: 143–154.

27. Panse VG, Sukhatme PV (1957) Statistical Methods for Agricultural Workers. ICAR (IARI), New Delhi.

28. Ayers AD, Brown JW, Wadleigh CH (1952) Salt tolerance of barley and wheat in soil plots receiving several salinization regimes. Agron J 44: 307-310.

29. Uprety DC (1970) Physiological studies on salt tolerance in two varieties of pea. Ph D Thesis, Agra Univ, Agra.

30. Richards LA (1954) Diagnosis and Improvement of Saline and Alkali Soils. USDA, Agric Hand Book No 60.

31. Ayers AD, Hayward HE (1948) A method for measuring the effects of soil salinity on germination with observations on several crop plants. Amer Proc Soil Sci 13: 224-226.

32. Ota K, Yasue T (1957) Studies on the salt injury to crops, XI the differences on the salt resistance in the young wheat varieties. Gifu U Facul Agric Res Bull 8: 14.

33. Wahhab A (1961) Salt tolerance of various varieties of agricultural crops at the germination stage, Salinity Problems in Arid Zone. Proc Tehran Symp UNESCO 185.

34. Dumbroff EB, Cooper AW (1974) Effects of salt stress applied in balanced nutrient solutions at several stages during growth of tomato. Bot Gaz 135: 219 – 224.

35. Eaton FM (1942) Toxicity and accumulation of chloride and sulphate salts in plants. J Agric Res 64: 357-399.

36. Ogra RK, Baijal BD (1978) Relative tolerance of some sorghum varieties to salt stress at early seedling stage. Indian J Agric Sci 48: 713-717.

37. Poljakoff-Mayber A, Gale J (1975) Plants in Saline Environments, Ecological studies. Springer-Verlag 15: 213.

38. Bhardwaj SN (1959) Influence of NaCl and Na_2CO_3 on some aspects of carbohydrate metabolism in wheat. Mem Indian bot soc 2: 75-78.

39. Sarin MN, Narayanan A (1968) Effects of soil salinity and growth regulators on germination and seedling metabolism of wheat. Physiol Plant 21: 1201-1209.

40. Ayers AD (1953) Germination and emergence of several varieties of barley in salinized soil cultures. Agron J 45: 68-71.

41. Sarin MN (1962) Physiological studies on salt tolerance of crop plants V. Use of IAA to overcome depressing effect of sodium sulphate on growth and maturity of wheat. Agra Univ J Res (Sci) 11: 187 – 196.

42. Bhumbla DR, Singh NT (1965) Effect of salt on seed germination. Sci Cult 31: 96-97.

43. Puntamkar SS, Mehta PC, Seth SP (1970) Note on the inducement of salt resistance in two wheat varieties by presoaking with different salts of varying concentrations. Indian J Agric Sci 41: 717 – 718.

44. Taylor RM, Young EF, Rivera RL (1975) Salt tolerance in cultivars of grain Sorghum. Crop Sci 15: 734-735.

45. Epstein E (1976) Genetic adaptation of crops to salinity; Proceedings, Workshop on Salt Effects on Plant Structures and Processes. Riverside Calif: 51-53.

46. Maas EV, Hoffman GJ (1977) Crop salt tolerance current assessment. J Irrig Drainage Div ASCE 103: 115-134.

47. Garrard A (1945) the effect of b-indolyl acetic acid on the germination and root growth of certain members of cruciferae. New Phytol 53: 165 – 176.

48. Sheoran IS, Garg OP (1978) Effect of salinity on the activities of RNase, DNase and protease during germination and early seedling growth of mung bean. Physiol Plant 44: 171 – 174.

23

Response of Sugarcane (*Saccharum officinarum L.*) Varieties to BAP and IAA on *In vitro* Shoot Multiplication

Belay Tolera[1]*, Mulugeta Diro[2] and Derbew Belew[3]

[1]Ethiopian Sugar Corporation, Research and Training Division, Variety Development Directorate, Biotechnology Research Team, Wonji Research Center, P.O.Box 15, Wonji, Ethiopia
[2]Capacity Building for Scaling up of Evidence-based Best Practices in Agricultural Production in Ethiopia (CASCAPE) Project, Addis Ababa, Ethiopia
[3]Jimma University, College of Agriculture and Veterinary Medicine, Jimma, Ethiopia

Abstract

In spite of its diverse limitations, the conventional propagation method is exclusively used for multiplication of sugarcane planting materials in the Ethiopian Sugar Estates since the establishment of the sugar industry in 1954. The present study was carried out to optimize *in vitro* shoot multiplication protocol for two selected sugarcane varieties (B41-227 and N14) widely grown in Ethiopian sugar estates to complement with the conventional propagation method. In the study, initiated aseptic shoot tip cultures of the two sugarcane varieties were treated with four concentrations (1.5, 2, 2.5 and 3 mgL^{-1}) of 6- benzylaminopurine (BAP) and Indole-3-acetic acid (IAA) (0.25, 0.5, 0.75 and 1 mgL^{-1}) while Plant growth regulator free medium was used as free check (control). The experiment was set up in a completely randomized design (CRD) with three factor factorial treatment combinations arrangements. Data were collected on number of shoots per explant, average shoot length and number of leaves per shoot after 30 days. Data were subjected to three way analysis of variance. The study verified that medium fortified with 1.5 mgL^{-1} 6-benzylaminopurine (BAP) and 0.5 mgL^{-1} indole-3-acetic acid (IAA) for B41-227 and 2 mgL^{-1} 6-benzylaminopurine (BAP) and 0.25 mgL^{-1} indole-3-acetic acid (IAA) for N14 resulted in optimum multiplication responses. On these media, B41-227 produced 15.5 ± 2.90 shoots per explant with 5.93 ± 0.57 cm average shoot length and 6.4 ± 1.49 leaves per shoot while N14 gave 11 ± 00 shoots per explant with 6.32 ± 0.23 cm average shoot length and 5.8 ± 0.06 leaves. Thus, the optimized protocol can be used for rapid *in vitro* mass multiplication of the sugarcane varieties and hence minimize the limitations sugarcane planting materials.

Keywords: Conventional propagation; *In vitro* shoot multiplication; Sugarcane; BAP; IAA

Introduction

Sugarcane is the most important cash crop widely grown in tropical and sub tropical regions of the world and is the major source of sugar [1-3]; accounting for 70% of the world's total sugar production [3-7]. Properties such as an efficient photosynthesis and efficient biomass production make this crop an excellent target for industrial processing and valuable alternative and prime candidate for ethanol and other byproducts production [8,9]. The Ethiopia sugar industry utilizes only sugarcane and has great contributions to the socio-economy of the country. It produces sugar for the household and industrial consumption, provides job opportunity for the nationals, produces ethanol and serves as a source of energy and co-products used for miscellaneous purposes [10].

However, the current sugar production of the Ethiopian Sugar Industry covers only 60% of the annual demand for domestic consumption while the deficient is imported from abroad. In order to make the country self sufficient in sugar and export the surplus sugar and produce ethanol and other by-products, the Federal government of Ethiopia is working to establish sugarcane plantation on 325,000 ha in addition to the vast expansion project of the previously established farms with erection of high crashing capacity 10 new sugar mills. However, availability of adequate amount of quality disease free planting materials of sugarcane within a short time is the major limiting factor. In addition, the yield of the existing few and old commercial cane varieties is declining sharply and some productive varieties were obsolete due to lack of alternative technologies for disease cleansing and rejuvenation. Moreover, commercialization of improved introduced and adapted sugarcane varieties took several years through the conventional propagation method, challenging the realization of the intended plan through conventional propagation method.

In conventional propagation method where stem cuttings with two to three nodes used as a planting material have various limitations. A bud produces 4 to 5 shoots [4] and the rate of propagation is 1:10 in a year [11,12]. In contrast, if estimated conservatively, micropropagation can produce 10,000 identical plants from a single bud in about 3 to 4 months [13] and the rate of propagation is 1:22 to 1:25 in 8 to 10 months [14]. Propagation from stem cuttings facilitates spread of pathogens with accumulation of disease over vegetative cycles leading to reduction in yield and quality [11,12]. Unlike the conventional propagation method, micropropagation using shoot tip or apical meristem culture have been widely used to produce virus-free plants [15-18] with rapid multiplication of new variety [19-21] and for rejuvenation and mass production of true to type and uniform planting materials from old diseased sugarcane plants [22,23]. Moreover, tissue culture raised sugarcane plants were reported to give superior cane and sugar yield as compared to their donors from conventional seed source under similar climatic conditions and agronomic management practices [24-28]. Thus, it is imperative to optimize *in vitro* propagation protocol to minimize the challenges of conventional propagation method and utilize the merits of micropropagation technology. In addition, there is no information on tissue culture study of sugarcane varieties grown in Ethiopian Sugar estates. Therefore, this study was carried out with the objective to evaluate the response of the two sugarcane varieties to different levels BAP and IAA on *in vitro* shoot multiplication.

*Corresponding author: Belay Tolera, Ethiopian Sugar Corporation, Research and Training Division, Variety Development Directorate, Biotechnology Research Team, Wonji Research Center, P.O.Box 15, Wonji, Ethiopia
E-mail: belaytolera@yahoo.com

Materials and Methods

The study was conducted at plant tissue culture laboratory of Jimma University College of Agriculture and Veterinary Medicine, Ethiopia. Two sugarcane varieties, B41-227 and N14, were used in this study. These varieties were collected from Metahara and Wonji-Shoa sugar estate seedcane nurseries of the Ethiopian Sugar Corporation. The stock plants were planted after hot water treatment (52°C for 2 hours) and grown under greenhouse conditions for two to three months. Then, preparation of explants was carried out according to the standard procedure of [12,29] with some modifications. Shoot tops were cut from actively growing sugarcane plants grown in the greenhouse. The leaves were removed and the shoot tops were taken to the laboratory. In the laboratory, surrounding leaf sheaths were removed carefully one by one until the inner white sheaths were exposed. Then, 10 cm long tops were collected by cutting off at the two ends, locating the growing point somewhere in the middle. The shoot tops were washed under running tap water for one minute with soap solution and treated with 0.3% Kocide (fungicide solution) for one and half an hour under laminar air flow cabinet. After decanting Kocide solution, shoot tops were washed three times with sterile distilled water and further immersed in 70% ethanol for one minute and rinsed three times each for five minute with sterile distilled water. Finally, the explants treated with 10% (v/v) Sodium hypochlorite solution (4% w/v active chlorine) for 20 minutes. After discarding the sodium hypochlorite solution, the explants were washed with sterile distilled water three times each for five minutes. The surface sterilized explants were excised and sized to 1 cm long and 0.5 cm diameter and cultured on initiation media. The initiated aseptic cultures were used to set up the multiplication experiment. Murashige and Skoog (1962), (MS) media, in full strength was used with different concentrations and combinations of 6-benzylaminopurine (BAP) and Indole-3-acetic acid (IAA). The medium contained 30 g/l sucrose as a carbon source and the pH was adjusted to 5.8 before gelling with 8 g/l agar and autoclaved at 121°C and 15 psi for 20 minutes. Molten medium of 40 ml was dispensed per each culture jar.

The experiment was carried out at growth room temperature range of 23-27°C under 16-hours light and eight hours dark photoperiod regimes maintained under fluorescent light having 2500 µmolm-2S-1 light intensity and 75-80% relative humidity. The experiment was laid out in a factorial treatment combination in a completely randomized design with a three factor factorial treatment combinations arrangement. Data were subjected to three way analysis of variance (ANOVA) using SAS statistical software *version 9.2* (SAS Institute Inc., 2008) and Treatments' means were separated using the procedure of REGWQ (Ryan-Einot-Gabriel-Welsch Multiple range test).

Results and Discussion

Analysis of variance (ANOVA) revealed that there was a very high significant interaction among the sugarcane varieties, BAP and IAA (vareity *BAP *IAA=$p < 0.0001$) on the number of shoots per explant, average shoot length and number of leaves per shoot in both sugarcane varieties. On MS media lacking plant growth regulators BAP and IAA (control), no multiple shoot formation occurred in any of the varieties tested. B41-227 gave the highest (15.5 ± 2.90) number of shoots per explant on MS medium supplemented with 1.5 mgL^{-1} BAP and 0.5 mgL^{-1} IAA (Table 1 and Figure 1) while N14 produced only 3.22 ± 1.67 shoots per explant on this medium composition. N14 produced maximum of 11 ± 0.00 shoots per explant on MS medium containing 2 mgL^{-1} BAP and 0.25 mgL^{-1} IAA (Table 1 and Figure 2) while the same medium composition resulted in only 5.83 ± 0.93 shoots per explant in B41-227. B41-227 also gave the highest average shoot length (5.93 ± 0.23 cm) with the largest number of 6.4 ± 1.49 leaves per shoot on MS medium containing 1.5 mgL^{-1} BAP + 0.5 mgL^{-1} IAA where only 2.92 ± 1.27 cm average shoot length with 3.5 ± 1.11 leaves per shoot observed in N14. In B41-227, maintaining BAP at 1.5 mgL^{-1} while increasing IAA levels from 0.25 to 0.5 mgL^{-1}, showed a marked increase in the number of shoots per explant from 4 ± 0.69 to 15.5 ± 2.90, average shoot length from 4.25 ± 1.02 to 5.93 ± 0.57 cm and number of leaves per explant from 3.52 ± 0.19 to 6.4 ± 1.49. However, further increase in the levels of IAA to 0.75 mgL^{-1} significantly reduced the number of shoot per explant, average shoot length and number of leaves per shoot to 7.0 ± 1.37, 5.17 ± 0.53 cm and 4.33 ± 0.11, respectively. The rate

PGRs (mg/l)		B41-227			N14		
BAP	IAA	Number of shoots per explant	Shoot length (cm)	Number of leaves per shoot	Number of shoots Per explant	Shoot length (cm)	Number of leaves per shoot
1.5	0.25	4.00^n ± 0.69	4.25^{gh} ± 1.02	3.52^i ± 0.19	2.80^t ± 0.88	2.35^{no} ± 0.54	3.33^m ± 0.42
	0.5	15.50^a ± 2.90	5.93^b ± 0.57	6.40^a ± 1.49	3.22^{qr} ± 1.67	2.92^i ± 1.27	3.50^l ± 1.11
	0.75	7.00^c ± 1.37	5.17^d ± 0.53	4.33^h ± 0.11	4.50^m ± 0.77	4.37^g ± 1.19	4.45^g ± 0.78
	1	6.21^d ± 0.21	4.74^e ± 0.27	4.50^{fg} ± 0.54	5.80^f ± 1.11	5.23^d ± 1.18	3.35^m ± 0.68
2	0.25	5.83^f ± 0.93	4.55^f ± 0.19	3.81^i ± 0.05	11.00^b ± 0.00	6.30^a ± 0.23	5.80^b ± 0.06
	0.5	10.90^b ± 0.82	5.26^d ± 1.35	5.20^d ± 0.08	5.41^{hi} ± 1.91	3.78^i ± 1.01	5.22^d ± 0.40
	0.75	6.02^e ± 1.26	5.85^b ± 0.00	4.50^{fg} ± 0.89	4.64^l ± 0.71	3.42^j ± 0.44	4.00^i ± 0.90
	1	5.52^{gh} ± 0.77	5.52^c ± 1.30	4.33^h ± 0.41	3.22^{rs} ± 0.92	2.88^l ± 0.83	3.54^l ± 0.78
2.5	0.25	6.00^e ± 0.38	4.17^h ± 1.46	4.51^f ± 0.91	3.33^q ± 1.22	4.15^h ± 0.64	2.90^o ± 0.62
	0.5	5.24^j ± 0.65	3.82^i ± 0.04	3.33^m ± 0.60	3.80^o ± 0.69	3.23^k ± 0.00	4.82^e ± 0.00
	0.75	5.20^j ± 0.82	3.52^j ± 0.31	5.51^c ± 1.27	4.10^n ± 0.54	2.25^o ± 0.47	3.00^n ± 0.81
	1	4.81^k ± 0.21	4.38^g ± 0.67	4.34^h ± 1.11	4.43^m ± 0.00	5.21^d ± 0.66	2.42^r ± 0.66
3	0.25	4.53^{lm} ± 0.55	3.19^k ± 1.02	3.50^l ± 0.00	3.52^p ± 0.00	4.53^f ± 0.31	3.84^i ± 0.11
	0.5	4.50^m ± 0.67	2.93^l ± 0.79	3.50^l ± 0.15	3.33^q ± 0.48	2.65^m ± 0.00	3.60^k ± 0.00
	0.75	3.17^{rs} ± 1.20	3.00^l ± 0.15	3.33^m ± 0.97	2.80^t ± 0.01	2.11^p ± 0.19	2.82^p ± 0.92
	1	3.05^s ± 0.55	2.44^n ± 0.44	2.92^o ± 0.24	2.52^u ± 0.51	2.46^n ± 0.00	2.50^q ± 0.71
CV (%)		8.33	5.27	7.91	8.33	5.27	7.91

PGRs=Plant growth regulators. *Values for number of shoots per explant, average shoot length and number of leaves per shoot given as mean ± SD. *Numbers with in the same column with different letter(s) are significantly different from each other at $p \leq 0.05$ according to REGWQ.

Table 1: Response of B41-227 and N14 to BAP and IAA on In vitro shoot multiplication

Figure 1: *In vitro* shoot multiplication of B41-227 at 1.5 mg/l BAP and 0.5 mg/l IAA

Figure 2: *In vitro* shoot multiplication of N14 at 2 mg/l BAP and 0.25 mg/l IAA

of sugarcane propagule multiplication depends upon auxin-cytokinin balance of culture medium [30]. A low concentration of auxin is often beneficial in conjunction with higher levels of cytokinin during shoot multiplication and exogenous auxin does not promote auxiliary shoot proliferation; however, their presence in culture medium may improve the culture growth [31]. Although cytokinins are known to stimulate cell division, but does not induce DNA synthesis. Nevertheless, the presence of auxin promotes DNA synthesis. Hence, the presence of auxin together with Cytokinin stimulates cell division and control morphogenesis thereby influences shoot multiplication. The present results in B41-227 are in line with the findings of [32] in terms of the number of shoots per explant. He found maximum of 16.5 shoots per explant on MS medium supplemented with 3 mgL^{-1} BAP with 1 mgL^{-1} IAA in sugarcane, but in contrast with the present result in N14. Murashige and Skoog (MS) medium supplemented with BAP and IAA for shoot induction in sugarcane showed the formation of profuse shoots on the MS medium containing 1 mgL^{-1} BAP along with 0.5 mgL^{-1} IAA [33,34].

Conclusion

Based on the current result, it is possible to deduce that, we have developed a rapid *in vitro* shoot multiplication protocol for the two sugarcane varieties which can use to complement the conventional propagation method. Murashige and Skoog (1962) (MS) media fortified with 1.5 mgL^{-1} benzylaminopurine (BAP) and 0.5 mgL^{-1} indole-3-acetic acid (IAA) for B41-227 and 2 mgL^{-1}benzylaminopurine (BAP) and 0.25 mgL^{-1} indole-3-acetic acid (IAA) for N14 can be used as a suitable medium combinations for rapid shoot multiplication *in vitro*.

Acknowledgements

We would like to thank Ethiopian Sugar Corporation for financing the research and Jimma University College of Agriculture and Veterinary Medicine for providing us plant Tissue Culture Laboratory with facilities.

References

1. Bahera KK, Sahoo S (2009) Rapid *in vitro* micropropagation of sugarcane (Saccharum officinarum L.cv-Nayana) through callus culture. Nature and science 7: 1545-0740.

2. Sajid GM, S Pervaiz (2008) Bioreactor mediated growth, culture ventilation, stationary and shake culture effect on *in vitro* growth of sugarcane. Pak J Bot 40: 1949-1956.

3. Suprasana P (2010) Biotechnological Interventions in sugarcane improvement: Strategies, methods and progress. Nuclear Agriculture and Biotechnology Division, Technology Development Article 316.

4. Khan S A, Rashid A, Chaudhary MF, Chaudhary Z, Afroz A (2008) Rapid Micropropagation of three elite Sugarcane (SaccharumofficinarumL.) varieties by shoottip culture. African Journal of Biotechnology 7: 2174-2180.

5. Anonymous (2005) The biology of Saccharum spp. (sugarcane). Department of health and Aging, office of the gene technology.

6. Ali A, Naz S, Iqbal J (2007) Effect of different explants and media compositions for efficient somatic embryogenesis in sugarcane (Saccharum officinarum L.). Pak J Bot 39: 1961-1977.

7. FAO (2013) Food and Agriculture Organization of the united States of America. World sugarcane production statistics.

8. Ather A, Khan S, Rehman A, Nazir M (2009) Optimization of the protocol for callus induction, regeneration and acclimatization of sugarcane cv.THATTA-10. Pak J Bot 41: 815-820.

9. Gallo-Meagher, R G. English, A Abouzid (2000)Thidiazuron stimulates shoot regeneration of sugarcane embryonic callus. *In vitro* cellular and developmental biology plant 36: 37-40.

10. Ambachew D, Firehun Y (2010) Cane sugar productivity potential in Ethiopia. Second Biannual Conference ESI 4-6

11. Biradar S, Biradar BP, Patil VC, Kambar NS (2009) *In vitro* plant regeneration using shoot tip culture in commercial cultivars of sugarcane. Karnataka Journal of Agric Scie 22: 21-24.

12. Jalaja NC, Neelamathi D, Sreenivasan TV (2008)Micropropagation for quality seed Production in sugarcane in Asia and the Pacific. Sugarcane pub 13-60.

13. Lee TSG (1987) Micropropagation of sugarcane (Saccharum spp). Plant cell Tissue Org Cul 10: 47-55.

14. Tawar PN (2004) Sugarcane Seed multiplication and Economics. National Training course on sugarcane micropropagation. VSI, Pune, India.

15. Hendre RR, Iyer RS, Kotwal M (1983) Rapid multiplication of sugarcane by tissue culture. Sugarcane 1: 58.

16. Parmessur Y, Aljanabi A, Saumtally S, Dookun-Saumtally A (2002) Sugarcane yellow leaf virus and sugarcane yellows phytoplasma: Elimination by tissue culture. Plant pathology 51: 561-566.

17. Fitch MMM, Leherer AT, Komor E, Moore PH (2001) Elimination of sugarcane yellow leaf virus from infected sugarcane plants by meristem tip culture visualized by tissue blot immunoassay.

18. Lal J, Pande HP, SK Awasthi (1996) A general micro propagation protocol for sugarcane varieties. New Bot 23: 13-19.

19. Anita PJ, Sehrawat RK, AR, Punia A (2000) Efficient and cost effective micropropagation of two early maturing varieties of sugarcane (Saccharum Spp.). India sugar 50: 611-618.

20. Sandhu SK, Gossal SS, Thind KS, Uppal SK, Sharma B, et al. (2009) Field performance of micrpropagated plants and potential of seed cane for stock yield and quality in sugarcane. Sugar tech research article11: 34-38.

21. Singh N, Kumar A, Garg GK (2006) Genotype influence of phytohormone combination and sub culturing on Micropropagation of sugarcane varieties. Indian Journal of biotechnology 5: 99-106.

22. Heinz DJ, Mee GW (1969) Plant differentiation from callus tissue ofSaccharum species. Crop sci 9: 346-348.

23. Lakshmanan (2012) Sugarcane tissue culture. Sugarcane for the future. Information sheet IS13034.

24. Anonymous(2002) Micropropagation: Tissue culture techniques in sugarcane. Indian Institute of sugarcane Research, Directorate of sugarcane Development1-2.

25. Comstock JC, Miller J (2004) Yield comparison: Disease free tissue cultures versus bud propagated planted sugarcane plants and healthy versus yellow leaf virus infected plants. USDA-ARS, Sugarcane field station, canal point, Florida 33438.

26. Geetha S, Padmanabhan D (2001) Effect of hormones on direct somatic embryogenesis in sugarcane. Sugar Tech 3: 120-121.

27. Nand L, Ram K (1997) Yield comparison in sugarcane crop raised from conventional and mericlone derived seedcane.Ind Sugar 47: 617-621.

28. Ramanand, Lal M, Singh S (2005) Comparative performance of micropropagated and conventionally raised crops of sugarcane. Sugar tech 7: 93-95.

29. Singh R (2003) Tissue culture studies of sugarcane. M.Sc. Thesis submitted to Thapar Institute of Engineering and Technology: 18-22.

30. Soodi N, Gupta PK, Srivastava RK, Gosal SS(2006) Comparative studies on field performance of micrpropagated and conventionally propagated sugarcane plants. Plant tissue cult & Biotech16: 25-29.

31. Kumari R, Verma DK(2001) Development of micropropagation protocol for sugarcane (Saccharum officinarum L)-A review. Agric Reve 22: 87-94.

32. George EF, Machakova I, Zazimalova E (2008) Plant propagation by tissue culture. 3rd edition, 175-205.

33. Dhumale ED, Engoe GL, Durge DV (1994) In vitro regeneration of sugarcane by tissue culture. Ann Pl Phys 8: 192-194.

34. Larkin PJ (1982) Sugarcane tissue and protoplast culture. Plant Cell Tissue Org. Cul 1: 149-164.

24

Microtuber Induction of Two Potato (*Solanum tuberosum* L.) Varieties

Miheretu Fufa[1]* and Mulugeta Diro[2]

[1]*Oromia Agricultural Research Institute, Adami Tullu Agricultural Research Center, Plant Biotechnology Team, P.O. Box 35, Zeway, Ethiopia*
[2]*Southern Agricultural Research Institute, Ethiopia*

Abstract

Two potato varieties namely, 'Hunde' and 'Ararsa' were tested for their microtuber induction under five levels of sucrose (40, 60, 80, 100 and 120 gl[-1]) in completely randomized design with 2×5 factorial combinations. The objective was to determine optimum concentration of sucrose for microtuber induction. In both varieties, among the five concentrations of sucrose, MS medium supplemented with 60 gl-1 sucrose exhibited a better response than the other concentrations in mean values of microtuber number, diameter, and weight and was found optimum. Accordingly, this medium produced, after 42.57 ± 0.58 days of culture, an average value of (1.97 ± 0.02) microtuber number, $(3.60 \pm 0.04$ mm) microtuber diameter, and $(0.08 \pm 0.002$ g) weight of microtuber in the variety Ararsa. On the other hand, it gave, after 35.67 ± 0.58 days of culture, mean value of (2.90 ± 0.031) microtuber number, $(2.95 \pm 0.01$ mm) microtuber diameter, and $(0.06 \pm 0.001$ g) weight of microtuber in the variety Hunde.

Keywords: *In vitro* tuberization; Microtuber; Potato; Sucrose

Introduction

The potato (*Solanum tuberosum* L.) is an important food and cash crop [1] having the first rank in the world from none grain crops to ensure food security [2]. It is a high biological value crop that gives an exceptionally high yield and more nutrients per unit area per unit time than any other major crops [3].

The tubers produced through the conventional propagation are characterized by low multiplication rate and susceptibility to pathogens [3]. Microtubers are an ideal propagating material for producing high-quality seed potatoes [4]. Microtubers, on the other hand, has several merits over *in vitro* plantlets due to their little size, reduced weight and vigorous nature.

Microtubers offer advantages of small space accommodation, ease of transport and storage for long time in addition to solving the problems of transplanting of plantlets [5,6]. Moreover, microtubers are utilized for minituber production in greenhouses and, less commonly, are directly field-planted [7]. These properties make microtubers an ideal propagating material [4] for producing high-quality seed potatoes [8].

Microtubers are also useful in other applications, including germplasm storage and exchange or as experimental research tools in the areas of plant metabolism, germplasm selection and evaluation, transformation, somatic hybridization or molecular farming, and for *in vitro* selection of agronomically important characters, such as maturity and abiotic stress tolerance [9].

Moreover, microtubers and field grown tubers have strong and consistent similarity in their morphology and biochemical features. This makes the induction, growth and development of microtubers a valuable model system [10].

In general, the use of microtuber technology appears to have enormous potential in seed tuber production, breeding programs, germplasm conservation and research. The technology helps to reduce the time necessary to supply seed tuber [7] of greatly improved quality in large scale with low cost.

In India, the advances in microtuber production is considered as second "green revolution" in agriculture and are expected to make farming more efficient, profitable and environmentally safe in addition to helping the farmers economically, socially and commercially. As a result, microtubers are used as an alternative in potato seed production [11].

Sucrose is the most critical stimulus for inducing microtubers at high concentration [6]. It is a cheap, safe and superior agent for microtuber induction [4]. No attempt was made for *in vitro* tuber induction in Ethiopia. Hence, the present study was initiated with the objective to determine optimum concentration of sucrose for microtuber induction.

Materials and Methods

Single nodal excision from one week old sprouts of the relatively clean tubers of 'Ararsa' and 'Hunde' potato varieties that were released by Sinana Agricultural Research Center in 2006 were used for microtuber induction experiment at Tissue Culture Laboratory of Jimma University College of Agriculture and Veterinary Medicine. The two varieties were tested for tuberization response under five levels of sucrose (40, 60, 80, 100 and 120 gl[-1]) in completely randomized design with 2×5 factorial combinations.

The pH of the medium was adjusted at 5.8, agar (8 gl[-1]) was added and then the medium was autoclaved at 121oC for 20 minutes at 15 psi. Murashige and Skoog (MS) basal medium containing gebberellic acid (0.1 mgl[-1]), naphthalene acetic acid (0.01 gl[-1]) and sucrose (30 gl[-1]) was used for initiation. In the case of microtuber induction, the MS basal medium was prepared for each treatment combination.

All the surface sterilization procedures were carried out under aseptic condition of laminar flow chamber following the procedure of Naik and Karihaloo [11]. One week old sprouts with buds were excised and used as initial explants. The excised explants were washed 3 times in running tap water with three drops of Tween-20. Then it was washed

*Corresponding author: Miheretu Fufa, Oromia Agricultural Research Institute, Adami Tullu Agricultural Research Center, Plant Biotechnology Team, P.O. Box 35, Zeway, Ethiopia, E-mail: miheretufufag@gmail.com

thoroughly three times with sterile distilled water and immersed in 70% ethyl alcohol for 10 seconds. The alcohol was removed by three times washing with sterile distilled water and sterilized with 10% NaOCl for 20 minutes.

The excised explants were dissected into single nodes (2cm) on a sterile plate after removing the leaves. Six explants were cultured into 40 ml of an initiation culture medium in culture jar and incubated under a 16 h photoperiod at 24oC with a light intensity of 2500 lux. For 3-4 weeks, the sprouts were allowed to grow into plantlets having nodal segments. After decanting the multiplication medium, the plantlets were kept in a conditioning medium before being used for microtuber induction medium to avoid the carryover effects of hormones. Forty milliliters of microtuber induction medium was dispensed into each culture jar. The culture were transferred to the growth room and kept at a temperature of 18oC under dark condition.

Data collection and analysis

The date of formation of microtuber (first, second and third) was followed carefully and days (50%) to microtuber formation was recorded and used for analysis. The number of microtubers produced by each explant was counted and all the microtubers produced were harvested. The diameter (mm) of each microtuber was measured by Digital Caliper. Immediately after harvest, each microtuber was weighed on sensitive balance to get the mean microtuber weight (g). After fifteen days of light exposure, the microtubers were treated with gibberellic acid (GA3) and incubated in the dark before planting in the green house. The number of the microtubers germinated and established was counted to get their percent survival under in vivo. The data were subjected to the analysis of variance (ANOVA) at 5% level of significance using SAS statistical software [12]. The REGWQ multiple comparison procedure was used for separating significant means.

Results and Discussion

Analysis of variance revealed that sucrose and variety interaction had very highly significant effect (α=5%) on days to microtuber induction and on the average number, diameter (mm) and weight (g) of microtubers (Table 1). This implied that there is interdependence of sucrose and genotype on induction of microtuber of potato. Thus, the microtuber induction of genotypes varies with the level of sucrose.

Effects of sucrose on days to tuberization, mean microtuber number and diameter

At 40 gl⁻¹ sucrose, both varieties did not produce microtubers. However, when 60 gl⁻¹ sucrose was added to growth media, 'Hunde' produced microtubers in 36 days, which is significantly earlier than that of 'Ararsa' (43 days). Increasing concentration of sucrose from

60 to 80 gl⁻¹ delayed microtuber formation in both varieties but more pronounced on Ararsa. This might be due to the marked variation in the responses of plant gene to changing sucrose status. Some genes are induced, some are repressed, and others are minimally affected [13].

Moreover, microtuber number and size get reduced, in both varieties, as the concentration of sucrose increased from 60 gl⁻¹ to 80 gl⁻¹ (Table 1). At 120 gl⁻¹ of sucrose, both genotypes did not produce microtubers (Table 1). The absence of microtuber formation at high sucrose concentration might be due to the effect of supra optimal level of sucrose that can result in an unfavorable osmotic condition for water uptake, and then affected microtuber formation of the seedlings.

Effects of sucrose on mean microtuber weight

A decreasing trend in mean weight (g) of microtuber was observed, in both varieties, as the level of sucrose increased (Figure 1). This might be, again, due to the effect of high sucrose level on osmotic condition of the culture for water uptake that affect cell turgidity [13] and hence microtuber weight.

In both varieties, among the five concentrations of sucrose, MS medium supplemented with 60 gram litre-1sucrose exhibited a better response than the other concentrations in mean values of microtuber number, diameter, and weight and was found optimum. Accordingly, this medium produced, after 42.57 ± 0.58 days of culture, an average value of (1.97 ± 0.02) microtuber number, (3.60 ± 0.04 mm) microtuber diameter, and (0.08 ± 0.002 g) weight of microtuber in the variety 'Ararsa'. On the other hand, it gave, after 35.67 ± 0.58 days of culture, mean value of (2.90 ± 0.031) microtuber number, (2.95 ± 0.01 mm) microtuber diameter, and (0.06 ± 0.001 g) weight of microtuber in the variety 'Hunde' (Table 1).

The present result is in agreement with that of Aslam et al. [9] who found that a medium containing 6% sucrose was optimal in terms of minimum time of induction (34), mean tuber number (1.2) and weight (0.03 g) of microtubers per single nodal explant in cultivar Desiree. Imani et al. [14] also reported that MS medium supplemented with 60 gl⁻¹ of sucrose as the best in producing the maximum number (4.20) and size (0.44 cm) of micro tubers. Iqbal et al. [4] also recorded similar results on mean numbers of tubers (4.8) on MS medium treated with 60 gl⁻¹ sucrose. Kanwal et al. [15], on the other hand, reported that MS medium supplemented with 30 and 40 gl⁻¹ sucrose did not produce microtubers.

Conclusion

A protocol for microtuber induction of potato varieties 'Ararsa' and 'Hunde' from single nodal explant has been developed. The result indicated that microtuber induction of potato was highly dependent on sucrose and genotype interaction.

ARARSA					HUNDE			
Sucrose (g/l)	DT	MTN	MTD	MTWT	DT	MTN	MTD	MTWT
40	0.00 ± 0.00ᶜ	0.00 ± 0.00ᶜ	0.00 ± 0.00ᶜ	0.00 ± 0.00ᵈ	0.00 ± 0.00ᵈ	0.00 ± 0.00ᵈ	0.00 ± 0.00ᵈ	0.00 ± 0.00ᵈ
60	42.67 ± 0.58ᵇ	1.97 ± 0.02ᵃ	3.60 ± 0.04ᵃ	0.08 ± 0.002ᵃ	35.67 ± 0.58ᶜ	2.90 ± 0.031ᵃ	2.95 ± 0.050ᵃ	0.06 ± 0.001ᵇ
80	45.00 ± 1.00ᵃ	1.30 ± 0.08ᵇ	3.07 ± 0.03ᵇ	0.05 ± 0.001ᵇ	40.00 ± 0.00ᵇ	2.06 ± 0.081ᵇ	2.81 ± 0.015ᵇ	0.04 ± 0.001ᵇ
100	0.00 ± 0.00ᶜ	0.00 ± 0.00ᶜ	0.00 ± 0.00ᶜ	0.02 ± 0.00ᵈ	44.67 ± 0.58ᵃ	1.30 ± 0.042ᶜ	2.56 ± 0.044ᶜ	0.03 ± 0.002ᶜ
120	0.00 ± 0.00ᶜ	0.00 ± 0.00ᶜ	0.00 ± 0.00ᶜ	0.00 ± 0.00ᵈ	0.00 ± 0.00ᵈ	0.00 ± 0.00ᵈ	0.00 ± 0.00ᵈ	0.00 ± 0.00ᵈ

Means with the same letters in a column are not significantly different from each other using the Ryan-Einot-Gabriel-Welsch Multiple Range Test (REGWQ) at α= 0.05.
DT= days to tuberization, MTN= microtuber number, MTD=microtuber diameter (mm), and MTWT= Microtuber Weight (g).

Table 1: Effects of sucrose on days to tuberization, number, diameter and weight (g) of microtuber.

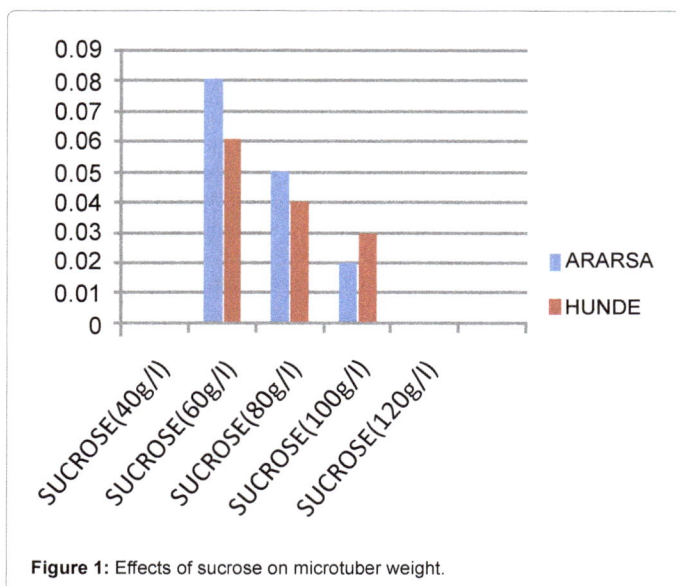

Figure 1: Effects of sucrose on microtuber weight.

MS medium supplemented with 60 gl⁻¹ exhibited fewer days to microtuber formation (35.67 ± 0.58/42.67 ± 0.58), better mean number (2.90 ± 0.031/1.97 ± 0.02), diameter (2.81 ± 0.015/3.60 ± 0.04), fresh weight (0.06 ± 0.001/0.08 ± 0.002) and dry weight (0.046 ± 0.001/0.063 ± 0.002) of microtubers in 'Hunde' and 'Ararsa' varieties, respectively. However, microtuber production needs further improvement as the size of microtubers produced was not large enough. Thus, trying different levels of BAP in combination with sucrose and extending the time of harvesting may be helpful to improve the size of the microtuber.

References

1. Lemaga B, Kakuhenzire R, Gildemacher P, Borus D, Gebremedhin WG, et al. (2009) Current status and opportunities for improving the access to quality potato seed by small frmers in Eastern Africa. Symposium 15th triennial of the Symposium of the International Society for Tropical Root Crops 2-6.

2. FAO (2008) Workshop on Opportunities and Challenges for Promotion and Sustainable Production and Protection of the Potato Crop in Vietnam and elsewhere in Asia.

3. Badoni A, Chauhan JS (2010) Conventional vis -a- visBiotechnological Methods of Propagation in Potato: A Review. Stem Cell 1: 1-6.

4. Hussain I, Zubeda C, Aish M, Rehana A, Naqvi S, et al. (2006) Effect of chlorocholine chloride, sucrose and BAP *in vitro* tuberization in Potato (*Solanumtuberosum* L. cv. cardinal). Pak J Bot 38: 275-282.

5. Hoque ME (2010) *In vitro* tuberization in potato (*Solanumtuberosum* L.).

6. Nistor A, Campeanu G, Atanasiu N, Chiru N, Karacsonyi D (2010) Influence of potato genotypes on "in vitro" production of microtubers. Romanian Biotechnological Letters 15: 1-8.

7. Badoni A, Chauhan JS (2009) A Note on Micro Tuber Seed Production of Potato: Necessitate Step for Uttarakhand Hills. Report and Opinion 1: 9-11.

8. Yu WC, Joyce PJ, Cameron DC, McCown BH (2000) Sucrose utilization during potato microtuber growth in bioreactors. Plant Cell Reports 19: 407-413.

9. Dobránszki J, Tábori KM (2010) Influence of Nitrogen Supply of Potato Plantlets on *In Vitro* Tuberization Pattern under Inductive and Non-inductive Conditions. Potato Research 53: 121-127.

10. Aslam A, Ali A, Naveed NH, Saleem A, Iqbal J (2011) Effect of interaction of 6-benzyl aminopurine (BA) and sucrose for efficient microtuberization of two elite potato (Solanum tuberosu L.) cultivars, Desiree and Cardinal. African Journal of Biotechnology 10: 12738-12744.

11. Naik PS, Karihaloo JL (2007) Micropropagation for production of quality potato seed in Asia-Pacific. Asia-Pacific consortium on agricultural biotechnology, New Delhi, India.

12. SAS Institute Inc. (2008) SAS/STAT ®9.2 user's guide.

13. George EF, Hall MA, De Klerk GJ (2008) Plant Propagation by Tissue Culture 3rd Edition: The Background 1:1-501.

14. Imani AA, Qhrmanzadeh R, Azimi J, Janpoor J (2010) The effect of various concentrations of 6-Benzylaminopurine (BAP) and Sucrose on *in vitro* potato (*Solanumtuberosum*L.) microtuber induction. American-Eurasian J Agric Environ Sci 8: 457-459.

15. Kanwal A, Shoaib K (2006) *In Vitro* Microtuberization of Potato (*Solanumtuberosum*L.) Cultivar Kuroda: A New Variety in Pakistan.

25

Study on the Reaction of Sugarcane Genotypes (CIRAD-2011) to Sugarcane Smut (*Sporisorium scitamineum*) in the Ethiopian Sugarcane Plantations

Abiyot Lemma[1]*, Hadush Hagos[1], Yohannes Zekarias[2] and Amrote Tekle[1]

[1]*Ethiopian Sugar Corporation Research and Training P.o.box 15, Wonji, Ethiopia*
[2]*Bayer Trade Representative Office P.o.box 1/1065, Addis Ababa, Ethiopia*

Abstract

Sugarcane smut disease caused by the fungus *Sporisorium scitamineum* significantly reduces the yield and quality of sugarcane. Study on the reaction of thirty one newly imported genotypes from CIRAD was conducted to evaluate their reaction to sugarcane smut at Metahara sugarcane plantation. The findings of this study were based on collection of sugarcane smut spores, testing the viability of the spore and inoculating the tested material by immersing in the spore suspension (5×10^6 spores ml^{-1}) for 30 minutes. During the course of the experiment, data on smut stool count begins six weeks after planting and continues at 10-day intervals till ten months after planting. Evaluation of genotypes for resistance to smut was made based on percentage of infected stools by using a 0-9 disease rating scale. Among the tested materials, no genotypes were shown an immune reaction to the Ethiopian sugarcane smut isolates. While about 81.3% of genotypes have shown from moderately resistant to very highly resistant reaction in which 21.9%, 9.4%, 21.9% and 28.1% were Very Highly Resistant (VHR), Highly Resistant (HR), Resistant (R) and Moderately Resistant (MR) reactions to sugarcane smut respectively. 18.7% of genotypes were shown Intermediate (I) (3.1%), Susceptible (S) (6.25%) and Very highly susceptible (VHS) (9.4%) reactions, so that these genotypes PSR 97 051 (I), FG 02 551 (S), FG 03 396 (S) FG 06 729 (VHS), FG 03 173 (VHS), and FG 03 526 (VHS), won't be considered as promising genotypes which can competently resist to sugarcane smut in the country.

Keywords: Reaction; Sugarcane genotypes; CIRAD-2011; Sugarcane smut; *Sporisorium scitamineum*

Introduction

Sugarcane smut is caused by *Sporisorium scitamineum*, a basidiomycetous fungus that exists in several physiologic races [1,2]. It is believed to be originated in Asia and spread slowly to other continents through Africa in the late nineteenth and early twentieth [3]. *S. scitamineum* is one of the most prevalent diseases and it has been responsible for the demise of several leading varieties [4-6]. This disease was first reported in Ethiopia with the commencement of commercial sugarcane plantation at Wonji Shoa in the early 1950's [7]. Since then, it spreads to other newer sugarcane commercial farms and is threatening the sugar industry due to its effect on cane yields.

Symptoms of sugarcane smut include black whip-like structures from terminal meristem or meristems of lateral buds of infected stalks [8]. On maturity, the whip raptures and frees millions of tiny black spores which are then disseminated by wind. Other symptoms include smutted side shoots, stem galls, grassy stools, bud proliferation, general reduction in plant size, and increased tillering [1,9,10]. Perpetuation of the smut pathogen occurs through planting diseased seed cane. In plant cane, smut infection usually remains latent in the buds of underground stubble. When such buds are harvested, the shoots that come up bear smutted whips [1]. Primary transmission of the smut fungus occurs through planting diseased seed cane. Secondary spread is through windblown spores. Spores in or on soil are carried to different fields by rain or irrigation water where they can cause new infections to cane [1].

Smut reduces the yield and quality of sugarcane. Reduction in yield and quality varies widely in different sugarcane growing areas of the world and is mainly dependent on the races of the pathogen present, the sugarcane genotypes and the prevailing environmental conditions [11]. Estimates of economic losses have ranged from negligible to levels serious enough to threaten the agricultural economy of an area [11].

Sandhu et al. [4] was reported, sugar cane smut can cause 12%–75% yield losses. Lee Lovick [11], was also mentioned, a total crop failure may possible if susceptible genotypes are used and conditions are favorable for infection. In Ethiopia sugarcane smut causes 19.3 to 43% in sugar yield and 30 to 43% in cane yield [12], the total monetary loss due to this disease in the older sugar estates of Ethiopia was estimated to about ten million birr per year [13]. According to Firehun et al. [13] Smut in Ethiopia led to the discontinued cultivation of varieties like Co 419 and NCo 310.

The most efficient and economic method for disease control, including sugarcane smut, is the use of resistant varieties [14-16]. The efforts made by the Research and Training of Ethiopian Sugar Corporation to attain high-yielding, disease-resistant *Saccharum* spp. which can be introduced for commercial production and to introduce *Saccharum* spp. hybrid clones with particular attributes for breeding program results with continuous use of introduced germplasm mainly from CIRAD and other organizations and institutes abroad. About 165 genotypes in four batches have been imported from CIRAD during 2011 – 2015. After the quarantine precautions are being exercised in closed quarantine at Worer center of EIAR [Ethiopian Institute of Agricultural Research], 31 genotypes from the first batch [CIRAD

*Corresponding author: Abiyot Lemma, Ethiopian Sugar Corporation Research and Training P.o.box 15, Wonji, Ethiopia
E-mail: alemma2009@gmail.com

2011] have evaluated for their reaction to smut at Metehara Sugarcane Plantation Farm.

Objective

To evaluate the reaction of sugarcane genotypes [CIRAD-2011] to sugarcane smut

Materials and Methods

Study on reaction of the newly imported genotypes to sugarcane smut was conducted in Metahara sugarcane plantation during 2014/15 cropping seasons. From the test genotypes, stalks were cut from 10 month-old seed cane nursery. The leaves were detached from the stalks to expose buds and then cut into single-budded setts. To swell up the buds and ensure their susceptibility, the tested materials were incubated for an overnight in a polythene bag filled with a liter of water following the procedures of Bock [17].

Sugarcane smut spore were collected from infected fields of Metehara sugarcane plantation commercial farm, and its viability were tested before it was used for inoculation. The incubated setts were immersed in the spore suspension [5 x 10^6 spores ml^-1] for 30 minutes (Lee-Lovick, 1978). To create favorable condition for infection, the inoculated setts were again incubated in a polythene bag filled with a liter of water just after inoculation [17].

A day after inoculation, the planting material were planted in Randomized Complete Block Design [RCBD] with three replications in the field number "OR 19C of Metehara Sugarcane Plantation". Plot size was six furrows of 6 m length each, i.e. 52.2 m^2. In the course of the experiment, all cultural practices remain the same as recommended for commercial sugarcane production.

During the course of the experiment, data on smuted stool count/observation begins six weeks after planting and continues at 10-day intervals till ten months after planting. Total number of stool count was taken at four months after planting. After recording smut-affected stools, they were uprooted and buried at the edge of the field. Then, evaluation of genotypes for resistance to smut was made based on percentage of infected stools by adopting the scale used by Latiza et al [18] (Table 1).

Result and Discussion

A total of 32, thirty one genotypes of CIRAD 2011 and one widely grown commercial check cultivar, which were planted during 2014, had shown variable reaction to sugarcane smut isolates at Metehara Commercial sugarcane plantation farm. Among these, no any material was shown an immune reaction to the Ethiopian sugarcane smut isolates. While about 81.3% of the tested materials have shown from moderately resistant to very highly resistant reaction in which 21.9%, 9.4%, 21.9% and 28.1% were Very Highly Resistant (VHR), Highly Resistant (HR), Resistant (R) and Moderately Resistant (MR) reactions to sugarcane smut respectively. Genotypes like FG 06 729, FG 03 526 and FG 03 173 which have imported to the country with import certificate information of resistant reaction to sugarcane smut have found to be very highly susceptible to the Ethiopian sugarcane smut isolates (Table 2). Likewise genotypes PSR 97 051 (intermediate), FG 02 551 (susceptible) and FG 03 396 (susceptible) had brought to the country with moderately resistant, intermediate and resistant reactions information respectively. Different sugarcane cultivars in the world may react differently to sugarcane smut isolates. For example cultivar H50-7209 is susceptible to smut in South Africa, but resistant in Taiwan

Infected stools (%)	Disease rating	Reaction group
0	0	Immune
0.1 – 2.5	1	Very highly resistant (VHR)
2.6 – 5.5	2	Highly resistant (HR)
5.6 – 7.5	3	Resistant (R)
7.6 – 12.5	4	Moderately resistant (MR)
12.6 – 15.5	5	Intermediate (I)
15.6 – 18.0	6	Moderately susceptible (MS)
18.1 – 22.5	7	Susceptible (S)
22.6 – 25.5	8	Highly susceptible (HS)
25.6 – 100	9	Very highly susceptible (VHS)

Table 1: Disease rating scale [18].

S/N	Genotypes	Infected Stool Incidence (%)	Disease rating	Reaction
1	FG 06 729	36.1	9	VHS
2	FG 03 204	1.8	1	VHR
3	FG 04 187	10.7	4	MR
4	FG 06-700	3.1	2	HR
5	CP 00 1252	6.9	3	R
6	FG 03 103	7.9	4	MR
7	DB 700 47	11.6	4	MR
8	TCP 93 4245	1.8	1	VHR
9	FG 03 318	11.5	4	MR
10	PSR 97 051	12.6	5	I
11	HO 95 988	11.0	4	MR
12	FG 03 173	40.4	9	VHS
13	FG 03 447	6.0	3	R
14	FG 02 551	18.6	7	S
15	CP 99 1534	5.8	4	MR
16	B52/298	8.2	4	MR
17	FG 02 553	8.9	4	MR
18	FG 03 418	2.8	2	HR
19	CP 99 1894	0.2	1	VHR
20	FG 03 396	19.2	7	S
21	PSR 97 087	4.7	3	R
22	FG 03 425	6.6	3	R
23	FG 03 214	1.3	1	VHR
24	PSR 97092	7.4	3	R
25	FG 03 372	5.4	3	R
26	FG 04 708	0.5	1	VHR
27	VMC 95 212	0.8	1	VHR
28	VMC 95 173	11.3	4	MR
29	FG 04 705	2.8	2	HR
30	FG 04 829	6.9	3	R
31	FG 03 526	31.4	9	VHS
32	FG 04 754	2.2	1	VHR

Table 2: Reaction of CIRAD 2011 genotypes to Smut.

[19]. NCo376 which has been considered as resistant cultivar in China [20], is out of production due to its high susceptibility to sugarcane smut in the Ethiopian Sugarcane plantation [13]. Unlike the varietal reaction information in the import certificate, this study has shown that about 16.1% of the test genotypes have not competently resist to the Ethiopian sugarcane smut isolates. In general, 18.7% of the tested materials in this study were shown intermediate (3.1%), susceptible (6.25%) and very highly susceptible (9.4%) reaction (Table 2) so that these genotypes will not be considered as promising genotypes which can competently resist to sugarcane smut in the country.

Conclusion and Recommendation

Thirty one genotypes of CIRAD 2011 and one widely grown commercial check cultivar, a total of 32 genotypes, which were planted during 2014, had shown variable reaction to sugarcane smut. Among these, no any material was shown an immune reaction to the Ethiopian sugarcane smut isolates. While about 81.3% of the tested materials have shown from moderately resistant to very highly resistant reaction in which 21.9%, 9.4%, 21.9% and 28.1% were very highly resistant (VHR), highly resistant (HR), resistant (R) and moderately resistant (MR) reactions to sugarcane smut respectively. 18.7% of genotypes were shown intermediate (PSR 97 051), susceptible (FG 02 551, FG 03 396) and very highly susceptible (FG 06 729, FG 03 173, FG 03 526) reaction, so that these genotypes won't be considered as promising genotypes which can competently resist to sugarcane smut in the country.

References

1. Agnihotri VP (1983) Diseases of Sugarcane. Oxford and IBH Publishing Company. 65-86.

2. Anon (1984) National Sugar Research Centre (NSRC) Annual Report. 41-42.

3. Elston DA, Simmonds W (1988) Models of sugarcane smut disease and their implications for testing variety resistance. Journal of Applied Ecology 25: 319-329.

4. Sandhu SA, Bhatti DS, Rattan BK (1969) Extent of losses caused by red (Physalosporatucumane NSis Speg.) and smut (Sporisorium scitamineum Syd) Journal of Research 6: 341-344.

5. Whittle AM (1982) Yield loss in sugar-cane due to culmicolous smut infection, Tropical Agriculture 3: 239-242.

6. Hoy JW, Hollier CA, Fontenot DB, Grelen LB (1986) Incidence of sugarcane smut in Louisiana and its effects on yield. Plant Disease 70: 59-60.

7. Anonymous (1986) A report on Agricultural Research and Services Project. Ethiopian Sugar Corporation Vol I, Agrima Project Engineering & Consultancy Services. Bombay, India.

8. Ferreira SA, Comstock JC (1989) In: Diseases of sugarcane. Major Diseases.

9. Ricaud CBT, Grishan MP (2001) an international project on genetic variability within sugarcane smut Proc. Int Soc Sugar Technol 24: 459-461.

10. Rott P, Bailey A, Comstock JC, Croft BJ, Sauntally AS (2000) A guide to sugarcane diseases. Published by CIRAD and ISSCT. 339.

11. Lee-Lovick G (1978) Smut of sugarcane-Sporisorium scitamineum. Review of Plant Pathology 147: 181-188.

12. Abera T, Mengistu H (1992) Effect of smut on yield of sugarcane in Ethiopia. Proceedings of the joint conference Ethiopia Phytopathological Comitee and Comitee of Ethiopian Entomologists.

13. Firehun, Abera YT, Yohannes Z, Leul M (2009) Hand book for Sugarcane Pest Management in Ethiopia, Ethiopian Sugar Development Agency Research Directorate.

14. Villalon B (1982) Sugarcane smut in lower Rico Grande Valley of south Texas. Plant Dis 66: 605-606.

15. Pruett CJH, Waller JM (1989) A report on sugarcane diseases in the Santa Cruz area of Bolivia. Sugar Azucar 84: 35-37.

16. Phelps RH, Donelan AF (1991) Disease and other problems affecting sugarcane cultivar change in Tridinad and Tobago 1975 to 1989. In: Proceedings of the 24th West Indies sugar Technologist' Conference, Kington, Bridgetown, Barbacaos, April 8-12 (Sugar Association of the Caribbean (1989) Caroni Research Station, carapichaina, Tridinad) 88-95.

17. Bock KR (1964) Studies on sugar-cane smut (Sporisorium scitamineum) in Kenya. Trans Br Mycol Soc 47: 403-417.

18. Latiza AS, Ampusta DC, Rivera JR (1980) Reaction of sugarcane clones to strain B of Sporisorium scitamineum syd. Int Soc Sugarcane Technology 2: 1456-1462.

19. Nzioki HS, Jamoza JE, Olweny CO, Rono JK (2010) Characterization of physiologic races of sugarcane smut (Sporisorium scitamineum) in Kenya. African Journal of Microbiology Research 4: 1694-1697.

20. You-Xiong Q, Jian-Wei L, Xian-Xian S, Xu Li-Ping, Ru-Kai C (2011) Differential Gene Expression in Sugarcane in Response to Challenge by Fungal Pathogen Sporisorium scitamineum Revealed by cDNA-AFLP. Journal of Biomedicine and Biotechnology 2011: 10.

26

In Vitro and *In Vivo* Effects of Aqueous Extract of *Rosmarinus officinalis* L. (Rosemary) in The Control of Late Blight Disease of Potato Caused by Phytophthora Infestans Mont. De Bary. in Algeria

Messgo Moumene S[1] *, Olubunmi OF[2], Laidani M[1], Saddek D[1], Houmani Z[1] and Bouznad Z[3]

[1]*Laboratory for Research on Medicinal and Aromatic plants, Science and life Faculty, University of Blida1, BP. 270, Soumaa road, 09100, Blida, Algeria*
[2]*Department of Crop Protection and Environmental Biology, University of Ibadan, Nigeria*
[3]*Laboratory of Phytopathology and Molecular Biology, National graduate school of Agronomy El Harrach, Algeria*

Abstract

The fungus *Phytophthora infestans* is known to develop resistance against the metalaxyl (fungicide), commonly used in the control of potato mildew disease. There is therefore urgent need to explore the potentials of alternative fungicides which are potent, affordable, readily available, easy to prepare and environment friendly. The study was carried out to test the fungicidal potential of aqueous extracts of *Rosmarinusofficinalis* (Rosemary), *in vitro* and *in vivo* on two isolates of *P. infestans* collected from two potato producing Algerian areas: Bourkika (Tipaza City) and El Abbadia (Aindefla City). Various concentrations of crude extracts of *Rosmarinus officinalis* applied by direct contact in the following dilutions: 5%, 10% and 20% on medium with pea-agar (PPA), allowed the inhibition of mycelial growth of *P. infestans* isolates. The observed rates of inhibition exceeded 85% and the inhibitive minimal concentration (CMI) was 5%. Parallel structural modifications, caused by mycelial lyses, as well as the deformation or, and the digestion of the contents of sporangia affected the morphology of both strains from the lowest concentration. The sporulation and the germination were inhibited by this aqueous extract (100%). Also, the absence of resumption of mycelial growth on medium PPA and absence of the mildew symptoms on detached Spunta potato leaves confirmed the fungicidal effect of the Rosemary aqueous extract. This also translated *in vivo* as significant reduction of the disease was observed. Disease reduction was recorded for the preventive application modes by spraying with the crude aqueous extract (86.2%) and by watering, while for the curative mode with crude extract (81%). On the other hand, Spunta variety was more marked for preventive mode by watering (85%) and the curative one (90%) also, A2 isolate was more inhibited for the application of *R. officinalis* aqueous extract by curative (83%), spraying mode (86%) and watering modes. Besides, treatments made in preventive modes by spraying and watering showed a total inhibition of the sporulation (100%), exceeding 85% in Spunta variety and 96% for A1 isolate was observed in the curative mode of application. This study thus confirms the antifungal potential of aqueous extract of *Rosmarinus officinalis* on *P. infestans* isolates. It is thus recommended for use as bio-fungicide in the management of potato mildew disease.

Keywords: *Rosmarinus officinalis*; *Phytophthora infestan*; Fungicide; *Solanum tuberosum*

Introduction

Late blight potato, caused by *Phytophthora infestans* is one of the most destructive diseases of potato. Until now, chemical control remains the most important control against the disease. However, the use of pesticides has several constraints such as high cost of fungicides, negative effects on the environment and the health of the consumers [1]. Also, the appearance of aggressive isolates of this fungus, mostly resistant to the current synthetic fungicides, has created new challenges for potato growers [1].

Various reports have highlighted the action of certain plant extracts and some essential oils against the phyto pathogenic agent of potato mildew disease [2].The experiment, therefore aimed at evaluation of the antifungal activities of crude aqueous extract of *Rosmarinus officinalis* on *Phytophthora infestans* isolates, while determining *in vitro* the effect of the extract on the mycelial growth, sporulation and germination, the determination of the minimal and lethal inhibitory concentrations, as well as the inhibition of their survivability after treatments. Also, observations were made *in vivo* on detached potato's leaves, its effect on the symptoms appearance period, disease reduction and sporulation inhibition.

Materials and Methods

Plant material: The plant material used includes the aerial parts composed of stalks, leaves and flowers of rosemary (*Rosmarinus officinalis* L.) as well as seed potato tubers. The collection of plants was made in May, 2011 in Medea city in M'sallah locality. After the harvest, the plant material was cleaned with tap water to clear it of fragments of soil, then it was left to dry away from direct sunlight, at ambient temperature and in open air.

Two approved varieties of potato, certified and widely cultivated in Algeria, Spunta and Kondor were collected and retained for *in vivo* study. Seed tubers were supplied by the National Center of Control and Certification of seeds and seedlings (C.N.C.C) of EL Harrach, Algeria.

Fungal isolates: Two purified fungal isolates of *Phytophthora*

*Corresponding author: Messgo-Moumene Saida, Laboratory for Research on Medicinal and Aromatic plants, Science and life Faculty, University of Blida1, BP. 270, Soumaa road, 09100, Blida, Algeria, E-mail: moumene_saida@yahoo.fr

infestans, identified respectively as A1 and A2, were selected for this study. The latter was taken from potato producing areas: El abadia of Ain defla city and Bourkika of Tipaza city. They were maintained by transplanting on pea- agar medium and incubated at 18°C during 20 days [3].

Preparation of rosemary aqueous extract: The aqueous extract was obtained by decoction of 100 g from dried plants in 1 L of distilled water and heated in the autoclave at 100°C for 30 minutes in well closed vials to avoid contamination.

The extract was filtered using Whatman's sterile filter paper in the laboratory. The obtained filtrate was collected in sterile glass vials hermetically closed and stored in a refrigerator at 4°C until its use in the following dilutions: 5, 10, 20 and 100% [4,5].

Potato varieties cultivation: Pre-germinated potato tubers were planted in pots previously prepared (at a rate of one tuber per pot) and depth of 4 to 5 cm, the substrate constituted a mixture of 2/3 of unused soil and 1/3 of peat [6].

Planting was done in 12 pots among which 6 were reserved as controls. The planting was replicated thrice for each variety. Aqueous extract were applied to the soil in 6 pots, to field capacity, every three days at a dilution of 20% from date of planting to pre-flowering. The control experiment was irrigated with clean tap water.

In vitro study: This part of study was based on inhibition of mycelial growth, sporulation and germination of both isolates of P. infestans.

Mycelial growth inhibition: Four concentrations: 5%, 10%, 20% and 100% of R. officinalis aqueous extract were treatments used for this study, and correspond respectively to treatments D1, D2, D3 and D4. Mycelial growth inhibition was based on direct contact method, described by Mishra and Dubey [7]. The microbiological procedures and the minimum inhibitory concentration (MIC) of the aqueous extract were determined according to Paranagama et al. [8] method For each treatment, 5 ml of plant extract was poured into Petri dishes of the same diameter (90 mm) using micropipettes. The Pea- agar medium was maintained in surfusion (45°C) then poured into Petri dishes containing aqueous extract. The latter were slightly shaken to homogenize the medium. Plant aqueous extract in the control experiment was substituted by sterile distilled water. Treatments were replicated five times for each P. infestans isolate.

Using sterile Pasteur pipettes, a disk of 50 mm in diameter of inoculum, for each isolate was taken and inoculated at the center of Petri dishes. Incubation of plated dishes in the hot air oven was done at a temperature of 18°C to evaluate the mycelial growth, which was daily observed for a period of 15 days. Readings were taken by calculating the average of two diameters measured on two perpendicular axes drawn on the reverse side of the plated petri dishes.

The minimum inhibitory concentration of mycelial growth (MIC) was determined for each isolate.

Mycelial growth inhibition rate was determined for each P. infestans isolate according to the formula described by Rollan et al. in Ibarra-Medina et al. [9].

$$I(\%)=\frac{(DT-Dt)}{DT}\times100$$

Where I is a percentage Inhibition rate of mycelial growth of P.

infestans isolate,

DT is a Mycelial growth (mm) of P. infestans isolates in control and

Dt is a Mycelial growth (mm) of P. infestans isolates developed in the medium in the presence of R. officinalis aqueous extract.

Antifungal activity of rosemary aqueous extract on morphology of P. infestans isolates: In evaluating the effects of rosemary aqueous extract on the phyto pathogenic isolates, a morphological description was done after 15 days of incubation, by direct observation of the treated cultures and controls of A1 and A2 of P. infestans isolates under photonic microscope at magnification (X125).

Sporulation and germination inhibition: After incubation for 21 days, at 18°C, each plated petri dish was collected, and 15 ml of sterile distilled water poured in, and then scraped with sterile Pasteur pipette to recover separately sporangial suspensions in sterilized test tubes. These were agitated using an agitator of tubes vortex. The sporangial suspensions prepared for A1 and A2 isolates were observed to determine the concentration of spores using a hemacytometer under optical microscope. Five repetitions were carried out for each fungal isolate, and each concentration to calculate sporulation inhibition rates according to the formula of Hibar et al. [10].

$$IS(\%)=\frac{(ST-St)}{ST}\times100$$

Where IS is a percentage Inhibition rate of P. infestans sporulation,

ST is a concentration in sporangia of control P. infestans isolates (sporangia.ml^{-1}) and

St is a concentration in sporangia of P. infestans isolates developed in the medium in the presence of R. officinalis aqueous extract (sporangia. ml^{-1}).

Parallel, germination inhibition rate (IG%) was calculated for each isolate, according to the formula described by Hill and Nelson [11].

$$IG(\%)=\frac{(NT-NPA)}{NT}\times100$$

Where IG is a percentage germination inhibition rate of P. infestans strain,

NT is a concentration of P. infestans sporangia germinated in control (sporangia.ml^{-1}) and

NPA is a concentration of P. infestans sporangia developed in medium in presence of R. officinalis aqueous extract (sporangia.ml^{-1}).

Survivability of treated P. infestans isolate: To evaluate the fungistatic and fungicidal effects of R. officinalis aqueous extract, in vitro and in vivo survivability of P. infestans isolates previously treated were monitored respectively on PPA medium and on leaf disks of Spunta potato's cultivar.

In vitro survivability study was based on the technique modified by Mahanta et al. [12]. The test was based on the resumption or the absence of mycelial growth on the isolates inhibited by R. officinalis aqueous extract. Explants were transplanted on fresh Pea- agar medium under conditions of incubation previously mentioned. Four explants of each P. infestans isolate were transferred to Petri dishes, with four replicates. Treatments were administered and observations made in comparison with the control. Readings were taken daily, for 7 days.

The lethal inhibitory concentrations (CIL) were estimated at the end of the experiment. The CIL was determined from the smallest concentration for which no mycelial growth and no resumption of the explant was observed on the PPA medium in the term of 7 days of incubation [8].

Besides, the *in vivo* survivability of *P. infestans* isolates beforehand treated with the plant extract at different concentrations was realized according to the method of Klarfeld et al. [13].

Healthy detached Spunta potato leaves having a diameter greater or equal to 50 mm were chosen and collected from healthy plants in Tipaza city. The leaves were cut to uniform disks using a punch, they were washed with clean tap water then disinfected in 2% of Sodium hypochlorite solution for 3 minutes, then rinsed in 3 changes of sterile distilled water.

Sterile filter paper moistened with sterile distilled water was deposited in transparent plastic and sterile boxes, a plastic mesh was also placed in, then 5 potato leaf disks were placed in the box, and the explants of isolates were introduced. Previously treated leaves along with the controls were also observed.

Incidence of the disease was defined by the number of leaf disks presenting typical symptoms of mildew, while disease severity was represented by expression of the symptoms in terms of percentage of surface infected by the mildew. Disease reduction rate was calculated using the formula proposed by Hill and Nelson [11].

$$Inf(\%) = \frac{(InfT\text{-}Inft)}{InfT} \times 100$$

Where Inf is a percentage infection rate of detached potato leaf disks,

Inf T is a% infection rate of positive controls detached potato leaf disks and

Inf t is a% infection rate of detached potato leaf disks treated by aqueous extract.

***In vivo* antifungal potential of *R. officinalis* aqueous extract:** *In vivo* antifungal potential evaluation was done by the application of potato leaf disks with treatments *in vivo* as well as the leaf disks controls inoculated by A1 and A2 of *P. infestans* isolates. Various modes of treatment were used for this study:

- Preventive application through spraying potato leaf disks with rosemary aqueous extract at concentration of 20% for few minutes. 24 hours after the treatment, 100 μl of sporangial suspension of 10^5 sporangia.ml^{-1} were deposited by means of a micropipette on the lower surface of potato leaf disk at 5 replications per fungal isolate.

- Curative application through the inoculation of potato leaf disks by depositing 100 μl of sporangial suspension on the lower leaf surface, then after 24 hours, application of droplets of 50 μl crude aqueous extract of *R. officinalis* at 20% concentration.

- Disks of detached potato leaves earlier treated with *R. officinalis* aqueous extract diluted at 20% were inoculated with 100 μl of sporangial suspension of *P. infestans* and incubated at 18°C for 10 days in the sterile transparent boxes.

The frequency of attacks was estimated two to four days later.

Both negative and positive controls were observed. Negative control, in which the disks of detached potato leaves were treated with sterile distilled water, and positive controls, where the detached leaf discs were inoculated with A1 and A2 of *P. infestans* isolates [14,15]. *In vivo* antifungal potential evaluation of *R. officinalis* aqueous extract on *P. infestans* isolates was done using the following parameters.

Period of appearance of the symptoms: It is the necessary time for the appearance of the infection by the phyto pathogenic agent on the inoculated foliar tissue.

Reduction of late blight disease: The reduction of the disease or (%DR) was translated by the product of the incidence of the disease (number of infected leaf disks) by the scale attributed to the infected foliar surface. It is determined by the formula proposed by Hill and Nelson [11].

RM (%)= CIP- CIPE/ CIP x 100

Where CIP is a coefficient of infection of controls (detached potato leaves inoculated with *P. infestans* isolates.),

CIPE is a coefficient of infection of treated detached potato leaves inoculated with the phyto pathogenic isolates.

Inhibition of the sporulation: After 10 days of incubation, the infected disks of detached potato leaves were carefully dipped into sterile tubes containing 10 ml of sterile distilled water then subjected to agitation by means of an agitator of tubes vortex to release the sporangia produced. The content of each tube was observed to determine the concentration of spores by means of a Hemacytometer under optical microscope at magnification (X125).

Sporangial production inhibition rate or IPC was calculated using the formula proposed by Hill and Nelson [11].

IPC (%)= NCP- NCPE/ NCP x 100

Where NCP is a number of sporangia produced on surface of detached potato's leaf disk inoculated by *P. infestans* isolates,

NCPE is a number of sporangia produced on surface of detached potato leaf disk treated with *R. officinalis* aqueous extract and inoculated with *P. infestans* isolates.

Statistical analysis: Data obtained was analyzed using Analysis of variance, ANOVA SYSTAT vers.7, variance calculated using the GLM (Generalized Linear Model), the differences were considered significant for $P < 0.05$.

Results and Discussion

In vitro antifungal potential

Evaluation of mycelial growth inhibition: The analysis of variance of mycelial growth inhibition showed statistically significant differences between the two isolates, but no significant difference between the various *R. officinalis* aqueous extract studied at different concentrations (Table 1). In GLM, the latter exceeded 85% for 5% concentration to evolve slightly to 20% concentration where, it was more pronounced on A1 isolate. Therefore, 5% concentration represents the minimum inhibitory concentration (CMI) of *R. officinalis* aqueous extract (Figures 1 and 2).

Antifungal effects of *R. officinalis* aqueous extract on *P. infestans* isolates: The inhibition of the mycelial growth of *P. infestans* isolate could have resulted from the effect of the extract causing lysis and vesiculation of the mycelium, as well as the deformation of sporangium and the digestion of their contents. These morphological modifications were also observed from the lowest concentration of this tested plant extract (Figure 3).

Factors	Sum-of-Squares	ddl	Mean- Square	F-ratio	P
Concentrations	2.635	3	0.878	1.000	0.500
Isolates	200.983	1	200.983	228.812	0.001

Table 1: Variance analysis of *P. infestans* mycelial growth inhibition rates according to rosemary aqueous extract concentrations and isolates.

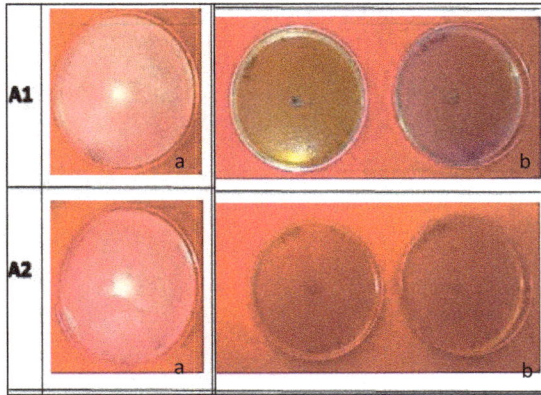

a: Controls strains
b: Treated strains respectively at 10 % and 20 %.

Figure 1: Mycelial growth variability of A1 and A2 *P. infestans* isolates under treatments of rosemary aqueous extract at different concentrations.

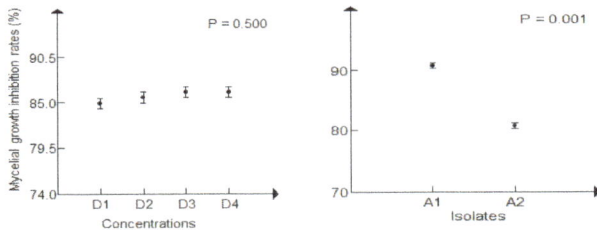

D1, D2, D3 and D4: Concentrations respectively at 5%, 10%, 20% and 100%

Figure 2: Variance analysis of mycelial growth inhibition rates of *P. infestans* in GLM according to rosemary aqueous extract concentrations and isolates.

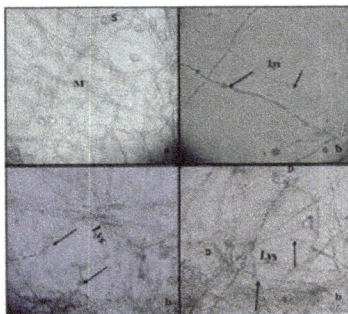

M: Mycelium; S: Sporangia; Lys: Lysis of Mycelium; D: Digest of content of sporangia; a:Control isolate; b:Treated isolate
Figure 3: Morphology of controls (a) and treated (b) isolates of *P. infestans* by rosemary aqueous extract

Sporulation and germination inhibition of *P. infestans* isolates: The sporulation, as well as the germination of the studied isolates were affected by aqueous extract of *R. officinalis* at 5%, concentration, where 100% inhibition was observed.

Effect of Rosemary aqueous extract on the survivability of *P. infestans* isolates: Concerning the *in vitro* study, the isolate A1 was more sensitive to the treatment and, the inhibition evolved with aqueous extract tested at different concentrations. Also, inhibition was observed *in vivo* for both isolates and four concentrations of the *R. officinalis* aqueous extract. However, the lethal inhibitory concentration (CIL) of the isolate A1 was higher than that of the A2 isolate.

In vivo antifungal potential

Determination of the symptoms appearance period: The analysis of variance of the symptoms appearance period did not show significant differences between the treatments application rates, potato varieties, *P. infestans* isolates and *R. officinalis* aqueous extract (Table 2).

Variability in the period of symptoms appearance was observed among the treatments. It was more in crude than in diluted extracts. However, it was approximately similar for both concentrations of aqueous extract applied with respect to the curative rate (2.6 days).

On the other hand, the periods of symptoms appearance by activity of both isolates of *P. infestans* showed that A1 isolate reproduced the symptoms more quickly than A2 isolate for various treatment application rates.

The symptoms appearance period extended till the 3rd day. The shortest period was marked for the preventive mode by watering while the mode of spraying with the crude aqueous extract showed the longest period. Therefore, the classification of the various application rates was established in the following decreasing order: preventive by spraying with the crude extract (SPR1: 3 days) and, in diluted extract to 20% (SPR2: 2.8 days), curative by use of crude and diluted extracts (CUR1, CUR2: 2.7 days), preventive by watering (WAT: 2.6 days) (Figure 4).

Antifungal potential of rosemary aqueous extract on disease reduction: The analysis of variance of disease reduction rates did not show significant difference between the application rates, on both isolates of *P. infestans*, concentrations of rosemary aqueous extract applied for all application rates and potato varieties for the preventive mode by spraying and watering (Table 3).

However, a significant difference was noticed between potato's varieties according to the curative application mode (F=10.662, P=0.031) (Table 3).

In GLM, the highest disease reduction was observed for the preventive mode in treatment with the crude extract (SPR1: 86.2%), while the lowest rate was observed for the preventive treatment by watering and curative with crude extract (WAT, CUR1: 81%).

Therefore, the classification of application rates was established in decreasing order.

- Preventive mode by spraying potato leaf disks with crude

Parameters	Application mode of treatments	Sum of squares	d.d.l	Mean square	F- Ratio	P
Modes of Application	All the modes	0.964	4	0.241	0.933	0.471
Varieties	Spraying	0.521	1	0.521	1.000	0.351
	Watering	0.125	1	0.125	1.000	0.374
	Curative	0.058	1	0.058	0.378	0.558
Phytophthora Infestans Isolates	Spraying	0.021	1	0.021	0.040	0.847
	Watering	0.125	1	0.125	1.000	0.374
	Curative	0.058	1	0.058	0.378	0.558
Concentrations	Spraying	2.882	2	1.441	2.766	0.130
	Curative	1.145	2	0.572	3.738	0.079

Table 2: Variance analysis of symptoms appearance periods of *P. infestans* according to the treatments application modes, concentrations, Potato's varieties and isolates

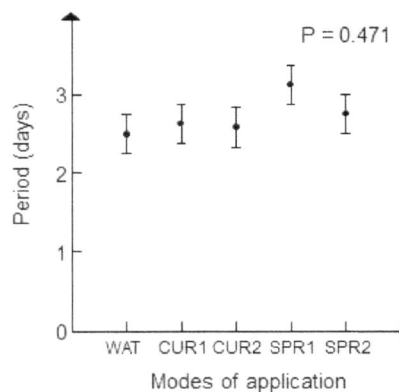

WAT: Watering by the crude extract SPR1: spraying with crude extract SPR2: spraying with diluted extract at 20% CUR1: curative treatment with crude extract CUR2: curative treatment with diluted extract at 20%.
Figure 4: Variance analysis of symptoms appearance period in GLM according to treatments application modes and concentrations.

Parameters	Application modes of treatments	Sum of squares	ddl	Mean square	F-Ratio	P
Modes of Application	All the modes	90.322	4	22.580	0.141	0.964
Varieties	Spraying	39.739	1	39.739	0.118	0.749
	Watering	52.345	1	52.345	5.395	0.259
	Curative	534.903	1	534.903	10.662	0.031
P. infestans isolates	Spraying	17.731	1	17.731	0.053	0.830
	Watering	22.515	1	22.515	2.320	0.370
	Curative	46.321	1	46.321	0.923	0.391
Concentrations	Spraying	6.643	1	6.643	0.020	0.895
	Curative	0.574	1	0.574	0.011	0.920

Table 3: Variance analysis of disease reduction rates according to treatments application modes, rosemary aqueous extract concentrations, potato's varieties and *P. infestans* isolates.

aqueous extract (SPR1: 86.2%)

• Preventive mode by spraying potato leaf disks with diluted aqueous extract at 20% and curative mode with diluted aqueous extract at 20% (SPR2, CUR2: 84%)

• Preventive mode of watering and curative mode with crude rosemary aqueous extract (WAT, CUR1: 81%).

On the other hand, the disease reduction rates registered for all treatment application modes were (up to 70%) on both varieties, but the Spunta variety showed a more significant disease reduction for

preventive mode treatment by watering (85%) and curative mode (90%) while it was more marked on kondor variety for the preventive mode by spraying (88%) (Figure 5).

Besides, it was important also for both isolates of *P. infestans* exceeding 75% for the various treatment application rates. On the other hand, the reduction affected much more A2 isolate for treatment application rates by spraying (86%), watering and curative (83%) (Figure 5).

This also confirms that the reduction of the disease was slightly

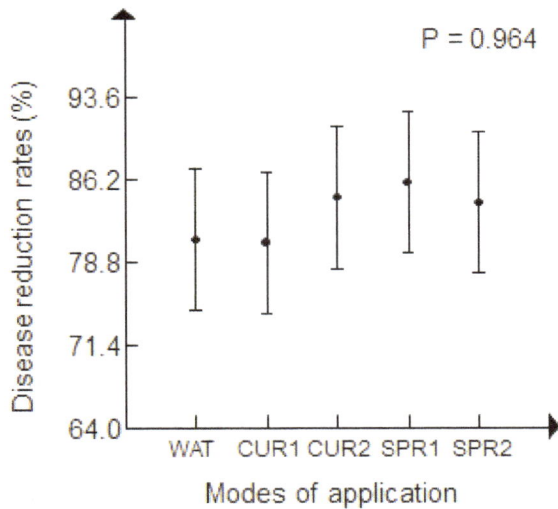

WAT: Watering by the extract SPR1: spraying with crude extract SPR2: spraying with diluted extract   CUR1: curative treatment with crude extract CUR2: curative treatment with diluted extract.

Figure 5: Variance analysis of disease reduction rates in GLM according to treatments application modes and concentrations.

higher in application of the crude treatments than with treatments diluted at 20% for both modes of application: preventive by spraying and curative.

Antifungal potential of R. officinalis aqueous extract on sporulation inhibition of P. infestans: The variance analysis of sporulation inhibition rates did not show significant difference between treatment application modes, potato's varieties, P. infestans isolates and, R. officinalis aqueous extract concentrations (Table 4).

In GLM, all the sporulation inhibition rates registered showed antifungal effect against P. infestans sporulation (rate exceeding 75% and reaching 100%) (Figure 6).

• Classification of treatments application rates was established in the following decreasing order.

• Watering by the crude aqueous extract and spraying with crude and diluted at 20% of rosemary aqueous extract (100%).

• Curative mode with a crude and diluted at (20%) of rosemary aqueous extract (77%).

The inhibition of sporulation was recorded on both varieties in curative application. Spunta variety showed higher inhibition (over 85%) than Kondor (bordering 55%).

The latter was variable on both isolates of Phytophtora infestans. Sporulation of A1 isolate was higher than A2 isolate for curative mode (over 96%), while the rate registered for A2 isolate borders 45%.

On the other hand, a slight variation of sporulation inhibition rates was noticed between both concentrations of aqueous extract applied for the curative mode (77%) for the crude extract and (68%) for the diluted extract at concentration of 20%.

It is very important to indicate that the treatments used in preventive

mode by spraying and watering showed a complete inhibition of the sporulation (100%) on P. infestans isolates and potato's varieties.

Discussion

Plants are able to produce various compounds. Besides the classic primary metabolites, they synthetize and accumulate secondary metabolites which the physiological function is not always obvious but represents a wide range of exploitable molecules in agriculture within the framework of phyto-protection.

R. officinalis antibacterial and antifungal activities can be summarized by the oil composition of these extracts [16]. The study revealed the phenolic compounds such as the terpenes, which include borneol, camphore, 1,8 cineole, pinene camphone, verbenonone and bornyl acetate [17].

This study as well as previous reports confirms the efficiency of certain extracts of plants in the control of potato mildew [18].

Several authors have also asserted that R. officinalis aqueous extract has a powerful antioxidant activity, associated with the presence of several di-terpenes phenolic as the carnosique acid, carnosol, rosmanol, rosmariquinone and rosmaridiphenol [19].

Besides, some reported that a number of compounds contained in the extracts of R. officinalis revealed antibacterial and antifungal properties [20,21].

So, the antifungal activity of Rosmarinus officinalis L. extracts was estimated against the fungal infections of wheat. Their inhibitory effect on the mycelial growth asserted that their use in low concentrations could have a significant potential for the biological control of phytopathogen fungi such as Alternaria alternata, Botrytis cinerea and Fusarium oxysporum [22].

On the other hand, Goussous et al. [23] reported important in vitro inhibitory effects of the crude extracts of various concentrations of R. officinalis against Alternaria solani. In addition, the microscopic observations of the fungal isolates treated by the extracts of R. officinalis revealed structural modifications leading to the lysis of the mycelium and, the digestion of sporangia contents, there by confirming the fungicidal effect of the plant extract. It is the toxic effect of its components on the structure and the physiology of the cellular membrane that is responsible for the antifungal effect [24,25]. In this sense, Omidbeygi et al. [26] suggested that the components of essential oil and extracts of plants cross cell wall and interact with enzymes and proteins, it is the production of a flux of protons towards the exterior cell which incites cellular changes and after all, the death of the microorganism. Cristani et al. [27] suggested that the antimicrobial activity is connected to the capacity of terpenes to affect not only the permeability but also the other functions of cell wall.

Our results revealed that all the extracts of R. officinalis even in the lowest concentration (5%), were excellent inhibitors of P. infestans sporulation and germination. Their inhibition rates were as high as 100%. This could have been caused by the deformation of sporangia and the lysis of the contents by the bioactive molecules contained in this aqueous extract.

It has further proved that the extracts of plants are rich in phenols also showed, an inhibitory activity particularly raised against the sporulation of fungi [28].

Duru et al. [29] confirmed that the antifungal effects of several extracts of plants seem to have a correlation with the incomplete

Para-meters	Application Modes of treatments	Sum of squares	d.d.l	Mean square	Test F	P
Modes of application	All the modes	2887.749	4	721.937	0.750	0.573
Varieties	Curative	1394.314	1	1394.314	0.678	0.457
P. infestans isoltates	Curative	5899.765	1	5899.765	2.867	0.166
Concentrations	Curative	127.417	1	127.417	0.062	0.816

Table 4: Variance analysis of sporulation inhibition rates according to treatments application modes, concentrations, varieties and P. infestans isolates.

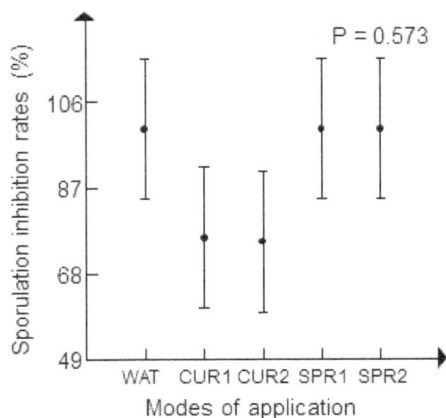

WAT: Watering by the crude extract SPR1: spraying with crude extract SPR 2: spraying with diluted extract at 20% CUR1: curative treatment with crude extract CUR 2: curative treatment with diluted extract at 20%
Figure 6: Variance analysis of sporulation inhibition rates in GLM according to treatments application modes.

development of conidiophores and morphological modifications of *Aspergillus fumigatus*.

In the same context, Blaeser and Steiner [30] showed that the extracts of various plants prevent the germination and affect the release and the mobility of *P. infestans* zoospores, in agreement with this study.

In vitro and in vivo survivability of *P. infestans* isolates showed a fungistatic effect of *R. officinalis* with low concentrations and a fungicidal effect at increasing concentrations. The latter was able to prevent completely the appearance of the late blight symptoms on leaf disk of potato. The CIL value allowed us to know from which concentration the extract of the rosemary becomes fungicidal. Based on the scale of Koba et al and Webster et al. [31,32], we can deduce that *R. officinalis* extract has an interesting inhibitory power for A1 isolate because the A2 CIL exceeds 70%.

These results could be connected to the phytochemical composition of the plant. They suggest that it contains molecules with fungicidal activity. This hypothesis coincides with several studies reported by the bibliography. Indeed, Banso et al. [33] revealed fungistatic antifungal substances in the extracts of plants in low concentrations but, which become fungicidal in higher concentrations.

Conclusion

Results from this study as certain that aqueous extract of *R. officinalis* powder was an excellent inhibitor of *P. infestans* mycelial growth, sporulation and germination in the lowest dilution (5%). This could have resulted from the action of bioactive molecules contained in the rosemary aqueous extract, that causes the lysis of the mycelia

and sporangia of the fungus. The fungicidal effects increased with increasing concentrations of this aqueous extract.

On the other hand, the aqueous extract of *R. officinalis* led to a reduction of the disease on leaf disk of potato. Important disease reduction rates (over 70%) were registered in both varieties and for both isolates of *P. infestans* exceeding 75%, while Spunta variety showed a more important reduction for treatment preventive application modes by watering (85%) and for the curative mode (90%). Also, the A2 isolate was greatly inhibited by treatments application rates of spraying (86%), watering and curative (83%). The sporulation inhibition was very pronounced *in vivo* (75% and 100%) and as the rates of preventive treatments made by spraying and watering. This present work thus confirms the bio-fungicidal potentialities of aqueous extract of *R. officinalis* on *P. infestans* isolates with the aim of its use in the bio control of late blight potato.

Acknowledgements

Authors are very grateful to National Plant Protection Institute (INPV) for providing necessary facilities.

References

1. Andrivon D, Lebreton L (1997) Mildiou de la pomme de terre, ou en sommes-nous après 150 ans. Phytoma 494: 24-27.

2. Rashid A, Ahmad I Iram S, Mirza JI, Rauf CA (2004) Efficiency of Different Neem (Azadirachta indica A. Juss) Products against Various Life Stages of Phytophthora infestans (Mont.) de Bary Pak J Bot 36: 881-886.

3. Gallegly ME, Galindo J (1958) Mating types and oospores of Phytophthora infestans in nature in Mexico. Phytopathology 48: 274-277.

4. Grainge M, Ahmed S (1988) Handbook of plant with pest control properties. Wiley, New York 2nd edition 470.

5. Krebs H, Dornand B, Forrer HR (2006) Fight against blight of potato with herbal preparations. Swiss magazine Agric 38: 203-207.

6. Compobello EWA, Drenth HH, Leifrink RS (2002) Professional culture of Potato, Plantation, 2nd edition, NIVVA, Dutch Institute for the promotion of markets for agricultural products 22.

7. Mishra AK, Dubey NK (1994) Evaluation of some essential oils for their toxicity against fungi causing stored deterioration of food commodities. Applied and environmental microbiology 60: 1101-1105.

8. Paranagama PA, Abeysekera KHT, Abeywickrama K, Nugaliyadde L (2003) Fungicidal and anti-aflatoxigenic effects of the essential oil of Cymbopogon citratus (DC.) Stapf. (lemon grass) against Aspergillus flavus Link. isolated from stored rice. Letter in Applied Microbiology 37: 86 - 90.

9. Ibarra-Medina VA, Ferrera-Cerrato R, Alarcón A, Lara-Hernández ME, Valdez-Carrasco JM (2010) Isolation and screening of Trichoderma strains antagonistic to Sclerotinia sclerotiorum and Sclerotinia minor. Rev Mex Mic 31: 53-63.

10. Hibar K, Daami-Remadi M, Khiareddine H, Mahjoub MEI (2005) In vitro and in vivo inhibitor effect of Trichoderma harzianum against Fusarium oxysporum f. sp. radicis-lycopersici. Biotechnol Agron Soc Environ 9: 163-171.

11. Hill JP, Nelson RR (1983) Genetic control of two parasitic fitness attributes of Helminthosporium maydis race T Phytopathology 73: 455-457.

12. Mahanta JJ, Chutia M, Bordoi M, Pathak MG, Adhikary RK, et al. (2007)

Cymbopogon citratus L. essential oil as a potential antifungal agent against key weed moulds of Pleurotus spp. Spawns. Flavour Fragrance Journal 22: 525-530.

13. Klarfeld S, Rubin AE, Cohen Y (2009) Pathogenic Fitness of Oosporic Progeny Isolates of Phytophthora infestans on Late-Blight-Resistant Tomato Lines. Plant Disease 93: 947-953.

14. Fontem DA, Olanya OM, Tsopmbeng GR, Owona MAP (2005) Pathogenicity and metalaxyl sensitivity of Phytophthora infestans isolates obtained from garden huckle berry, potato and tomato in Cameroon. Crop Protection Journal 24: 449-456.

15. Abd- El- Khair H, Haggag WM (2007) Application of Some Egyptian Medicinal Plant Extracts Against Potato Late and Early Blights. Res J Agric Biol Sci 3: 166-175.

16. Pinto E, Vaz CP, Salgueiro L, Goncalves MJ, Costa-de-Oliveira SC, et al. (2006) Antifungal activity of the essential oil of Thymus pulegioides on Candida, Aspergillus and dermatophyte species. Journal of Medical Microbiology 55: 1367-1373.

17. Makhloufi A, Moussaoui A , Lazouni HA, Hasnat N, Abdelouahid DE (2011) Antifungal activity of essential oil of Rosmarinus officinalis L. and its impact on the conservation of a local variety of dates during storage. Medicinal Plants-International Journal of Phytomedicines and Related Industries 3: 129-134.

18. Ashrafuzzaman MH, Khan AR, Howlide AR (1990) In vitro effect of lemon grass oil and crude extracts of some higher plants on Rhizoctonia solani. Bangladesh. J Plant Pathol 6: L 17-18.

19. Georgantelis D, Ambrosiadis I, Katikou P, Blekas G, Georgakis SA (2007) Effect of rosemary extract, chitosan and α-tocopherol on microbiological parameters and lipid oxidation of fresh pork sausages stored at 4°C. Meat Sci 76: 172-181.

20. Del Campo J, Amiot MJ, Nguyen C (2000) Antimicrobial effect of rosemary extracts. Journal of Food Protection, 63: 1359-1368.

21. Djenane D, Sánchez-Escalante A, Bel-trán JA, Roncalés P (2002) Ability of α-tocopherol, taurine and rosemary, in combination with vitamin C, to increase the oxidative stability of beef steaks packaged in modified atmosphere. Food Chemistry 76: 407-415.

22. Centeno S, Calvo MA, Adelantado C, Figueroa S (2010) Antifungal activity of extracts of Rosmarinus officinalis and Thymus vulgaris against Aspergillus flavus and A. ochraceus. Pakistan. Journal of Biological Sciences 13: 452-455.

23. Goussous SJ, Abu-El-Samen FM, Mas'adb IS, Tahhan RA (2013) In vitro inhibitory effects of rosemary and sage extracts on mycelial growth and sclerotial formation and germination of Sclerotinia sclerotiorum. Archives of Phytopathology and Plant Protection 46: 1745-1757.

24. Bouchra C, Achouri M, Hassani LMI, Hmamouchi M (2003) Chemical composition and anti-fungal activity of essential oils of seven Moroccan Labiatae against Botrytis cinerea Pers: Fr J Ethnopharmacol 89: 165-169.

25. Yoshimura H, Sawai Y, Tamotsu S, Sakai A (2011) 1,8-cineole inhibits both proliferation and elongation of BY-2 cultured tobacco cells. Journal of Chemical Ecology 37: 320-328.

26. Omidbeygi M, Barzegar M, Hamidi Z, Naghdibadi H (2007) Antifungal activity of thyme, summer savory and clove essential oils against Aspergillus flavus in liquid medium and tomato paste. Food control 18: 1518-1523.

27. Cristani M, Arrigo MD, Mandalari G, Castelli F, Sarpietro MG, et al. (2007) Interaction of four monoterpenes contained in essential oils with models membranes: application for their antibacterial activity. J Agric Food Chem 55: 6300-6308.

28. Inouye S, Watanabe M, Nishiyama Y, Takeo K, Akao M, et al. (1998) Anti-sporulating and respiration-inhibitory effects of essential oils on filamentous fungi. Mycoses 41: 403-410.

29. Duru ME, Cakir A, Kordali S, Zengin H, Harmandar M, et al. (2003) Chemical composition and anti-fungal properties of essential oils of three Pistacia species. Fitoterapia 74: 170-176.

30. Blaeser P, Steiner U (1999) Antifungal activity of plant pathology extracts against potato late blight (Phytophthora infestans), In: Lyr H, Russel PE, Dehne HW, Sisler HD (eds.). Book Modern Fungicides and Antifungal Compounds II, Thuringia, Germany 491-499.

31. Koba K, Sanda K, Raynaud C, Nenonene Y A, Millet J, et al. (2004) Activités antimicrobiennes d'huiles essentielles de trois Cymbopogon sp. africains vis-à-vis de germes pathogènes d'animaux de compagnie. Annales de Médecine Vétérinaire 148: 202-206.

32. Webster D, Taschereau P, Belland RJ, Rennie RP (2008) Antifungal activity of medicinal plant extracts; preliminary screening studies. Journal of Ethno pharmacology 115: 140-146.

33. Banso ASO, Adeyemo, Jeremiah P (1999) Antimicrobial properties of Vernonia amygdalina extract. J Appl Sci Manage 3: 9-11.

27

Genotype and Environment Interaction and Marketable Tuber Yield Stability of Potato (*Solanumtuberosum L*) Genotypes Grown in Bale Highlands, Southeastern Ethiopia

Fufa Miheretu*

Sinana Agricultural Research Center, Horticulture division, Ethiopia

Abstract

Twelve potato (*Solanumtuberosum L.*) genotypes were evaluated using randomized complete block design (RCBD) with three replications to evaluate their genotype x environment interaction (GEI) and marketable tuber yield stability across nine environments during 2009-2011 at highlands of Bale, Southeastern Ethiopia. The combined analysis of variance (ANOVA) revealed that there was significant ($p < 0.05$) variation in genotype x environment interaction in marketable tuber yield. Genotype, environment and genotype x environment interaction respectively explained 18.86%, 51.88% and 29.26% of the total sum of squares in marketable tuber yield (t/ha). Most of the total sum of squares in marketable tuber yield is contributed by environment. The AMMI analysis for marketable tuber yield (t/ha) indicated that IPCA1, IPCA2 and IPCA3 were highly significant ($p < 0.01$) while IPCA4 showed non-significant interaction. The first and second principal component axis captured 40.37% and 30.8% of the GEI sum of squares in marketable tuber yield. Genotype 394640-539 gave high mean marketable tuber yield that is the most stable across environments. It was, thus, selected and recommended for wide production across locations.

Key words: AMMI stability; GEI; Potato (*Solanumtuberosum L.*)

Introduction

Potato (*Solanumtuberosum L.*), belonging to the family Solanaceae, is an important food and cash crop ranking fourth after maize, wheat and rice in annual production in the world [1,2]. It is the world's number one none-grain crop to ensure food security due to its growing demand [3]. It is a high biological value crop that gives an exceptionally high yield with more nutritious content per unit area per unit time than any other major crops. Thus, it can play a remarkable role in human diet as a supplement to other food crops such as wheat and rice [4]. Furthermore, the contribution of potato to the diversification of the cereal mono-cropping in Bale is great.

Despite the importance of potato in the country agriculture, its productivity has shown a decreasing trend even if its production is expanding steadily [5,6].One of the major factors contributing to reduction in yield of potato is inadequacy of improved cultivars with wide adaptability and stability in tuber yield. Thus, evaluating genotypes across various environments for their stability of performance and range of adaptation is crucial and is an important component of the research activity of the national as well as regional research program.

Evaluating genotypes over diverse environments is universal practice to ensure the stability of performance of the genotypes [7]. Stability in performance is one of the most desirable properties of a genotype to be released as a variety for wide cultivation [8]. However, the activity of identification, selection and recommendation of superior genotypes is complicated and severely limited by genotype × environment interaction that is inevitable in multi-environmental trails [9-13]. The presence of genotype x environment interaction may confound the genotypic performance with environmental effects [14].

Several statistical models and procedures have been developed and exploited for studying the genotype x environment interaction effects, stability of genotypes and their relationships in varietal development process [9-11,15]. A combined analysis of variance (ANOVA) can quantify the interactions and describe the main effects. However, it is uninformative for explaining genotype x environment interaction. To increase accuracy, additive main effects and multiplicative interaction (AMMI) is the model of first choice when main effects and interaction are both important [13]. It is a powerful tool for effective analysis and interpretation of multi-environment data structure in breeding programs and is useful for understanding genotype x environment interaction [7,9]. Plant breeders frequently apply AMMI model for explaining genotype x environment interaction and analyzing the performance of genotypes and test environments [16,17]. Therefore, this paper assesses genotype x environment interaction and marketable tuber yield stability of potato genotypes under Bale highlands, Southeastern Ethiopia.

Materials and Methods

Twelve genotypes of potato were evaluated for their adaptability and stability in marketable tuber yield across locations in Bale highlands at Sinana, Shallo and Dinsho during 2009, 2010 and 2011. Sinana is located at an altitude of 2400 m.a.s.l. with a range of mean annual rainfall of 563-1018 millimeter and minimum and maximum temperature of 7.9 0C and 24.30C, respectively. The soil type is dark-brown with slightly acidic reaction [18].

The experiment was laid out in randomized complete block design with three replications. The genotypes were planted on a plot area of 9m2with spacing of 75 cm and 30cm between rows and plants respectively. All agronomic and cultural practices were followed as per the general recommendation: the fertilizer rate of 90Kg/ha P2O5 and 110 Kg/ha N was used without fungicide application. At physiological maturity, the tubers were harvested from two middle rows and washed with clean tap water to remove soils. The clean tubers were sorted and graded into large, medium and small based on their size. The weight of

***Corresponding author:** Fufa Miheretu, Sinana Agricultural Research Center, Horticulture division, Ethiopia, E-mail: miheretufufag@gmail.com*

the tubers per plot (kg) was recorded and their mean was subjected to analysis.

Statistical Analysis

Analysis of variance (ANOVA) was carried out on marketable tuber mean (t/ha) on plot basis and pooled over locations and seasons using the Generalized Linear Model (GLM) procedures of the Statistical Analysis System version, 9.2 [19]. The Additive Main Effects and Multiplicative Interactions (AMMI) statistical model and biplot were produced using Irristat software [20].Furthermore, AMMI's stability value (ASV) was calculated in order to rank genotypes in terms of stability using the formula suggested by Purchase [21] as shown below:

$$\text{AMMI stability value (ASV)} = \sqrt{\left[\frac{SSIPCA1}{SSIPCA2}(IPCA1score)\right]^2 + \left[IPCAscore2\right]^2}$$

where, SS = Sum of squares; IPCA1 = interaction principal component analysis axis 1and IPCA2 = interaction principal component analysis axis 2. Genotype by environment interaction

Results and Discussion

The combined analysis of variance indicated that there is significant variation (p<0.05)in genotype x environment interaction for marketable tuber yield (t/ha) (Table 1). The significant variation in genotype x environment interaction indicate that there is a need to undertake additive main effects and multiplicative interaction(AMMI) analysis to distinguish which genotypes are stable in their marketable tuber yield. The analysis of variance for additive main effect and multiplicative interaction model of marketable tuber yield (t/ha) of the 12 potato genotypes was indicated in Table 2.

Genotypes, environment and genotype x environment interaction respectively explained 18.86%, 51.88% and 29.26%of the total sum of squares in marketable tuber yield (t/ha). Most of the total sum of squares are contributed by environment indicating that environment is diverse, with large difference among the environmental means causing most of the variation in marketable tuber yield (t/ha). The magnitude of genotype x environment interaction sum of squares was 1.551 times larger than that of genotypes in marketable tuber yield, implying that there was difference among genotypic response across environments (Table 1).This variability may be due to the variability of soil and rainfall across locations. AMMI stability analysis of marketable tuber yield

The AMMI analysis for marketable tuber yield (t/ha) indicated that IPCA 1, IPCA2 and IPCA3 were highly significant (p<0.01) while IPCA 4 showed non-significant variation (Table 2). The first principal component axis captured 40.37 % of the interaction sum of squares while second principal component axis explained 30.80% of the GEI sum of squares. IPCA 1 and IPCA 2 together had greater contribution (71.17%) to the total sum of squares than that of genotypes.

The mean and AMMI stability values of marketable tuber yield were indicated in Table 3. The highest (28.42t/ha) and the lowest (9.94t/ha) mean marketable tuber yield were recorded by genotype 387967-3 and the local cultivar respectively. Genotypes 387967-3, 394640-539, 90147-41, Jalane, Ararsa, 90170-37, 390012-2 and Hunde gave mean marketable tuber yield higher than the grand mean. On the other hand, 90147-15, 392637-500, 90147-46 and local gave mean yield lower than the grand mean. The most stable genotype in marketable tuber yield was 394640-539 based on AMMI stability value while the local, 90147-41 and Jalane are the most unstable genotypes. Based on the AMMI stability value, genotype 394640-539 was selected for wide production as it gave high mean marketable tuber yield that is stable across environments.

AMMI biplot analysis of marketable tuber yield

Both genotypes and environments differed in their interaction as well as main effects for marketable tuber yield (Figure 1). Genotype3 (Local) was the lowest in its mean marketable tuber yield while genotype 5 (387967-3) and 2 (394640-539) were higher in their marketable tuber yield. Environment E and B were highly productive while environment D was poor in marketable tuber yield. Genotypes 3 (Local), 6 (90147-46), 10 (90170-37), 2 (394640-539) and 5 (387967-3) showed positive interaction with environment B, D, E and F. On the other hand, genotypes 4 (392637-500), 7 (90147-15), 1 (Hunde), 8 (390012-2), 9 (Ararsa), 11(Jalane) and 12 (90147-41)showed negative interaction with environment G, C, A and H. Genotype 3 (Local) was found adaptable to poor environment (D). On the other hand, genotypes 5 and 2 were found suitable to productive environments.

Figure 2 indicates the interaction pattern of the 12 potato genotypes with 9 environments for their marketable tuber yield (kg/

Source	DF	Mean Square
GEN	11	738.09687***
LOC	2	8953.56433***
Year	2	1347.07762***
REP (LOC*Year)	18	68.25173ns
GEN*LOC	22	135.37407***
GEN_*Year	22	255.59237***
LOC*Year	4	432.03913***
GEN*LOC*Year	44	90.71289**
Error	198	61.32870
CV	34.77	
R2	0.79	

Ns, ** and ***= non-significant, significant and highly significant at 0.05 and 0.01 level of significance respectively.
Table 1. Mean squares of combined analysis of variance of marketable tuber yield (t/ha) of 12 genotypes evaluated across location (2009-2011)

Source of variation	D.F	S.S	M.S	F	% explained
Genotypes	11	2706.36	246.032		18.86
Environments	8	7443.15	930.393		51.88
Genotypes X environments	88	4197.54	47.6994		29.26
AMMI COMPONENT 1	18	1694.65	94.1471	2.633***	40.37
AMMI COMPONENT 2	16	1292.99	80.8117	3.607***	30.80
AMMI COMPONENT 3	14	575.534	41.1096	2.592***	13.71
AMMI COMPONENT 4	12	224.574	18.7145	1.279ns	5.35
GXE RESIDUAL	28	409.800			
TOTAL	107				

Ns, ** and *** = non-significant, significant and highly significant at 0.05 and 0.01 level of significance respectively.
Table 2 Analysis of variance for Additive Main effect and Multiplicative Interaction (AMMI) model of marketable tuber yield of potato genotypes grown at highlands of Bale, South eastern Ethiopia (2009-2011)

Genotype	Mean	AMMI1	AMMI2	ASV	R
387967-3	28.42	4.535***	1.382 ***	1.905	6
394640-539	28.10	0.5389***	0.5713***	0.908	1
90147-41	26.29	-2.652***	0.7525	3.556	11
Jalane	25.87	-1.490***	2.961	3.547	10
Ararsa	24.82	-1.158***	0.8290	1.729	4
90170-37	23.77	1.674	0.8806	2.364	6
390012-2	23.19	-1.372***	-0.9574***	1.798	5
Hunde	25.72	-0.8035***	1.154***	1.562	2
90147-15	20.34	-1.102	-2.435 ***	2.831	9
392637-500	20.26	-0.3904	1.583	1.665	3
90147-46	16.69	0.5864	-2.256	2.383	8
Local	9.94	1.634***	-2.899	3.604	12
Mean	21.11				

ASV= AMMI stability value

Table 3: Mean and AMMI stabilityof marketable tubers of 12 genotypes evaluated over locations (2009-2011)

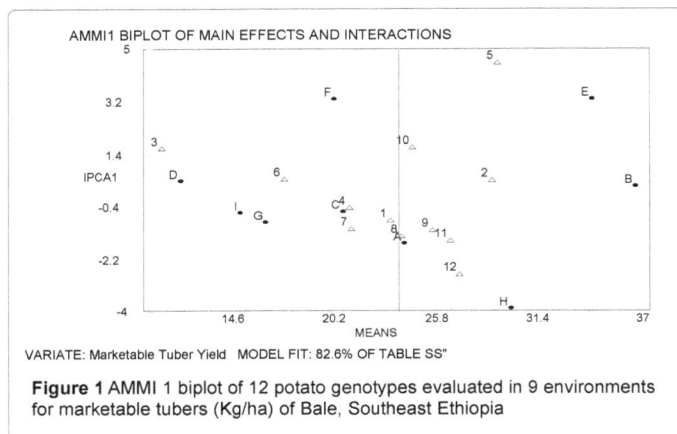

Figure 1 AMMI 1 biplot of 12 potato genotypes evaluated in 9 environments for marketable tubers (Kg/ha) of Bale, Southeast Ethiopia

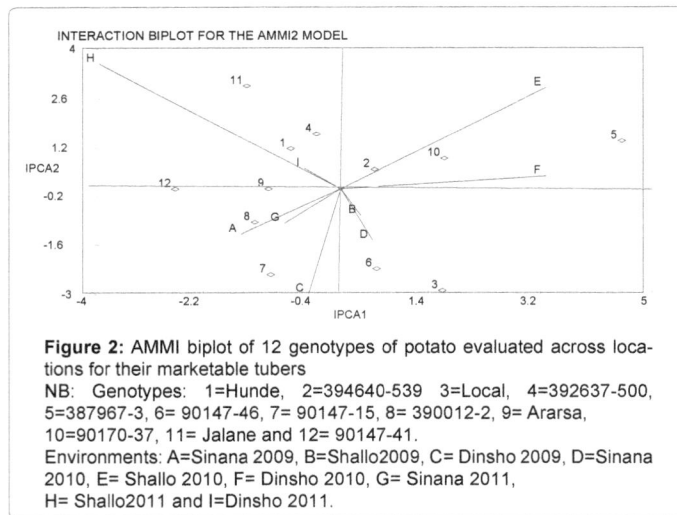

Figure 2: AMMI biplot of 12 genotypes of potato evaluated across locations for their marketable tubers
NB: Genotypes: 1=Hunde, 2=394640-539 3=Local, 4=392637-500, 5=387967-3, 6= 90147-46, 7= 90147-15, 8= 390012-2, 9= Ararsa, 10=90170-37, 11= Jalane and 12= 90147-41.
Environments: A=Sinana 2009, B=Shallo2009, C= Dinsho 2009, D=Sinana 2010, E= Shallo 2010, F= Dinsho 2010, G= Sinana 2011, H= Shallo2011 and I=Dinsho 2011.

ha). The distance from the origin (0,0) is indicative of the amount of interaction that was exhibited by genotypes either over environments or environments over genotypes [22].The genotypes 5(387967-3), 11 (Jalane), 3 (Local), 12 (90147-41), 7 (90147-15) and 6 (90147-46)

expressed a highly interactive behavior (positively or negatively) while genotype 2 (394640-539) show low interaction and thus stable in its marketable tuber yield (Kg/ha) (Figure 2). The environments H, E and F were highly interactive while B, I and D showed low interaction.

Conclusion

The combined analysis of variance of marketable tuber yield (t/ha) indicated that there was significant (p<0.05) genotype x environment interaction. Most of the total sum of squares in marketable tuber yield was explained by environment than genotype. The Local cultivar was found adaptable to poor environment while genotypes 394640-539 and387967-3were found suitable to productive environments. Genotype 394640-539 was selected for wide production for it had stable and high mean marketable tuber yield across the environments.

Acknowledgment

I express my appreciation to Guta Leggesse, Mulugeta Gadif, Adam Abdul hamid and Asnakech Ishetu for their cooperation in the collection of data.

References

1. FAO, 2008

2. Lemaga B, Kakuhenzire R, Gildemacher P, Borus D, Gebremedhin WG, et al. (2009) Current status and opportunities for improving the access to quality potato seed by small farmers in Eastern Africa. symposium 15th triennial of the symposium of the international society for tropical root crops, 2-6.

3. FAO (2008) Workshop on opportunities and challenges for promotion and sustainable production and protection of the potato crop in Vietnam and elsewhere in Asia, 25-28.

4. Badoni A, Chauhan JS (2010) Potato seed production of cultivarKufriHimalini, In vitro. stem cell, 1: 1-6.

5. CSA (2003) Ethiopian agricultural sample enumeration, 2001/02 (1994 E.C.) Statistical report on socio-economic characteristics of the population agricultural households, land use and are production of crops. Part 1. Addis Ababa.

6. CSA (2008) Agricultural Sample Survey 2007/2008 (2000 EC) volume1 report on area and production of crops (private peasant holdings, Meher season) Addis Ababa.

7. Sadeghi SM, Samizadeh H, Amiri E, Ashouri M (2011) Additive main effects and multiplicative interactions (AMMI) analysis of dry leaf yield in tobacco hybrids across environments. Afr. J. of biotechnology 10: 4358-4364.

8. Singh RK, Chaudhary BD (1977) Biometrical methods in quantitative genetic analysis. Kalyani publishers, India.

9. Asfaw A, Alemayehu F, Gurmu F, Atnaf M (2009) Ammi and sreggebiplot analysis for matching varieties onto soybean production environments in Ethiopia. Scientific research and essay 4: 1322-1330.

10. Eberhart SA, Russell WA (1966) Stability parameters for comparing varieties. Crop Sci, 6: 36-40.

11. Finlay KW, Wilkinson GN (1963) The analysis of adaptation in a plant breeding program. Aust. J. Agr. Res. 14: 742-754.

12. Shafii B, Price WJ (1998) Analysis of genotype-by-environment interaction using the additive main effects and multiplicative interaction model and stability estimates. J. agr.biol. environ. stat. 3: 335-345.

13. Zobel RW, Wright MS, Gauch HG (1988) Statistical analysis of a yield trial. Agron J. 80: 388-393.

14. Thillainathan M, Fernandez GCJ (2002) A novel approach to plant genotypic classification in multi-site evaluation. Hort. science 37: 793-798.

15. Crossa J (1990) Statistical analysis of multi-location trials. Adv. agron.44: 55-85.

16. Gauch HG (2006) Statistical analysis of yield trials by AMMI and GGE. Crop Sci. 46: 1488-1500.

17. Yan W, Kang MS, Ma B, Wood S, Cornelius PL (2007) GGE biplot vs. AMMI analysis of genotype-by-environment data. Crop sci. 47: 643-655.

18. SARC (1998) A decade of research experience. Geremew E Tilahun G AliyH(eds.). Bulletin no. 4. Sinana agricultural research center, agricultural research coordination service, Oromia agricultural development bureau.

19. SAS (2008) SAS software version 9.2. SAS Institute INC, Cary. NC. USA.

20. IRRI Stat (2003) International rice research institute. Philippines.

21. Purchase JL (1997) Parametric stability to describe G x E interactions and yield stability in winter wheat. South Africa.

22. Voltas J, Van EF, Igartua E, García del Moral LF, Molina-Cano JL, et al. (2002) Genotype by environment interaction andadaptation in barley breeding: Basic concepts and methods ofanalysis.

28

Survey of Rust and *Septoria* Leaf Blotch Diseases of Wheat in Central Ethiopia and Virulence Diversity of Stem Rust *Puccinia graminis* f. sp. *tritici*

Endale Hailu* and Getaneh Woldeab

Ethiopian Institute ofAgricultural Research,Ambo Plant Protection Research Center, P.O. Box 37,Ambo, West Shewa, Ethiopia

Abstract

Wheat is one of the most important cereal crops in Ethiopia and produced across large area of the country. Production of the crop constrained by several infection diseases including rust and Septoria leaf blotch diseases which are the major bottle neck of wheat production in Ethiopia. The objective of this study was to ass distribution, incidence and severity of wheat rusts and Septoria leaf blotch in west and South West Shewa zones and identification of Puccinia graminis f.sp. *tritici* virulences in Ethiopia. The survey was made in 2013 main cropping season (from September to October) following the main roads and accessible routes in each survey district, and stops were made at every 5 km intervals based on vehicles odometers as per wheat fields available. Five stops were made in each wheat field by moving "W" fashion at each stop interval using quadrants and data were collected from each. Race analysis was carried out by inoculating single uredinial isolates on to the 20 differential hosts. The result indicated that, stem, leaf and yellow rust mean incidence value 54.7%, 19.4% and 7.7% were recorded in the surveyed areas, respectively and mean severity value of 7.0%, 9.7% and 5.5% in the same order. Septoria leaf blotch was the most prevalent disease with100%. Mean incidence of 83% and 0.44 disease index of Septoria leaf blotch were recorded. The most widely grown Varieties Digelu and Kakaba showed susceptible reaction to stem and leaf rust whereas Meda wolabu were free both rusts. Variety Kubisa were susceptible to the three rusts (stem, leaf and yellow). Out of 20 isolates, two races namely TTKSK and TKTTF were identified. Race TTKSK was the most frequent with 95%. Stem rust resistance genes *Sr36*, *SrTmp* and *Sr24* were effective against TTKSK and while, *Sr8a*, *Sr24* and *Sr31* were effective against TKTTF (Digelu race). Most of the genes possessed by the differentials were ineffective against one or more of the tested isolates except *Sr24*. *Sr24* which confer resistance against most of the races detected and prevalent in Ethiopia can be used in breeding for resistance to stem rust in the country.

Keywords: Race; *Puccinia graminis* f. sp. *Tritici*; *Sr* genes

Introduction

Bread wheat (*Triticum aestivum* L. em. Thell) is the world's leading cereal grain where more than one-third of the population of the world uses as a staple food. It is one of the most important cereal crops of Ethiopia [1,2]. It ranked fourth in land coverage and total production after tef, maize and sorghum [3]. Wheat is produced across a wide range of agro ecological and crop management regime. The most suitable area for wheat production falls between 1900-2700 m.a.s.l [1]. Despite the large area under wheat in Ethiopia the national average yield is 2.11 t/ha [3], which is far below the average of African and world yield productivity. The low productivity is attributed to a number of factors including biotic (diseases, insect pest and weeds) and a biotic (moisture, soil fertility, etc) and adoption of new agricultural technologies [4]. Among these factors, diseases play a significant role in yield reduction.

Wheat is susceptible to many diseases including the highly destructive ones like rusts (*Puccinia spp.*), *Septoria* leaf blotches (*Septoria tritici*), *Fusarium* head blight (*Fusarium graminearum*), tan spot (*Pyrenophora tritici repentis*), smut (*Ustilago tritici*) and powedery mildew (*Erysiphe graminis* f.sp.*tritici*) [5]. Over 30 diseases have been reported on wheat in Ethiopia [6]. Of these, fungal diseases like rusts (stem, stripe and leaf rust), *Fusarium* head blight (FHB), *Septoria* blotch, *Helminthosporium spp.*, and tan spot are the dominant ones that were reported over time [6-9]. Rusts are the most important disease of wheat worldwide, in spite of great progress made in their control in many countries [10] and considered the major diseases of wheat since no other wheat disease could result in greater loss over large area in a given year [11]. Rusts can cause up to 60 percent of yield loss for leaf or stripe (yellow) rust and 100 percent loss for stem rust. The persistence of rust as a significant disease in wheat can be attributed to specific characteristics of the rust fungi. These characteristics include a capacity to produce a large number of spores which can be wind-disseminated over long distances and infect wheat under favorable environmental conditions and the ability to change genetically, thereby producing new races with increased aggressiveness on resistant wheat cultivars.

In West and South West Shewa zones, wheat is highly exposed to wheat rust and *Septoria* blotch damages. Most probable farmer of the area use the most susceptible wheat cultivars like Kubsa and Dashen, the involvement of new wheat rust races and most of the farmer use double cropping system can be used as green bridge for wheat rusts.

Therefore, disease monitoring and surveillance are of paramount significant for sustainable wheat production and tackle food insecurity. Hence, the present research aims to survey of rusts and *Septoria* leaf blotch of wheat in West and South West Shewa zones in addition to

*Corresponding author: Endale Hailu, Ethiopian Institute of Agricultural Research, Ambo Plant Protection Research Center, P.O. Box 37, Ambo, West Shewa, Ethiopia, E-mail: sebhailuabera@yahoo.com

identify *Puccinia graminis* f.sp. *tritici* virulences in the surveyed areas of Ethiopia.

Materials and Methods

Survey of wheat diseases

Wheat diseases survey was conducted in 2013 cropping season in two major wheat producing districts of South West and West Shewa zones. A total of 65 fields were assessed in 10 districts in two zones. The surveys were made following the main roads and accessible routes in each survey district, and stops were made at every 5-10 km intervals based on vehicles odometers. Three to five stops were made in each wheat field along a diagonal move at each stop interval. Disease prevalence, incidence and severity were recorded for three rusts (stripe, leaf and stem) and *Septoria* leaf blotch. The disease prevalence was calculated using the number of fields affected divided by the total number of field assessed and expressed in percentage. Incidence was calculated by using the number of plants infected and expressed as percentage of the total number of plants assessed. Severity was scored visually using the modified Cobb's Scale [12] for the three rusts, and the double digit scale (00-99) for *Septoria* leaf blotch [10,13]. The first scale (0-9) in the double digit scale represents the blotch development up the plant height (5 if the disease reached at the middle height, 8 for flag leaf and 9 for spike), and the second digit stands for severity (1=10%, 2=20% and 9=90%). For each score, the disease severity percentage was calculated based on the following formula [14].

% Disease severity (DS)=(D1/9) × (D2/9) × 100.

The mean incidence and severity of each field was computed from three to five stops. The results of the survey were summarized by districts and varieties. The geographic coordinates (latitude and longitude), and altitude were recorded using Geographic Positioning System (GPS) unit. The latitude and longitude coordinates were used to map the distributing of the three rusts and *Septoria* leaf blotches of wheat in the survey areas using the Environmental Systems Research Institute (ESRI) Arc View 3.0.

Stem rust race analysis

Race analysis was conducted only for stem rust. Samples were collected from South West and West Shewa zone of Oromia region. A total of 20 samples were collected and analyzed. Rusted stems were cut into small pieces of 5 to 10 cm in length using scissors and placed in paper bags after the leaf sheath was separated from the stem in order to keep the leaf sheath dry.

Urediniopores were collected from samples using atomizer collector in capsule and suspensions were prepared by mixing spores with lightweight mineral oil (Soltrol). For each isolate, the prepared spore suspension was inoculated using atomized inoculator on seven-days-old seedlings of the universally rust susceptible variety "Morocco" which does not carry known stem rust resistance to get enough amounts of spores to inoculate on stem rust differentials. Greenhouse inoculations were done using the methods and procedures developed by Stakman et al. [11]. The mono-pustule (single uredinial isolate) was further multiplied to get enough spores for the differentials. Seedlings were moistened with fine droplets of distilled water produced with an atomizer and placed in dew chamber for 18 h dark at 18 to 22°C followed by exposure to light for 3 to 4 h to provide condition for infection and seedlings were allowed to dry their dew for about 2 h. Then, the seedlings were transferred from the dew chamber to glass compartments in the greenhouse where conditions was regulated at

12 h photoperiod, at temperature of 18 to 25°C and relative humidity (RH) of 60 to 70%.

After two weeks of inoculation, urediniospores of each single pustule were collected in separate capsule and inoculated on the twenty standard differential sets. Five seeds of the twenty wheat stem rust differentials with known resistance genes (Sr5, Sr6, Sr7b, Sr8a, Sr9a, Sr9b, Sr9d, Sr9e, Sr9g, Sr10, Sr11, Sr31, Sr17, Sr21, Sr30, Sr36, Sr38, Sr24, SrTmp, and SrMcN) and one susceptible variety Morocco were grown in 3 cm diameter pots separately in greenhouse. The single pustule derived spores was suspended in soltrol inoculated onto seven-day-old seedlings using atomizers and/or an air pump. After inoculation, the formal procedure was repeated in dew chamber room. Upon removal from the dew chamber, plants were placed in separate glass compartments in greenhouse.

Infection types were scored after 14 days from inoculations based on 0-4 scale as described by Stakman et al. [11] where 0-2 stands for low infection and 3-4 for high infection. Five latters race code nomenclature system was used according to Roelfs and Martens and Jin et al. [15,16].

Results and Discussion

Survey of wheat stem, leaf and yellow rust

The survey result showed that the overall stem rust, leaf rust and yellow rust prevalence in the surveyed areas (West and South West Shewa) were 76.7%, 33.7% and 13%, respectively (Table 1).

Stem, leaf and yellow rust mean incidence of 6.4%, 26.9% and 5.3% were recorded in South West Shewa zone whereas 5.4, 19.6 and 7.7 was recorded in West Shewa zone in this order. The maximum incidence with 20%, 67% and 30% of stem rust, leaf rust and yellow rust, were recorded in Dawo, Seden Sodo and Chelia districts, respectively. In contrary, no leaf rust and yellow rust were observed in Ameya and Bacho districts (Table 1).

Over all mean stem, leaf and yellow rust severity with 6.5%, 9.7% and 5.5%, respectively were recorded in the surveyed areas. The maximum mean stem and leaf rust severity of 15% and 25%, respectively were recorded in Dawo district while, 20% severity of yellow rust was recorded in Chelia district (Table 1).

In general, the distribution of stem, leaf and yellow rust in both zones were less. This may be due to the unfavorable weather condition during 2013 cropping season. However, districts like Dawo, become hot spot for stem and leaf rust and attention should be given for such areas in the feature to manage loss due to the rust diseases. Chalia district was also affected by yellow rust which may became main treat for the feature unless resistant wheat varieties used.

Reaction of wheat varieties to rust diseases

Wheat varieties Digelu, Bonde, Dashen, Danda'a, Emer wheat, Gisso, Kekeba, Kubsa, Kumute and Medawelabu were grown in the surveyed areas. Among the wheat varieties, Digelu was the most popular and widely grown variety in two zones followed by Kubisa. Stem rust severity up to 20MS and 15S were recorded on variety Digelu and Kubisa, respectively. Zero stem rust severity was recorded on Danda'a, Gisso and Medawelabu Varieties. Varieties Digelu, Kekeba and Kumute showed moderately susceptible reactions while, Bonde, Emer wheat and Kubsa showed susceptible reactions to stem rust. On contrary, variety Dashen and Medawelabu showed moderately resistant reactions to stem rust (Table 2).

All varieties grown in the areas except Danda'a were affected by leaf

Zone	Districts	Stem rust			Leaf Rust			Yellow rust		
		Prevalence (%)	Incidence (%)	Severity (%)	Prevalence (%)	Incidence (%)	Severity (%)	Prevalence (%)	Incidence (%)	Severity (%)
South west Shewa	Ameya	67	5	5	0	0	0	0	0	0
South west Shewa	Bacho	100	5	10	0	0	0	0	0	0
South West Shewa	Dawo	100	20	15	75	40	25	50	25	10
South west Shewa	Seden Sodo	100	5	5	67	67	20	0	0	0
South west Shewa	Wolisso	100	5	10	40	54	17	0	0	0
south west Shewa	Wenchi	60	5	5	40	22	10	10	2	5
Mean		87.8	7.5	8.3	37.4	30.5	12	10	4.5	2.5
West Shewa	Ambo	100	5	5	40	5	10	20	10	10
West Shewa	Chelia	20	1	5	65	3	5	30	30	20
West Shewa	Dendi Toke	100	2	5	10	1	5	10	5	5
West Shewa	kutaye	20	1	5	40	2	5	20	5	5
Mean		60.0	2.3	5	38.8	2.8	6.3	20	12.5	10
G mean		76.7	54.7	7.0	37.7	19.4	9.7	14	7.7	5.5

Table 1: Distribution of Stem, Leaf and Yellow rust of wheat in South West and West Shewa zones.

Variety	Altitude (masl)	Stem rust				Leaf Rust				Yellow rust			
		Incidence		Severity		Incidence		Severity		Incidence		Severity	
		Range	Mean	Range	Mean	Range	Mean	Range	Mean	Range	Mean	Range	Mean
Digelu	1896-2843	0-20	5	0-20MS	5.5MS	0-5	1	05MR-Ms	1	0-5	1	0-5R	0.5R
Bonde	2306	2	2	5 S	5S	5	5	10MS	10	5	5	5MR	5MR
Dashen	1967-2346	0-5	4	0-10R	5R	0-100	98	0-60mss	54	0	0	0	0
Danda'a	2679	0	0	0	0	0	0	0	0	0	0	0	0
Emer wheat	2148	1	1	5S	5S	100	100	80S	80	0	0	0	0
Gisso	2378	0	0	0	0	5R	5	5MS	5	5R	5	5MS	5MS
Kekeba	2132-2361	5-10	7.5	5MS	5MSS	0-100	50	0-40MS	20	0	0	0	0
Kubsa	2016-2817	0-60	10	0-15S	10S	0-100	25	0-40mss	12	0-100	33	0-70MSS	20mss
Kumute	2612	5	5	5MS	5MS	5R	5	5R	5	0	0	0	0
Medawelabu	2758-2870	0	0	0	0	0-5	3	0-5R	2.5	0	0	0	0

Table 2: Response of improved wheat varieties to stem, leaf and yellow rust.

rust. The maximum leaf rust severity 80S were recorded on Emer wheat. Leaf rust severity with 60S, 40S, 40S were recorded on variety Dashen, Kekeba and Kubisa. Varieties Digelu showed moderately resistant to moderately susceptible reaction to leaf rust. Bonde, Dashen, Gisso, Kekeba and Kubisa wheat varieties showed moderately susceptible to susceptible reaction while, varieties Kumute and Medawelabu showed resistant reactions (Table 2).

Varieties Digelu, Bonde, Gisso and Kubsa were affected by yellow rust while, the rest varieties were free yellow rust. The highest yellow rust severity 70MS were recorded on Kubsa.Kekeba and Kubissa varieties, showing moderately susceptible reaction. On other hand Bonde and Digelu wheat varieties showed moderately resistant and resistant reaction, respectively (Table 2).

Varieties Danda'a and Meda Walabu were resistant against the three rusts (stem, leaf and yellow) while, the widely grown and popular variety Digelu showed susceptible reaction to stem rust. So wheat growers should be advised to grow resistance varieties Danda'a and Meda walabu replacing to the susceptible varieties like Digelu.

Survey of *Septoria* leaf blotch of wheat

Septoria leaf blotch was found to be among the most destructive disease observed during the growing season across surveyed areas (Table 3). The overall distribution/prevalence of the disease in the ten districts reached 100%. The disease was prevalent in all surveyed areas of the zones. The highest mean incidence (100%) was recorded

in Wolisso and Toke kutaye districts followed by the mean (90%) in Chelia. The overall mean incidence 83% was recorded in the surveyed areas in which mean incidence 81% and 85% were recorded in South West and West Shewa zone, respectively. The disease index ranges from 0.19-0.6 in the surveyed areas. The maximum severity index value 0.6 was obtained in Wanchi districts of South west Shewa zone. The overall severity index was 0.44 (Table 3). This indicated that *Septoria* leaf blotch became the most important disease in the surveyed areas.

Stem rust race analysis

Out of 20 stem rust samples collected and analyzed, 2 races were identified. Ten *P. graminis f. sp. tritici* isolates collected from West Shewa zone were assigned to one race and the rest 10 from South West Shewa zone were assigned to 2 races (Table 3). The highly virulent race called Ug99 (TTKSK) was the most abundant and widely distributed race across both zones with a frequency of 95%. The race TKTTF was detected in South west Shewa zone with frequency value 5%. The identification of two races from 20 samples is a clear indication of high virulence diversity within the *P. graminis* f. sp. *tritici* populations in Ethiopia. Admassu and Fekadu reported that there is high variability of *P. graminis* f. sp. *tritici* populations in Ethiopia [17].

The two races were virulent to one or more of the resistance genes (Table 5). For instance, the differential host carrying the resistance gene 5, 21,9e, 7b, 11, 6, 9g, 9b, 30, 17, 9a, 9d, 10, 31, 38 and McNair (SrMcN) were susceptible to both of the races. Race TTKSK was virulent on resistance gene 8a and 31 whereas TKTTF was virulent on resistance

Zone	Woreda	Altitude (m.a.sl)	Prevalence (%)	Septoria leaf blotch		
				Incidence (%)	Severity (%)	Disease index
	Ameya	1896-2006	100	87	45	0.25
	Bacho	2161-2293	100	80	76	0.52
	Dawo	2148-2306	100	82	76	0.52
South west Shewa	Seden Sodo	2292-2413	100	83	54	0.25
	Wolisso	1967-2415	100	100	76	0.52
	Wenchi	2100-2817	100	78	77	0.60
	Mean	**1967-2817**	**100**	**85**	**66**	**0.44**
	Chelia	2383-2870	100	90	65	0.37
	Dendi	2227-2497	100	64	84	0.40
West Shewa	Toke kutaye	2247-2413	100	100	54	0.25
	Ambo	2073-2679	100	70	35	0.19
	Mean	**2073-2679**	**100**	**81**	**65**	**0.37**
Grand mean		**1967-2817**	**100**	**83**	**66**	**0.44**

Table 3: Distribution of *Septoria* leaf blotch of wheat in South West and West Shewa zones.

Pgt-code					
	Set 1	5	21	9e	7b
	Set 2	11	6	8a	9g
	Set 3	36	9b	30	17
	Set 4	9a	9d	10	Tmp
	Set 5	24	31	38	McN
B		Low	Low	Low	Low
C		Low	Low	Low	High
D		Low	Low	High	Low
F		Low	Low	High	High
G		Low	High	Low	Low
H		Low	High	Low	High
J		Low	High	High	Low
K		Low	High	High	High
L		High	Low	Low	Low
M		High	Low	Low	High
N		High	Low	High	Low
P		High	Low	High	High
Q		High	High	Low	Low
R		High	High	Low	High
S		High	High	High	Low
T		High	High	High	High

*Low: Infection types 0, 1, and 2 and combinations of these values. **High: Infection types 3 and 4 and a combination of these values [15,16]

Table 4: Nomenclature of *Puccinia graminis* f. sp. *tritici* (*Pgt*) based on 20 differential wheat hosts.

Race	Virulence spectrum (ineffective Sr resistance genes)	No	%
Oromia TTKSK	*5,21,9e,7b,11,6,8a,9g,9b,30,17,9a,9d,10, 31, 38,MCN*	19	95
TKTTF	*5,21,9e,7b,11,6,9g,36,9b,30,17,9a,9d,10,TMP,38,MCN*	1	5
	Total	20	100

Table 5: Races of *P. graminis* f.sp. *tritici* identified and their virulence spectrum in west and south west Shewa zones, Oromia region, Ethiopia in 2013.

gene 36 and *Tmp*. Only three of the differential lines carrying resistance gene *Sr36*, *SrTmp* and *Sr24*, were effective against the most dominate race TTKSK (Ug99) whereas only *Sr8a*, *Sr24* and *Sr31* were effective against the most virulent race TKTTF. *Sr24* gene was found to be effective to all races detected in this study and hence can be considered as source of resistance (Table 5).

In general out of two races identified the most dominant and virulent race were TTKSK. Most of the genes were ineffective except *Sr36*, *SrTmp* and *Sr24* against TTKSK race but *Sr36* and *SrTmp* susceptible to TKTTF. The discovery of the race Ug99 with Virulence to *Sr31* in Uganda in 1999 [18] represented a real threat to wheat production in the world, including Ethiopia where stem rust epidemics

had not occurred since the resistant cultivar 'Enkoy, lost its resistance in 1993. In Ethiopia Ug99 was first detected in 2003 at six dispersed sites [16]. In this study also this race is widely distributed in the central part of the country. Previous study also indicated that Ug99 were predominantly distributed in the southern and central parts of the country than in northern west of Ethiopia [19].

Sr24 was effective against most of the isolates tested in Ethiopia. Admassu et al., also indicated that no virulent race detected against *Sr24* gene in Ethiopia [19]. Use of this gene for breeding in Ethiopia is permanent [20,21]. Countries like Ethiopia in which stem and yellow rust severely occur every year and the majority of wheat grown by subsistence farmers, for whom use of chemical fungicide against stem rust is not economical, continuous supply of resistance varieties decidedly needed to avoid wheat rust epidemics.

Acknowledgment

We would like to offer a great thanks to wheat rust research team of Ambo plant protection research Center for their valuable encouragement and technical support during the whole period of the study.

References

1. Hailu G, Tanner DG, Mengistu H (2011) Wheat research in Ethiopia: A Historical perspective, IARI and CIMMYT, Addis Ababa.

2. Bekele, Verkuiji H, MWangi W, Tanner D (2000) Adoption of improved wheat technologies in Adaba and Dodola woredas of the Bale high lands, Ethiopia, Mexico.

3. CSA (centeral statistics Authority) (2013) Report on area and crop production forecast for major grain crops, Addis Ababa, Ethiopia: statistical bulletin.

4. Zegeye T, Taye G, Tanner D, Verkuijl H, Agidie A, et al. (2001) Adoption of improved bread wheat varieties and inorganic fertilizer by small-scale farmers in yelmana Densa and Farta districts of North western Ethiopia. EARO and CIMMYT. Mexico city, Mexico.

5. Prescott JM, Burnett PA, LeSaari EE, Ranson J, Bowman J, et al. (1986) Wheat diseases and pests: a guide for field identification. CIMMYT, Mixico city. DF. Mexico. pp. 135.

6. Bekele E (1985) A review of research on diseases of berley, tef and wheat in Ethiopia. In: Tsedeke Abate (ed.), A review of crop protection research in Ethiopia. Institute of Agricultural Research (IAR), Ethiopia, pp 79-107.

7. Yirgu D (1967) Plant diseases of economic importance in Ethiopia. Haile Selassie I university, college of Agriculture, Experimental station bulletin no.50, Addis Ababa, Ethiopia.

8. Badebo A (2002) Breeding Bread Wheat with Multiple Disease resistance and high the yielding for the Ethiopian highlands: broadening Genetic basis of yellow rust and tan spot resistance. Gottingen, Germany: Gottingen University, Ph.D thesis.

9. CIMMYT (2005) Sounding the alarm on global stem rust: an assessment of race Ug99 in Kenya and Ethiopia and potential for impact in neighboring countries and beyond. Mexico City, Mexico.

10. Saari EE, Prescott JM (1975) A scale for appraising the foliar intensity of wheat disease. Plant Disease Reporter 59: 377-380.

11. Stakman EC, Stewart DM, Loegering WQ (1962) Identification of physiologic races of Puccinia graminis var. tritici.' USDA ARS, E716. United States Government Printing Office: Washington, DC.

12. Peterson RF, Campbell AB, Hannah A (1948) A diagrammatic scale for estimating rust intensity on leaves and stems of cereals. Canadaian Journal Research 26: 496-500.

13. Eyal, Z, Scharen AL, Prescott JM, Van Ginkel M (1987) The Septoria diseases of wheat: Concepts and methods of disease management. CIMMYT, Mexico DF, Mexico.

14. Sharma RC, Duveiller E (2007) Advancement toward new Spot Blotch resistant wheat in south Asia. Crop Science 47: 961-968.

15. Roelfs AP, Martens JW (1988) An international system of nomenclature for P. graminis f. sp. tritici. Phytopathology 78: 526-533.

16. Jin Y, Szabo U, Pretorius ZA, Singh RP, Ward R, et al. (2008) Detection of virulence to resistance gene Sr24 within race TTKS of Puccinia graminis f. sp. tritici. Plant Dis. 92: 923-926.

17. Admassu B, Fekadu E (2005) Physiological races and virulence diversity of Puccinia graminis f.sp. tritici on wheat in Ethiopi. Phytopathologia Mediterranea 44: 313-318.

18. Pretorius ZA, Singh RP, Wagoire WW, Payne TS (2000) Detection of virulence to wheat stem rust resistance gene Sr31 in Puccinia graminis f. sp. tritici in Uganda. Phytopathology 84: 526-533.

19. Admassu B (2010) Genetic and virulence diversity of Puccinia graminis f. sp. tritici population in Ethiopia and stem rust resistance genes in wheat. Gottingen, Germany: Gottingen University, Ph.D thesis.

20. Teklay A, Getaneh W, Woubit D (2012) Analysis of pathogen virulence of wheat stem rust and cultivar reaction to virulent races in Tigray, Ethiopia. African Journal of Plant Science 6: 244-250.

21. Teklay A, Woubit D, Getanh W (2013) Physiological races and virulence diversity of Puccinia graminis pers. f. sp. tritici eriks. & e. Henn. On wheat in Tigray region of Ethiopia. ESci J Plant Pathol 2: 1-7.

29

Management of Weeds in Maize (*Zea mays* L.) through Various Pre and Post Emergency Herbicides

Amare Tesfay, Mohammed Amin* and Negeri Mulugeta

Department of Plant Science, College of Agriculture and Veterinary Science, Ambo University, Ethiopia

Abstract

Field experiments were conducted in 2013 during main cropping season at Ambo and Guder to determine the effect of different post and pre emergency herbicides application on weed dynamics in maize (*Zea mays* L.) variety, BH-660 in randomized complete block design with three replications. Six treatments including Nicosulfuron (Arrow 75 WDG) at 0.09 kgha^{-1}+ silwet gold (adjuvant) at 0.10%, $_s$-metolachlor 290 + Atrazine (Primagram) at 3.00 kgha^{-1}, s-metolachlor (dual gold) 1.5 kgha^{-1}, hand weeding as standard check and weedy check as control were used. Effect of different herbicides on weed density was significant. The lowest weed density was recorded in plot treated with hand weeding and hoeing (3.12 m^{-2}) followed by Nicosulfuron (18.67 m^{-2}) and Primagram (3.88 m^{-2}). But, the maximum was recorded in weedy check (14.16 m^{-2}). However, no significant difference was observed between Nicosulfuron and Primagram. The minimum dry weight of weeds (0.77 gm^{-2}) was observed in hand weeding and hoeing followed by Nicosulfuron which is not significantly different from s-metolachlor. Moreover, those treatments also significantly increased the yield and yield component of maize. This is an indication of the reliability and promise as well as the exhibition of the great potential of the Nicosulfuron is the effective control of the weeds and enhancing yield of maize in Guder and Ambo, Ethiopia.

Keywords: Atrazine; Nicosulfuron; Primagram; Silwet gold; *Zea mays*

Introduction

The major constraints of maize production in Ethiopia include both biotic (weeds, plant pathogens, insect pests, rodents, wild animals) and abiotic factors (drought, hailstorm, flood, nutrient deficiency, soil type, topographic features) [1]. Weed infestation is supreme importance among biotic factors that are responsible for low maize grain yield. Worldwide maize production is hampered up to 40% by competition from weeds which are the most important pest group of this crop [2]. Generally weeds reduce crop yields by competing for light, nutrients, water and carbon dioxide as well as interfering with harvesting and increasing the cost involved in crop production. Overall, weeds impose the highest loss potential (37%), which is higher than the loss potentials due to animal pests (18%), fungal and bacterial pathogens (16%) and viruses (2%) [3]. Kebede [4] reported that most farmers in Ethiopia commonly lose up to 40, 30, 35, 18 and 30% of yield in maize, sorghum, wheat, barley and teff, respectively, due to weed infestations.

Weeds have a more direct influence on human beings than any other pest in developing countries like Ethiopia. Weeds not only cause severe crop losses but also compete with farmers and their families to spend a considerable amount of their time on weeding [5]. More than 50% of labor time is devoted to weeding, and is mainly done by the women and children in the farmer's family [6,7]. In the hand hoe system, weeding alone accounts for 40-54% of the total labor input in farming in Ethiopia, Ghana, Malawi, *Nigeria, Sierra Leone, Tanzania and Zambia, requiring 300-400 man-hours per hectare [8]. In most cases, farmers are unable to do their weeding on time due to limitations on family labor.* According to Unger [9], the taller and more numerous the weeds are in relation to the crop, the stronger is the competition. Weed competition in a cereal generally reduces crop vigor, tillers, head size, kernel weight and, consequently, grain yield.

Control of weeds in the fields of maize is, therefore, very essential for obtaining good crop-harvest. Weed control practices in maize resulted in 77 to 96.7% higher grain yield than the weedy check. Different weed control methods have been used to manage the weeds but mechanical and chemical methods are more frequently used for the control of weeds than any other control methods. Mechanical methods including hand weeding are still useful but are getting expensive, laborious and time-consuming. In the less developed countries, the situation still exists where the peak labor requirement is often for hand weeding [10]. Herbicides weed control is an important alternative to manual weeding because it is cheaper, faster and gives better weed control [3]. Chemical control is a better alternative to manual weeding because it is cheaper, faster, and gives better control [2,11]. Weed control in maize with herbicides has been suggested by researchers [12,13]. Ali et al. [14] also reported that herbicides significantly increased maize yield and decreased the weed density. However, continuous application of currently registered herbicides caused changing weed flora, poor controlling, and evolution of some herbicide resistant weed biotypes. This necessitates the introduction of some other new herbicide options with different modes of action. Therefore, this research work was carried out to evaluate the effect of new herbicide (Nicosulfron) on weeds and yield and yield components of maize under field condition at Guder and Ambo district, West Shoa, Ethiopia.

Materials and Methods

Location of study areas

Field experiments were conducted at two different areas viz. Guder and Ambo in maize cultivated field, West Showa, Ethiopia during

---

*Corresponding author: Mohammed Amin, Department of Plant Science, College of Agriculture and Veterinary Science, Ambo University, Ethiopia
E-mail: yonis_1986@yahoo.com

the main cropping season of 2013, for the management of weeds. The altitude of the study areas are between 1900 and 3100 m. a. s. l, geographical positions of N 08° 43.423-N 10° 12.082 and E 037° 28.902-040° 62.590. Guder and Ambo district has total geographical area of 78887 sq.km and is located at 8° 57 'North latitude and 38° 07 'East longitude at an average elevation of 1800-2300 m. a. s. l. The district lay under different climatic zones, which are 23% of highland, 60% of middle altitude and 17% is low land. In addition, the district has bi-model rainfall distribution with small amount of rainfall during autumn season and much rainfall during summer season. Heavy rain observed from onset of July to the end of August. The annual rainfall ranges from 1000 -1588.06 mm and the temperature of the district ranged between 9.44°C and 21.86°C with average of 15.65°C. The soil of the experimental site is light red in color, clay loam in texture and pH value of 6.8.

Treatments and experimental design

After determining the appropriate rate of application, field experiments were undertaken at both Guder and Ambo to compared the newly introduced herbicide (Nicosulfuron) with the traditional method (hand weeding and hoeing) and with already introduced herbicides (Primagram and s-metolachlor) and weedy check. Field experiment consists of six treatments, $_s$-metolachlor 290. + Atrazine(Primagram) at 3 kg/ha, s-metolachlor (dual gold) 1.5 kg/ha, Nicosulfuron (Arrow 750 WDG) at 0.09 kg/ha + silwet gold (adjuvant) at 0.10%, hand weeding and hoeing and weedy check were carried out and arranged in a randomized complete block design with three replications. Herbicides was applied 2 days after sowing as pre-emergence and 30 days after planting for post emergence using Knapsack/ Backpack sprayer. The spray volume was 600 L of water per ha. The size of each plot was 1.5 m×2.4 m. The distance between adjacent replications (blocks) and plots were 1 m and 0.5 m, respectively.

Agronomic practices for both locations

The experimental plots were ploughed twice to get fine seed bed, by oxen and plots were leveled manually before the field layouts were made. Variety BH-660 was used as a planting material. The maize seeds were planted manually in the month of May at both sites. During planting time, two maize seeds were placed at each hole and thinned to one plant per hill 20 days after sowing. The recommended amount of Nitrogen and phosphorus was applied. The source of nutrients was Urea and DAP, respectively. Half of N and the whole phosphorus were drilled in rows at the time of sowing. The remaining half N was applied at knee high growth stage of the plant.

Data Collected

Weeds data

Population: The weed population was counted before first 45 days after planting and at tasseling. The population count was taken with the help of 0.25 m×0.25 m quadrate thrown randomly at two places in each plot and was identified and converted to population/density per m².

Dry weight: While recording weed population the biomass was harvested from each quadrate. The harvested weeds were placed into paper bags separately and drying in oven at a 65°C temperature for 24 hours till constant weight and subsequently the dry weight was measured and converted in to gm⁻².

Weed Control Efficiency (WCE): It was calculated from weed control treatments in controlling weeds.

$$WCE = \frac{WDC - WDT}{WDC} X100;$$ Where WDC= weed dry matter in weedy check,

WDT= weed dry matter in a treatment

Maize data

Plant height (cm): Plant height was measured from 8 randomly selected (pre tagged) plants at the middle four rows, from the ground level to the apex of each plant at dough stage of the plant.

Number of cobs per plant: The number of productive ears was counted in each sample plants. Eight randomly selected tagged plants from the four central rows were used for counting productive ears.

Ear length (cm): The diameter of eight randomly taken ears was measured at mid length using caliper and the averages was recorded.

Hundred kernels weight (g): Thousand kernels were counted from each plot and their weight was recorded and adjusted to 12.5% moisture content.

Grain yield (kg/ha): The final produce was measured and adjusted to 12.5% moisture content with the help of formula:

$$Adjusted\ grain\ yield \left(kg\,ha^{-1} \right) = \frac{Actual\ yield\ X100 - M}{100 - D}$$

Where, M is the measured moisture content in grain and D is the designated moisture content.

Relative yield loss: Crop yield loss was calculated based on the maximum yield obtained from a treatment /treatment combination i.e. interaction as follows:

$$Relative\ Yield\ loss = \frac{MY - YT}{MY} X100\ ,$$

Where, MY= maximum yield from a treatment, YT = yield from a particular treatment.

Statistical analysis

Population density of weed was subjected to square root transformation $\left(\sqrt{(X + 0.5)} \right)$ to have data normal distribution using scientific calculator. Data were subjected to the analysis of variance. Mean separation was conducted for significant treatment means using Least Significance Differences (LSD) at 5% probability level.

Results and Discussion

Weed floral composition of the experimental sites

At Ambo, maize was infested with different weed species belongs to different family. 12 weeds species belongs to 8 families were identified. Out of the total weeds, 91.7% were broad leaved whereas the remaining 8.3% were grasses weeds (Tables 1 and 2). These indicate that species-rich weed community in the experimental field. Similarly at Guder site, 11 weeds species belongs to 9 families were identified. Out of the total weeds 72.7% were broadleaved weeds whereas the remaining 9.09% and 18.19% were grasses and sedges weeds respectively. This result is in agreement with Mehmeti et al. [15] who found that different weeds species in a single experimental site.

Common name	Trade Name	Rate	Time of Application
Nicosulfuron +silwet gold (adjuvant) at 0.10%	Arrow 75WDG	0.90kgha-1	Post
s-metolachlor	Dual Gold	1.50 kgha-1	Pre
Primagram	Primagram Gold 660EC	3.00 kgha-1	Pre
Hand weeding and hoeing	-	-	Post
Weedy check	-	-	-

Table 1: Description of treatments used in the experimental sites.

S.No	Guder		Ambo	
	Botanical name	Family name	Botanical name	Family name
1	*Amarathushybridus* L.	Amaranthaceae	*Amarathushybridus* L.	Amaranthaceae
2	*Commelinabanghalensis* L.	Commelineae	*Bidensbiternate*	Asteraceae
3	*Corrigiolacapensis* L.	Caryophyllaceae	*Canyzaboniersis*	Asteraceae
4	*Cynodondactylon* L.	Poaceae	*Daturastramorium*	Solanaceae
5	*Cyprus esculentus* L.	Cyperaceae	*Digitariaabysinca.*	Poaceae
6	*Cyprus rotundus* L.	Cyperaceae	*Erucastrumarabicum*Fisch and May	Brassicaceae
7	*Erucastrumarabicum*Fisch and May	Brassicaceae	*Galinsogaparviflora* cav.	Asteraceae
8	*Galinsogaparviflora* cav.	Asteraceae	*Ipomeaariocarpa*	Convolvulaceae
9	*Oxalis comiculate*L.	Oxalidaceae	*Launaeacornuta*	Asteraceae
10	*Oxalis latifolia* L.	Oxalidaceae	*Oxalis comiculate*L.	Oxalidaceae
11	*Polygonumnepalense*Meisn	Polygonaceae	*Polygonumnepalense*Meisn	Polygonaceae
12			*Tribulusterrestris*	Convolvulaceae

Table 2: Weed floral composition of at Guder and Ambo study sites.

Treatment	Guder		Ambo	
	weeds Density (m-2)	Dry weight(gm-2)	weeds Density (m-2)	Dry weight(gm-2)
Nicosulfuron	3.68(13.33)[d]	2.13[bc]	5.92(34.67)[c]	65.60[c]
s-metolachlor	5.45(29.33)[b]	21.33[bc]	12.87(168.00)[b]	105.07[b]
Primagram	4.65(21.33)[c]	26.67[bc]	11.99(144.00)[b]	93.33[b]
Hand weeding +hoeing	0.71 (0.00)[e]	0.00[c]	4.90(24.00)[c]	26.67[d]
Weedy check	14.16(200.00)[a]	170.93[a]	24.24(589.33)[a]	382.13[a]
LSD (0.05)	0.49	25.9	2.81	26.16
CV (%)	4.6	31.1	12.4	10.3

Table 3: Effect of different herbicides on density (m-2) and dry weight of weeds (gm-2). Figures or numbers in the parenthesis are original value, LSD= least significant difference, CV= coefficient of variation, means within a column followed by the same letter are not significantly different at the 0.05 probability level using Fisher's protected LDS test.

Density and dry weight of weeds

Effect of different herbicides on weed density both at 45 days after planting and tasseling stage was significant. As described on Table 3, the lowest weed density (0.71) was recorded in plot treated with hand weeding followed by Nicosulfuron (3.68) whereas the maximum was recorded in weedy check (14.16 m-2). Similar finding was reported Mehmeti et al. [15] who found that highest weed density in weedy check.

Moreover, the effect of herbicides application significantly affected the dry weight of weeds at both stage. The lowest of weight of weeds (0.0 gm-2) was recorded in plot treated with hand weeding followed by Nicosulfuron (2.13 gm-2) however, non-significant difference was existed among them, whereas the highest was observed in weedy check (170.93 gm-2). These results are in agreement with those reported by Hassan et al. [16] who reported reduced weed biomass due to use of selective pre-emergence and post emergences herbicides best for controlling different maize weed species.

Weed control efficiency

Weed control efficiency at both crop stages was also significantly affected. The minimum weed control efficiency was observed in weedy check (0.00%) whereas the highest (100.0%) was recorded in a plot treated with hand weeding and hoeing which is not significantly

different Nicosulfuron (98.8). This result further indicates that herbicides are more effective in reducing density and dry weights of weeds next to hand weeding and hoeing as compared to weedy check. This result was in accordance with Mehmeti et al. [15] who reported that it is evident herbicides reduced the weed infestation and control better than in the maize crop in comparison to the control plots (Table 4) [16].

Maize yield and yield components

At Guder except plant height, cobs number per plant, ear length and diameter were significantly affected by weed control methods. According the result showed in Table 5, plant height was not significantly affected. The maximum number of cobs per plant (1.9) was observed in hand weeding and hoeing followed by Nicosulfuron (1.8) however no significant were exist between them, whereas the lowest was recorded weedy check (0.47). Similarly at Ambo, effect of weed control methods was also significantly affecting the yield component of maize [17-19].

As described in Table 6, the maximum hundred seed weight was scored on combination of hand weeding and hoeing and the minimum was recorded on weedy check both at Guder and Ambo. Moreover, the highest grain yields were obtained from hand weeding + hoeing and followed by plot treated with Nicosulfuron at both study sites. While, the lowest grain yields were scored on weedy check [20-22].

Treatment	Weed Control Efficiency (%)	
	Guder	Ambo
Nicosulfuron	98.81[a]	83.02[b]
s-metolachlor	87.08[b]	72.48[c]
Primagram	83.91[b]	75.48[c]
Hand weeding + hoeing	100.00[a]	92.98[a]
Weedy check	0.00[c]	0.00[d]
LSD (0.05)	7.95	4.08
CV	5.71	3.35

Table 4: Effect of various herbicides on weed control efficiency (%). LSD= least significant difference, CV= coefficient of variation, means within a column followed by the same letter are not significantly different at the 0.05 probability level using Fisher's protected LDS test.

Treatments	Guder				Ambo			
	PH (cm)	Cobs /plant	EL(cm)	ED (cm)	PH (cm)	Cobs /plant	EL(cm)	ED(cm)
Nicosulfuron	150.47[a]	1.87[a]	18.0[a]	7.1[b]	175.5a[b]	1.9[a]	19.5[a]	7.1[b]
s-metolachlor	148.00[a]	1.20[b]	17.1[ab]	7.1[b]	160.7[ab]	1.4[b]	18.8[b]	7.2[b]
Primagram	157.00[a]	1.33[b]	16.8[ab]	7.2[b]	175.5[ab]	1.5[ab]	19.2[a]	7.1[b]
Hand weeding + hoeing	152.73[a]	1.93[a]	16.3[ab]	8.2[a]	179.1[a]	1.9[a]	19.7[a]	8.1[a]
Weedy check	147.87[a]	0.47[c]	12.2[c]	6.5[b]	144.3[b]	0.8[c]	12.9[a]	6.1[c]
LSD (0.05)	NS	0.29	2.28	0.8	31.39	0.42	1.90	0.8
CV	3.44	11.70	7.23	5.92	9.98	15.11	5.60	5.81

Table 5: Effect of various herbicides on plant height, cobs per plant, ear length and diameter (cm) in Guder and Ambo. PH=plant height, EL=ear length, ED=ear diameter, LSD= least significant difference, CV= coefficient of variation, means within a column followed by the same letter are not significantly different at the 0.05 probability level using Fisher's protected LDS test.

Treatments	Guder			Ambo		
	HSW	GY	RYL	HSW	GY	RYL
Nicosulfuron	41.53[a]	6883.3[a]	4.737[cd]	44.667[b]	6883.3[ab]	6.314[d]
s-metolachlor	42.633[a]	5026.4[b]	30.15[b]	41.167[c]	5026.4[c]	29.368[b]
Primagram	42.833[a]	6159.2[a]	14.519[c]	41.30[c]	6159.2[b]	11.803[c]
Hand weeding + hoeing	45.333[a]	6989.8[a]	0.000[cd]	49.667[a]	7223.1[a]	0.00[e]
Weedy check	33.80[b]	2312.4[c]	63.655[a]	29.80[d]	2612.4[d]	75.712[a]
LSD (0.05)	5.19	921.28	9.79	3.29	812.36	5.32
CV	6.68	8.84	23.01	4.24	7.73	11.47

Table 6: Effect of various herbicides on 100 seed Weight (g), Grain Yield (kgha⁻¹), and Relative Yield Loss (%). HSW =hundred seed weight, GY=grain yield, RYL=relative yield loss, LSD= least significant difference, CV= coefficient of variation, means within a column followed by the same letter are not significantly different at the 0.05 probability level using Fisher's protected LDS test.

Conclusions

In Ethiopia, maize has been selected as one of the national commodity crops to satisfy the food self-sufficiency program of the country to feed the alarmingly increasing population. Control of weeds in the fields of maize is very essential for obtaining good crop-harvest. From the result it can be stated that effect of different pre and post emergency herbicides on weed density, weed dry weight and weed control efficiency were significant. The lowest weed density was recorded in plot treated with hand weeding and hoeing followed by Nicosulfuron whereas the maximum was recorded in weedy check. Like density, dry weight of weeds the minimum was observed in hand weeding and hoeing followed by Nicosulfuron. Moreover, those treatments also significantly increased the yield and yield component of maize. Therefore from this field experiment, hand weeding and hoeing is most effective measure of weed control and increasing yields of maize however, due to labor shortage; herbicides are the most effective in terms of time and cost. Even though herbicides are more effective in time and cost, the candidate herbicide (Nicosulfuron+silwet gold (adjuvant) at 0.10%) is the outstanding for weed control in maize as compared to the already registered herbicides (Primagram and ₛ-metolachlor).

References

1. Ransom JK, Short K, Waddington S (1993) Improving productivity of maize under stress conditions. 30-33. *In*: Benti T. and Ransom J.K. ed. Proceeding of the First National Maize Workshop of Ethiopia, May 5-7 1992, IAR/ CIMMYT, Addis Ababa, Ethiopia.

2. Chikoye D, Schulz S, Ekeleme F (2004) Evaluation of integrated weed management practices for maize in the northern Guinea savanna of Nigeria. Crop Protection 23: 895-900.

3. Chikoye D, Udensi UE, Fontem A, Lum (2005) Evaluation of a new formulation of atrazine and metolachlor mixture for weed control in maize in Nigeria. Crop Protection 24: 1016-1020.

4. Kebede D (2000) Weed control methods used in Ethiopia. 250-251.*In*: Starkey, P. and Simalenga, T. (eds.). Animal power for weed control. A resource book of the Animal Traction Network for Eastern and Southern Africa (ATNESA). Technical Centre for Agricultural and Rural Cooperation (CTA), Wageningen, the Netherlands.

5. Fasil R, Matias M, Kiros M, Kasshun Z, Rezene F, et al. (2006) Weed research in sorghum and maize. pp. 303-323. *In*: Abraham Tadesse (eds.) Proceeding of the 14th annual conference of the plant protection society of Ethiopia (PPSE) 19-22 December 2006, Addis Ababa, Ethiopia 6.

6. Ellis-Jones J, Twomlow S, Willcocks T, Riches C, Dhliwayo H, et al. (1993) Conservation tillage/ weed control systems for communal farming areas in Semi-Arid Zimbabwe. Brighton Crop Protection Conference-Weeds 3: 1161-1166.

7. Akobundu IO (1996) Principle and practices of integrated weed management in

developing countries. *In: Proceeding 2nd International Weed Control Congress.* Copenhagen. Denmark 591-600.

8. Akobundu IO (1987) *Weed science in the tropics*, Principles and Practices. John Wiley and Sons, Chichester 522.

9. Unger J (1984) Principles and practice of weed management. College of Agriculture, Alemaya, Ethiopia 185.

10. Chikoye D, Ellis-Jone J, Riches C, Kanyomeka L (2007) Weed management in Africa : Experiences, Challenges and Opportunities. *In* : 16th International plant protection Congress 652-653.

11. Chikoye D, Manyong VM, Carsky RJ, Ekeleme F, Gbehounou G, et al. (2002) Response of speargrass [*Imperatacylindrica* (L.) Raeusch.] to cover crops integrated with handweeding and chemical control in maize and cassava. *Crop Protection* 21: 145-156.

12. Correa AJA, De La Rosa Mora MM, Dominguez YJA (1990) Demonstration plots for chemical weed control in rain fed maize *(Zea mays* L.) sown with minimum tillage in Acolmn Mexico RevistaChapingo 15: 164-166.

13. Owen MDK, Hartzler RG, Lux J (1993) Wooly cup grass *(Eriochloaviosa)* control in corn *(Zea. mays* L.) with Chloroacetamide herbicides. Weed Technology 7: 925-929.

14. Ali R, Khalil SK, Raza SM, Khan H (2003) Effects of herbicides and row spacing on maize (*Zea mays* L.). Pakistan Journal of Weed Science Research 9: 171-178.

15. Mehmeti AA, Demaj I, Demelezi, Rudari H (2012) Effect of Post-Emergence Herbicides on Weeds and Yield of Maize. Pak J Weed Sci Res 18: 27-37.

16. Hassan G, Tanveerl S, Khanand NU, Munir M (2010) Integrating cultivars with reduced herbicide Rates for weed management in maize. Pak J Bot 42: 1923-1929.

17. Akinyemiju AO (1988) Chemical weed control in maize (*Zea mays* L.) and cowpea [*Vignaunguiculata* (L.) Walp.] in the rainforest zone of southwestern Nigeria. Nigeria Journal of Weed Science 1: 29-41.

18. Gomez KA, Gomez AA (1984) Statistical procedure for agricultural research. 2nd edition. A Wiley Interscience Publications, New York 122-123.

19. Khan MA, Marwat KB, Khan N (2003) Efficacy of different herbicides on the yield and yield components of maize. Asian Journal of Plant Science 2: 300-304.

20. Lyon, Alex R, Martin R, Klein N (2006) Cultural practices to improve weed control in winter wheat. The University of Nebraska-Lincoln.

21. Radosevich SR, Todies SH, Ghersa C (1996) Weed Ecology: Implication for 2nd edition. John Wiley and Sons, Inc. New York 589.

22. Takele A (2008) Witch weed (*Striga hermonthca* (Del.)) Benth. Infestation and Component crop productivity as Influenced by Different Intercropping Patterns of Groundnut Varieties in Maize at Metekel Zone, North-western Ethiopia. M.Sc. Thesis, Haramaya University, Ethiopia 1-3.

30

Hydraulic Redistribution from Wet to Drying Roots of Potatoes (*Solanum tubersosum* L.) During Partial Rootzone Drying

Hamad Saeed[1]*, Ivan G. Grove[2], Peter S. Kettlewell[2], Nigel W. Hall[2], Ian J. Fairchild[3] and Ian Boomer[3]

[1]*Department of Agriculture, Wiltshire College, Lacock, Chippenham, Wiltshire, UK*
[2]*Crop and Environment Research Centre, Harper Adams University College, Newport, Shropshire, UK*
[3]*School of Geography, Earth and Environmental Sciences, University of Birmingham, UK*

Abstract

Hydraulic redistribution, redistribution of water upward or downward within a soil profile through roots as a consequence of root-soil water potential gradients, can be an important mechanism in transporting chemical signals (i.e. abscisic acid) to the shoot for stomatal closure or in maintaining the root system during dry periods of partial rootzone drying (PRD). PRD involves alternate irrigation to two sides of a plant root system. The study reported here investigated the occurrence and magnitude of hydraulic redistribution in glasshouse-grown potatoes (*Solanum tuberosum* L.) under PRD. Deuterium labelled water was applied to only one half of the root system to field capacity at tuber initiation. The roots from the drying side of the dual pot were extracted at 3, 6, 12, 18 and 24 h following watering by the dry sieving method. Water from the roots was extracted by azeotropic distillation and analysed for hydrogen isotope ratios. Hydraulic redistribution occurred the most at night when stomatal conductance was considerably lower and leaf water potential was higher (less negative). The magnitude of the redistributed water, however, did not exceed 3.5%, indicating limited water redistribution under PRD. The observed water redistribution would probably be of little significance for the survival of roots present in the upper drier portion of the soil under higher water demanding conditions but its role in sending the chemical signals to the shoot to conserve water by reducing transpiration would be of particular significance during drying periods of partial rootzone drying.

Keywords: Potatoes; Partial rootzone drying; Hydraulic redistribution; Water utilization; Stomatal conductance; Leaf water potential

Introduction

Water is the most important factor controlling plant growth [1]. Plants shift growth in favour of roots under water deficit conditions [2]. Roots under water deficits continue growth by adjusting the minimum pressure in cells required for the expansion and by regulating solute transport within the elongation zone by altering cell wall elasticity or cell size [3,4]. This adoptive response to a water deficit results in vertical root penetration reaching the moist soil layers. Roots in the deeper moist soil layers often help plants to overcome drought stress by extracting and supplying more water to the shoot [5,6]. These roots can also redistribute water to the upper drier soil layers at night by a process known as hydraulic lift [7,8]. Although the direction of water movement is typically upward towards the shallower soil layers, it has been demonstrated that roots can also redistribute water from the surface to deep soil layers along water potential gradients [9,10]. The process is thought to be largely passive, requiring only a gradient in soil water potential, a more positive water potential in the root xylem than in the surrounding dry soil layers, and a relatively low resistance to reverse flow from the roots [11]. Because of the bi-directional and passive nature of the phenomenon, Burgess et al. [9] have proposed 'hydraulic redistribution' as a more comprehensive term for the phenomenon. This downward movement of water has also been described as 'downward siphoning' [12], 'inverse hydraulic lift' [13] or 'reverse flow' [10] in the literature. Scholz et al. [14] found that the rate of reverse flow was linearly related to soil-leaf water potential gradient, with the greatest reverse flow rates occurring when this potential gradient was at its most negative values. Hultine et al. [15] reported greater magnitude of hydraulic redistribution in roots when night-time vapour pressure deficit was low. The onset and the magnitude of hydraulic redistribution is thought to be regulated by the development of water potential gradients within the plant parts, between the plant parts and soil, and the nocturnal demand for water by the plant [16]. Hydraulically redistributed water may buffer plants against water stress during a water deficit by replenishing up to 28–35% of the soil water removed each day by plants from the upper soil layers [17]. The redistributed water can contribute positively in transporting chemical signals (i.e. abscisic acid) to the shoot for stomatal closure [18] or in prolonging or enhancing the activity (e.g. growth and solute uptake) and life span of fine roots in a dry soil profile [19,20], Hydraulic redistribution has been reported to be a common phenomenon in numerous plant species including trees, shrubs and grasses from deserts to tropical forests [21]. Little attention has been paid, however, to the phenomenon in agricultural crops. There has been a report documenting the occurrence of hydraulic redistribution in grapevines under partial rootzone drying (PRD) [18]. PRD is an irrigation practice in which one half of the plant root system is irrigated as in standard irrigation whilst the other half is kept in a drying state [22]. Results of this study are, however, inconclusive in terms of the extent to which

*Corresponding author: Hamad Saeed, Department of Agriculture, Wiltshire College, Lacock, Chippenham, Wiltshire, UK, E-mail: hamadsaeed@yahoo.com

hydraulic redistribution occurs to roots growing in the drying soil. Due to inadequate information on this subject, the degree of dependency of roots of the drying soil on roots of the wet soil under partial rootzone drying (PRD) is unknown. To date, there is no published work describing the detailed insight of the water-supplying characteristics of the wet roots to the roots of the drying soil under PRD. Further, until now hydraulic redistribution phenomenon has not been studied in potatoes (*Solanum tuberosum* L.) under partial rootzone drying. The present study investigated the hydraulic redistribution mechanism and its magnitude in potatoes (*Solanum tuberosum* L.) under partial rootzone drying using the stable isotope of hydrogen (deuterium). The hypothesis tested was that hydraulic redistribution does occur in potatoes during PRD and at an increased rate when transpiration is low. The objectives were to investigate (1) whether hydraulic redistribution occurs in potatoes during PRD, (2) the time of water movement from wet to the drying half of the root system, and (3) the magnitude of the redistributed water in the roots growing in the drying soil.

Materials and Methods

Potatoes (cv. Estima) with a split root system were grown in John Innes No. 2 compost in a dual flexible pot system under glasshouse conditions at Harper Adams University College, Shropshire, UK during April to June 2006. The dual flexible pot system was made by joining two flexible plant pots (LBS Polypot, LBS Horticultural Ltd., UK), each 18 cm wide and 30 cm tall with a volume of 6.11 L. Seed tubers of the potato cv. Estima were placed in a suitably-sized hole created on the inner sides of the pot, 10 cm from the top, in such a way that half of the sprouts laid in each side of the pot. A 2 cm layer of gravel (6 mm diameter) was added at the surface after planting to prevent evaporation. Plant emergence was completed at 13 ± 1 days ($n=40 \pm$ standard deviation; S.D). All plants were thinned to a single shoot growing in the middle of the dual pot a week after plant emergence. The aim was to minimise root growth variability between the split root system and between the plants due to variable number of stems. Volumetric water content (%) of the compost from both sides of a dual pot was monitored regularly with time domain reflectometry using Trime FM (Imko, Germany). Both sides of a dual pot were irrigated close to field capacity until tuber initiation. At tuber initiation (four weeks after plant emergence), plants were randomly assigned to different treatments in a complete block experiment. Treatments were the combination of two water types and five root sampling times with four replications of each treatment. Water types were tap water and deuterated water. Tap water was the normal irrigation water with an isotopic composition ($\delta2H$) of -51 parts per thousand (%). Deuterated water was prepared by mixing 10 ml of deuterium oxide (99.96% deuterium, Merck KGaA, Germany) in 40 litres of tap water. The $\delta2H$ value of the deuterated water obtained was 1217 (%). The obtained $\delta2H$ value was in the range of enrichment used for grapevines [18] and Douglas-fir trees [17]. Both types of water were applied in the morning between 08:00-09:00 h at a slow trickle, away from the stem base, to only one side of the dual pot to FC. The other side was kept in a drying state over the treatment period. Plants irrigated with tap water were denoted as 'control' plants whilst those irrigated with deuterated water were denoted as 'treated' plants. On average, control and treated plants received 1332 and 1353 ml of tap water and deuterated water, respectively. Shortly after watering, the irrigated side was covered with aluminium foil to prevent deuterium fractionation due to surface evaporation. The movement of water from the base of irrigated to the drying side of a dual pot was eliminated by placing each side of the dual pot in a pot saucer. Soil water content was close to 25% by vol. in both sides of a dual pot at the time of water application. This corresponded to approximately 50% of FC, or to a soil matric potential of approximately -150 kPa according to the moisture release curve for this compost. Roots from drying side of a dual pot were extracted at 3, 6, 12, 18 and 24 h following watering. Root sampling times fell at local time 12:00, 15:00, 21:00, 03:00, and 09:00 h, respectively, allowing to determine the time and rate of water influx from irrigated to drying side of the root system under high (day) and low (night) evaporative demand conditions.

Physiological measurements

Physiological influences of PRD on the movement and magnitude of water redistribution were evaluated by measuring abaxial stomatal conductance and leaf water potential for each sampling time. Abaxial stomatal conductance was measured from the terminal leaflet of the 4th fully expanded young leaf from the apex [23,24] using a portable porometer (Delta-T AP4, Delta-T devices, Cambridge, UK). Leaf water potential was measured immediately after the stomatal conductance measurements on the same leaf within a minute of its excision using a Scholander portable pressure chamber [25]. The physiological measurements were completed within half an hour at any sampling time.

Root extraction

Roots from the drying side of a dual pot were extracted from compost by the dry sieving method [26]. Roots retained on the sieve (5 × 5 mm mesh size) were collected, washed with tap water to remove any compost traces, excess water removed immediately with filter paper, placed in self-seal plastic bags and stored at 0°C in a water bath to prevent isotopic fractionation due to evaporation [6]. Root samples were transported to the laboratory after each sampling time where they were stored at -30°C until water extraction [6,27]. The root extraction time did not exceed one hour for any sampling time. Water extraction, purification and hydrogen isotope analysis Water from all root samples was extracted by azeotropic distillation with kerosene (liquid paraffin; boiling point >23°C, VWR Ltd. UK) as the solvent [28,29] using the methods described by Revesz and Woods [30]. The water extraction process took 2-3 h to complete. Water samples were purified with powder paraffin wax (solidifying point 63-66°C) for any impurities [30-32]. The purified water samples were stored in 2 ml vials at −30°C [6,27]. All water samples were sent to Isotope and Luminescence Laboratory, School of Geography, Earth and Environmental Sciences, University of Birmingham, UK for hydrogen isotope analysis. Water samples were analysed for hydrogen isotope ratios on a continuous flow isotope ratio mass spectrometer (Isoprime™, GV Instruments, Manchester,UK) interfaced with an elemental analyser (Eurovector, GV Instruments) and a autosampler. The hydrogen isotope ratios were expressed as $\delta2H$ (delta values) in parts per thousand (%) relative to V-SMOW (Vienna Standard Mean Ocean Water):

$$\delta2H\ (\%)=[(R_{sample}/R_{VSMOW})-1]1000$$

Where R_{sample} and R_{VSMOW} are the ratios of deuterium to hydrogen atoms ($^2H/^1H$) of the isotope sample and the standard VSMOW, respectively. The total analytical uncertainty of the instrument was $\pm$ 1%.

Proportion of deuterated water uptake (%)

The proportion of deuterated water taken up by the treated plants relative to the control plants was calculated using the following formula (pers. communication: Prof. Ian Fairchild, School of Geography, Earth and Environmental Sciences, University of Birmingham, UK):

$$\text{Proportion of water uptake} = \frac{\text{Rtreated plant} - \text{Rcontrol plant}}{\text{Rdeuterated water} - \text{Rtap water}}$$

Where $R_{\text{treated plant}}$ and $R_{\text{control plant}}$ are the ratios of deuterium to hydrogen atoms ($^2H/^1H$) of the water extracted from a plant that received either deuterated or tap water, respectively. $R_{\text{deuterated water}}$ and $R_{\text{tap water}}$ are the ratios of deuterium to hydrogen atoms ($^2H/^1H$) of deuterated and tap water applied to a plant, respectively. These proportions were expressed as percentages after multiplying by 100.

Water utilisation

Water utilisation from either side of a dual pot was calculated by the following

Formula:

Water use (%, vol.) of a side of a dual pot was calculated as follows

Water use (%, vol.)=$WC_{(s)}$-$WC_{(E)}$

Where:

$WC_{(S)}$ is the water content (%, vol.) measured from a side of a dual pot when treatments began. WC(s) (%, vol.) for the wet side of the pot was taken as 50.33% as this side was only irrigated to field capacity. $WC_{(E)}$ is the water content (%, vol.) measured at the end of a sampling time.

Air temperature

Air temperature during the treatment period was recorded by positioning a Tinytag® data logger (Gemini Data Loggers (UK) Ltd., Chichester, Sussex, UK) one metre above the plant canopy. The changes in temperature over each sampling time are shown in Figure 1.

Statistical data and analysis

During water application, approximately 500 ml of deuterated water (δ^2H=1217‰) accidentally moved to the dry side of the pot (supposed not to receive this water) in one of the 3 h root sampling time plants, thus omitted from the experiment. Additionally, one water sample bottle of 12 h root sampling time damaged during storage, so was discarded in the isotope analysis. Data were subjected to polynomial analysis of variance with water types and sampling times as factors for all measured variables except for percent deuterated water in the drying roots water and soil water content measured at the end of each sampling time. Percent deuterated water in the drying roots water was analysed with sampling times as the main factor. Soil water content data was analysed by considering pot sides another factor with water types and sampling times. Data were analysed using Genstat 8th edition (PC/Windows XP), Lawes Agricultural Trust (Rothamsted Experimental Station). Treatments means were considered significantly different at the 5% level of probability using Tukey's HSD test.

Results

Water content (%, vol.)

The volumetric water content measured at the end of each sampling time was not significantly different between control and treated plants (P=0.907) (Table 1a) but differed significantly between the wet and the dry side of the pot (P<0.001) (Table 1b). The polynomial analysis of variance revealed a significant decrease in the water content over time (P=0.004), with a quadratic effect (P=0.002). The non-linear relationship was mainly due to high soil water content in one of the control plants harvested after 24 h of water application.

Of the possible interactions, pot side x sampling time interaction significantly affected the water content of the pots (P=0.005), with a linear effect (P<0.001) (Table 1b). Tukey's test revealed that water content on the wet side was significantly higher from the drying side of the pot for all sampling times. Although water content of the drying side of the pot was not significantly different between sampling times, the water content of 3 h sampling time was significantly higher from the water content of 12, 18 and 24 h sampling times on the wet side of the dual pot.

Water utilisation from drying side of the pot (%)

The percent of water utilisation from drying side of the pot was not significantly different between control and treated plants (P=0.563), between sampling times (P=0.354), and between control and treated plants over the treatment period (P=0.942) (Table 2).

Stomatal conductance

Stomatal conductance was not significantly different between

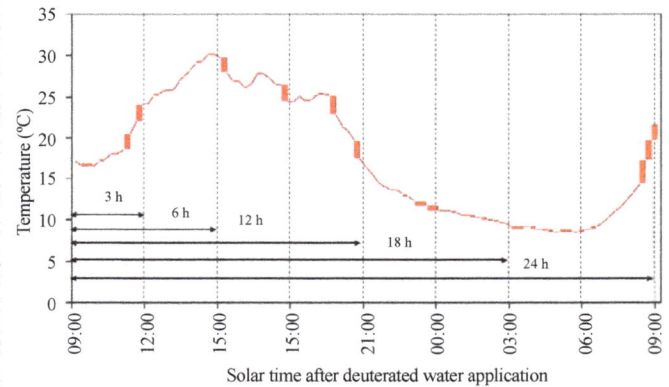

Figure 1: Changes in temperature over the treatment period (2-3 June 2006) following deuterated water application to potatoes (cv. Estima) under glasshouse conditions. Arrows and numbers indicate air temperature experienced by the potato plants during each sampling time (h). Experiment was started at 09:00 am local time.

a) Water type × sampling time interaction.

Water type	Sampling time after watering (h)					Mean
	3	6	12	18	24	
Tap water (control plants)	31.31	30.71	27.98	28.58	30.16	29.75
Deuterated water (treated plants)	32.13	30.04	28.53	29.24	29.15	29.82
Mean	31.72	30.38	28.25	28.91	29.65	

b) Pot side × sampling time interaction.

Pot side	Sampling time after watering (h)					Mean
	3	6	12	18	24	
Wet	40.08	37.55	34.29	34.54	34.73	36.24
Dry	23.36	23.20	22.21	23.28	24.59	23.33

C) Water type × pot side × sampling time interaction.

Water Type	Pot side	Sampling time after watering (h)					Mean
		3	6	12	18	24	
Tap water (control pants)	Wet	39.88	37.98	34.58	23.05	34.93	36.07
	Dry	22.75	23.45	21.38	24.18	25.40	23.43
Deuterated water (treated plants)	Wet	40.28	37.13	34.00	36.10	34.53	36.41
	Dry	23.98	22.95	23.05	22.38	23.78	23.23

Factor	P value	sed (df=55)
Water type	0.907	0.576
Pot side	<0.001	0.576
Sampling time	0.004	
Linear	0.015	
Quadratic	0.002	0.910
Deviations	0.645	
Water type × pot side	0.638	0.814
Water type × sampling time	0.785	
Water type. linear water	0.600	
type. quadratic	0.599	1.288
Deviations	0.561	
Pot side × sampling time	0.005	
Pot side linear	<0.001	
Pot side quadratic	0.277	1.288
Deviations	0.842	
Water type × pot side × sampling time	0.349	
Water type × pot sides Linear	0.244	
Water type × pot sides Quadratic	0.900	1.821
Deviations	0.217	
CV	8.6%	

Table 1: Volumetric water content (%) of dual pots measured at the end of each sampling time along with all possible interactions of glasshouse grown potatoes of the cv. Estima.

control and treated plants (*P*=0.802) (Table 3) but differed significantly over the treatment period (*P*<0.001) (Figure 2) with a linear and quadratic relationship; both with *P*<0.001. Although the deviation remained significant (*P*<0.001), the quadratic relationship described the stomatal conductance response better as it reflected the biological response of plants with the time of the day. Stomatal conductance measured during daytime at 12:00 h after 3 h of water application was significantly higher from all other sampling times. The lowest stomatal conductance (23.2 mmolm-2 s-1) was, however, measured in plants harvested after 18 h of water application at dawn. There was no significant water type x sampling time interaction effect on stomatal conductance (*P*=0.226) (Table 3).

Leaf water potential

There were no significant differences between control and treated plants for leaf water potential (*P*=0.910) (Table 4). Leaf water potential, however, differed significantly between the sampling times (*P*<0.001), with a cubic effect (P<0.001) (Figure 3). Sampling times 3h and 6h were statistically similar to each other for leaf water potential but were significantly different from rest of the sampling times. The higher (less negative) leaf water potential of –222.7 kPa measured in plants harvested at dawn after 18 h of water application was only found to be non-significant with leaf water potential of the plants harvested in the morning at 09:00 am after 24 h of water application. Water type x sampling time interaction effect was found to be non- significant on leaf water potential of the plants (*P*=0.344) (Table 4).

Isotopic composition (δ2H,%) of water of 'drying roots'

The rate and pattern of water redistribution cannot be described from variations in the δ^2H values between the sampling times as mean δ^2H value of each sampling time is an average of control and treated plants δ^2H values. The δ^2H values of control plants (received tap water) predominantly reflect natural changes in isotope ratios of root water over time and cannot be used to describe the water redistribution

pattern over the treatment period, only δ^2H values of the treated plants appear to be more applicable and reliable in describing such trends. Changes in δ^2H values of the water extracted from 'drying roots' of control and treated plants are, therefore, described separately over the treatment period.

The isotopic composition (δ^2H, %) of water extracted from 'drying roots' was significantly different between control and treated plants (*P*<0.001) (Table 5), water of the treated plants being 34.9% isotopically heavier relative to the control plants. The effect of water type x sampling time interaction on δ^2H values of water extracted from 'drying roots' was found close to the level of significance (*P*=0.052) with a quadratic relationship (*P*=0.015) (Figure 4). The deuterium concentration-time curves for control and treated plants show that deuterated water moved to the drying half of the root system after 3 h of water application, indicated by relatively higher 'drying roots' water δ^2H values of the treated plants than the control plants (Figure 4).

The deuterium concentration, however, reached a peak after 12 h of water application, which remained relatively constant until 18

Water type	Sampling time after watering (h)					Mean
	3	6	12	18	24	
Tap water (control pants)	18.3	17.3	15.9	14.9	15.2	16.3
Deuterated water (treated plants)	18.8	17.2	15.4	16.3	17.1	16.9
Mean	18.5	17.3	15.7	15.6	16.1	

Factor	P value	sed (df=26)
Water type	0.563	1.048
Sampling time	0.354	
Linear	0.110	
Quadratic	0.186	1.657
Deviations	0.984	
Water type × sampling time	0.942	
Water type. Linear	0.524	
Water type. Quadratic	0.679	2.343
Deviations	0.921	
CV	19.9%	

Table 2: Water utilisation (%) by glasshouse-grown potatoes (cv. Estima) from drying side of the pot over treatment period.

Water Type	Sampling time after watering (h)					Mean
	3	6	12	18	24	
Tap water (control pants)	158.8	86.5	93.2	25.0	61.5	85.0
Deuterated water (treated plants)	199.2	87.0	58.2	21.5	71.5	87.5
Factor	P value				sed (df=26)	
Water type	0.802				9.86	
Sampling time	<0.001					
Linear	<0.001					
Quadratic	<0.001				15.58	
Cubic	0.592					
Water type × sampling time	0.226					
Water type. Linear	0.473					
Water type. Quadratic	0.045					
Water types. Cubic	0.339					
Deviations	0.707				22.04	
CV	36.1 %					

Table 3: Stomatal conductance (mmolm^{-2}s^{-1}) over treatment period of potato plants (cv. Estima) received tap water and deuterated water under glasshouse conditions.

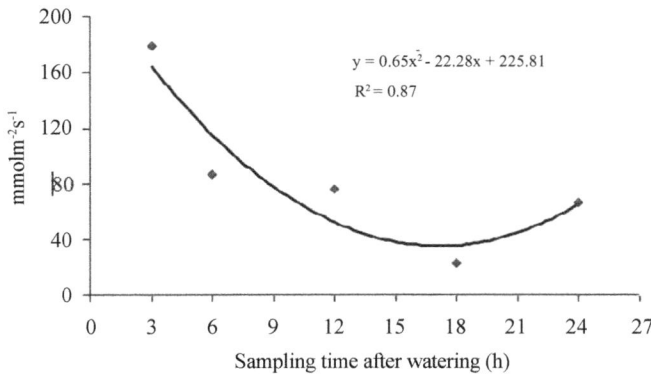

Figure 2: Diurnal changes in stomatal conductance (mmolm^{-2}s^{-1}) following deuterated water application to glasshouse grown potatoes (cv Estima) at 09:00 am local time.

Water type	Sampling time after watering (h)					Mean
	3	6	12	18	24	
Tap water (control pants)	−429.1	−483.3	−380.4	−218.5	−291.9	−360.7
Deuterated water (treated plants)	−492.8	−475.8	−350.1	−226.9	−266.6	−362.4
Factor			*P* value		sed (df = 26)	
Water type			0.910		15.59	
Sampling time			<0.001			
Linear Quadratic			0.008			
Cubic			<0.001		24.65	
Deviations			0.925			
Water type × sampling time			0.344			
Water type. Linear			0.201			
Water type. Quadratic			0.365			
Water type. Cubic			0.156		34.87	
Deviations			0.967			
CV			13.6%			

Table 4: Leaf water potential (kPa) measured over time following deuterated water application to glasshouse grown potatoes of the cv. Estima.

h and then showed a declining trend, with the lowest δ^2H value of −33.4% in plants harvested after 24 h of water application. The 'drying roots' water δ^2H values of the control plants were not significantly between the sampling times and varied from −53.3 to −63.7% (Figure 4), with a mean of −57.6% (Table 5). Root water of the control plants was 6.6% isotopically lighter than the source irrigation water δ^2H value of −51%, indicating the liberation of organically bound hydrogen into the bulk root water due to the breakdown of either plant tissues or sap carbohydrates to some extent at high distillation temperature [33].

Percent deuterated water in the 'drying roots' water

The proportion of deuterated water in the water extracted from 'drying roots' of the treated plants was significantly different over the time course of the study (P=0.046) (Figure 5). Figure 5 shows a steady increase in the redistribution of deuterated water from the roots in the wet soil to the roots in the drying soil from 6 to 18 h of water application, with the highest deuterium concentration of 3.48% in plants harvested

after 18 h of water application at dawn. The percent deuterated water, however, declined to 1.57% in plants harvested in the morning at 09:00 am after 24 h of water application. The small proportion of the deuterated water in 'drying roots' of the treated plants indicates that

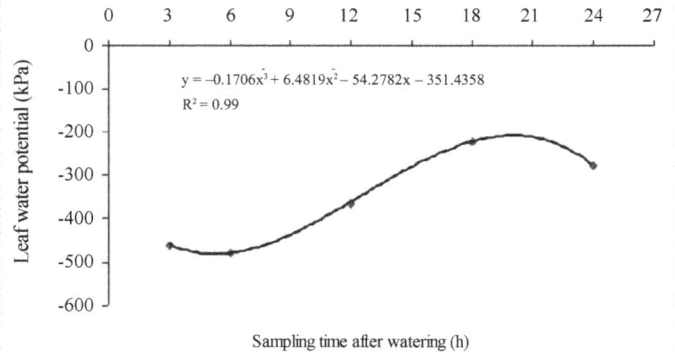

Figure 3: Diurnal changes in leaf water potential (kPa) following deuterated water application to glasshouse grown potatoes (cv Estima) at 09:00 am local time.

Water type			Δ^{2}H (%)
Tap water (control plants)			−57.6
Deuterated water (treated plants)			−22.7
Factor		*P* value	sed (df = 26)
Water type		<0.001	2.66
Sampling time		0.653	
Linear		0.275	
Quadratic		0.350	4.21
Deviations		0.852	
Water type × sampling time		0.052	
Water type. Linear		0.302	
Water type. Quadratic		0.015	5.96
Deviations		0.248	
CV		21.0%	

Table 5: Mean δ^2H values (%) of water extracted from 'drying roots' of the control (received tap water) and the treated (received deuterium-enriched water) plants of potato cv. Estima.

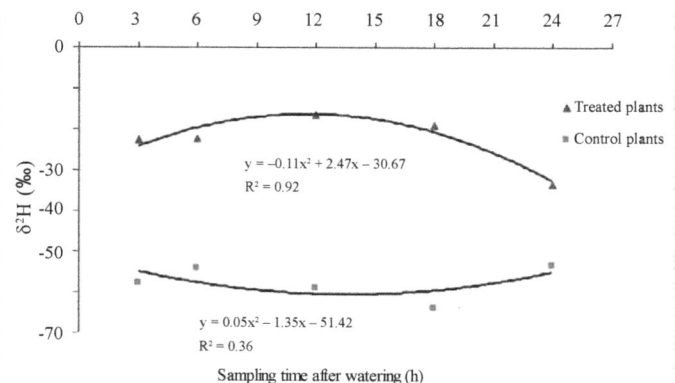

Figure 4: Time course changes in δ^2H (%) values of water extracted from 'drying roots' of control and treated potato plants (cv. Estima) grown under glasshouse conditions. Experiment was started at 09:00 am local time.

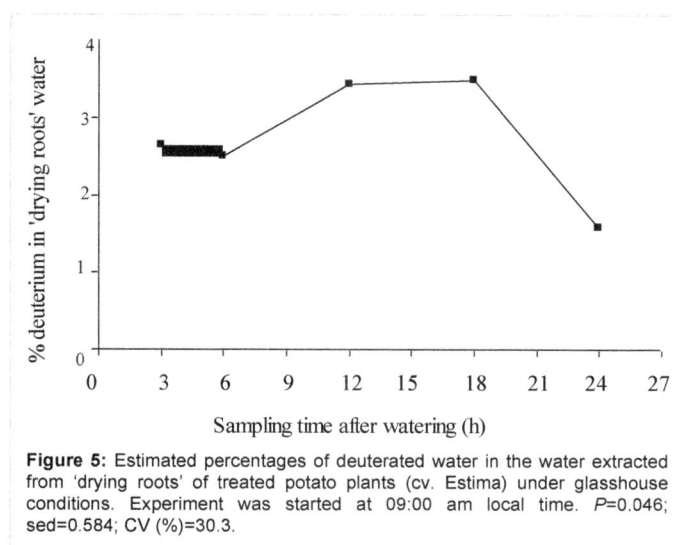

Figure 5: Estimated percentages of deuterated water in the water extracted from 'drying roots' of treated potato plants (cv. Estima) under glasshouse conditions. Experiment was started at 09:00 am local time. $P=0.046$; sed=0.584; CV (%)=30.3.

the 'drying roots' were less dependent on the water of the wet soil but were largely relying on the water existing in the drying soil (Figure 5).

Discussion

Irrespective of the sampling time, δ^2H values of the 'drying roots' of the treated plants remained negative relative to the standard (VSMOW). There could be several reasons. Firstly, the negative δ^2H values might be associated with the root sample size. Whole root system growing in the drying soil was used to extract water for isotope analysis. Although the whole root system was used to reduce bias in the results, the existing water in the roots, absorbed from the soil previously watered with tap water ($\delta^2H=-51\%$), might have diluted the deuterium concentration of the redistributed water. Secondly, there is a possibility that the quantity of water absorbed by roots from the irrigated side of the pot was perhaps sufficient to hydrate the above-ground plant parts but not sufficient enough to hydrate all roots in the drying soil given the relatively short time period over which the samplings were taken. Thirdly, the liberation of hydrogen (H) from breaking down of plant tissues or sap carbohydrates at high temperature of the azeotropic distillation technique might have modified the δ^2H values of water [31]. Finally, deuterium might have been retained in the tissue matrix, or mixed and exchanged with hydrogen atoms of cell tissues in different compartments of the above-ground plant parts [34], which could have resulted in either loss of deuterium or slowed down the deuterium content reaching the roots growing in the drying soil, thereby reducing deuterium concentration. Schiegl and Vogel [35] also reported depletion in the D/H ratio (δ^2H) of several percent during the conversion of water to organic matter in living plants. The results, therefore, need to be read cautiously.

The results indicate that water moved from roots in the wet soil to roots in the drying soil as the root water of the treated plants was isotopically heavier relative to the control plants. Although there was an indication of water redistribution, the magnitude of this redistribution, however, was less distinct between sampling times. Surprisingly, the treated plants harvested after 3h and 6h of water application at 12:00 and 15:00 h had higher δ^2H values of the 'drying roots' water relative to the control plants, showing water movement of 2.63% and 2.50% from wet soil roots to 'drying roots'. Physiologically, plants of these sampling times had higher stomatal conductance and lower leaf water potential (more negative) (Figures 2 and 3), indicating high evaporative demands

of the plants relative to other sampling times. Under high evaporative demands, water evaporation in the stomatal chamber develops a highly negative pressure (i.e. tensions up to –1000 kPa) in the xylem vessels, which draws water from the roots up into the aerial parts [36] and is distributed to the cells that are losing water, predominantly by the apoplastic pathway [37].

During this unidirectional water movement, the reasons for this higher δ^2H values in the 'drying roots' are not clear. Since the wet and the drying roots had originated from the same stem base, there is a possibility that deuterated water entering the stem might have moved across (or around) the stem laterally from the root-stem interface [20] and then transported downwards into 'drying roots', contrary to the direction of the transpiration stream. Brooks et al. [17] studied hydraulic redistribution of woody perennials and pointed out that the traverse flow across (around) the trunk and then reversal into roots and soils on the non-irrigated side was possible provided resistances to hydraulic conductance across the tissues of the trunk were greater that roots. Lateral movement of deuterated water between the sides of the pot at the time of application could be the other possibility but this potential source of error was eliminated by applying water away from the stem base at a slow trickle. This method of water application effectively restricted water movement within the wet soil column of the pot as there were no visible signs of lateral water movement between the pot sides.

Several researchers have successfully demonstrated that hydraulic redistribution usually occurs at night when transpiration diminished sufficiently to allow water potential of the roots to exceed that of the other plant parts or the drier portions of the soil profile [7,17,29,38-40]. In this study, the plants harvested after 18 h of water application at dawn (03:00 am) had the lowest stomatal conductance and highest leaf water potential (less negative) but the amount of water redistributed to the drying roots was only 3.48% (Figure 5). This suggests that the hydraulic resistances encountered by water being redistributed from roots in the wet soil to roots in the drying soil through shoots were probably high [29], thereby resulting in reduced water transport into the 'drying roots'. It is also postulated that reduced stomatal conductance and higher leaf water potential (less negative) coupled with warmer and drier conditions at night (increased vapour pressure deficit) were perhaps still high enough to sustain a water potential gradient between the soil and roots to extract water to refill the above-ground plant storage compartments [29]. The above-ground plant parts refilling demand may have indirectly limited or reduced the magnitude of hydraulic redistribution by creating strong sinks for water within the transpiring foliage than existed in the roots growing in the drying soil. Other possible explanation could be that hydraulic lift was also occurring simultaneously. Thus, the 'drying roots' were probably being rehydrated both from above-ground plant parts due to hydraulic redistribution and from deeper moist soil layers due to hydraulic lift, thereby neutralising the deuterium concentration to some extent. Hultine et al. [41] found that the magnitude of hydraulic redistribution was greater in roots when night-time vapour pressure deficit was low.

The isotopic composition of the root water declined after 24 h of water application, which corresponded to daytime 09:00 am in the morning. The reasons for this decline are not clear. It is likely that the 'drying roots' were also supplying water to the shoot to meet the plant transpirational demand on a bright and hot morning with an air temperature above 20°C (Figure 1). This is supported by the increased stomatal conductance and a lower leaf water potential (more negative) of this sampling time relative to the 18 h sampling time (Figures 2 and 3). As a result of this, the deuterium concentration (δ^2H) of over nightly redistributed water might have been diluted by water absorbed from the drying soil or a portion of the redistributed water might have been supplied to the shoot before sampling (Table 2), thereby showing a decline in water redistribution from 3.48% to 1.57% relative to the 18 h sampling time. Alternatively, the decline in deuterium concentration could partly be due to inherent uncertainties associated with the sampling, distillation or analytical technique. Soil water content was measured at the end of each sampling time from both sides of the dual pot to quantify its effect on hydraulic redistribution. On average, the soil water content remained above 36% on the irrigated side and above 23% on the dry side of the pot (Table 1b). Theoretically, substantial water redistribution should have occurred due to the unequal soil water content between the wet and the dry part of the root system but this was not observed even during periods of low transpiration (i.e. night-time) (Figure 2). The magnitude of the hydraulically redistributed water was perhaps largely dependent on the hydraulic resistances encountered by water during the cell-to-cell pathway [42].

The water utilisation pattern indicated that the plants predominantly utilised water from the irrigated side of the dual, as expected. Despite the wet side being the dominant source of water use, the small changes in δ^2H values of the 'drying roots' water between the sampling times suggest less dependency of the 'drying roots' on the wet side of the root system. Further, water supplying ability of the 'drying roots' to the shoot (Table 2) may have deterred water redistribution substantially or water absorbed by these roots from the soil may have altered the isotope ratio (δ^2H) of root water upon mixing with the redistributed water. Based on the results, we suggest the occurrence of hydraulic redistribution in potatoes during partial root zone drying at a limited rate. The limited water redistribution would probably be of little significance in the survival of roots present in the upper drier portion of the soil under higher water demanding conditions but its role in sending the chemical signals to the shoot to conserve water by reducing transpiration would be of particular significance during drying periods of partial rootzone drying. It is planned to study this mechanism over days by observing the movement of deuterium from wet roots to stems, leaves and then to the roots growing in the drying soil column for better understanding of the hydraulic redistribution mechanism under partial rootzone drying.

Acknowledgements

The first author thanks Harper Adams University College, Shropshire, UK, for funding this research project.

References

1. Wesseling JG, Feddes RA (2006) Assessing crop water productivity from field to regional scale. Agricultural Water Management 86: 30-39.

2. Jefferies RA (1993) Response of potato genotypes to drought. I. Expansion of individual leaves and osmotic adjustment. Annals of Applied Biology 122: 93-104.

3. Frensch J (1997) Primary responses of root and leaf elongation to water deficits in the atmosphere and soil solution. Journal of Experimental Botany 48: 985-999.

4. Heuer B, Nadler A (1998) Physiological response of potato plants to soil salinity and water deficit. Plant Science 137: 43-51.

5. Sharp RE, Davies WJ (1985) Root growth and water uptake by maize plants in drying soils. Journal of Experimental Botany 36: 1441-1456.

6. Zegada-Lizarazu W, Iijima M (2004) Hydrogen stable isotope analysis of water acquisition ability of deep roots and hydraulic lift in sixteen food crop species. Plant Production Science 7: 427-434.

7. Richards JH, Caldwell MM (1987) Hydraulic lift: substantial nocturnal water transport between soil layers by Artemisia tridentate roots. Oecologia 73: 486-489.

8. Caldwell MM, Richards JH (1989) Hydraulic lift: water efflux from upper roots improves effectiveness of water uptake by deep roots. Oecologia 79: 1-5.

9. Burgess SSO, Adams MA, Turner NC, Ong CK (1998) The redistribution of soil water by tree root systems. Oecologia 115: 306-311.

10. Sakuratani T, Aoe T, Higuchi H (1999) Reverse flow in roots of Sesbania rostrata measured using the constant power heat balance method. Plant, Cell and Environment 22: 1153-1160.

11. Meinzer FC, Clearwater MJ, Goldstein G (2001) Water transport in trees: current perspectives, new insights and some controversies. Environmental and Experimental Botany 45: 239-262.

12. Smith DM, Jackson NA, Roberts JM, Ong CK (1999) Reverse flow in tree roots and downward siphoning of water by Grevillea robusta. Functional Ecology 13: 256-264.

13. Schulze ED, Caldwell MM, Canedell J, Mooney HA, Jackson RB, et al. (1998) Downward flux of water through roots (i.e. inverse hydraulic lift) in dry Kalahari sands. Oecologia 115: 460-462.

14. Scholz FG, Bucci SJ, Goldstein GH, Meinzer FC, Franco AC (2002) Hydraulic redistribution of soil water by neotropical savanna trees. Tree Physiology 22: 603-612.

15. Hultine KR, Williams DG, Burgess SSO (2003) Contrasting patterns of hydraulic redistribution in three desert phreatophytes. Oecologia 135: 167-175.

16. Brooks JR, Meinzer FC, Warren JM, Domec J (2006) Hydraulic redistribution in a Douglas-fir forest: lessons from system manipulation. Plant, Cell and Environment 29: 138-150.

17. Brooks JR, Meinzer FC, Coulombe R, Gregg J (2002) Hydraulic redistribution of soil water during summer drought in two contrasting Pacific Northwest coniferous forests. Tree Physiology 22: 1107-1117.

18. Loveys BR, Dry PR, Stoll M, McCarthy MG (2000) Using plant physiology to improve the water use efficiency of horticultural crops. Acta Horticulturae 537: 187-197.

19. Caldwell MM, Dawson TE, Richards JH (1998) Hydraulic lift: consequences of water efflux from the roots of plants. Oecologia 113: 151-161.

20. Smart DR, Carlisle E, Goebel M, Nunez BA (2005) Transverse hydraulic redistribution by a grapevine. Plant, Cell and Environment 28: 157-166.

21. Jackson RB, Sperry JS, Dawson TE (2000) Root water uptake and transport: using physiological processes in global predictions. Trends in Plant Science 5: 482-488.

22. Loveys BR, Dry PR, Stoll M, McCarthy MG (2000) Using plant physiology to improve the water use efficiency of horticultural crops. Acta Horticulturae 537: 187-197.

23. Firman DM (1988) Field measurement of the photosynthetic rate of potatoes grown with different amounts of nitrogen fertilizer. Journal of Agricultural Science 111: 85-90.

24. Basu PS, Sharma A, Garg ID, Sukumaran NP (1999) Tuber sink modifies photosynthetic response in potato under water stress. Environmental and Experimental Botany 42: 25-39.

25. Scholander PF, Hammel HT, Bradstreet ED, Hemmingsen EA (1965) Sap pressure in vascular plants. Science 148: 339-346.

26. Böhm W (1979) Methods of Studying Root Systems. (1st edn) Springer-Verlag, Berlin and New York.

27. Peñuelas J, Fielella I (2003) Deuterium labelling of roots provides evidence of deep water access and hydraulic lift by Pinus nigra in a Mediterranean forest of NE Spain. Environmental and Experimental Botany 49: 201-208.

28. Thorburn PJ, Mensforth LJ (1993) Sampling water from alfalfa (Medicago sativa) for analysis of stable isotopes of water. Communications in Soil Science and Plant Analysis 24: 549-557.

29. Stoll M, Loveys B, Dry P (2000) Hormonal changes induced by partial rootzone drying of irrigated grapevine. Journal of Experimental Botany 51: 1627-1634.

30. Revesz K, Woods P (1990) A method to extract soil water for stable isotope analysis. Journal of Hydrology 115: 397-406.

31. Allison GB, Gat JR, Leaney FWJ (1985) The relationship between deuterium and oxygen-18 delta values in leaf water. Chemical Geology 58: 145-156.

32. Leaney FW, Osmond CB, Allison GB, Ziegler H (1985) Hydrogen-isotope

composition of leaf water in C3 and C4 plants: its relationship to the hydrogen-isotope composition of dry matter. Planta 164: 215-220.

33. Walker CD, Lance RMC (1991) The fractionation of 2H and 18O in leaf water of barley. Australian Journal of Plant Physiology 18: 411-425.

34. Calder IR (1991) Implications and assumptions in using the "total counts" and convection-dispersion equations for tracer flow measurements – with particular reference to transpiration measurements in tress. Journal of Hydrology 125: 149-158.

35. Schiegl WE, Vogel JC (1970) Deuterium content of organic matter. Earth and Planetary Science Letters 7: 307-313.

36. Steudle E (2001) The cohesion-tension mechanism and the acquisition of water by plant roots. Annual Review of Plant Physiology and Plant Molecular Biology 52: 847-875.

37. Steudle E, Peterson CA (1998) How does water get through roots? Journal of Experimental Botany 49: 775-788.

38. Dawson TE (1993) Hydraulic lift and water use by plants: implication for water balance, performance and plant-plant interactions. Oecologia 95: 565-574.

39. Yoder CK, Nowak RS (1999) Hydraulic lift among native plant species in the Mohave Desert. Plant and Soil 215: 93-102.

40. Millikin IC, Bledsoe CS (2000) Seasonal and diurnal patterns of soil water potential in the rhizosphere of blue oaks: evidence for hydraulic lift. Oecologia 125: 459-465.

41. Hultine KR, Williams DG, Burgess SSO (2003) Contrasting patterns of hydraulic redistribution in three desert phreatophytes. Oecologia 135: 167-175.

42. Steudle E (2001) The cohesion-tension mechanism and the acquisition of water by plant roots. Annual Review of Plant Physiology and Plant Molecular Biology 52: 847-875.

31

Quantitative Assessment of Wheat Pollen Shed by Digital Image Analysis of Trapped Airborne Pollen Grains

Katja Kempe[1], Anastassia Boudichevskaia[1], Robert Jerchel[1], Dmitri Pescianschi, Renate Schmidt[1], Martin Kirchhoff[2], Ralf Schachschneider[2] and Mario Gils[1]*

[1]Leibniz Institute of Plant Genetics and Crop Plant Research (IPK), Germany
[2]Nordsaat Saatzucht GmbH, Böhnshauser Straße 1, D-38895 Langenstein, OT Böhnshausen, Germany

Abstract

The objective of the present study was to develop a technique for quantifying the dynamics of wheat pollen shed under field conditions. Pollen traps with an adhesive film were used to assess the relative pollen shed of 12 winter wheat lines. Quantitative measurements were performed in 2012 and 2013 over a period of up to 20 days. The amounts of trapped pollen were automatically determined using a customized image analysis program. We demonstrated that this method is suitable for the assessment of wheat pollen shed. The possible impact of the technical advances revealed in this study for the selection of pollinators in routine wheat breeding programs is discussed.

Keywords: Pollen shed; Pollen traps; Digital image analysis; Field trials; Hybrid wheat

Introduction

Wheat (*Triticum aestivum* L.) is predominantly a self-pollinating crop. Still, outcrossing via pollen dispersal is possible at variable rates [1]. The rate of outcrossing depends upon genotypes, populations and environmental conditions and directly correlates to the amount of wind-borne pollen [2-4].

In the vast majority of studies, the outcrossing capability of wheat plants was measured by their ability to fertilize crossing partners. Such examinations were conducted either on male-sterile crossing partners [5-8] or under natural conditions of pollen competition [3,9,10].

However, in order to evaluate the potential of wheat plants for out-crossing, an efficient method for quantifying the dynamics of wheat pollen shed would be highly advantageous. An approach to the measurement of wheat pollen shed may include a device suitable for capturing the pollen that is shed over a certain period of time ("Pollen trap") coupled with a tool for counting of pollen. Pollen traps may be active or passive, depending on whether they actively suck in air containing pollen or whether they rely on the passive transport of pollen by wind. Previously, the use of active pollen traps [11] or passive pollen traps [12,13] was described to measure wheat pollen dispersal under open- field. The pollen number was determined by counting the pollen grains in microscopic images through visual inspection. The use of image analysis software for the quantification of pollen shed was reported for maize [14].

In this technical report, we describe the combined use of a new type of passive adhesive pollen trap with image analysis software that was specially tailored to the needs of wheat pollen analysis. We assume that this method offers the potential to facilitate a reliable quantitative assessment of wheat pollen shed and that this technology may be utilized for the identification of wheat lines with high pollen shed. Since high amounts of air-borne pollen are a prerequisite for successful commercial hybrid wheat production, we believe that the described method has a direct value for wheat breeding programs.

Material and Methods

Field trials were conducted in 2012 and 2013 at the breeding station of Nordsaat Saatzucht GmbH located in Langenstein, Germany (51°53'N, 10°59'E). A total of 12 winter wheat lines were analyzed. All of the lines were bred in identical field plots of 7 by 9 m at a plant density of 300/m².

Pollen capture

For pollen capture, two slides of optically clear adhesive seal sheets for polymerase chain reaction (PCR) multiwell plates (AB-1170, Thermo Fisher Scientific, Schwerte, Germany) were fixed with clamps on a metal base (14 by 10 cm) and mounted at a 45° vertical angle and 40 cm height on a metal pole (Figure 1a). In the experiments carried out in 2012, the pollen traps were placed at each end and in the center of each field plot as depicted in Figure 1b. The distance to the pollen source was 60 cm. The slides were exchanged twice daily at 7:00-8:00 a.m. and 5:00-6:00 p.m. In the experiments that were conducted in 2013, the number of traps was reduced from five to three and pollen capture was carried out only over the day (7:00 a.m. to 6:00 p.m.) while the remaining experimental parameters were identical to the field trials of the previous year. We have chosen these periods of time on the basis of a case study in which the maximum release of pollen was found to occur from 8 to 11 a.m. and a less pronounced peak was found at 3 to 6 p.m. [15].

Sealing/conservation of traps

Immediately after collection, the adhesive films with captured pollen were covered with a transparent sheet to facilitate sealing of the pollen trap for archiving and subsequent analysis.

*Corresponding author: Dr. Mario Gils, Leibniz Institute of Plant Genetics and Crop Plant Research (IPK), Germany, E-mail: gils@ipk-gatersleben.de

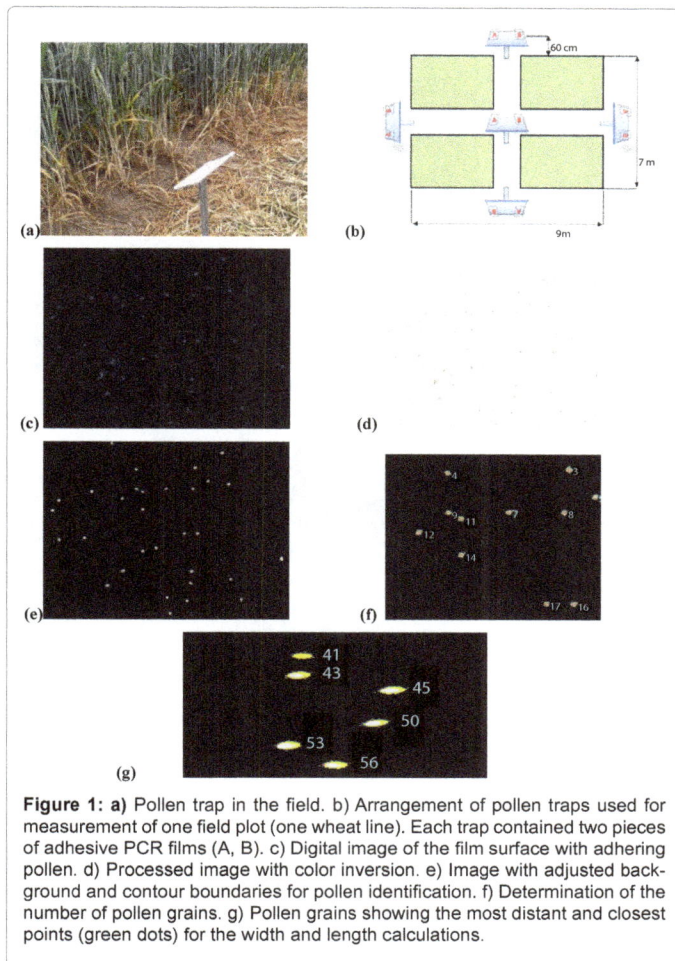

Figure 1: a) Pollen trap in the field. b) Arrangement of pollen traps used for measurement of one field plot (one wheat line). Each trap contained two pieces of adhesive PCR films (A, B). c) Digital image of the film surface with adhering pollen. d) Processed image with color inversion. e) Image with adjusted background and contour boundaries for pollen identification. f) Determination of the number of pollen grains. g) Pollen grains showing the most distant and closest points (green dots) for the width and length calculations.

Selection of adhesive surfaces for pollen capture

Pilot experiments were conducted in June 2011. Different adhesive films were tested for their ability to collect wheat pollen under field conditions and then assessed for their suitability for use in microscopic analyses. Due their easy handling and high quality standards (homogenous dispersion of the adhesive, stiffness, transparency) adhesive PCR seal foils were applied. The particular films tested were E2796-9793 and E2796-9795 from Starlab GmbH (Hamburg, Germany), AB-1170 and AB-0558 from Thermo Fisher Scientific (Schwerte, Germany) and UC-500 from Axygen (Union City, CA, USA). Four samples of each film were fixed on metal plates and arranged around a wheat field (pollen source). The films were exposed from 8 to 11 a.m. and then sealed for microscopic analysis. All the films had a comparable capability to collect pollen, but we observed variation in their suitability for microscopic analysis.

In general, the thinner seal films used for real-time PCR applications (E2796-9795, AB-1170 and UC-500) were more suited for microscopic examination because of their higher transparency and more homogenous texture, which results in a higher accuracy of the automated pollen assessment. In comparison to the real-time PCR seals, use of E2796-9793 and AB-0558 led to an error rate of 25-30% when pollen was counted automatically (i.e. 25-30% of the pollen that was detected by visual detection was not identified by the software EVALUATOR). In case of the real-time PCR seals, an average deviation

of ± 2% between automatically counting and visual inspection was measured.

In addition, all of the seals were tested for their performance at rain. For this purpose, three of each seal were exposed to a flowering wheat field from 7 a.m. to 6 p.m. at a rainy day. Whereas the adhesives of the seals E2796-9793 and AB-0558 were negatively affected by humidity (tendency to cloudiness and in-transparency) the real-time PCR seals could be dried easily during the sealing without adverse effects. Among the real-time PCR foils, AB-1170 was chosen for further analysis because of its favorable price and availability in higher quantities.

To evaluate whether any unequal distribution of pollen occurs on the surface, up to 10 images were analyzed per trap. We did not find any obvious bias in pollen deposition on the trap surface at any pollen shed density. The average of the pollen numbers collected on two randomly chosen areas of 0.25 cm² was used as the measured value for each pollen trap.

Digital image analysis

The digital images were produced using a Zeiss AxioCam digital camera system with AxioVision software in combination with a Stemi 2000 microscope (Zeiss, Jena, Germany). The images were captured at 16X magnification (Figure 1c). The camera exposure time was 0.7 s.

The program EVALUATOR was used following the production of the digital image to generate an automated color-inverted image (Figure 1d). Subsequently, the software algorithm isolates the pollen grains from the background based on differences in pixel intensities and creates a contour boundary for the precise identification of each grain and its orientation, along with a new background for the grains with definable RGB values. The pollen grains appear as bright spots (Figure 1e). The customized software determines the number of grains and numbers them consecutively in the digital image (Figure 1f). Furthermore, the two most distant and the two closest points on opposite sides of the grain are identified, and the linear distances between the two sets of points are defined as the length and the width, respectively (Figure 1g). The determination of the pixel number inside the boundaries enables the calculation of the two-dimensional area of the pollen grain.

The particles whose shapes and sizes are not within a selected range are reliably removed from the image. In most cases, touching or misshaped grains are recognized and excluded from the analysis (not counted) because their size is out of a defined range that is set by the operator. All filters and threshold parameters can be manually set by the operator.

As an alternative option, the program offers the opportunity to check questionable cases by visual inspection of the screen picture. Structures can be added or subtracted to the count if appropriate. This approach was used in some cases in which the adhesive surface was scratched or excessively covered with dust.

Notably, the software allows the filter and threshold conditions optimized for one image to be automatically applied to all other images of the same project without further adjustments, which enables high-throughput image analysis. The images (processed and unprocessed) are stored along with the report document in BMP format. The summarized results of the analysis can be exported to an SCV file, and the measured values (including the number of pollen grains, the number of rejected objects and the pollen length, width and area) are stored as an Excel file. The program EVALUATOR works under Windows.

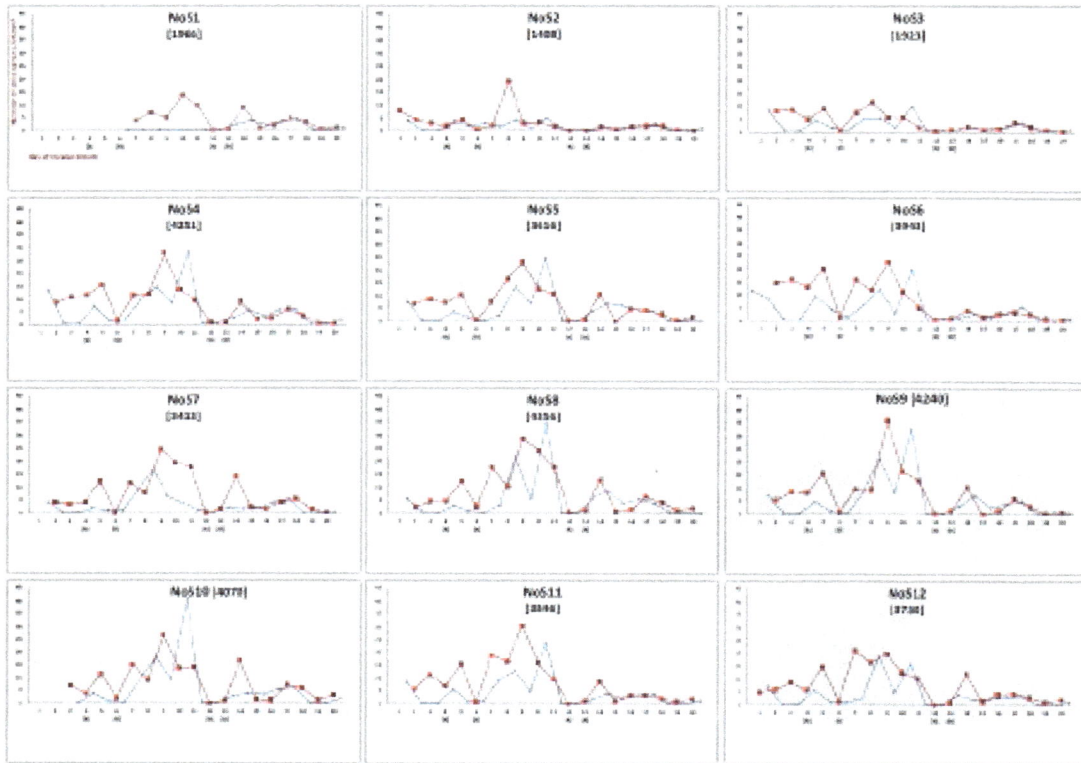

Figure 2: Pollen shed curves of 12 wheat lines, with red and blue curves showing the measurements for the day and night periods, respectively. The values for pollen density/25 mm² are the averaged results from 20 image analyses (5 traps per line x 2 slides per trap x 2 image analyses per slide). The cumulative pollen shed (total number of pollen collected) is given in brackets. Note that the decline of pollen shed on days 4, 6, 12 and 13 is most likely due to rain [R].

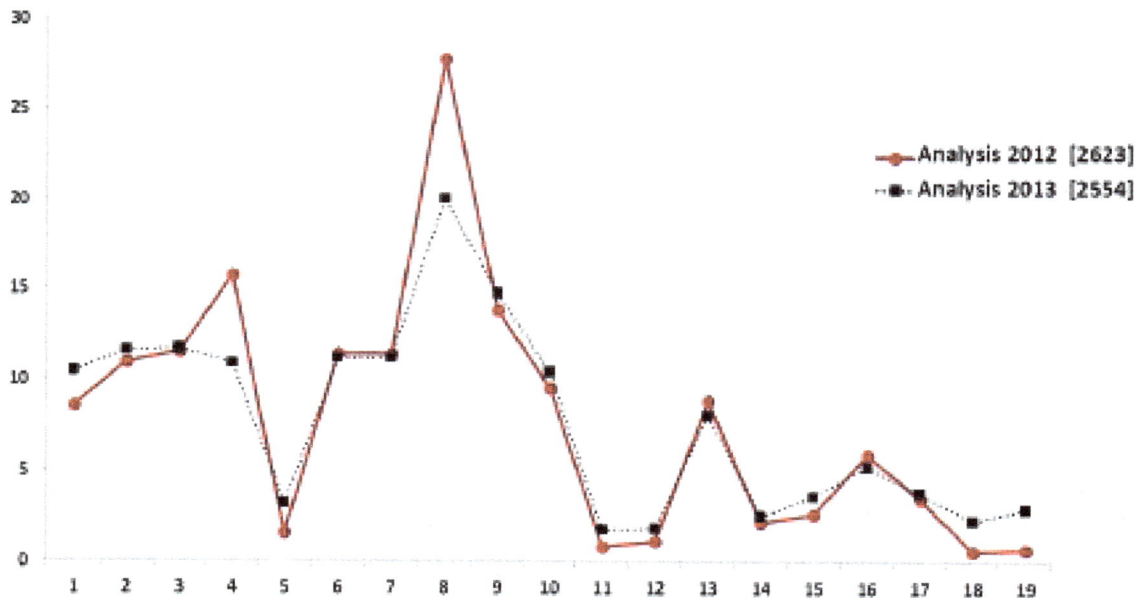

Figure 3: Results from two independent pollen counts of line NoS4. The measurements were conducted immediately after the slides were harvested in 2012 (solid line) and after 14 month of storage (dashed line). Note that non-identical areas of the identical slides were chosen for determining the number of captured pollen and those only measurements over the day were included. The cumulative values of captured pollen are given in brackets.

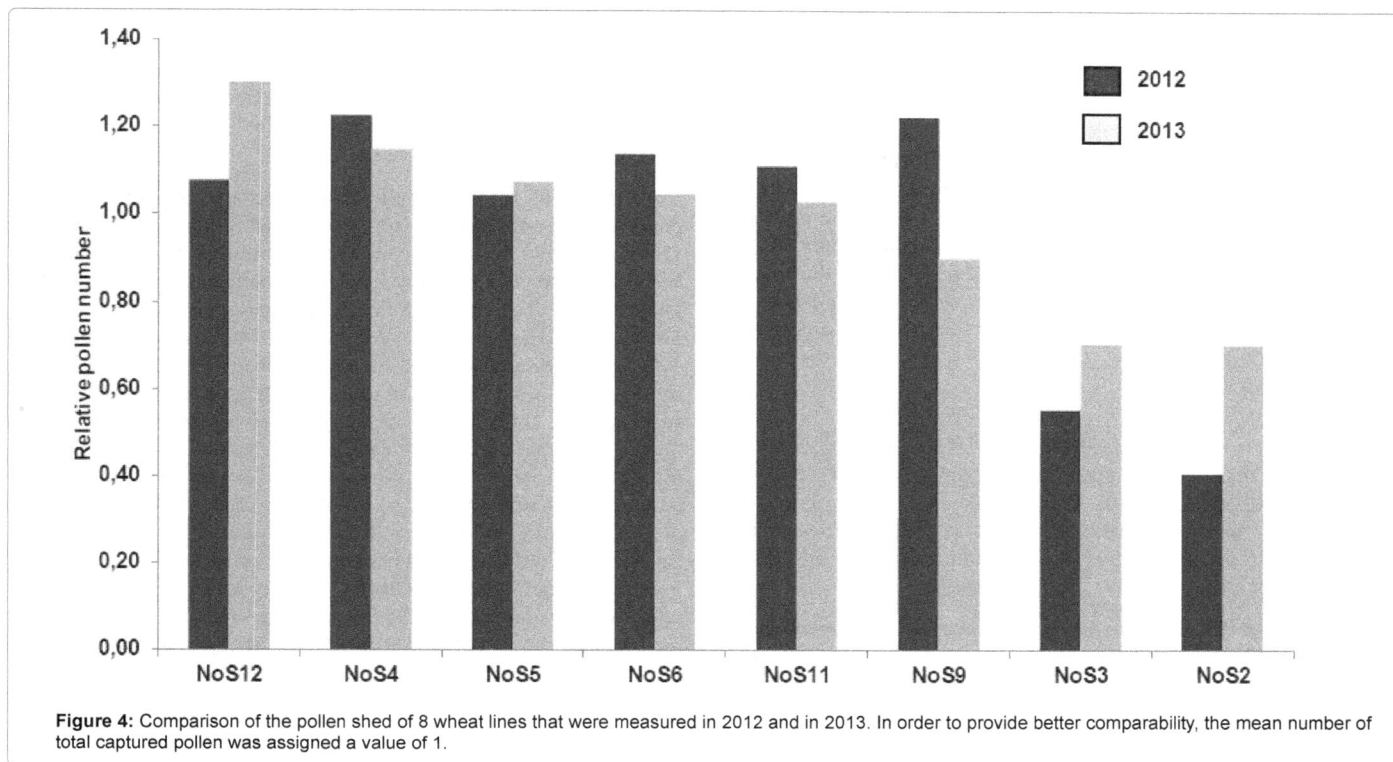

Figure 4: Comparison of the pollen shed of 8 wheat lines that were measured in 2012 and in 2013. In order to provide better comparability, the mean number of total captured pollen was assigned a value of 1.

For this study, the Delphi software EVALUATOR was adapted according to the specific requirements for the examination of wheat pollen. Other versions of the EVALUATOR software were recently used to analyze the size of *Arabidopsis thaliana cotyledons* [16] and seeds (Boudichevskaia and Schmidt, unpublished results).

Results

Field experiments in 2012 (Experiment I)

The relative dynamics of pollen shed for 12 winter wheat lines was measured (Figure 2). Measurements were initiated for each line at the individual start of anthesis. The data points for day and night values were expressed as the mean number of pollen grains counted in 20 image analyses (5 traps/line×2 slides per trap×2 images per slide; mean value=total number of counted pollen/20). Most of the wheat lines displayed a similar temporal pattern of pollen shed in which the shedding increased until day 9, when the number of pollen grains reached a maximum of between 21 and 36 per 25 mm² of trap surface. The lines NoS1, NoS2 and NoS3 exhibited lower pollen shed than the other lines and a later initiation of anthesis. As expected, pollen shed was reduced during the night, with the exception of night 10, when large amounts of pollen were collected. However, this outlier can be attributed to exceptionally strong nocturnal winds.

On several days (4, 6, 12 and 13), it was raining during the period of measurement, and the amount of captured pollen decreased significantly. This decrease can be explained by either the removal of pollen from the trap, a lower adhesive force of the trap, reduced pollen shed by the plants or a combination of these factors.

The image analysis results were confirmed on more than 200 slides by visually counting the pollen using unprocessed images. Overall, a high accordance between the counting methods was found (average deviation of ± 2%). The counting errors appeared predominantly at

higher pollen densities when the grains tended to touch each other. Small particles, such as dust, and larger objects such as insects or anthers were reliably removed during image processing. From our results we concluded that that the automated pollen counts were highly accurate.

Pollen counting was conducted immediately after harvesting the traps in 2012, and, in a second analysis, after storage of 14 months (Figure 3). The close match between the two sets of data indicates that the traps can be stored for an extended period of time under ambient conditions without significantly affecting the counting results.

Field experiments in 2013 (Experiment II)

In 2013, the pollen shed measurements of eight of the 12 winter wheat lines were repeated. Lines were chosen that represented both classes of high and low-pollen shedding genotypes. Among the selected lines were two lines that showed low pollen shed in 2012 (lines NoS2 and NoS3; line NoS1 was excluded because of the late flowering phenotype), two lines with the highest cumulative amounts of pollen counted in 2012 (NoS4 and NoS9) and four lines with medium pollen shed in 2012.

Figure 4 compares the relative values of the total pollen count measured for both sets of experiments. The relative values for each particular genotype in the chart are calculated as (total number of pollen captured from particular genotype per year)/(average pollen number captured from the all genotypes per year). The results obtained from both experiments display a significant correlation (correlation coefficient=0.77; P<0.01). From these data we deduce that the described method is in general a suitable tool for a comparative assessment of winter wheat pollen shed. A certain variation between the pollen shed in 2012 and 2013 is expected since the weather conditions were not identical; however, from our results it is possible to confirm the categorization of the lines into classes with high pollen shed and such lines that display relative low levels of pollen shed (NoS2 and NoS3).

Pollen size

The program EVALUATOR also allows for the measurement of pollen size. In total, 16,000 pollen grains were analyzed. The mean size was 84 μm (± 2.8 μm). An ANOVA revealed that there was no statistically significant difference in pollen size among the different wheat lines investigated (P=0.94). Nevertheless, the results of the measurements have to be interpreted with care. The repositioning of the samples during image production may move the pollen grains out of the optimal optical focus plane which might contribute to differences in the measured object sizes.

Discussion

Techniques for estimating the degree of pollen grain shed are important for many topics related to pollination biology. The role of pollen in pathogenesis of allergic diseases is well established [17]. Also, in recent years, pollen dispersal in wheat has gained increasing attention due to the potential of GM crops to transfer foreign genes from GM plants to related non-GM plants [18,19].

The method described in this article appears to be suitable for characterizing the temporal pattern of wheat pollen shed under field conditions. An important practical application of the technique for plant breeding might be the identification of wheat lines that have the prerequisites to constantly and reliably pollinate females in hybrid production. Previous studies on pollen dispersal focused on the commercial production of hybrid wheat, where reaching high levels of genetic purity and sufficient seed set on male sterile plants were essential [20].

Hybrids often display a yield increase, enhanced yield stability and improved abiotic and biotic stress resistance due to the exploitation of heterosis (hybrid vigor) [21,22]. When applied on a large scale, hybrid wheat is the product of the controlled crossing of two pure lines (a sterile "mother line" and a "male" pollinator) in a propagation field. Despite its outstanding performance, hybrid wheat currently occupies only a niche sector in commercial wheat production. One major reason for this is that the production of "good males" may require significant breeding efforts [20,23] because excellent pollinator characteristics (e.g., amount of pollen shedding, anther extrusion and length, duration of pollination, pollen viability) have to be available in combination with high breeding values. It is estimated that suitable pollinators constitute only 1-2% of the modern wheat varieties (Schachschneider, unpublished results). The reason for this rareness is that a number of characteristics, particularly the pollen and floral biology, render wheat ill-adapted to cross-fertilization [1,5,23,24]. Moreover, compared with other grass species, wheat produces a small amount of pollen, and the pollen settles quickly because it is relatively heavy [25]. Thus, wind is required to move wheat pollen over an appreciable distance [2,26]. Furthermore, wheat pollen is viable for a relatively short period of time. In a previous study, its viability was completely lost within 65-70 min [27].

In contemporary hybrid wheat breeding programs, pollinator lines are selected by testing their ability to cross-fertilize male-sterile females in so-called "experimental crossing blocks" [8,28]. This method presents several disadvantages, such as high operation costs, the risk of failure of the chemical hybridization agent and the necessity of using male and female lines that have an overlapping flowering period [29,30]. The technology described in this study appears to be advantageous for

the precise selection of pollinator candidates because only the male component is involved. The removal of bottlenecks to progress in the identification of pollinators may lead to considerable cost savings. Nevertheless, in the future, it will be important to determine to what extent the measured differences in pollen shed correlate with different rates of seed set on the female lines.

It is difficult to compare the absolute numbers of captured wheat pollen with those obtained in previous studies because the experiments varied regarding the distance of the traps from the pollen source and the type of trap used. Moreover, the measurements were made at other locations and different wheat lines were analyzed. Yet, the maximum number of pollen grains per area of slide surface was comparable with the values published in a study of wheat [12] and, as expected, significantly lower (~30x) than those reported for efficient pollinators, such as maize [14]. The quantity of captured pollen was considerably reduced by rain, and additional environmental factors and their interactions, such as the wind speed and direction, air turbulence, temperature and humidity, may influence the amount of trapped pollen. Therefore, to further substantiate claims about the pollination capabilities of different wheat lines, future trials should be carried out at multiple locations. Still, since it can be assumed that rainfall efficiently inhibits the overall pollen flow, the low values measured on rainy days may represent a realistic scenario. Nevertheless, future pollen traps might be equipped with devices that protect the slides from rain in order to prevent that already captured pollen is washed off the surface.

The proven storability of the pollen-covered slides might offer a major technical advantage over traps that collect pollen in isotonic solutions [31]. Several morphological flower and plant characteristics may influence the amount of pollen shed [1,23,26]. While observations of several characteristics, for example, anther extrusion or plant height, are readily obtainable, the analysis of others, such as pollen number or size, is more difficult. For breeders, it would be highly interesting to specifically correlate plants characteristics with the dynamics of pollen shed in order to enable a selection of pollinators via "indirect" traits. The availability of an efficient technology for assaying the pollen would strongly facilitate correlation studies. We hope that the described method will be a valuable tool for such on-field plant phenotyping.

Acknowledgements

The authors are grateful to Iris Röhlich, Karoline Buller, Corinna Schollmeier and Manja Franke for their excellent technical support. We thank the federal ministry of education and research for funding the project "KMU-Innovativ" (funding code 0315889) as a joint project between the IPK Gatersleben and the Nordsaat GmbH (Langenstein, Germany). The work was also supported by a BMBF grant to Renate Schmidt (funding code 0315053G).

References

1. Waines JG, Hedge SG (2003) Intraspecific gene flow in bread wheat as affected by reproductive biology and pollination ecology of wheat flowers. Crop Science 43: 451-463.

2. Beckie HJ, Hall LM (2008) Simple to complex: Modelling crop pollen-mediated gene flow. Plant Science 175: 615-628.

3. Hucl P (1996) Out-crossing rates for 10 Canadian spring wheat cultivars. Can J Plant Sci 76: 423-427.

4. Matus-Cadiz MA, Hucl P, Dupuis B (2007) Pollen-mediated gene flow in wheat at a commercial scale. Crop Science. 47: 573-581.

5. Arya RK, Sethi SK (2005) Studies on floral traits influencing outcrossing in wheat (Triticum aestivum L.). National Journal of Plant Improvement 7: 73-76.

6. de Vries AP (1974) Some aspects of cross-pollination in wheat (Triticum aestivum L.). 4. Seed set on male sterile plants as influenced by distance from

the pollen source, pollinator: male sterile ratio and width of the male sterile strip. Euphytica 23: 601-622.

7. Lu AZ, Zhao H, Wang TY, Wang HB (2002) Study of possibility of target gene introgression from transgenic wheat into non-transgenic plants through pollens. Acta Agric Bor Sin 17: 1-6.

8. Wilson JA, Ross WM (1962) Cross Breeding in Wheat, Triticum aestivum L. II. Hybrid Seed Set on a Cytoplasmic Male-Sterile Winter Wheat Composite Subjected to Cross- Pollination. Crop Science 2: 415-417.

9. Hucl P, Matus-Cadiz MA (2001) Isolation distances for minimizing outcrossing in spring wheat. Crop Science 41: 1348-1351.

10. Matus-Cadiz MA, Hucl P, Horak MJ, Blomquist LK (2004) Gene flow in wheat at the field scale. Crop Science 44: 718-727.

11. Alcázar P, Galán C, Cariñanos P, Domínguez-Vilches E (2003) A new adhesive for airborne pollen sampling in Spain. Aerobiologia 19: 57-61.

12. Loureiro I, Concepción Escorial M, González-Andujar JL, García-Baudin JM, Chueca MC (2007) Wheat pollen dispersal under semiarid field conditions: potential outcrossing with Triticum aestivum and Triticum turgidum. Euphytica 156: 25-37.

13. Sapra VT, Hughes JL (1975) Pollen production in hexaploid Triticale. Euphytica 24: 237-243.

14. Fonseca AE, Westgate ME, Doyle RT (2002) Application of fluorescence microscopy and image analysis for quantifying dynamics of maize pollen shed. Crop Science 42: 2201- 2206.

15. de Vries AP (1972) Some aspects of cross-pollination in wheat (Triticum aestivum L.) 1. Pollen concentration in the field as influenced by variety, diurnal pattern, weather conditions and level as compared to the height of the pollen donor. Euphytica 21: 185-203.

16. Meyer RC, Witucka-Wall H, Becher M, Blacha A, Boudichevskaia A, et al. (2012) Heterosis manifestation during early Arabidopsis seedling development is characterized by intermediate gene expression and enhanced metabolic activity in the hybrids. Plant J 71: 669-683.

17. Sofiev M, Bergmann KC (2013) Allergenic Pollen: A Review of the Production, Release, Distribution and Health Impacts. Springer 247.

18. Conner AJ, Glare TR, Nap JP (2003) The release of genetically modified crops into the environment. Part II. Overview of ecological risk assessment. The Plant J 33: 19-46.

19. Singh OV, Ghai S, Paul D, Jain RK (2006) Genetically modified crops: success, safety assessment, and public concern. Appl Microbiol Biotechnol 71: 598-607.

20. Pickett AA (1993) Hybrid wheat—results and problems. In: Advances in Plant Breeding, Paul Parey Scientific Publishers, Berlin 15.

21. Longin CF, Gowda M, Muhleisen J, Ebmeyer E, Kazman E, et al. (2013) Hybrid wheat: quantitative genetic parameters and consequences for the design of breeding programs. Theor Appl Genet 126: 2791-2801.

22. Mühleisen J, Piepho HP, Maurer HP, Longin CF, Reif JC (2013) Yield stability of hybrids versus lines in wheat, barley, and triticale. Theor Appl Genet.

23. Whitford R, Fleury D, Reif JC, Garcia M, Okada T, et al (2013) Hybrid breeding in wheat: technologies to improve hybrid wheat seed production. J Exp Bot.

24. Gatford KT, Basri Z, Edlington J, Lloyd J, Qureshi JA, et al. (2006) Gene flow from transgenic wheat and barley under field conditions. Euphytica 151: 383-391.

25. Lelley J (1966) Observation on the biology of fertilization with regard to seed production in hybrid wheat. Der Züchter 36: 314-317.

26. Gustafson DI, Horak MJ, Rempel CB, Metz SG, Gigax DR, et al. (2005) An Empirical Model for Pollen-Mediated Gene Flow in Wheat. Crop Science 45: 1286-1294.

27. Fritz SE, Lukaszewski AJ (1989) Pollen longevity in wheat, rye and triticale. Plant Breeding 102: 31-34.

28. Porter KB, Lahr KA, Atkins IM (1965) Cross-Pollination of Male-Sterile Winter Wheat (Triticum aestivum L.) Having Aegilops caudata L. and Aegilops ovata L. Cytoplasm. Crop Science 5: 161-163.

29. Curtis BC, Rajaram S, Gomez Macpherson H (2002) Bread wheat: improvement and production. FAO Plant Production and Protection Series.

30. Kempe K, Gils M (2011) Pollination control technologies for hybrid breeding. Plant Breeding 27: 417-437.

31. Carre S, Tasei JN (1997) Use of a flow-cytometer for pollen counting-An application to Vicia faba. L. Acta Hortic 437: 369-372.

32

Integrated Sugarcane Trash Management: A Novel Technology for Sustaining Soil Health and Sugarcane Yield

Suma R[1]* and Savitha CM[2]

[1]University of Horticultural Sciences, Bagalkot, Karnataka, India
[2]KVK, V.C. Farm, Mandya, Karnataka, India

Abstract

With the raising concern on soil conservation and health in the context of depleting traditional organic manures, efforts are required to harness the potentiality of crop biomass wastes effectively. Sugarcane is one such crop that produces 7-12 t ha^{-1} of trash, which is a rich source of organic carbon and plant nutrients. The burning of trash would lead to environment pollution besides depleting the soil biological properties and fertility. In this context, integrated sugarcane trash management (ISTM) that conserves and decomposes trash using microbial enriched (*Trichoderma viridae*) farm yard manure and urea (75 kg/ha) serve as a novel technology in sustaining soil health and sugarcane yield.

The results revealed that intense heat generated due to trash burning has reduced the germination of sugarcane to an extent of 68 percent compared to 82 percent in ISTM. The ISTM has increased the organic carbon content, available nitrogen, phosphorus and potassium in soil to an extent of 11.2, 3.6, 8.5 and 11.2 percent respectively in three years. The increased average cane yield was 12.8 percent over trash burning. The economic analysis showed that the gross income increased to 18.2 percent with the benefit of 2.63 rupees per rupee invested over three years. Farmers surveyed indicated that the trash management technology increased soil moisture and number of earthworms, and reduced weed incidence. Farmers also expressed that buds germinated 15 days earlier in ISTM practice and that ISTM increased cane yield and did not hinder ratoon practices.

Keywords: Sugarcane trash; Frontline demonstration; Organic carbon; Nitrogen

Introduction

The depleting soil health and crop productivity in the sugarcane cultivating area of Mandya district of Karnataka is a major concern because of reduced yields. This can be clearly visualized from the static average productivity hovering close to 98 ton/hectare in last five years compared to its potential yield of 150 ton/hectare. Although soil fertility is closely linked to the physical and chemical characteristics of the environment, it is strongly influenced by human management practices. One such practice followed by farmers is burning trash after the harvest of sugarcane. About 7-12 tons of trash can be obtained from 1 ha of sugarcane [1]. Every ton of sugarcane trash contains about 5.4 kg N, 1.3 kg P_2O_5, 3.1 kg K_2O and small quantities of micronutrients [2]. However, when sugarcane trash is burnt, most of the organic matter and nutrients in the trash are lost, leading to environmental pollution [3]. Farmers usually burn the trash with the opinions that its management is laborious, will reduce germination and hinders routine ratoon cultivation practices. On the other hand farmers apply huge quantity of fertilizers to meet the nutrient requirement of crop. Hence, the present study was taken up as a frontline demonstration in the farmer's field by Krishi Vigyan Kendra, V.C. Farm, Mandya, Karnataka, India, with the objective to educate farmers on conservation of crop residue and to assess its impact on soil fertility and sugarcane yield.

Materials and Methods

The study on integrated sugarcane trash management (ISTM) was conducted as a frontline demonstration (FLD) at Mallanayakanakatte village of Mandya district, Karnataka for three successive ratoon sugarcane crops in 1.0 hectare area. The FLD was implemented after the harvest of plant-cane crop during rabi 2007 and continued for successive three crops. The following technologies were implemented for in-situ management of sugarcane trash;

1. Irrigation of sugarcane plot for complete soaking of trash; this would soften trash and help for easy handling.

2. Mulching of sugarcane trash in alternate rows; this would help in following ratoon cultivation practices in un-mulched rows.

3. Broadcasting of 75 kg/ha urea on sugarcane trash; enhancing N narrows the wider C:N ratio of trash and helps for faster decomposition.

4. Application of 500 kg of farm yard manure (FYM) enriched with 25 kg microbial culture (*Trichoderma viridae*) on sugarcane trash; this would help in enhancing decomposition rate.

5. Stubble shaving, shoulder breaking, gap filling and following recommended ratoon sugarcane cultivation practices

The farmers practice (1.0 ha) of burning the trash after the harvest of sugarcane crop was considered as check plot. The summary of FLD and check plot is depicted in Table 1.

Collection and analysis of soil samples

Composite soil sample was collected after the harvest of each crop, from a depth of   0-15 cm, between the crop rows in check plot and

---

*Corresponding author: Suma R, University of Horticultural Sciences, Bagalkot, Karnataka-587104, India, E-mail: sumassac@gmail.com

Tangential contact behavior		Parameter
Variety	CO-62175	CO-62175
Planting date	October 2006	October 2006
Plant-cane harvest	November 2007	November 2007
First ratoon harvest	First week of October 2008	Second week of October 2008
Second ratoon harvest	Third week of October 2009	First week of October 2009
Third ratoon harvest	First week of October2010	Fourth week of October 2010
Trash management practice	As detailed in FLD	Trash burning after each harvest
Initial soil properties		
Texture	Clay loam	Clay loam
pH (1:2.5)	8.23	8.21
EC (dS/m)	1.67	1.73
OC (%)	0.42 %	0.44
Available N (kg/ha)	312.4	324.8
Available P_2O_5 (kg/ha)	31.8	31.2
Available K_2O (kg/ha)	232.5	244.6
Nutrient application (N: P_2O_5: K_2O- kg/ha)	250:100:125	350:125:170
Gap filling	Practiced	Practiced

Table 1: Summary of the FLD and check plot.

between the trash mulched rows in FLD plot. Soil pH and electrical conductivity (EC) were determined in 1:2.5, soil: water suspension by using digital pH meter and Conductivity Bridge (expressed in dS m^{-1} at 25°C) respectively. Chromic acid oxidation method [4] and alkaline permanganate method [5] were followed for determining organic carbon (OC) and available nitrogen content in soil. Available phosphorus from soil was extracted using Olsen's extractant and blue colour was developed by ascorbic acid method and the intensity was read at 660 nm using spectrophotometer and calculated referring to P-standard curve [6]. Available potassium was extracted from soil using neutral normal ammonium and determined using Flame photometer [6].

Collection and analysis of trash sample

Sugarcane trash and ash (after trash burning) samples were collected after the harvest from the field and nitrogen content was determined by Kjeldhal distillation method. A known weight of sample was digested with conc. H_2SO_4 and digestion mixture, further distilled for estimating N-content. The total phosphorus and potassium were determined by digesting with di-acid ($HClO_4$+ HNO_3) and analysed by phospho-vanado-molybdate complex and flame photometer methods respectively [7].

Other data collection

The number sets giving healthy tillers to number of sets planted in five randomly selected rows was used for measuring ratoon germination and expressed in per cent. The yield data were recorded after the harvest of each crop. Farmers usually sell the sugarcane to sugar factories and its price will be fixed by the government of Karnataka. It was Rs. 900/ton during 2007-08 and Rs. 1200/ton during 2009-10. The same prices were used for determining gross income. The cost incurred in inputs, machinery and labour were accounted for calculating cost of production. The difference in gross income and cost of cultivation was used for determine net income. The ratio of net income to cost of cultivation was used for calculating benefit to cost ratio (B: C). The other data on relevant parameters were collected and pooled for scientific interpretation.

Results and Discussion

Nutrient content of sugarcane trash

The samples of raw trash and the ash of the burnt trash were collected randomly after the harvest. The analysis results (Table 2) showed that the trash contain 0.51% total N that signifies the need of external nitrogen source for reducing its wider C:N ratio and for enhancing decomposition rate [8]. The trash ash samples showed more total phosphorus and potassium compared to raw trash, but the quantity of ash produced after the burning is quite low (~0.75 - 1.0 ton/ha) compared to raw trash (7-12 ton/ha) which, ultimately results in low input for one hectare area.

Impact of ISTM on soil properties

Organic carbon: Organic carbon (OC) content in soil is a key factor for its health and fertility. The impact of ISTM on soil organic carbon is presented in Table 3. The ISTM resulted in increased OC content of soil from 0.42 to 0.58 per cent over the three years, which amounts to an average increase of 11.2 per cent. Further, intervention of application of N and lignolytic microbial culture (*Trichoderma viridae*) might have enhanced the faster decomposition of trash resulting buildup of organic carbon in ISTM plots [9]. Trash burning decreased the organic carbon content, such that at the end of third year the organic carbon content was to 0.40 per cent compared to 0.44 percent in the initial period. This may be due to loss of dry matter and carbon during the burning processes of trash. According to Mitchell et al., depending on the severity of the fire, 77-97 per cent of the dry matter and carbon may be lost by burning sugarcane trash. On contrary, the retention of trash in the field will increase the organic carbon through decomposition process in the long term [10].

Available nitrogen: The available nitrogen in the soil depends mainly on the sources of nitrogen supply, crop removal and organic carbon content of the soil. The increase in the available nitrogen in ISTM soil was low (3% on an average), while, the trash burnt soil showed decreased N content over the years even after excess N application through fertilizers (350 kg/ha) (Table 3). This may be due to loss of N during burning of trash. The inorganic N supplied through fertilizer is prone to more loss than retention. The low buildup of N in ISTM

Season	Sample	Total N (%)	Total P_2O_5 (%)	Total K_2O (%)
Rabi 2008	Raw trash	0.53	0.09	0.42
	Trash ash	Traces	0.11	0.56
Rabi 2009	Raw trash	0.49	0.13	0.51
	Trash ash	Nil	0.21	0.62
Rabi 2010	Raw trash	0.51	0.14	0.50
	Trash ash	Nil	0.22	0.58
Average	**Raw trash**	**0.51**	**0.12**	**0.48**
	Trash ash	-	**0.18**	**0.59**

Table 2: Major nutrient content of sugarcane trash and burnt trash ash.

Year	Organic carbon		% difference in OC over years		Available N		% difference in Avail. N over years	
	ISTM	Check	ISTM	Check	ISTM	Check	ISTM	Check
Before	0.42	0.44	-	-	312.4	324.8	-	-
Rabi 2008	0.45	0.42	6.3	-4.3	319.6	306.4	2.3	-5.7
Rabi 2009	0.50	0.40	11.3	-4.1	330.5	297.3	3.4	-3.0
Rabi 2010	0.58	0.40	16.0	-1.5	347.3	284.7	5.1	-4.2
Average	**0.49**	**0.41**	**11.2**	**-3.3**	**327.4**	**303.3**	**3.6**	**-4.3**

Table 3: Impact of ISTM on soil organic carbon and available nitrogen.

may be attributed to wider C:N ratio in trash and might have resulted in immobilization by microbes. The immobilization was higher during the initial year and in the later years the N builds up was comparatively higher. Robertson and Thorburn [11] reported similar results of slow N buildup in wider C:N crop residue management.

Available phosphorus: The impact of ISTM on available phosphorus (P) content of soil is presented in Table 4. The results revealed that there was a trend of increased available P content in soil in ISTM plot, while, the check plot recorded decreasing trend. This increased available P may be attributed to inoculation of *Trichoderma viridae,* which mobilizes the unavailable P content in the soil through production of organic acids [12] and also P from sugarcane trash. Even though trash ash had higher P content and farmers usually applied more P than the recommendation (125 kg/ha), there was decrease in the P content in check plot. This might be due to alkaline soil which turns P into unavailable form.

Available potassium: The available potassium (K) also recorded the similar trend as that of the P (Table 4). The effective management of trash results in increased potassium content as the trash is a rich source of K. The results are in line with findings of Graham et al., [13]. Though the trash ash contained more potassium, the volume of ash generated through burning was not sufficient to meet the potassium requirement of the sugarcane crop.

Impact of ISTM on sugarcane germination and yield

Sugarcane germination: The pre-knowledge test conducted on sugarcane trash burning by farmers revealed that the farmers had an opinion that trash mulching will reduce germination and come in the way of routine ratoon cultivation practices. However, the results showed higher germination percentage in ISTM compared to trash burning (Table 5). The excess heat generated during burning might have resulted in reduced germination per cent and time taken for germination. Also, the maturation of sugarcane was 15-20 days earlier in ISTM than the burnt plot (Table 1).

Sugarcane yield: The average yield of first ratoon sugarcane crop of Mallanayakanakatte village, Mandya district, Karnataka was taken as base data to assess the impact of ISTM on sugarcane yield, as the yield of main crop will be always higher than the ratoon crops. The yield data revealed that the cane yield increased substantially in second and third ratoon crop compared to first ratoon and base year in ISTM and the average increase in yield was 8.5 per cent. While in the check plot, the yield showed the decreasing trend. The increased cane yield might be attributed to increased germination percent, increased soil fertility and over all positive effect of trash mulching on soil health. Research results on trash mulching in sugarcane revealed that the trash mulching had an added advantage of moisture conservation, weed control, increased soil biological activity and increased number of earthworms, which eventually resulted in increased yield [14]. Thorburn et al., [11] also determined that the trash mulching sustains the sugarcane productivity and soil health.

Impact of ISTM on sugarcane economics: The economics of sugarcane cultivation are described in Table 6 and 7. The results revealed that though the average cost involved in ISTM was marginally higher (3.2 %) than the check, the gross income realized was 13.9 per cent greater than the check. The net income obtained in ISTM plot over the years trended higher, while the net income in the check plot showed a decreasing trend. This could be attributed to increased cane yield and price per ton over the years (Rs. 900/ton during 2007 to Rs. 1200/ton

Year	Available P$_2$O$_5$		% difference in Avail. P$_2$O$_5$ over years		Available K$_2$O		% difference in Avail. K$_2$O over years	
	ISTM	Check	ISTM	Check	ISTM	Check	ISTM	Check
Before	31.8	31.2	-	-	232.5	244.6	-	-
Rabi 2008	34.2	29.6	7.4	-5.1	248.3	236.8	6.8	-3.2
Rabi 2009	37.0	29.2	8.3	-1.4	277.1	228.4	11.6	-3.5
Rabi 2010	40.6	27.6	9.8	-5.5	319.2	230.3	15.2	0.8
Average	**35.9**	**29.4**	**8.5**	**-4.0**	**269.3**	**235.0**	**11.2**	**-2.0**

Table 4: Impact of ISTM on soil available phosphorus and potassium.

Year	% S. cane germination		S. cane yield (ton/ha)		% difference in yield over years	
	ISTM	Check	ISTM	Check	ISTM	Check
Before*			104.3	104.3		
Rabi 2008	86	73	110.9	97.3	6.3	-6.7
Rabi 2009	73	65	121.0	95.4	9.1	-2.0
Rabi 2010	87	66	133.2	91.4	10.1	-4.2
Average	**82**	**68**	**117.3**	**97.1**	**8.5**	**-4.3**

*The average yield of first sugarcane ratoon crop of Mallanayakanakatte village, Mandya ditrict, Karnataka.

Table 5: Impact of ISTM on sugarcane germination and yield.

Year	Gross cost (Rs.)		% difference in GC over years		Gross income (Rs.)		% difference in GI over years	
	ISTM	Check	ISTM	Check	ISTM	Check	ISTM	Check
Before*	41720	39113			96750	96750		
Rabi 2008	44348	40380	6.3	3.2	110871	97300	14.6	0.6
Rabi 2009	48384	42930	9.1	6.3	133056	104940	20.0	7.9
Rabi 2010	53271	45700	10.1	6.5	159813	109680	20.1	4.5
Average	**46931**	**42031**	**8.5**	**5.3**	**125122**	**102168**	**18.2**	**4.3**

* Economics calculated for average first ratoon crop.

Table 6: Impact of ISTM on sugarcane economics-Gross cost and income.

Year	Net income (Rs.)		% difference in NI over years		B:C Ratio		% difference in B:C over years	
	ISTM	Check	ISTM	Check	ISTM	Check	ISTM	Check
Before*	55030	55030			2.32	2.32		
Rabi 2008	66523	53515	20.9	-2.8	2.50	2.22	7.8	-4.2
Rabi 2009	84672	57240	27.3	7.0	2.75	2.20	10.0	-1.0
Rabi 2010	106542	54840	25.8	-4.2	2.96	2.00	7.6	-9.1
Average	**78192**	**55156**	**24.7**	**0.0**	**2.63**	**2.19**	**8.5**	**-4.8**

*Economics calculated for average first ratoon crop.

Table 7: Impact of ISTM on sugarcane economics-Net income and B:C ratio.

during 2009). The increase in the Benefit to cost ratio (B:C) was 8.5 per cent in ISTM plot, while in the check plot the difference was negative. ISTM recorded a benefit of Rs. 2.63 over rupee invested.

Conclusion

The present study, indicated that ISTM technology of trash mulching in alternate rows along with application of urea (75 kh/ha) and microbial enriched (*Trichoderma viridae*) farm yard manure is effective in enhancing the soil health and sugarcane yield. This technology was popularized to farmers through field visits, trainings, demonstration, group discussion and field days. The Farmers stated that ISTM is highly beneficial and helped in conserving the soil moisture which decreased the number and frequency of irrigations to the sugarcane crop. The trash mulching helped in improving soil health which can be realized with increased number of earthworms in soil. This technology can be adopted without hampering the ratoon sugarcane cultivation practices. Also, ISTM resulted in increased cane yield and allowed farmers to harvest the crop 15 days earlier compared to their usual practice.

References

1. Robertson FA, Thorburn PJ (2000) Trash management - consequences for soil carbon and nitrogen. Proceedings of the Australian Society of Sugar Cane Technologists 22: 225-229.

2. Singh JB, Soleman S (1995) Sugarcane Agro-industrial Alternatives. Oxford & IBN Publishing company, New Delhi.

3. Mitchell RDJ, Thorburn PJ, Larsen P (2000) Quantifying the loss of nutrients from the immediate area when sugarcane residues are burnt. Proceedings of the Australian Society of Sugar Cane Technologists 22: 206-211.

4. Walkley AJ, Black CA (1934) An examination of the method for determining soil organic matter and a proposed modification of the chromic acid titration method. Soil Sci 37: 29-38.

5. Subbiah BV, Asija GL (1956) A rapid procedure for the estimation of available nitrogen in soil. Curr Sci 25: 259-260.

6. Jackson ML (1973) Soil Chemical Analysis, Prentice Hall of India Pvt. Ltd., New Delhi.

7. Piper CS (1966) Soil and Plant Analysis. Han's Publication, Bombay.

8. Gentile R, Vanlauwe B, Chivenge P, Six J (2008) Interactive effects from combining fertilizer and organic residue inputs on nitrogen transformations. Soil Biology and Biochemistry 40: 2375-2384.

9. Yadav RL, Shukla SK, Suman A, Singh PN (2009) Trichoderma inoculation and trash management effects on soil microbial biomass, soil respiration, nutrient uptake and yield of ratoon sugarcane under subtropical conditions. Biology and Fertility of Soils 461-468.

10. Chan KY, Heenan DP, Oates A (2002) Soil carbon fractions and relationship to soil quality under different tillage and stubble management. Soil and Tillage Research 63: 133-139.

11. Thorburn PJ, Robertson FA, Lisson SN, Biggs JS (2001) Modeling decomposition of sugarcane surface residues and the impact on simulated yields. In: 'Sustainable Management of Soil Organic Matter 74-82.

12. Allison MF, Killham K (1988) Response of soil microbial biomass to straw incorporation. Journal of Soil Science 39: 237-242.

13. Graham MH, Haynes RJ, Meyer JH (2000) Changes in soil fertility induced by trash retention and fertiliser applications on the long-term trash management trial at Mount Edgecombe. Proceedings of the South African Sugar Technologists Association 74: 109-113.

14. Anonymous (1994) Crop production-Sugarcane. Proceedings of Annual Review Workshop Meeting, Zonal Agriculture Research Station, VC Farm, Mandya.

33

Test Cross Mean Performance and Combining Ability Study of Elite Lowland Maize (*Zea mays L.*) Inbred Lines at Melkassa, Ethiopia

Tessema Tamirat[1], Sentayehu Alamerew[1], Dagne Wegary[2] and Temesgen Menamo[1*]

[1]Jimma University, Collage of Agriculture and Veterinary Medicine, P.O.Box: 307, Jimma, Ethiopia
[2]CIMMYT–Ethiopia, ILRI Campus, P.O. Box 5689, Addis Ababa, Ethiopia

Abstract

In Ethiopia, there are several maize production constraints, among which shortage of improved varieties is the major one. The objective of this study was to observe the mean performance of crosses and estimate combining abilities for grain yield and other agronomic traits in thirty six maize inbred lines using Line x Tester mating design. Seventy-two lines x test crosses (L1xT1, L1xT2, L2xT1, L2xT2... L36xT1 and L36xT2) and three standard checks were evaluated for 11 traits in alpha lattice design at Melkassa, Ethiopia. Analyses of variances showed significant mean squares due to crosses for all traits. Among the crosses, L34xT2, L36xT2, L30xT1, L19xT1 and L33xT2 crosses were the top five highest grain yield mean performance. While L35xT1, L23xT1, L13xT1, L24xT2 and L15xT1 crosses were lowest mean grain yield performance. GCA Mean squares were significant, but SCA mean squares were non-significant for all traits. The ratio of GCA/SCA mean square was high which exhibited the preponderance of additive gene effects in the inheritance of all traits. Furthermore, the GCA sum of square component was greater than the SCA sum of square for all traits, supported that variation among crosses was mainly due to additive rather than non-additive gene effect. Inbred lines L15, L25, L27, L33, L34 and L35 were the best general combiners for grain yield, and hence were promising parents for hybrids as well as for inclusion in breeding programmes for yield improvements. Inbred lines L2, L3, L4, L6, L10, L27, L30, L31 and L32 had negative and significant GCA effects for the days to anthesis, indicating that the lines had gene combinations that can enhance early maturity.

Keywords: Cross mean performance; SCA; GCA; Maize inbred lines; Ethiopia

Introduction

Maize [1] is one of the oldest food grains of the world and it is believed to be originated in Mexico [2]. It belongs to the grass family *Poaceae (Gramineae)*, tribe *Maydeae* and is the only cultivated species in this genus [3]. Maize grain is recognized worldwide as a strategic food and feed crop that provides an enormous amount of protein and energy for humans and livestock, like Quality protein maize [4].

Several million people in the developing world consume maize as an important staple food and derive their protein and calorie requirements from it [5]. Maize is a potential source of protein for humans and animals [5]. It holds a great promise for increasing production [5]. Maize [1] was introduced to West Africa in the early 1500s by Portuguese traders and then to Ethiopia during the 1600s and 1700s [6]. Today, maize is becoming one of the most important food crops throughout Ethiopia. In terms of area, it is the second most important commodity next to tef (*Eragrostis tef*) [7]. In Ethiopia, maize is one of the top priority food crops selected to achieve food security. It is the staple food and one of the main sources of calorie, particularly in the major maize producing regions [7]. Overall, the area allocated and the productivity level of maize has been increasing since 1994. The area allocated in 1994 was about one million ha, which has increased to about 1.96 million ha of land in 2010/11 production season. Similarly, the average national productivity of maize has increased from about 1.5 t/ha in 1994 to about 2.54 t/ha in 2010/11 mainly due to the strong public push of improved seed and fertilizer [7].

Maize grows in different agro-ecological zones of Ethiopia ranging from sea level up to 2800 m a.s.l [8]. It is grown in areas with light to heavy soils, wide ranges of temperatures and rainfall, indicating that maize has good adaptability to different arrays of environmental variables. However, in spite of its wide adaptation and efforts made to develop improved maize technologies for different maize agro-ecological zones still many biotic (such as common leaf rust (CLR), maize streak virus (MSV), stalk borer, termite, etc.) and abiotic constraints (such as drought, low soil fertility, etc.) limit maize production and productivity in different maize producing agro-ecological zones [9]. Global coarse grains (such as maize) production in 2012 is expected to fall by 2.5 percent from 2011 2.5 percent. Severe droughts this year in the United States and across a large part of Europe and into central Asia have been the main cause of the reduced coarse grain crops [10].

Combining ability is an effective tool which gives useful genetic information for the choice of parents in terms of their performance in a series of crosses [11]. Line x tester [12] is useful in deciding the relative ability of female and male lines to produce desirable hybrid combinations. It also provides information on genetic components and enables the breeder to choose appropriate breeding methods for hybrid variety or cultivar development programmes. Information on combining ability effects helps the breeder in choosing the parents with the high general combining ability and hybrids with high specific combining. In the maize breeding program, analysis of

Corresponding author: Temesgen Menamo, Jimma University, Collage of Agriculture and Veterinary Medicine; P.O.Box: 307, Jimma, Ethiopia
E-mail: temesgen2008@hotmail.com

general combining ability (GCA), specific combining ability (SCA) and heterosis would help to identify best inbred lines for hybrid development and hybrid combinations for better specific combining ability. Combining ability analysis helps in identifying potential inbred lines for producing hybrids and synthetic varieties in maize breeding. Luchsinger and Violic (1972) examined eight characters in a maize diallel cross analysis and observed that additive effects were more important than non-additive ones for all the variables studied. Piovarci [13] Also determined combining ability and its components for maize grain yield in an eight inbred diallels and concluded that additive gene effects were more important than non-additive effects. Currently, large numbers of introduced elite maize inbred lines from CIMMYT-Zimbabwe are available in Ethiopian; Maize Improvement Project at the Melkassa Agricultural Research Center (MARC). Thirty-six lines selected for this study. These lines were highest resistance for drought and saline soils. Rift valley areas have drought and saline problems. MARC is one of the best maize growing areas in the Rift Valley of Ethiopia [14]. Therefore, the objectives of this study were to estimate crosses mean performance and combining abilities for grain yield and other agronomic traits in thirty six maize inbred lines using Line x Tester mating design.

Material and Methods

Experimental materials

The experimental material for the study consisted of seventy two lines by test crosses (L1xT1, L1xT2, L2xT1, L2xT2... L36xT1 and L36xT2) and three standard checks. The parents were crossed in Line x Tester mating design to generate 72 F_1 hybrids at Melkasa Agricultural Research Center, Ethiopia at main cropping season 2011/2012. The experiment was conducted during the main cropping season of 2012/13 at the Melkasa Agricultural Research Center (MARC) on F_1 (LineXTester crosses). This location is one of the best maize growing areas in the Rift Valley of Ethiopia [14]. The inbred lines were introduced from CIMMYT-Zimbabwe and were bred for resistance to various biotic and abiotic stresses of Africa. The two testers are single crosses of commercial CIMMYT inbred lines of known heterotic groups; CML312/CML442 (Tester A) and CML202/CML395 (Tester B). MHQ138 (check) is a three way hybrid, high yielding and tolerance/resistance to maize diseases. MH130 is also check which is a double top cross hybrid. Melkassa_2 is the 3rd check; it is early maturing and drought tolerant variety. All checks are released by the Melkassa Agriculture Research Center.

Experimental design and field managements

The experiment was designed 7×15 alpha lattice design [15]. The experiment had two replications. Each plot consisted of one row of 5 m length with 75 cm and 25 cm spacing between rows and plants, respectively. As recommended by the center, 50 kg P_2O_5 ha^{-1} and 25 kg N ha^{-1} was applied at planting, followed by a side dressing of 25 kg N ha^{-1} 35 days later. Urea and Dominium phosphate (DAP) was used as sources of N and P_2O_5, respectively. Other crop management practices such as land preparation, weeding, disease and insect control were applied following research recommendations

Statistical analysis and procedures

Data collection and Analysis of variance (ANOVA): Data were recorded on eleven quantitative characters. Data related to days to 50% anthesis, 50% days to silking, 1000-kernel weight, grain yield and anthesis silking interval were recorded on the plot basis while data related to other characters were recorded on five randomly selected plants leaving border plants of each row. The mean values were subjected to line x tester analysis. Analyses of variances (ANOVA) were computed for grain yield and other agronomic traits by using SAS 9.2 software.

Combining ability analysis: Line x tester analysis was done for traits that showed statistically significant differences among the crosses using the adjusted means based on the method described by Kempthorne [12]. General combining ability (GCA) and specific combining ability (SCA) effects for grain yield and other agronomic traits were calculated using the line x tester model. The F-test of mean square due to lines, testers and their interactions were computed against mean square due to error [16]. Significances of GCA and SCA effects of the lines and hybrids were determined by an F - test using the standard errors of GCA and SCA effects.

Results and Discussion

Analysis of variance

Analysis of variance showed a highly significant difference (P<0.01) among the crosses for the anthesis data, plant height, ear height silking day and ear length. While the mean square due to crosses were significant (P<0.05) for the grain yield, anthesis silking interval, number of ears per plant, 1000 kernel weight, number of kernels per row and ear diameter (Table 1). Mean square due to check were highly significant for the anthesis day and the number of days to silking. Mean squares due to check and check Vs cross were significant for the anthesis day, plant height, and ear length. This indicated the presence of inherent variation among the lines, which makes selection possible. Similarly, several previously studied reported significant differences among crosses for grain yield and related trait in different sets of maize genotypes [17-22] (Table 1).

Mean performance of F_1 crosses (Line x Tester)

Mean values for the 11 characters are shown in Table 2. Forty-six of 72 crosses had a higher grain yield than the best hybrid check MHQ 138 (5.40t/ha). The overall mean grain yield of the crosses were 5.45 t/ha, ranging from 4.13 t/ha to 6.27 t/ha. Cross L34xT2 (6.27 t/ha), L25xT2l (6.24), and L18 xT2 (6.00 t/ha) had a higher grain yield, whereas crosses L23xT1 (4.856t/ha), L24xT2 (4.88t/ha) and L35xT1 (4.13t/ha) were lower grain yield. Cross L12xT1 (78.5) was the latest anthesis day while cross L2xT2 (70.2) was earliest athesis day.

The mean anthesis silking interval was ranging from 0.7 (L31xT2) to 1.7 (L26xT1, L26xT2, L28xT1) day (Table 2). Crosses L9xT2 (258.13 cm), L13xT2 (216.3 cm) and L33 x T2 (215.3 cm) had highest plant height while crosses L7 x T1 (173.3 cm), L17x T1 (175 cm) and L35xT21 (177.4 cm) showed lowest plant height. Crosses L11 x T2 (118.8 cm), L16 x T1 (117.0 cm) and L16 x T2 (115.5cm) had highest ear height while L31xLT2 (87.2 cm), L7xT1 (90.5 cm) and L10xT2 (92.5 cm) had lowest ear height. All crosses were showed less thousand kernels weight than check of MHQ 138 (242 gm).Overall means ear diameter of 45.3 mm with ranged from 41.2 mm (L11 x T1) to 48.6 mm (L13xT2). Cross L13xT2 (48.8 mm) followed by cross L5xT1 (48.1 mm) and L25xT2 (48.2 mm) (Table 2) had a highest ear diameter while crosses L11xT1 (41.2 mm), L21xT1 (42.6 mm) and L10xT1 (41.7 mm) showed a lowest ear diameter. MHQ 138 (35.88) standard check had a higher number of kernels per row than all crosses.

Crosses L17xT2, L29xT2 and L13xT1 had the lowest number of ears per plant while crosses L1xT1, L11xT2, and L16xT2 had showed the maximum number of ears per plant. Mean ear length was 16.6 cm

SoV	Df	GY(t/ha)	AD(days)	ASI(day)	PH(cm)	EH(cm)	ED(cm)	EL(cm)	EPP(#)	KPR(#)	SD(days)	TKWT(gm)
Rep	1	4.15*	2.4*	1.3	60.16*	218.4*	.0.92	1.02	0.13*	76.32*	3.8*	13949.1*
Entry	74	1.59*	10.9**	1.15*	435.6**	178.3**	12.12*	3.1**	0.04*	8.4*	11.1**	1741.8*
Crosses(Cr)	71	1.62*	9.51**	1.14*	440.7**	184.4**	12.42*	3.05**	0.04*	8.15*	10.78**	1680.9*
GCA line	16	1.82*	12.76**	1.33*	697.4**	306.8**	16.3*	4.11**	0.06**	11.4*	13.1**	2103.4*
GCA tester	1	4.71*	112**	5.84*	2425.5**	31.2	31.1*	14.95*	0.002	5.14	117.2**	9236.8*
SCA	35	1.33	3.33	0.82	127.3	66.3	8.04	1.6	0.02	4.9	4.36	1042.5
Ck	2	0.99	10.6**	0.5	460.6*	43.1	7.56	6.14*	0.018	0.04	11.5**	4303.0
Ck vs Cr	1	1.62	16.6*	1.147	440.7*	120.2	0.044	1680.9*	0.01	0.04	11.8	1680.0
Error	74	0.96	2.87	0.73	83.2	45.1	0.03	0.68	___	4.72	3.72	806.9
% contribution. GCA		60.87	82.79	65.02	85.79	82.64	68.12	73.49	76.4	73.36	85.2	72.62
% contribution. SCA		39.13	17.21	34.98	14.21	17.36	31.88	26.51	23.54	26.64	14.8	27.38
Ratio GCA/SCA		1.56	4.81	1.86	4.14	4.76	2.14	2.77	3.24	2.75	5.76	2.65

**=significant at P<0.01 level of probability, * = Significant at P<0.05 Level of probability, GY= grain yield AD = anthesis, date, ASI=anthesis silking interval, Df = degrees of freedom, ED = ear diameter, EH = ear height, EL = ear length, EPP= number of ears per plant, KPR = number of kernels per row, SD=silking days, PH = plant height, Rep= replication, TKWT = thousand kernels weight.

Table 1: Analysis of variance for grain yield and other agronomic traits of line by tester crosses involving 36 lines and two testers evaluated at Melkassa 2012

Crosses	GY (t/ha)	AD (Days)	ASI	PH (cm)	EH (cm)	TKWT (gm)	ED (mm)	SD (days)	KPR	EL (cm)	EPP
L1xT2	5.71ab	71.9 u-y	1.2ab	204.2c-o	95.6i-v	221.1a-g	45.4a-k	73.1j-o	35.9a-e	17.2a-g	1.33a-i
L2xT2	5.45ab	70.2xy	0.9ab	195.4j-u	97.1g-v	227.4a-e	46.2a-i	71.1no	36.7a-e	17a-h	1.08i-r
L3xT2	5.75ab	72.8o-w	0.7b	198.7h-s	95.9i-v	188.5g	46.9a-g	73.5i-o	36.7a-e	17.1a-h	1.4a-f
L4xT2	5.19abc	72.3s-y	0.7b	190.8n-v	91r-v	210.1a-g	43.8f-m	73k-o	35.8a-e	16.9a-i	1.25b-l
L5xT2	5.57ab	75c-r	1.5ab	192.2m-v	94.1k-v	214.9ab-g	43.9e-m	76.5b-k	34.5de	16.9a-i	1.07i-r
L6xT2	5.02bc	73.6k-v	0.7b	200.6f-r	100e-s	197.9d-g	44.5d-m	74.3f-o	35.2a-e	16.8a-j	1.22c-n
L7xT2	5.14abc	74.6e-s	0.8ab	187.6p-x	93.5m-v	204.9b-g	44.9b-l	75.4c-l	35.4a-e	17a-h	1.22c-n
L8xT2	5.37ab	75.2b-p	1ab	201.8e-q	101.3d-s	212.2a-g	45.1a-l	76.2b-k	37.7b	17.3a-f	1.23b-m
L9xT2	5.52ab	75.1b-q	1.4ab	235.6a	111.2a-d	208.2a-g	44.6c-m	76.5b-k	36.8a-e	17.2a-g	1.4a-f
L10xT2	5bc	72.7p-x	1.1ab	202.6c-p	92.5o-v	202.1b-g	43.1h-m	73.8i-o	36a-e	16.2d-l	1.13g-q
L11xT2	5.61ab	73.9i-u	1.3ab	220.1b	118.8a	195.5e-g	42.3k-m	75.2c-l	35.9a-e	16.9a-i	1.49ab
L12xT2	5.51ab	74.5e-t	1.4ab	214.4b-g	96.4h-v	204.7b-g	44.8b-l	75.9b-k	35.8a-e	17.4a-f	1.27a-k
L13xT2	5.95ab	75.4b-n	1.2ab	216.3b-e	108.1a-g	223.3a-g	48.6a	76.6a-j	36.7a-e	16.9a-i	1.13g-q
L14xT2	5.74ab	74.1g-u	0.8ab	207.4b-m	101.4d-s	221.6a-g	44.5d-m	74.9c-l	36.2a-e	17.2a-g	1.25b-l
L15xT2	4.99bc	76.2a-j	1.6ab	200.2f-r	98.2g-v	220.9a-g	44.9b-l	77.8a-f	35.6a-e	16.3c-l	1.1h-r
L16xT2	5.65ab	75.6b-l	1ab	215.1b-f	115.2a-c	208.8a-g	47.3a-f	76.6a-j	35.8a-e	15.8f-l	1.18d-o
L17xT2	5.19abc	73.7j-v	1.1ab	197.8h-t	102.6d-p	224.2a-g	44.2d-m	74.8d-m	35b-e	16e-l	0.91p-r
L18xT2	6ab	73.7j-v	0.8ab	201.4e-r	99.2e-t	233.5a-d	46.5a-h	74.5e-n	36.6a-e	16.1e-l	1.01k-r
L19xT2	4.9bc	74.7d-s	1.3ab	205.9b-n	102.8d-p	220.2a-g	46.4a-h	76b-k	36.2a-e	17.9a-c	1.03j-r
L20xT2	5.83ab	76.1a-k	0.9ab	197i-u	96.4h-v	226.9a-e	46.7a-g	77a-i	36.6a-e	17.9a-c	1.09i-r
L21xT2	5.65ab	74.9d-r	1.5ab	209.3b-k	105.1c-k	225.4a-f	45.3a-k	76.4b-k	36.3a-e	17.1a-h	1.11g-r
L22xT2	5.68ab	72.3s-y	1.4ab	212.9b-h	104.7c-l	218.9a-g	45.8a-k	73.7i-o	36.9a-e	16.9a-i	1.19c-o
L23xT2	5.72ab	73.3l-v	0.8ab	213b-h	107.3b-h	217.8a-g	46.2a-i	74.1g-o	35.2a-e	16.6b-k	1.21c-n
L24xT2	4.88bc	73m-v	0.9ab	202.3d-p	99.4e-t	213a-g	45.2a-l	73.9i-o	35.9a-e	16.7a-k	1.11g-r
L25xT2	6.24a	73.8j-v	1.1ab	206.4b-n	104.8c-l	237.9ab	48.2ab	74.9c-l	36a-e	17.2a-g	1.01k-r
L26xT2	5.23abc	75.3b-o	1.7a	208.3b-l	104.4c-m	199.2c-g	47.4a-e	77a-i	36.3a-e	15.3i-l	1.07i-r
L27xT2	5.66ab	72t-y	1.3ab	184.3s-x	96.2i-v	202.3b-g	47.1a-f	73.3j-o	36a-e	17a-h	1.13g-q
L28xT2	5.91ab	72.8o-w	1.1ab	182.4t-x	95.4i-v	225.4a-f	44.4d-m	73.9i-o	37.2a-d	16e-l	1.27a-k
L29xT2	5.66ab	73.8j-v	1ab	186.4q-x	88.1uv	226.6a-e	46.8a-g	74.8d-m	36.5a-e	17.4a-f	0.96n-r

L30xT2	5.66ab	72.7p-x	0.8ab	201.3e-r	98.7f-v	204.3b-g	46.9a-g	73.5i-o	36.4a-e	17.8a-d	0.96n-r
L31xT2	5.51ab	71.3v-y	0.7b	186.2r-x	87.2v	222.3a-g	46.4a-h	72l-o	36.3a-e	17.3a-f	1l-r
L32xT2	5.1abc	73.1l-v	1.2ab	196.4j-u	96i-v	219.6a-g	44.5d-m	74.3f-o	36.7a-e	17.3a-f	1.1h-r
L33xT2	5.87ab	75.5b-m	1.1ab	215.4b-f	107.4b-h	227.7a-e	47.5a-d	76.6a-j	36.2a-e	16.7a-k	1.23b-m
L34xT2	6.27a	77a-e	1.1ab	207.9b-l	100.4e-s	225.3a-f	45.6a-k	78.1a-d	36.1a-e	16.1e-l	1.13g-q
L35xT2	5.13abc	75.1b-q	0.9ab	192.1m-v	100e-s	208.4a-g	46.3a-h	76b-k	36a-e	15.8f-l	1.03j-r
L36xT2	5.99ab	74.7d-s	1.5ab	208.9b-k	104.3c-m	216.1a-g	44.7b-m	76.2b-k	35.5a-e	16.3c-l	1.13g-q
L1xT1	5.18abc	76.6a-g	1ab	195.6j-u	92.4o-v	214.5a-g	45.2a-l	77.6a-g	36.3a-e	17.1a-h	1.53a
L2xT1	5.3abc	72.6q-x	1.1ab	186r-x	97.8g-v	215.1a-g	46.2a-i	73.7i-o	36.9a-e	17.3a-f	1.05j-r
L3xT1	5.49ab	73m-v	1.3ab	189.4o-w	92.8o-v	208.8a-g	44.1d-m	74.3f-o	36a-e	15.5h-l	1.42abcd
L4xT1	5.48ab	73.6k-v	1.1ab	192.9l-v	94.5k-v	213a-g	42.7i-m	74.7d-m	34.6de	16e-l	1.36a-h
L5xT1	4.97bc	77.5a-c	0.9ab	200.8e-r	99.7e-s	193.3e-g	48.1a-c	78.4a-c	34.5de	15.6g-l	1.27a-k
L6xT1	5.22abc	72.9n-v	0.9ab	191.4n-v	99.1e-u	219.5a-g	43.4g-m	73.8i-o	35.6a-e	16.7a-k	1.16d-p
L7xT1	5.36ab	74.5e-t	1.5ab	173.3x	90.5s-v	213.2a-g	44.6c-m	76b-k	34.8de	15.1kl	1.04j-r
L8xT1	5.51ab	75c-r	1.3ab	205.7b-n	98.1g-v	197.9d-g	44.1d-m	76.3b-k	35.5a-e	16.5b-l	1.41a-e
L9xT1	5.17abc	75.2b-p	1.1ab	220.8ab	105.9c-i	214.2a-g	43.9e-m	76.3b-k	35.6a-e	16e-l	1.25b-l
L10xT1	5.76ab	73.7j-v	0.8ab	188.3p-x	92.1o-v	193.5e-g	41.7lm	74.5e-n	35.6a-e	15.6g-l	1.28a-j
L11xT1	5.61ab	74.3f-u	1.4ab	217.5b-d	107.5b-g	203.9b-g	41.2m	75.7b-k	36.6a-e	16.8a-j	1.37a-g
L12xT1	4.98bc	78.5a	1.6ab	214.9b-f	104.6c-l	197.2d-g	46.5a-h	80.1a	36.8a-e	17.5a-e	1.01k-r
L13xT1	4.88bc	77.6ab	1.6ab	198.2h-s	104.8c-l	214.4a-g	43.9e-m	79.2ab	35.6a-e	15.5h-l	0.93op-r
L14xT1	5.66ab	75.2b-p	0.9ab	218.4bc	111.8a-d	207.2a-g	46.5a-h	76.1b-k	35.9a-e	17a-h	1.15e-p
L15xT1	4.91bc	77a-e	1.2ab	181.6u-x	94.1k-v	201.6b-g	45.3a-k	78.2a-d	34.3e	15.3i-l	1.01k-r
L16xT1	5.55ab	78.1a	1.1ab	200.6f-r	117ab	189.2fg	46.9a-g	79.2ab	35.3a-e	14.9l	1.45a-c
L17xT1	5.2abc	76.5a-h	1.3ab	175wx	97.3g-v	197e-g	44.2d-m	77.8a-f	34.8de	16.1e-l	0.88qr
L18xT1	5.79ab	74.6e-s	1.1ab	195j-u	102.3d-q	234.8a-c	44.9b-l	75.7b-k	37.6a-c	18.3a	1.06j-r
L19xT1	5.9ab	77.2a-d	1.2ab	194k-u	100.9d-s	203.6bc-g	45.5a-k	78.4a-c	37.1a-d	16.9a-i	1.09i-r
L20xT1	5.56ab	75.4b-n	1.5ab	201.9d-q	103.1d-o	199.6c-g	45.4a-k	76.9a-i	37.1a-d	17.2a-g	1.27a-k
L21xT1	5.44ab	76.7a-f	1.2ab	198.8g-s	103d-o	202.8b-g	42.6j-m	77.9a-e	37.9a	16.9a-i	0.97m-r
L22xT1	4.98bc	76.8a-f	1.1ab	210.1b-j	109.1a-f	206.3a-g	44.3d-m	77.9a-e	35.6a-e	16.6b-k	1.26b-l
L23xT1	4.85bc	75.5b-m	1ab	209.4b-k	107.4b-h	210.6a-g	44.6c-m	76.5b-k	35.7a-e	16.8a-j	1.23b-m
L24xT1	5.56ab	75.9b-k	1.1ab	196.3j-u	103.9d-n	209.8a-g	45.2a-l	77a-i	37a-e	17.3a-f	1.25b-l
L25xT1	5.65ab	72.5r-x	1.5ab	195.5j-u	101.7d-r	207.9a-g	44.8b-l	74h-o	35.8a-e	17.3a-f	1.13g-q
L26xT1	5.66ab	77.5a-c	1.7a	200.4fg-r	105.6c-j	200.1c-g	46.4a-h	79.2ab	35b-e	14.9l	1.24b-l
L27xT1	5.79ab	74h-u	1.4ab	189.3o-w	96i-v	205.2b-g	45.8a-k	75.4c-l	37a-e	16.4c-l	1.25b-l
L28xT1	5.49ab	74.5e-t	1.7a	189.2o-w	93n-v	219.3a-g	46.7a-g	76.2b-k	35.9a-e	16.5b-l	1.02j-r
L29xT1	5.58ab	75.2b-p	1.3ab	194.2k-u	96.4h-v	224.8a-g	46.3a-h	76.5b-k	36.9a-e	16.1e-l	1.07i-r
L30xT1	5.96ab	73.3l-v	1.1ab	203c-p	91.5q-v	222.4a-g	45.8a-k	74.4e-n	36.6a-e	18.1ab	1.1h-r
L31xT1	5.04bc	72.9n-v	1.5ab	182.1u-x	92.1o-v	213.9a-g	45.1a-l	74.4e-n	36.5a-e	15.2j-l	1.2c-n
L32xT1	5.87ab	73m-v	1ab	184.2s-x	88.4t-v	222.3a-g	44.5d-m	74h-o	37.2a-d	17.4a-f	1.14f-q
L33xT1	5.67ab	76.4a-i	1.3ab	188.1p-x	91.9p-v	220.8a-g	47.3a-f	77.7a-f	35.3a-e	15.8f-l	1.24b-l
L34xT1	5.52ab	74.8d-s	1.4ab	205.7b-n	100.9d-s	225.6a-e	46.6a-h	76.2b-k	35.1b-e	16e-l	0.93o-r
L35xT1	4.13c	76.5a-h	1.6ab	177.4v-x	94.1k-v	207.6a-g	45b-l	78.1a-d	34.9c-e	15.2j-l	0.86r
L36xT1	5.36ab	76.4a-i	1.1ab	200.7e-r	110a-e	208.1a-g	45.4a-k	77.5a-h	37a-e	16.4c-l	0.93o-r
MH 130	5.18abc	69.9y	0.9ab	206.1b-n	102.3d-q	210.3a-g	46.1a-j	70.8o	35.6a-e	16.5b-l	1.02j-r
Melkassa 2	5.23abc	73.9i-u	1.1ab	212.3b-i	98.4g-v	202.3b-g	44.2d-m	75c-l	36a-e	15.3i-l	1.2c-n
MHQ138	5.4ab	70.5w-y	0.8ab	185.9r-x	93.8l-v	242a	45.4a-k	71.3m-o	37.9a	17.4a-f	1.02j-r

Cr mean	5.46	74.6	1.2	199.5	100	212.4	45.3	75.8	36.1	16.6	1.15
Ck mean	5.27	71.3	0.9	120.9	98.2	218.2	45.22	72.2	36.5	16.4	1.08
Grand Mean	5.45	74.5	1.2	199.6	100	212.7	45.3	75.7	36.1	16.6	1.15
LS (5%)	1.2	2.6	1	15.7	11.1	36.4	3.6	3.6	2.8	1.7	0.27
CV (%^)	18.12	2.2	2.9	4.6	7	14.4	5.4	5.1	6.2	6.4	11.8
Min	4.13	69.9	0.7	173.3	87.2	188.5	41.2	71.1	34.3	18.3	0.9
Max	6.27	78.5	1.7	235.6	118.8	242	48.6	80.2	37.9	14.9	1.53

GY=grain yield AD=anthesis date, ASI=anthesis silking interval, Df=degrees of freedom, ED=ear diameter, EH=ear height, EL=ear length, EPP=number of ears per plant, , KPR=number of kernels per row, PH=plant height, Rep=replication, SD=silking day ,TKWT=thousand kernels weight . Numbers within the same column with different letter/s are significantly difference from each other according to LSD test. Some letters represent as interval e.g. Cross L6xT2 had K-v in anthesis date; it included from k to v letters alphabetically.

Table 2: Estimates of mean values for grain yield and related traits with LSD comparisons.

ranged from 14.9 cm (L26xT1) to 18.3 cm (L18xT1). Crosses L18xT1, L19xT2 and L20xT2 had a higher ear length while crosses L26xT1, L16xT1 and L7xT1 had showed lower ear length. Among the 72 crosses, L34xT2 (6.27t/ha), L36xT2 (5.99t/ha), L30xT1 (5.96t/ha), L19xT1 (6.27t/ha), and L33xT2 (5.87/ha) were shown best yield (Table 2). The least yield was obtained from the crosses L35xT1 (4.13t/ha), L23xT1 (4.85t/ha), L13xT1 (4.88t/ha), L24xT2 (4.88t/ha), and L15xT1 (4.91t/ha). The overall yield mean was 5.45t/ha. High grain yield was founded 45 out of 72 crosses than the best hybrid MHQ138 which yielded 5.40t/ha (Table 2). However, they are non-significant when compared to check MHQ138 (Table 2). Cross combination that had a high grain yield could be used in the breeding program to improve the grain yield.

A number of crosses showed mean performance for more than one trait as compared to the best hybrid check (Table 3). Cross combinations that were earliest in anthesis date, shorter in the ear and plant height could be used as source of gene for the development of early maturing and shorter statures variety. Similar, Dagne et al. and Gudeta [22,23] identified experiment variety performing better than the best check for most yield and related traits. Most of the crosses had a comparable grain yield to checks. Table 3 showed top 5 crosses which showed good and poor grain yield performance compared to checks. Statistically, the top five crosses didn't show significant differences from the check, but they were comparable (Table 2).

Combining ability analysis

A GCA Significant difference was observed among crosses for all traits while non-significant difference for SCA (Table 1). This indicated that additive gene action was more important than non-additive gene action. The ratio of GCA/SCA mean square further exhibited the preponderance of additive gene effects in the inheritance of all traits (Table 1).

Grain yield

GCA means squares were significantly difference, whereas SCA mean squares were non-significant for grain yield (Table 1). Pswararyi and Vivek [24] Carried out diallel analysis, among cimmyt's early maturing maize germplasm and reported significant GCA mean square and non-significant SCA for grain yield. This study was indicated that the importance of both additive and non-additive gene action for this trait. Piovarci [25] Analyzed a diallel cross and noted that GCA was more important than SCA for yield. In the line with this study Pswararyi and Vivek, Dange et al., [17,24] previously reported dominate the role SCA gene action in grain yield and Strube [26] reported that SCA effects were greater than GCA effects for yield.

Line GCA effects for grain yield ranged between -0.98 t/ ha (L6) to 1.88 t/ ha (L35) (Table 4). Even though a total of 18 of the inbred lines showed positive GCA effects for grain yield, only 6 inbred lines (L15, L25, L27, L33.L34 and L35) were found to be the best general combiner for grain yield as these lines had positive and significant GCA effect. Inbred lines with positive and significant GCA are desirable parents for hybrid development as well as for inclusion in the breeding program, as the lines may contribute favorable alleles in the synthesis of new varieties. Inbred lines with negative and significant GCA effect were L6 and L17, indicating that these were poor general combining ability for grain yield (Table 4). Similarly, Teshale [18] recorded significant positive and negative GCA effects for grain yield and Mandefro [17] also found significant positive GCA effects for grain yield. The results of this study are also in agreement with the findings of Amiruzzaman et al., Legesse et al. Gudeta, Hadji and Dagne et al. [17,20-22,27,28] who reported significant positive and negative GCA effects for grain yield in maize germplasm.

Number of days to tasseling and silking

For a number of days to anthesis and silking, mean squares due to line GCA were highly significant (p<0.01) (Table 1) but SCA were not significant. Shewangizaw and Leta et al. [28,29] Reported GCA is more important than SCA for days to anthesis and silking. Ahmad and Saleem (2003) [30] also reported that the preponderance of additive gene action in inheritance of days to anthesis days and silking. Further Leggess et al. [28] reported a predominance of additive gene action in inheritance of days to tussling and silking.

Line GCA effect for days to anthesis ranged between - 4.20 (L2) to 2.79 (L13) (Table 4). Among 17 inbred lines with negative GCA effects, eight inbred lines had a significant GCA effect for a number of days to tasseling. These inbred lines had gene combinations that enhance early maturity. Ten inbred lines exhibited positive and significant GCA effect for number of days to tasseling, indicating that these lines were undesirable as they showed a tendency to increase late maturity. Seventeen of the inbred lines showed negative GCA effects for number of days to Silking, out of which L2, L3, L6, L23, and L30 were highly significant (p<0.01) and L4, L10 and L32 was significant (P<0.05). On the other hand, inbred lines L5, L6, L12, L13, L15, L19, L26 and L31and L35 expressed positive and significant GCA (Table 4). This result is in agreement with Demissew et al. and Gudeta [22,23] who reported significant positive and negative GCA effect for a number of days to tasseling and silking. Similar finding were also reported by Teshale Dagne et al. [1,18].

Good performing genotype for grain yield		poor performing genotype for grain yield	
Genotype	GY t/ha	Genotype	GY t/ha
L34 XT2	6.27	L35 XT1	4.13
L36 XT2	5.99	L23 XT1	4.85
L30 XT1	5.96	L13XT1	4.88
L19XT1	5.90	L24 XT2	4.88
L33 XT2	5.87	L15 XT1	4.91

Table 3: Summary of five good and poor performing crosses for grain yield evaluated at Melkassa in 2012/2013

line	GY(t/ha)	AD(days)	SI(days)	PH(cm)	EH(cm)	ED(cm)	EL(cm)	EPP	KPR(#)	SD(days)	TKWT(gm)
L1	0.39	-0.45	-0.43	1.38	-8.52*	0.65	0.74	0.26**	-0.03	-0.88	19.94
L2	-0.54	-4.20**	-0.68	-11.36*	-3.52	1	1.24*	-0.09	2.16*	-4.88**	21.94
L3	0.20	-2.45**	-0.43	-6.61	-8.02*	0.05	-0.45	0.24**	0.56	-2.88**	-43.05**
L4	-0.27	-1.70*	-0.68	-9.76*	-12.0**	-3.47**	-0.3	0.15*	-2.23*	-2.38*	01.94
L5	-0.63	2.79**	0.06	-5.86	-4.27	0.6	-0.55	0.01	-3.83	2.85**	-27.80
L6	-0.98*	-1.95*	-0.93*	-5.11	1.22	-2.58*	0.29	0.03	-1.43	-2.85**	-13.05
L7	-0.51	2.54**	0.06	-23.3**	-10.0**	-0.77	-0.85	-0.02	-2.53*	2.6	-01.30
L8	-0.18	-0.20	0.06	17.13**	-1.02	-1.09	0.29	0.16*	1.26	-0.14	-11.80
L9	-0.04	0.29	0.31	34.88**	12.72**	-1.62	-0.1	0.17**	0.26	0.6	-03.90
L10	-0.41	-1.70*	-0.68	-5.61	-12.0**	-4.82**	-0.8	0.05	-0.53	-2.38*	-33.05*
L11	0.53	-0.95	0.65	23.38**	17.47**	-5.67**	0.29	0.27**	0.46	-0.3	-22.80
L12	-0.45	1.79*	1.16**	16.63**	0.72	0.22	1.29*	-0.02	0.56	2.95**	-49.3**
L13	-0.17	2.79**	0.82	12.13**	11.22**	1.4	-0.65	-0.13*	0.06	3.61**	02.99
L14	0.48	0.04	-0.94*	4.88	11.72**	-0.09	0.84	0.05	-0.08	-0.9	-15.30
L15	1.2**	2.54**	0.81	-11.36*	-6.27	-0.42	-1.35*	-0.09	-2.73*	3.35**	15.69
L16	0.35	2.29**	-0.19	11.63*	22.2**	3.22*	-2.05**	0.17**	-1.38	2.1*	-27.30
L17	-0.97*	0.54	0.06	-14.8**	1.22	-1.59	-1	-0.27**	-2.98**	0.6	-10.55
L18	0.77	-0.2	-0.68	-0.36	-0.52	0.82	0.94	-0.13*	2.66*	-0.88	56.19**
L19	-0.27	1.79*	0.31	-1.11	5.97	1.07	1.24*	-0.09	1.61	2.1*	-03.05
L20	0.75	1.04	0.06	-0.61	-0.52	1.07	1.34*	0.02	1.86	1.1	-05.05
L21	0.21	1.04	0.31	3.88	3.72	-2.64*	0.54	-0.13*	2.51*	1.35	15.19
L22	-0.43	0.29	0.06	15.63**	10.72**	-0.92	0.44	0.05	0.66	0.35	-05.80
L23	-0.25	-0.2	-0.94*	12.38**	8.72*	-0.32	-0.008	0.05	-1.38	-1.1	13.94
L24	-0.32	0.04	-0.44	-0.36	1.47	0.55	0.59	0.03	0.86	-0.4	-02.10
L25	1.21**	-1.2	0.31	2.13	1.97	2.56*	0.99	-0.08	-0.23	-0.89	39.44**
L26	-0.12	2.29**	0.56	7.13	3.97	2.32	-2.40**	-0.02	-1.18	2.89**	-19.65
L27	0.95*	-2.28*	1.31**	-16.1**	-5.77	2.05	-0.05	0.05	2.46*	-0.97	-10.75
L28	0.64	-0.7	0.56	-16.8**	-8.02*	0.7	-0.7	-0.03	0.21	-0.14	20.84
L29	0.50	1.79*	0.06	-9.51*	-9.52**	1.27	0.29	-0.14*	1.21	1.85	29.44*
L30	0.73	-2.28**	-0.68	3.88	-6.27	1.5	2.14**	-0.11	1.16	-2.96**	-3.05
L31	-0.44	-3.2**	-0.19	-19.1**	13.52**	0.8	-0.55	-0.05	1.41	3.39**	5.44
L32	-0.07	-1.95*	-0.43	-9.86*	-9.27**	-1.09	1.19*	-0.06	1.66	-2.38*	10.69
L33	1.00*	1.04	0.06	2.88	-0.77	3.55**	-0.6	0.09	-0.93	1.1	29.94*
L34	0.98*	1.04	0.31	10.63*	1.22	1.0	-0.45	-0.11	-1.08	1.35	29.54*
L35	1.88**	2.04*	0.31	7.38	8.47*	0.62	-1.5	-0.2	-2.18*	2.35*	-12.97

L36	0.33	-0.2	0.31	-21.8**	-5.02	0.05	-0.2	-0.13*	1.06	0.11	8.44
SE	0.48	0.84	0.42	4.56	3.35	1.27	0.54	0.06	1.08	0.97	14.2
SED	0.69	1.19	0.6	6.45	4.75	1.8	0.76	0.09	1.53	1.37	20.08

GY=grain yield AD=anthesis, date, ASI=anthesis silking interval, DF=degrees of freedom, ED=ear diameter, EH=ear height, EL=ear length, EPP=number of ears per plant, KPR=number of kernels per row, PH=plant height, Rep=replication, SD=silking day, TKWT=thousand kernels weight.

Table 4: Estimates of general combining ability effects for eleven traits of 36 maize inbred lines used in a Line by tester study at Melkassa, Ethiopia during 2012 cropping season

Ear length

Line GCA mean squares were highly significant (P<0.01) while non-significant difference for SCA (Table 1). Similarly Dagne et al. [17] reported significant mean square due to GCA for ear length. Mandefro [27] reported no importance of non- additive gene action for ear length.

Eighteen inbred lines showed a positive GCA effect for ear length. Among which five inbred lines (L2, L12, L19, L20 and L32) had positive and significant (P<0.05) and one inbred line (L30) had positive and highly significant (P<0.01) GCA effect. These inbred lines had a tendency to increase ear length. Nineteen inbred lines showed a negative GCA effect for ear length, among which two inbred lines (L16 and L26) had negative and highly significant GCA effect. These two inbred lines found to be poor general combiners as they showed negative and significant GCA effect. Similarly, [17,20,31] Amiruzzaman et al., Jumbo and Carena and Dagne et al. reported positive and negative significant GCA effect for ear length. The positive GCA effect is desirable as to indicate the tendency to increase ear length, which directly contributes to increase grain yield maize.

Ear diameter

Line GCA means squares were significant difference in ear diameter (Table 1). Both additive and non-additive gene effects were important as reported by Dagne, Hadji and Gudeta [1,21,22]. Line GCA effects were ranged from -5.67 (L11) to 3.55 (L33). Twenty-one inbred lines showed positive GCA effects. Three inbred lines (L16, L25 and L33) showed significant and positive GCA effects (Table 4). These lines were the best general combiners for ear diameter as they had significant and positive GCA effects. L33 (3.55) had highest positive GCA effects. On the other hand, five inbred lines had highly significant and negative GCA effects were poor general combiners for this trait. The present study is in agreement with Amiruzzman et al., Jumbo and Carena, Dagne et al. and Gudeta [17,20,22,31] who reported significant positive and negative GCA effects for ear diameter.

Anthesis silking interval

Anthesis silking interval, lines GCA were significant difference (P<0.05). Line GCA effects ranged from -0.94 (L14 and L23) to 1.31 (L27) (Table 4). Fourteen inbred lines expressed negative GCA effects in the desired direction, among these L6, L14 and L23 showed negative and significant (p<0.05) GCA effects. On the other hand, L12 and L27 expressed positive GCA effects and highly significant in the undesired direction. Similar to this result Mwambula [32] reported that anthesis-silking interval was one of the most useful secondary traits for selecting for better yields under drought stress condition.

Thousand-kernel weight

For thousand kernels weight means squares due to line GCA were significant difference (P<0.05). But mean squares due to SCA were not significant 0(Table 1). Similarly, Gudeta [22] reported non-significant SCA mean square. In contrast, Dagne et al. [17] reported

that the importance of both additive and non-additive gene action for the trait. Line GCA effects of thousand-kernel weight was ranged from -49.3 g (L12) to 56.19 (L18) (Table 4). Sixteen lines expressed positive GCA effects for thousand-kernel weight, out of which L18 and L25 showed positive and highly significant (p<0.01) GCA effects in the desired direction. Alternatively, L3, L10 and L12 expressed a negative and significant difference of GCA effects in the undesired direction. The present result is in agreement with the findings of several researchers who reported significant positive and negative GCA effects for thousand-kernel weight. [17,18,20,22,33]

Number of kernel per rows

Numbers of kernels per row mean square due to SCA were not significant while significant difference due to GCA. This result is similar to the report of Dagne and Gudeta [1,22] on maize. Line GCA effects for number of kernels per rows were ranged from -2.98 (L17) to 2.66 (L18) (Table 8). A total of 20 inbred lines showed positive GCA effects for number of kernels per rows, only three inbred lines (L2, L18, L21 and L27) showed positive and significant (P<0.05) GCA effects for number of kernels per rows. Five inbred lines (L4, L7, and L15and L35) exhibited significant negative GCA effects for number of kernels per rows. Significantly positive and negative GCA effects were obtained for number of kernels per row. Inbred lines with significant difference and positive GCA effect, suggesting the presence of divergence to improve this trait. This result is in agreement with the finding of Dagne et al., Gudeta and Amiruzzaman et al.,[17,20,22] who reported both positive and negative significant GCA effects for number of kernels per row.

Plant height

Mean squares due to line GCA were highly significant difference (p<0.01), Whereas, Mean squares due to SCA was not significant (Table 1). Line GCA effects of plant height was ranged from -23.36 (L7) to 34.88 (L9). Seventeen inbred lines expressed positively and the remaining 19 expressed negative GCA effects. Among these, L8, L9, L11, L12, L13, L22 and L23 expressed positive and highly significant (p<0.01) GCA effects for increased plant height while six of the inbred lines L7, L17, L27, L28, L31 and L36 had negative and highly significant (p<0.01) GCA effects for reduced plant height (Table 4). L7 was the most general combiner for reduced plant height; whereas L9 and L11 were the best general combiners for increased plant height. Significantly positive and negative GCA effects were obtained for plant height, suggesting the presence of divergence to improve this trait. Teshale, Dagne and Hadji [1,18,21] recorded significant positive and negative GCA effects for plant height.

Ear height

Mean squares due to line GCA were highly significant (p<0.01) for and plant ear height, whereas, mean squares due to SCA was not significant (Table 1). Line GCA effects for ear height ranged from 12.02 (L4 and L10) to 22.22 cm (L16) (Table 4).Among a total of 18 of the inbred lines that showed the positive GCA effect for ear height, only six inbred lines (L9, L13, L14, L16, L22 and L31) showed positive and highly significant (p<0.01) GCA effects. On the other hand, six (L4, L7, L10, L11, L29 and L32) showed negative and highly significant (p<0.01) GCA effects for ear height. Dagne [1] Recorded significant positive and negative GCA effects for ear height. Further, Gudeta [22] found significant positive and negative GCA effects for ear height.

Number of ear per plant

For number ear per plant (EPP), mean squares due to line GCA were highly significant (p<0.01) while mean squares due to SCA were not significant (Table 1). Seventeen inbred lines showed positive GCA effects among which 5 inbred lines (L1, L3, L9, L11 and L16) had positive and highly significant (p<0.01) GCA effects. On the other hand, five inbred lines (L17, L13, L18, L21, L29 and L36) showed negative and significant (p<0.05). Similarly, Manda and Mwambula [32] reported that ears per plant were the most useful secondary traits for selecting for better yields under drought stress conditions. Additionally, Dagne et al. [17] who reported significant positive and negative GCA effect for number of ears per plant.

Conclusion and recommendation

The present study consisted of 72 crosses and 3 checks which were evaluated at the Melkassa Agricultural Research Centre; Ethiopia with the objectives of observing of mean crosses performance and estimating combining abilities maize elites for 11 characters. The analysis of variance revealed that the crosses were significantly different from most of the characters. Further, significant differences were also recorded among the checks and checks vs crosses for anthesis day, Plant height and ear length. Results of Line Tester analysis showed that the lines GCA mean squares were significant for all studied traits. The Testers GCA mean squares were significant for all traits except ear height, number of ears per plant and kernels per row. SCA mean squares were non-significant for all traits. Significant lines in terms of GCA mean squares for all traits indicated the predominant role of additive gene actions in determining the inheritance of these traits. In this study, the GCA sum of squares component was greater than the SCA sum of squares for the studied traits, suggesting that variations among crosses were mainly due to additive rather than non-additive gene effects; and hence, the selection would be effective in improving grain yield and other agronomic traits.

Based on combining ability analysis L15, L25, L27, L33, L34 and L35 were the top general combiners for grain yield and these inbred lines can be used for variety development in the future lowland maize improvement program. Inbred lines L2, L3, L4, L6, L10, L27, L30, L31 and L32 were the best general combiners for days to anthesis, indicating these lines had a favorable allele frequency for earliness and can be used to develop early maturing varieties. Inbred lines L2, L4, L7, L15, L17, L27, L28, L29, L31, L32 and L36 were best general combiners for shorter plant height, which are desirable for lodging resistance. Inbred lines L2, L18, L21 and L27 were the best general combiners for kernels per row. These lines had favorable allele to improve the number of seeds per cob.

For ear diameter L16, L25and L33 lines were good general combiners, indicating these lines had the tendency to increase ear diameter. For ear length L2, L12, L19, L20 and L30 lines were good general combiner indicating these lines had the tendency to increase ear length. For thousand-kernel weight L18, L25, L29, L33 and L34 were the top general combiners as such line had the tendency to increase thousand kernel weights. Among the crosses, L34xT2 (6.27 t/ha),

L36xT2 (5.99t/ha) and L30xT1 (5.96 t/ha) crosses were highest grain yield (t/ha). These hybrids could be included in further investigation for grain yield and related traits and could be possible candidates of future releases.

From these finding better performing test crosses, inbred lines with desirable GCA effects for grain yield and other grain yield related traits were successfully identified. These germplasm constitute a source of valuable genetic material that could be successively used for future breeding work. In general, the results of this study could be useful for researchers who need to develop high yielding varieties of maize, particularly adapted to the rift valley areas of Ethiopia as well as sub-Saharan Africa. However, the present study was conducted at one location and the result is only an indication and we cannot reach at a definite conclusion. Therefore, it is advisable to continue with this study over many years and locations.

References

1. Dagne Wegary (2002) Combining ability analysis for traits of agronomic importance in Maize (Zea mays L.) inbred lines with different levels of resistance to grey leaf spot (Cercospora Zea maydis) M.Sc. Thesis. School of Graduate studies, Alemaya University, Ethiopia

2. Piperno DR and Flannery KV (2001)"The earliest archaeological maize (Zea mays L.) From highland Mexico: New accelerator mass spectrometry dates and their implications." PNAS 98: 2101-2103

3. Aldrich SR, Scott WO (1986) Modern Corn Production. A & L Publications

4. CIMMYT. 2001. 2000/2001 world maize facts and trends. Mexico D.F.

5. Vasal SK (1999) Improving Human Nutrition through Agriculture: The Role of International Agricultural Research. Quality Protein Maize Story, CIMMYT

6. Dawswell, CR, Paliwal RL, RP Cantrell (1996) Maize in the third world Westview press, Colorado, USA

7. CSA (Central Statically Agency) Federal Democratic Republic of Ethiopia 2011. Agricultural sample survey (2009/2010) Report on area and production of crops Volume I. Addis Ababa, Ethiopia

8. CIMMYT 1(993/1994) World maize facts and trends. Mexico DF.

9. EARO/CIMMYT (2002) Proceedings of the Second National Maize Workshop of Ethiopia. EARO/CIMMYT, Addis Ababa, Ethiopia.

10. FAO (2012) World cereals production statistics

11. Sprague GF, Tatum LA (1942) General versus specific combining ability in single crosses of corn. J Amer Soc Agron 34: 923-932

12. Kempthorne (1957) An introduction to genetic statistics John Wiley and Sons Inc., New York 453-471

13. Piovarci WS (1973) An analysis of the combining ability of inbred lines of maize (Zea mays L.) in a diallel cross system. Plant Breeding Abstract 43: 6812.

14. Seboksa G, Nigussie M, Bongale G (2001) Stability of drought tolerant maize genotypes in the drought stressed areas of Ethiopia. In: DK Friesen and AFE Palmer (eds.), Integrated Approaches to Higher Maize Productivity in the New Millennium: Proceedings of the Seventh Eastern and Southern Africa Regional Maize Conference, 5-11 February 2002. Nairobi Kenya. pp. 301-304.

15. Patterson HD, Williams ER (1976) A new class of resolvable incomplete block designs. Biometrika 63: 83-92

16. Singh RK, Chaudhary BD (1999) Biometrical methods in quantitative genetic analysis Kalyani Pub Ludhina, New Delhi, Revised Ed 92-101.

17. Dagne Wegary, Zeleke H, Labuscagne MT, Hussien T and Singn H, (2007) Heterosis and combining ability for grain yield and its component in selected maize inbred line. S Afr J Plant Soil 24: 133-137

18. Teshale Asefa (2001) Analysis of tropical High land Maize (Zea May L) inbred lines top crossed with three east African population M.Sc. Thesis school of Graduate studies, Alemaya University, Ethiopia.

19. Abdurrahman J (1999) Heterosis and combining for yield and related trait in maize M.Sc. Thesis School of Graduate studies, Alemaya University

20. Amiruzzaman M, Islam MA, Hassan L and Rohman MM (2010) Combining and heterosis for yield and component character in maize. Academic journal of plant Sciences 3: 79-84

21. Hadji Tuna, (2004) Combining Ability Analysis for yield and yield related traits in quality Protein Maize (QPM) inbred lines M.Sc. Thesis School of Graduate studies, Alemaya University.

22. Gudeta Nesir, (2007) Heterosis and combining ability in QPM version of early generation high land maize (Zea mays L) inbred line M.Sc. Thesis School of Graduate studies, Alemaya University

23. Dagne Wegary, Vivek BS, Birhnau Tadesse, Koste Abdissa, Mosisa Worku, et.al. (2010) Combining ability and heterotic relationship between CIMMYT and Ethiopia inbred line Ethiopia J Agric Sci 20: 82-93

24. Pswararyi A, Vivek BS (2008) Combining ability amongst CIMMYT's early maturing maize (Zea mays L.) germplasm under stress and non-stress conditions and identification of testers. Eupytica (in print).

25. Piovarci WS (1973) An analysis of the combining ability of inbred lines of maize (Zea mays L.) in a diallel crosses system. Plant Breeding Abstract, 43: 6812.

26. Strube HG (1967) Correlation between characters in hybrid maize and their importance in selection. Dissertation Landw Hochsch. Hohenheim, 61.

27. Mandefro Nigusie, 1999. Heterosis, Combining ability and correlation in 8 x 8 diallel crosses of drought tolerant maize (Zea mays L) population. M.Sc. Thesis. School of Graduate studies, Alemaya University, Ethiopia.

28. Leta Tulu, Legesse Wolde and Tasew Gobezayew, 1999. Combining ability of some traits in seven- parent diallel crosses of selected maize (Zea mays L.) Populations. pp.78-80. In: Maize production Technology for future: Challenges and opportunities, proceeding of the 6th eastern and southern Africa regional maize conference, Addis Ababa, Ethiopia, 21-25 Sept.1998.CIMMYT and EARO.

29. Shewangizaw Abebe. 1983. Heterosis and combining ability in 7x7 diallel cross of selected inbred lines of maize (Zea mays L.) M.Sc. Thesis, Addis Ababa University, Ethiopia.

30. 30.   Ahmad SA, El-Shouny KA, Olfat HE, Ibrahim KIM (2003). Heterosis and combining ability in yellow maize (Zea mays L.) crosses under two planting dates Ann Agric Sci Cairo,   49(2): 531-543

31. Jumbo MB, Carena MJ (2008) Combining ability maternal and reciprocal effect of elite early-maturing maize population hybrid .Eupytica 162: 325-333

32. Manda THE, Mwambula C (1999) Screening and selection of tropical maize genotypes for drought tolerance using primary and secondary traits. Maize production technology for the future: challenges and opportunities, Proceedings of the Eastern and Southern Africa Regional Maize Conference, 21-

33. Demissew Abakemal, Habtamu Zelleke, Kanuajia KR, Dagne Wegary, (2011) Combining ability in maize lines for agronomic trait and resistance to Weevil. Ethiopia J Agr Sci 2: 37-48

34

Virulence Spectrum of *Puccinia hordei* of Barley in Western and Central Highlands of Ethiopia

Getaneh Woldeab[1], Endale Hailu[1]*, Teklay Ababa[2] and Teklu Negash[1]

[1]*Ethiopian Institute of Agricultural Research, Ambo Plant Protection Research Center, P.O. Box 37, Ambo, West Shewa, Ethiopia*
[2]*Tigray Agricultural Research Institute, Alamata Agricultural Research center, P.O. Box 56, Alamata, Ethiopia*

Abstract

Virulence surveys of Puccinia hordei of barley were conducted in the main and off crop seasons of 2010/11 and 2011/12 in West Shewa, Wellega (western part of the country) and Arsi (central part of the country) zones of Oromiya region, Ethiopia to determine the virulence spectrum of the pathogen, and identify the effective resistance genes to the pathotypes. In the two crop seasons, 56 leaf (brown) rust samples in the main and 32 in the off-season were collected. From each barley field, single pustule descent spores were multiplied and inoculated onto the seedlings of 12 leaf rust differentials carrying Rph1 - Rph12 genes to designate the pathotypes. A total of 88 leaf rust isolates were processed and based on infection phenotype on the resistance genes, 7 pathotypes (ETPh6631, ETPh6611, ETPh6671, ETPh7671, ETPh7631, ETPh7611 and ETPh7651) were identified. The most frequently isolated pathotype was ETPh6631 with 43.2% followed by ETPh6611 with 19%. Moreover, virulence spectrum of *P. hordei* pathotypes identified in this study was diverse. Resistance genes with Rph1 (Sudan), Rph4 (Gold), Rph8 (Egypt4), Rph9 (Hor 2596), Rph11 (Clipper BC68) and Rph12 (Triumph) were non-effective to all pathotypes identified whereas genes Rph5 (Magnif), Rph6 (Bolivia) and Rph10 (Clipper BC 8) were effective to 26.1, 73.9 and 78.4% of the isolates, respectively. Virulence against Rph2 (Peruvian), Rph3 (Estate) and Rph7 (Cebada Capa) was absent. Therefore, the effective major genes to the existing leaf rust populations could be utilized as sources of resistance in the barley breeding program.

Keywords: Barley; *Hordeum vulgar;* Leaf/brown rust; *Pathotypes*; *Puccinia hordei*; Pathotype specific resistance

Introduction

Barley (*Hordeum vulgarae*) is one of the most important staple food crops in the highlands (2,000-3,000 meters above sea level) and ranks fifth in area and production among the cereals in Ethiopia. It was cultivated on 1.02 million hectares and total annual grain production in 2012/13 cropping season was about 1.78 million tons [1]. Barley is produced mainly in Shewa, Gojam, Bale, Arsi, Gondar, Wello, Wellega and Tigray zones [2]. It is grown in main rainy, residual and belg/off-seasons, the largest production being in the main rainy season (June– October). Productivity of barley in Ethiopia is low (only about 1.8 ton ha⁻¹) as compared to some major barley producing countries. Diseases are one of the main biotic constraints to barley production [2,3].

Barely leaf rust caused by a fungus *Puccinia hordei* is prevalent on cultivated and indigenous wild barleys. It is particularly important in areas where the crop matures late, occurring extensively in barley areas of the Eastern, Mid-Western United States and in North Africa, Europe, New Zealand, Australia, some part of Asia and in the Andes region of south America [4,5]. The disease was first reported in Ethiopia by Stewart and Dagnatchew [6]. Since then, it has become widespread in all barley growing regions of Ethiopia [7]. It is favoured by a relatively warm and moist climate [8]. Barley leaf rust causes serious yield losses in the countries of North Africa and in Pakistan [9]. Losses due to this disease reached 23% on a susceptible variety Trompllo at Ambo Plant Protection Research Centre experimental field [10] and 28% on white seeded barley variety on farmer's fields at Tikur Inchini and Shenen districts of West Shewa zone [11].

The fungus is a heterocious pathogen with Hordeum species as primary hosts and *Orinthgalem umbellatum* as alternate host [12]. It completes its life cycle on alternate host and other species. The alternate host also supports genetic-recombination of the leaf rust fungus resulting in the evolution of broad spectra of the pathogen variability [13]. At present, many of the leaf rust resistance genes derived from barley have limited value for plant breeding because *P. hordei* pathotypes with virulence on them have evolved [14]. More than 52 physiological pathotypes of *P. hordei* have been reported to infect barley in the world [15]. Earlier investigations of barley leaf rust isolates indicated the presence of pathotypes 77 and 184 in a few locations of Ethiopia [3]. In a recent study, seven pathotypes were identified from 381 isolates collected from the barley growing highlands of Gojam, Shewa and Bale zones [16]. However, the study made in 2006 did not cover other major barley growing areas like west Shewa, Wellega and Arsi zones. Therefore, these surveys were conducted to determine the virulence spectrum of *P. hordei* pathotypes occurring in three zones of Oromiya region, Ethiopia.

Materials and Methods

Barley leaf rust surveys were conducted in 2010/11 and 2011/12 cropping seasons. The major barley growing areas of Toke Kutaye district in west Shewa; Jima Arjo, Abay Chomen, Jima Geneti and Diga Leka districts in Wellega (western part of Ethiopia) and Robe, Sude, Hitosa and Meraro districts in Arsi zone (central part of Ethiopia) were surveyed in the main (September – October) seasons. Barley fields

Corresponding author: Endale Hailu, Ethiopian Institute of Agricultural Research, Ambo Plant Protection Research Center, P.O. Box 37, Ambo, West Shewa, Ethiopia, E-mail: sebhailuabera@yahoo.com

were also assessed for leaf rust infection in the off season (December – January) in Wellega zone. During the surveys, rust infected leaf samples were collected from the barley fields and the rust spores from each field were inoculated with an atomizer on the seedlings of susceptible cultivar L94 in the greenhouse at Ambo Plant Protection Research Centre to get isolated pustules. Each pustule/isolate extracted from the leaf rust population to represent a field was multiplied until sufficient spores were collected by tapping the rusted leaves in the test tube/ watch glass and this was inoculated on the differential hosts.

Cereal introduction/plant introduction accession number assigned by United States Department of Agriculture (USDA).

Twelve barley leaf rust differential hosts indicated on Table 1 and a universally susceptible check L94 were used to analyse *Puccinia hordei* populations collected from the three zones. Barley seeds were grown in greenhouse at temperature of 20-24°C in suitable 10 cm diameter clay pots filled with the ratio of 2:1:1 soil: sand: farm yard manure. Inoculation was carried out when the seedlings were between 5 and 10 cm long or at two leaf stages when the first leaf was fully extended. Before inoculation, the leaf surfaces were gently rubbed with moistened fingers to remove the outer waxy coating of the leaves and then the seedlings were sprayed/misted with distilled water.

Urediniospores of each isolate were suspended in distilled water with two drops of tween 20 as spreading and adhering agent. The suspension was sprayed onto seedlings of the differentials using atomizers. After that, the plants were moistened with distilled water and placed in an incubation chamber for 16-20 hr dark period at 20°C. The seedlings were then taken out onto the growth room where florescent lighting was provided for 16 hours at 20°C and RH of 60 - 70%. Within 7 to 10 days after inoculation, data on the infection types were recorded.

The infection types of barley leaf rust were read for each differential host using Levine and Cherewick [17] in 0-4 scoring scale. The infection types 0, 0; 1 and 2+ were taken as incompatible/ resistance (low) and 3- and 4++ were taken as compatible/susceptible (high) reactions. Octal race designation system was used to name the *pathotypes*. In this system, *pathotypes* are determined by adding values that correspond to each differential to which the isolate was virulent [18].

Differential hosts	CI/PI number	Resistance gene	Octal value
Sudan	CI6489*	Rph1	1
Peruvian	CI935	Rph2	2
Estate	CI34102	Rph3	4
Gold	CI1145	Rph4	10
Magnif	CI13806	Rph5	20
Bolivia	CI1257	Rph6 +Rph2	40
Cebada Capa	CI6193	Rph7	100
Egypt 4	CI6481	Rph8	200
Hor 2596	CI1243	Rph9	400
Clipper BC8	-	Rph10	1000
Clipper BC67	-	Rph11	2000
Triumph	PI2668 180	Rph12	4000

Table 1: *Puccinia hordei* differential hosts with CI/PI number, resistance genes and octal values.

For example: ETPh6611 is virulent on *Rph1+Rph4+Rph8+Rph9+ Rph11+Rph12= 1+10+200+400+2000+4000=6611* and ETPh means Ethiopian *Puccinia hordei*.

Results and Discussion

Pathotype and virulence surveys of *Puccinia hordei* were carried out in one district of West Shewa; four districts of Wellega and four districts of Arsi zones during the main season of 2010/11 and 2011/12. In the first main season, 40 isolates and in the second, 16 isolates collected from the three zones were processed and seven pathotypes were identified (Table 2). In the two off-seasons in the four districts of Wellega zone, 32 isolates were collected and analyzed and four pathotypes (ETPh6611, ETPh6631, ETPh6671 and ETPh7671) were isolated. From the whole of 88 leaf rust samples, seven pathotypes (ETPh6631, ETPh6611, ETPh6671, ETPh7671, ETPh7631, ETPh7611 and ETPh7651) were identified. The numbers of pathotypes isolated were 5 5, and 6 from Wellega, West Shewa and Arsi zones, respectively. In the off-season, of the seven, pathotypes ETPh7631 and ETPh7651 were absent. In the main and off seasons in Wellega, four pathotypes each were identified, of which three of them were identical while the other two were different. The off season barley comes right after the main season barley matures, so late maturing varieties as well as volunteer barleys can carry the rust disease from the main to the off season barley. The similarity of the pathotypes might be due to the movement of

Year	Season	Zone	No. of isolates tested	No. of pathotypes identified	Pathotypes
2010/11	1*	West Shewa	13	3	ETPh6631, ETPh7631, ETPh6671
	1	Arsi	21	6	ETPh6611, ETPh6631, ETPh6671 ETPh7611, ETPh7631, ETPh7671
	1	Wellega	6	3	ETPh6611, ETPh6631, ETPh7631
	Sub		40	6	
2011/12	1	West Shewa	3	2	ETPh7611, ETPh7651
	1	Arsi	8	4	ETPh6611, ETPh6631, ETPh6671, ETPh7671
	1	Wellega	5	2	ETPh6611, ETPh7611
	Sub		16	6	
	Total		56	7	
2010/11	2	Wellega	26	3	ETPh6631, ETPh6671, ETPh7671
2011/12	2	Wellega	6	1	ETPh6611
	Sub total		32	4	
	Grand total (1+2)		88	7	ETPh6611, ETPh6631, ETPh6671, ETPh7611, ETPh7631, ETPh7651, ETPh7671

*1 is main season, 2 off season

Table 2: *Puccinia hordei* pathotypes identified from barley growing areas of West Shewa, Wellega and Arsi zones, 2010/11-2011/12 seasons.

spores from one to the other season. The number of pathotypes in Arsi zone was higher than the races in West Shewa and Wellega. This could probably be because in Arsi zone the diversity of local and improved barley varieties grown is high. Moreover, the pathotypes identified in the two different seasons were similar indicating that carry over of the pathogen from one to the other season occurs in the areas. Besides, many of the pathotypes found during this study were similar with those identified from Bale, and north west Shewa zone of Oromiya region as well as from Gojam and north east Shewa zones of Amhara region in 2003 and 2004 [16]. Thus, earlier and this studies demonstrate the heterogeneity of the leaf rust pathogen in Ethiopia.

Virulence spectrum and frequency of *Puccina hordei* pathotypes in the three zones and in the main and off-seasons of 2010/11 – 2011/12 are shown in Table 3. Thirty-eight (43.2%) of the 88 isolates were pathotype ETPh6631. It was distributed in all three zones in both seasons and years. In the first main season, pathotype ETPh6631 was isolated from all three zones, while in the second, it was observed in Arsi zone. In the off season of 2010/2011 pathotype ETPh6631 was recorded, but not in the second season. This pathotype was not isolated in the virulence study of *P. hordei* in 2003 and 2004. This pathotype was identified for the first time in Ethiopia, however, theses zones were not covered in the previous study. ETPh6611 was the third frequently isolated pathotype in the previous study. Pathotype ETPh6671 was not isolated in earlier studies. Except ETPh6631 and ETPh6671, the rest of the pathotypes were isolated in Ethiopia before. The second commonly found pathotype was ETPh6611 with a frequency of 17 isolates (19.3%), followed by pathotype ETPh6671 with 14 isolates (15.9%). ETPh6611 was isolated from Wellega and Arsi, while ETPh6671 was from west Shewa and Arsi zones. The rest of the pathotypes had distribution below 10%, with a minimum frequency of 2.3% for pathotype ETPh7651. This pathotype was identified twice in the main season of 2011/12 in west Shewa zone. Previously, Woldeab et al. reported that pathotype ETPh6611 and ETPh7651 were the third and the fourth frequently isolated pathotypes, respectively in 2003 and 2004 cropping seasons [16].

Virulence spectrum of *Puccinia hordei* pathotypes identified from the three zones was diverse (Table 3). The least virulence was observed for pathotype ETPh6611, while the highest was recorded for pathotype ETPh7671. Six and nine out of the 12 *Puccinia hordei* resistant genes were defeated by these pathotypes, respectively. The second highest virulence spectrum was recorded for pathotypes ETPh6671, ETPh7631 and ETPh7651 with virulence to eight resistant genes.

Among the resistant genes, *Rph1* (Sudan), *Rph4* (Gold), *Rph8* (Egypt4), *Rph9* (Hor 2596), *Rph11* (Clipper BC68) and *Rph12* (Triumph) were non-effective to the 88 isolates tested. *Rph5* (Magnif), *Rph6* (Bolivia) and *Rph10* (Clipper BC8) were also ineffective to 73.9,

26.1 and 21.6% of the isolates, while *Rph2* (Peruvian), *Rph3* (Estate) and *Rph7* (Cebada Capa) genes were effective/resistant to all isolates identified (Table 3).

In earlier studies, *Rph3* (Estate) and *Rph7* (Cebada Capa) genes were also effective to 381 leaf rust isolates collected from different regions and production systems of Ethiopia [16]. Before 2006, similar results were reported by Alemayehu from Ethiopia and Park from Australia [7,19]. However, there is a report of virulence to *Rph3* from Europe [20] pathotypes virulent for *Rph7* have been reported in Israel [21], but not in Europe and Australia [14]. Furthermore, gene *Rph9* has been deployed in commercial cultivars worldwide, [22] but this gene has been defeated by the pathogen in many locations, including North Africa, the Middle East, and Australia [19,23]. At present, most of the Ethiopian barley landraces are susceptible to leaf rust, and one such landrace, Abyssinian, known to contain *Rph9*, is very susceptible.

The present study demonstrated pathogenic heterogeneity of the barley leaf rust populations in West Shewa, Wellega and Arsi zones. If barley with known major gene resistance were to be deployed, then the use of *Rph3* and *Rph7*, and perhaps *Rph2*, would be expected to give the best protection against leaf rust in Ethiopia.

References

1. CSA, Central Statistical Authority (2013) Agricultural sample survey 2012/2013, Report on area and production of crops meher season. Statistical Bulletin No 532, 1:15 CSA, Addis Ababa.

2. Lakew B, Gebre H, Alemayehu F (1996) Barley production and research. In: Gebre H, van Leur J (eds) Barley research in Ethiopia: past work and future prospect. IAR, Addis Ababa pp. 1-8.

3. Yitbarek S, Bekele H, Getaneh W, Dereje T (1996) Disease survey and loss assessment studies on barley. In: Hailu Gebre and Joop van Leur (eds) Barley research in Ethiopia, past work and future prospects. pp. 105-115, Addis Ababa, Ethiopia.

4. Dickson JC (1956) Diseases of field crops. Mc-Crow-Hill Publishing Co.Ltd, Bombay-New Delhi, India.

5. Mathre DE (1987) Compendium of barley diseases. American Phytopathological Society, pp. 78.

6. Stewart RB, Yirgu D (1967) Index of plant diseases in Ethiopia. College of Agriculture Haile Selase I. University Experimntal Station Bulletin No, 30.

7. Alemayehu F (1995) Genetic variation between and within Ethiopian barley landraces with emphasis on durable disease resistance. The Netherlands: Wageningen Agricultural University. PhD Thesis.

8. Bokelman HE, Sharp EI, Sands DC, Schoren AL, Mathre DE (1981) Field manual of common barley diseases including a section on Breeding disease resistant barley. Department of Plant Pathology, Montana State University, Bozeman, Montana 59717, USA. pp. 3-56.

9. Stubbs RW, Prescott JM, Saari EE, Dubin HJ (1986) Cereal disease methodology manual. CIMMYT, Mexico. pp. 45-46.

10. Getaneh W (1998) Yield loss due to leaf rust on barley at different sowing dates. Pest Management Journal of Ethiopia 1: 79-84.

11. Getaneh Woldeab, Fekadu Alemayehu (2001) On-farm yield loss due to leaf rust (Puccinia hordei Otth) on barley. Pest Management Journal of Ethiopia. 5: 29-35.

12. Clifford BC (1985) Barley leaf rust. In: Roelfs AP, Rashnell WH (eds), The cereal rusts II Disease, distribution, epidemiology and control. Orlando Press, Baton Rouge, pp. 173-305.

13. Anikster Y, Abraham C, Greenberger Y, Wahl I (1971) A contribution to the taxonomy of Puccinia brown leaf rust of barley in Israel. Israel journal of Botany 20: 1-12.

14. Dreiseitl A, Steffenson BJ (2000) Postulation of leaf rust resistance genes in Czech and Slovak barley cultivars and breeding lines. Plant Breeding 119: 211-4.

Races	Virulence/avirulence of the pathotypes	Frequency (No. & percent)
ETPh6631	*Rph*1, 4, 5, 8, 9, 11, 12 / 2, 3, 6, 7, 10	38 (43%)
ETPh6611	*Rph*1, 4, 8, 9, 11, 12 / 2, 3, 5, 6, 7, 10	17 (19.3%)
ETPh6671	*Rph*1, 4, 5, 6, 8, 9, 11,12 / 2, 3, 7, 10	14 (15.9%)
ETPh7671	*Rph*1, 4, 5, 6, 8, 9, 10, 11, 12 / 2, 3, 7	7 (7.9%)
ETPh7631	*Rph*1, 4, 5, 8, 9, 10, 11,12 / 2, 3, 6,7	6 (6.8%)
ETPh7611	*Rph*1, 4, 8, 9, 10, 11, 12 / 2, 3, 5, 6, 7	4 (4.5%)
ETPh7651	*Rph*1, 4, 6, 8, 9,10, 11, 12 / 2, 3, 5, 7	2 (2.3%)

Table 3: Virulence and avirulence formula and the frequency of the seven pathotypes identified on the 12 *Rph* genes from west Shewa, Wellega and Arsi zones, 2010/11 – 2011/12 cropping seasons.

15. Fadeev UM (1977) Metodicheskie recommendasi po izucheniyu racavova sostava vozbuditelei rrzavchini hlebnih zlakov, Moskva pp. 132-140.

16. Woldeab G, Fininsa C, Singh H, Yuen J (2006) Virulence spectrum of Puccinia hordei in barley production systems in Ethiopia. Plant Pathology 55: 351-357.

17. Levine MN, Cherewick WJ (1952) Studies on Dwarf Leaf Rust of Barley. Department of Agriculture. (USDA Technical Bulletin 1056), Washington: United States.

18. Gilmour J (1973) Octal notation for designating physiological races of plant pathogens. Nature 242: 260.

19. Park RF (2003) Pathogenic specialization and pathotype distribution of Puccinia hordei in Australia, 1992-2001. Plant Disease 87:1311-6.

20. Niks RE, Walther U, Jaiser H, Martinez F, Rubiales De, et al. (2000) Resistance against barley leaf rust (Puccinia hordei) in West European spring barley germplasm. Agronomie 20: 769-82.

21. Brodny U, Rivadeneira M (1996) Physiologic specialization of Puccinia hordei in Israel and Ecuador: 1992–94. Canadian Journal of Plant Pathology 18: 375-8.

22. Brooks WS, Griffey CA, Steffenson BJ, Vivar HE (2000) Genes Governing Resistance to Puccinia hordei in Thirteen Spring Barley Accessions. Phytopathology 90: 1131-1136.

23. Yahyaoui AH, Sharp EL (1987) Virulence spectrum of Puccinia hordei in North Africa and the Middle East. Plant Disease 71: 597-8.

35

Yield and Macronutrient Accumulation in Grain of Spring Wheat (*Triticum aestivum* ssp. *vulgare* L.) as Affected by Biostimulant Application

Szczepanek M* and Grzybowski K

Department of Agrotechnology, UTP University of Science and Technology

Abstract

Plant growth and development, resistance to stress, nutrient uptake and yield can be supported by the use of biostimulants. The aim of this study was to determine the influence of the seaweed biostimulant Kelpak and its combined application with the preparation Lithovit on yield components, grain yield and accumulation of macroelements in spring wheat grain. The study was based on field experiments located in Poland, (53°13'N; 17°51'E), conducted in 2010-2011 on a typical Alfisol (USDA). In the one-factorial field experiment with the spring wheat (*Triticum aestivum* ssp. *vulgare*) cultivar Katoda, we compared the effects of the application of the biostimulant Kelpak (seaweed extract) (T1) alone at a dose of 2 l ha^{-1}, to a mixture of two preparations, Kelpak 1.5 l ha^{-1}+Lithovit (finely milled limestone) 1.5 kg ha^{-1} (T2) with the control (without preparations) (T3). The study showed that foliar application of the biostimulant Kelpak at shooting increased spring wheat yield, but there was no additional increase in yield after its combined application with Lithovit. Application of the biostimulant Kelpak increased the number of generative tillers, and the combined application of prepared Kelpak and Lithovit had a favourable effect on the number of grains per ear. Application of biostimulant Kelpak and its mixture with Lithovit caused increased accumulation of N, P and K in grain.

Keywords: Grain yield; Straw weight; Harvest index; Macronutrient content; Nutrient accumulation

Introduction

Biostimulants are organic substances which, even when used in small amounts, stimulate growth and development of plants, and this response cannot be attributed to traditional nutrients [1]. Favourable effects of the application of extracts from algae indicate the benefit of their use in cultivation of horticultural and agricultural crops [2-4]. Seaweed biostimulants contain many active substances, including growth hormones: auxins, cytokinins and also polyamines and brassinosteroids [5,6]. Auxins have an effect on root formation, cytokinins on stem elongation, and polyamines on growth and development [7]. Due to the presence of phytohormones, marine algae-based seaweed extracts can increase plant biomass and reproductive yields [8,9]. In marine algae preparations, the presence of alginate, fucans, laminaran were also found, which are necessary in defensive activity of plants against diseases and pests [3,4,10]. Active substances of algae extracts applied in field crops can also reduce the effect of drought stress [11] or P and K deficiency [12]. Algae biostimulants may be used alone or combined with synthetic preparations and growth regulators [1]. They are applied mostly on leaves and may be used several times during the growth period [4]. The effect of such preparations is dependent on the time of their application and the dose [13,14].

One of the seaweed biostimulants is Kelpak, which is obtained from marine alga (*Ecklonia maxima* Osbeck), belongs to the class of brown algae (*Phaeophyta*), and is harvested on the coast of Africa. Kelpak contains phytohomones: auxins and cytokinins (11 and 0.031 mg l^{-1}, respectively), alginians, amino acids, as well as small amounts of macro and microelements [5]. In studies of cereal crops, the application of seaweed biostimulants caused root and shoot growth [15-17]. Increased accumulation of macro- and micro-nutrients was also indicated [18-20]. In another group of biostimulants is Lithovit, which is a finely milled limestone (a particle size smaller than the size of stomata) consisting mainly of (Ca, Mg)-CO$_3$ as well as different micronutrients (Mn, Cu, Zn, Ni, Fe) relevant for plant physiology (Patent DE202006011165 U1). However, there are no

of recommendations for dose and time of biostimulator Kelpak and Lithovit application in spring wheat. According to Sharma et al. [1] further studies concerning the use of biostimulants in cereals crop should be focused on the evaluation of crop trial protocols to reduce the impact of biotic and abiotic stresses as well as assessing synergistic responses to mixtures including macroalgal extracts and other biostimulants.

The aim of this study was to determine the influence of the seaweed biostimulant Kelpak and its combined application with the preparation Lithovit on yield components, grain yield and accumulation of macroelements in spring wheat grain yield.

Materials and Methods

Experimental site

The study was based on field experiments located in Poland, (53°13'N; 17°51'E), conducted in 2010-2011 on a typical Alfisol (USDA). The topsoil at the experimental site was characterized by medium content of available potassium 95-150 mg kg^{-1} and phosphorus 190-210 mgkg^{-1} (both determined with Egner-Riehm method), very low content of magnesium <20.0 mg kg^{-1} (Schetschabel method), and was slightly acidic (pH in 1M KCL 5.7-6.1) (with the use of potentiometry). The content of total nitrogen 0.69-0.75 g kg^{-1} and organic carbon 7.55-7.80 g kg^{-1} in soil was relatively low.

---

*Corresponding author: Małgorzata Szczepanek, Department of Agrotechnology, UTP University of Science and Technology, Al. prof. S. Kaliskiego 7, 85-796 Bydgoszcz, Poland, E-mail: Malgorzata.Szczepanek@utp.edu.pl

Experimental design

This study was based on a strict, one-factorial field experiment with the spring wheat (*Triticum aestivum ssp. vulgare*) cultivar Katoda, in which the effects of the application of the biostimulant Kelpak (T1) alone in a dose of 2 l ha^{-1}, or a mixture of two preparations: Kelpak 1.5 l ha^{-1}+Lithovit 1.5 l ha^{-1} (T2), were compared with the control (with no preparations) (T3). Preparations were applied at the shooting stage (4-5 leaves), after dissolving in water (300 l ha^{-1}).

Crop management

The spring wheat was sown 02-04 of April at a density of 500 no m^2 (at amount of 230 kg ha^{-1}) on plots with an area of 12 m^2, in four replications. Presowing fertilization of 31 kg ha^{-1} P (superphosphate), 66 kg ha^{-1} K (potassium chloride) and 80 kg ha^{-1} N (ammonium nitrate) was applied. At the beginning of the shooting state, the other rate of N (ammonium nitrate) was applied at a rate of 60 kgha^{-1} N. For weed control, triasulfuron 118.6 gha^{-1}+dicamba 7.4 gha^{-1} was applied at BBCH 22-24. To protect against diseases, epoxiconazole 93 g ha^{-1}+fenpropimorph 300 g ha^{-1}, and metrafenone 112.5 g ha^{-1} were applied at BBCH 34-39 and fusilazole 125 g ha^{-1}+karbendazim 250 g ha^{-1} at BBCH 51-59. Pest control was performed as a single application at BBCH 59, using dimethoat 200 g ha^{-1}. Wheat harvest was performed over the first ten days of August with the plot combine harvester Wintersteiger.

Sampling, measurement and chemical analysis

The number of generative tillers in the area of 1 m^2 was calculated at the flowering stage (BBCH 75). The number of grains per ear was determined at the full maturity stage on 30 randomly selected ears from each plot. The grain yield was determined directly after harvest and straw yield 7 days after. Presented grain and straw yields were calculated per the set humidity 14%. The 1000-kernel weight was assessed one month after harvest based on 200 grains from each plot. Harvest index was calculated for each plot as dry matter of grain yield divided by the sum of dry matter of grain yield and straw yield. Wheat grain was ground prior to chemical analyses. Mineralization was obtained by wet burning of fragmented material with perhydrol and sulfuric acid. Analyses were made with the following methods: total nitrogen content with the Kjeldahl method, phosphorus content with the molybdenum-vanadium test, and potassium with flame photometry. N, P and K uptake was calculated for each plot as the product of grain dry mass yield and the contents of individual macroelements.

Statistical analysis

The obtained results were analysed statistically using the statistical program Analysis of Variance for orthogonal experiments by the University of Technology and Life Sciences in Bydgoszcz, Poland. The differences between values were verified with Tukey's test on the significance level P ≤ 0.05.

Results and Discussion

The weather conditions in the study region were not very favorable for spring wheat (Table 1). The average precipitation of 275.1 mm from March to August limits plant growth and yield. Thermal conditions are mostly favorable, but periodic heat waves in conditions of low total precipitation may enhance drought stress. The course of the weather conditions was different in the successive years of the study (Table 1). In 2010, from March to August, the total precipitation was higher by 18.3% than in 2011. In 2010, during the period from sowing (April)

to the flag leaf stage (May), rainfalls were higher than in 2011, but at ear formation and flowering stages (June), there was only 18.1 mm of rainfall. The amount of rainfall during kernel development and maturation (July-August) did not limit plant growth. The years 2010 and 2011 were slightly warmer than the long-term average. In 2010 it was cooler in the successive months of growth from April to June than in 2011. In contrast, July and August were cooler in 2011.

In the present study, the number of generative shoots was on average 492 pieces m^{-2} and was similar in both years of the study. In 2010 and on average from the two years of the study, the application of the biostimulant Kelpak caused an increase in the number of generative shoots as compared with the control (Table 2). Increased wheat tillering after the use of algae extracts was noted also in another study [14,20]. In the study by Kumar and Sahoo [15], a positive effect of soaking seeds in an extract from algae was proved both for the number of branches and for the tiller length.

In our study of 2011, the number of grains per ear was higher, and the thousand grain weight was almost two times lower, as compared to 2010 (Table 3). In 2011 as well as on average in the long-term period, an increase in the number of grain per ear was observed after the combined application of Kelpak and Lithovit. A favorable effect of algae extracts on the yield components of spring wheat was also reported by Beckett and van Staden [21]. In this study the preparation applied at the 1-4 leaf stage in conditions of K deficit (moderate and heavy) increased both the number of grains per ear and the weight of a single grain. Zodape et al. [14] also indicated an increase in the 1000-kernel weight of wheat grains, but they did not find an effect on the number of grains per ear. In the study by Matysiak and Adamczewski [22], the 1000-kernel weight of spring wheat grain was higher after the application of the biostimulant Kelpak in a dose of 1.5 l ha^{-1} as compared with the variant without the application.

The spring wheat grain yield of 3.5 t ha^{-1} obtained in the study (Table 3) was similar to that obtained in the study by Matysiak and Adamczewski [22]. In the present study, despite different weather conditions, no considerable differences in yield were recorded in the harvest years 2010 and 2011. The applied biostimulants had a significant effect (P ≤ 0.05) on grain yield quantity. In 2010 and 2011, after the application of the biostimulant Kelpak, the spring wheat grain was higher than in the control. After the combined application of the preparations Kelpak and Lithovit the yield was similar to that obtained in the treatment with Kelpak alone and to the control. On average of the two years of the study, the application of the seaweed extract alone as well as its combined application with the biostimulant Lithovit brought a significant increase (P ≤ 0.05) in grain yield as compared with the variant without the application. This increase was obtained due to the increase in the number of generative shoots after the application of biostimulant Kelpak alone or as a result of an increase in the number of grains per ear, after combined application of Kelpak and Lithovit (Table 2). Similarly, Zodape et al. [14] indicate an increase in the number of generative shoots and thousand seed weight and the wheat grain yield after the application of an algae preparation, whereas this response was dependent on the dose applied. Shah et al. [20], when applying an algae extract, obtained an increase in wheat yield by 2.5-10%, depending on the dose. Favorable effect of the biostimulant Kelpak was also shown in another study by Matysiak and Adamczewski [22] when the application of doses 1.5 or 2 l ha^{-1} at the shooting stage resulted in an increase in the wheat grain yield by 13% as compared with the control. In contrast, the study by Beckett and van Staden [21], Kelpak did not affect the weight of grains per ear, in

Month	Precipitation (mm)			Air temperature °C		
	2010	2011	1949-2011	2010	2011	1949-2011
March	28.6	11.7	24.5	2.4	2.2	1.9
April	33.8	13.5	27.4	7.8	10.5	7.4
May	92.6	38.4	43.2	11.5	13.5	12.7
June	18.1	100.8	53.7	16.7	17.7	16.3
July	107.4	132.5	73.1	21.6	17.5	18
August	150.7	67.7	53.2	18.4	17.7	17.5
Total/mean	431.2	364.6	275.1	13.1	13.2	12.3

Table 1: Weather conditions at experimental locality.

Year	Treatment			Mean	LSD
	Control	Kelpak 2 l ha^{-1}	Kelpak 1.5 l ha^{-1} +Lithovit 1.5 kg ha^{-1}		
Number of generative tillers [no m^{-2}]					
2010	471.0	529.0	464.7	488.2	54.3
2011	496.0	498.5	493.0	495.8	ns
Mean	483.5	513.8	478.8	492.0	30.0
Number of grain per ear [no]					
2010	19.1	19.3	19.8	19.4	ns
2011	35.0	35.3	37.5	35.9	2.16
Mean	27.1	27.3	28.7	27.7	1.05
1000-kernel weight [g]					
2010	50.2	50.4	49.5	50.0	ns
2011	26.3	26.1	26.6	26.3	ns
Mean	38.3	38.3	38.1	38.2	ns

Table 2: Effect of biostimulant application on number of generative tillers, number of grain per ear and 1000-kernel weight of spring wheat.

conditions of the optimal K, but it significantly (P ≤ 0.05) increased the yield in conditions of stress caused by K deficit. This increase resulted from increasing the number of grains per ear and the weight of a single grain. This study also indicated an increase in the leaf area and root weight after the application of Kelpak in conditions of K deficit. Also a significant (P ≤ 0.05) influence of Lithovit on foliar area growth and development was shown [23].

In the present study, the straw weight in 2010, characterized by high total precipitation at the shooting stage (May), was two times higher than in 2011. The application of biostimulant Kelpak in 2010 or its combined application with Lithovit in 2011 caused a significant reduction (P ≤ 0.05) in straw weight as compared with the control. Straw weight, on average in the two years of the study, was lower after the application of the seaweed extract alone in comparison with the variant with no preparations. Literature reports show an increase in the number of shoots per plant and their height after the application of seaweed extracts [15,20,24]. In our study of spring wheat the reduction of straw weight after the application of biostimulants resulting from different distribution of assimilates than in the control. An increased accumulation of assimilates after the application of biostimulant was noted only in the grain. Because of this, both in successive years of the study and on average from the years, the harvest index was higher in the control as compared with the treatments with biostimulant Kelpak as well as Kelpak+Lithovit.

Significant effect (P ≤ 0.05) of the studied preparations on the average content from two years of the study of N, P, K Mg in spring wheat grain was indicated (Table 4). Wheat grain contained less N and Mg after the application of Kelpak as compared with the control. The

Year	Treatment			Mean	LSD
	Control	Kelpak 2 l ha^{-1}	Kelpak 1.5 l ha^{-1} +Lithovit 1.5 kg ha^{-1}		
Grain yield [t ha^{-1}]					
2010	3.27	3.56	3.54	3.45	0.272
2011	3.46	3.65	3.62	3.57	0.171
Mean	3.36	3.60	3.58	3.51	0.210
Straw weight [t ha^{-1}]					
2010	6.53	5.88	6.29	6.23	0.472
2011	3.09	2.95	2.88	2.97	0.173
Mean	4.81	4.42	4.59	4.60	0.380
Harvest index [%]					
2010	33.4	37.7	36.1	35.7	1.85
2011	52.8	55.3	55.8	54.6	1.66
Mean	43.1	46.5	45.9	45.1	2.1

Table 3: Effect of biostimulant application on grain yield, weight of straw and harvest index of spring wheat.

Year	Treatment			Mean	LSD
	Control	Kelpak 2 l ha^{-1}	Kelpak 1.5 l ha^{-1} +Lithovit 1.5 kg ha^{-1}		
N					
2010	1.66	1.62	1.64	1.64	ns
2011	2.41	2.30	2.36	2.35	ns
Mean	2.03	1.96	2.00	1.99	0.047
P					
2010	0.299	0.296	0.306	0.300	ns
2011	0.371	0.369	0.369	0.370	ns
Mean	0.335	0.333	0.337	0.335	0.003
K					
2010	0.187	0.199	0.195	0.193	ns
2011	0.498	0.490	0.485	0.491	ns
Mean	0.342	0.345	0.340	0.342	0.003
Mg					
2010	0.136	0.130	0.129	0.131	ns
2011	0.183	0.183	0.186	0.184	ns
Mean	0.160	0.157	0.158	0.158	0.003

Table 4: Effect of biostimulant application on macronutrient content in spring wheat grain.

content of P after the application of seaweed extract was lower than after the combined application of Kelpak and Lithovit. Wheat grain contained more K when only biostimulant Kelpak was applied, as compared with its combined application with the preparation Lithovit, as well as in relation to the control. N, P and K uptake in wheat grain, on average in the two-year period of the study, was higher both after the application of the biostimulant Kelpak and its combined application with the Lithovit compared with the control (Table 5). For N and K such relationship was also shown in 2010. The use of biostimulants did not affect Mg uptake in wheat grain. Shah et al. [20] report an increase in K and N uptake in some variants of seaweed extract application and mostly the lack of effect on P uptake. Zodape et al. also indicated an increasing accumulation of N, P, K under the influence of the seaweed biostimulant, particularly at the highest dose. Similarly, in the study by Beckett and van Staden [18], liquid feed (hydroponic) improved the yield of nutrient stressed wheat allowing better nutrient uptake.

Base on the study we can concluded that foliar application of the biostimulant Kelpak (seaweed extract) at the shooting stage increased spring wheat yield, but there was no additional growth in yield after

Year	Treatment			Mean	LSD
	Control	Kelpak 2 l ha^{-1} BBCH 22	Kelpak 1.5 l ha^{-1} +Lithovit 1.5 kg ha^{-1}		
N					
2010	45.9	49.5	49.9	48.4	2.84
2011	71.6	72.1	73.5	72.4	ns
Mean	58.8	60.8	61.7	60.4	1.375
P					
2010	8.37	9.06	9.31	8.91	ns
2011	11.0	11.6	11.7	11.4	ns
Mean	9.7	10.31	10.5	10.2	0.595
K					
2010	5.26	6.08	5.94	5.76	0.661
2011	14.8	15.3	14.8	15.0	ns
Mean	10.0	10.7	10.4	10.4	0.320
Mg					
2010	3.82	3.96	3.92	3.90	ns
2011	5.43	5.72	5.80	5.65	ns
Mean	4.63	4.84	4.86	4.78	ns

Table 5: Effect of biostimulant application on accumulation of macronutrient in spring wheat grain.

its combined application with Lithovit (finely milled limestone). Application of the biostimulant Kelpak increased the number of generative shoots and combined use of prepared Kelpak and Lithovit had a favorable effect on the number of grains per spring wheat spike. Application of the biostimulant Kelpak as well as its mixture with the prepared Lithovit caused increased accumulation of N, P, and K in wheat grain.

References

1. Sharma SH, Fleming C, Selby Ch, Rao JR, Trevor M (2014) Plant biostimulants: a review on the processing of macroalgae and use of extracts for crop management to reduce abiotic and biotic stresses. J Appl Phycol 26: 465-490.

2. Calvo P, Nelson L, Kloepper JW (2014) Agricultural uses of plant biostimulants. Plant Soil 383: 3-41.

3. Craigie JS (2011) Seaweed extract stimuli in plant science and agriculture. J Appl Phycol 23: 371-393.

4. Khan W, Rayireth U, Subramanian S, Jithesh M, Rayoreth P, et al. (2009) Seaweed extracts as biostimulants of plant growth and development. J Plant Growth Regul 28: 386-399.

5. Stirk W, Tarkowska D, Turecova V, Strand M, van Staden J (2014) Abscisic acid, gibberellins and brassinosteroids in Kelpak and commercial seaweed extract made from Ecklonia maxima. J Appl Phycol 26: 561-567.

6. Stirk WA, van Staden J (2014) Plant growth regulators in seaweeds: Occurrence regulation and functions. Adv Bot Res 71: 125-159.

7. Tarakhovskaya ER, Maslov YI, Shishova MF (2007) Phytohormones in algae. Russ J Plant Physiol 54: 163-170.

8. Kurepin LV, Zaman M, Pharis RP (2014) Phytohormonal basis for the plant growth promoting action of naturally occurring biostimulators. J Sci Food Agric 94: 1715-1722.

9. Rayorath P, Jithesh MN, Farid A, Khan W, Palanisamy R, et al. (2008) Rapid bioassays to evaluate the plant growth promoting activity of Ascophyllum nodosum (L) Le Jol. using a model plant, Arabidopsis thaliana (L.) Heynn. J Appl Phycol 20: 423-429.

10. Stadnik MJ, de Freitas MB (2014) Algal polysaccharides as source of plant resistance inducers. Trop Plant Pathol 39: 111-118.

11. Zhang XZ, Ervin EH (2004) Cytokinin-containing seaweed and humic acid extracts associated with creeping bentgrass leaf cytokinins and drought resistance. Crop Sci 44: 1737-1745.

12. Papenfus HB, Kulkarni MG, Stirk WA, Finnie JF, van Staden J (2013) Effect of a commercial seaweed extract (Kelpak®) and polyamines on nutrient-deprived (N, P and K) okra seedlings. Sci Hortic 151: 142-146.

13. Matysiak K, Adamczewski K (2006) Influence of bioregulator Kelpak on yield of cereals and other crops. Prog Plant Prot 46: 102-108.

14. Kumar G, Sahoo D (2011) Effect of seaweed liquid extract on growth and yield of Triticum aestivum var. Pusa Gold. J Appl Phycol 23: 251-255.

15. Zodape ST, Mukherjee MP, Reddy DR, Chaudhary DR (2009) Effect of Kappaphycus alvarezii (Doty) Doty ex silva extract on grain quality, yield and some yield components of wheat (Triticum aestivum L.) Int J Plant Prod 3: 97-101.

16. Nelson WR, van Staden J (1986) Effect of seaweed concentrate on the growth of wheat. S Afr J Sci 82: 199-200.

17. Steveni CM, Norrington-Davies J, Hankins SD (1992) Effect of seaweed concentrate on hydroponically grown spring barley. J Appl Phycol 4: 173-180.

18. Beckett RP, van Staden J (1990) The effect of seaweed concentrate on the yield of nutrient stressed wheat. Bot Mar 33: 147-152.

19. Shaaban MM, El-Saady AM, El-Sayed AB (2010) Green microalgae water extract and micronutrients foliar application as promoters to nutrient balance and growth of wheat plants. J Am Sci 6: 631-636.

20. Shah MT, Zodape ST, Chaudhary DR, Eswaran K, Chikara J (2013) Seaweed SAP as an alternative liquid fertilizer for yield and quality improvement of wheat. J Plant Nutr 36: 192-200.

21. Beckett RP, van Staden J (1989) The effect of seaweed concentrate on the growth and yield of potassium stressed wheat. Plant Soil 116: 29-36.

22. Matysiak K, Kaczmarek S, Leszczynska D (2012) Influence of liquid seaweed extract of Ecklonia maxima on winter wheat cv. Tonacja. J Res Appl Agric Eng 57: 44-47.

23. Posta D (2013) Research concerning the use of some seed and material preparation methods in the production of biological material in generative Koelreuteria paniculata LAXM. J Hort For Biot 17: 185-188.

24. Muhammad S, Anjum AS, Kasana MI, Randhawa MA (2013) Impact of organic fertilizer, humic acid and sea weed extract on wheat production in Pothowar region of Pakistan. Pak J Agri Sci 50: 677-681.

36

The Effects of Sucrose on *in vitro* Tuberization of Potato Cultivars

Miheretu Fufa[1]* and Mulugeta Diro[2]

[1]*Adami Tullu Agricultural Research Center; Plant Biotechnology Team, P.O. Box 35, Zeway, Ethiopia*
[2]*Cascape Project Coordinator, Addis Ababa, Ethiopia*

Abstract

Two potato varieties namely, 'Hunde' and 'Ararsa' were tested for *in vitro* Tuberization response under five levels of sucrose (40, 60, 80, 100 and 120gram litre[-1]) in completely randomized design with 2×5 factorial combinations. The objective was to determine optimum concentration of sucrose for *in vitro* Tuberization. In both varieties, among the five concentrations of sucrose, Murashige and Skoog (MS) medium supplemented with 60 gram litter[-1] sucrose exhibited a better response than the other concentrations in mean values of microtuber number, diameter, and weight and was found optimum. Accordingly, this medium gave an average value of (1.97 ± 0.02) microtuber number, (3.60 ± 0.04 mm) microtuber diameter, and (0.08 ± 0.002 gram) weight of microtuberin variety 'Ararsa' after 42.57 ± 0.58 days of culture. On the other hand, it gave mean value of (2.90 ± 0.031) microtuber number, (2.95 ± 0.01 mm) microtuber diameter, and (0.06 ± 0.001 gram) weight of microtuber in variety 'Hunde' after 35.67 ± 0.58 days of culture.

Keywords: *In vitro* tuberization; Microtuber; Potato; Sucrose

Introduction

Potato (*Solanum tuberosum* L.) is an important food and cash crop [1]. It ranks first in the world from none-grain crop to ensure food security [2]. It is a high biological value crop that gives an exceptionally high yield, more protein and calories, vitamins, minerals, carbohydrates and iron per unit area per unit time than any other major crops [3].

The conventional propagation of potato is characterized by low multiplication rate and susceptibility to pathogens. Susceptibility to pathogens often leads to poor quality and yields due to degeneration [3]. In Ethiopia, the production of potato is expanding steadily. However, its productivity has shown a decreasing trend [4,5] due to the unavailability of good quality clean seed tubers [6]. This can be avoided through *in vitro* tuberization, a better alternative to conventional propagation that can produce uniform and identical materials in large scale within a short time [7].

The most critical stimulus influencing *in vitro* Tuberization is sucrose at high concentration [8]. Sucrose is a cheap, safe and superior agent for *in vitro* Tuberization [9]. Hence, the present study was initiated with the objective to determine optimum concentration of sucrose for *in vitro* Tuberization.

Materials and Methods

Single nodal excision from one week old sprouts of the relatively clean tubers of 'Ararsa' and 'Hunde' potato varieties was used for *in vitro* Tuberization experiment at Tissue Culture Laboratory of Jimma University College of Agriculture and Veterinary Medicine. The two varieties were tested for Tuberization response under five levels of sucrose (40, 60, 80, 100 and 120 gram liter[-1]) in completely randomized design with 2×5 factorial combinations.

The pH of the medium was adjusted at 5.8, agar (8 gram liter[-1]) was added and then the medium was autoclaved at 121°C for 20 minutes at 15 psi. MS basal medium containing gibberellic acid (0.1 milligram liter[-1]), naphthalene acetic acid (0.01 mg/l) and sucrose (30 g/l) was used for initiation. In the case of *in vitro* Tuberization, the Murashige and Skoog (MS) (1962) basal medium was prepared for each treatment combination.

All the surface sterilization procedures were carried out under aseptic condition of laminar flow chamber, following the procedure of Naik and Karihaloo [10]. One week old sprouts along buds were excised and used as initial explants. The excised explants were washed 3 times in running tap water with 0.25 millliliter of Tween-20, and then washed thoroughly three times with sterile distilled water and immersed in 70% ethyl alcohol for 10 seconds. The alcohol was removed by three times washing with sterile distilled water. Finally, the sprouts were sterilized with 10% sodium hypochlorite (NaOCl) for 20 minutes before dissection.

After removing the leaves, the excised explants were dissected into single nodes (2 cm long) on a sterile plate. Six explants were cultured into 40 milliliter of an initiation culture medium in culture jar and incubated under a 16 hour photoperiod at 24°C with a light intensity of 2500 lux. The sprouts were allowed to grow into plantlets having nodal segments for 3 to 4 weeks. The multiplication medium was decanted and the plantlets were kept in a conditioning medium before being used for *in vitro* Tuberization to avoid the carryover effects of hormones. Forty milliliters of *in vitro* Tuberization medium was dispensed into each culture jar before transferring the culture to the growth room. Finally, the culture was kept at a temperature of 18°C under dark condition.

Data collection and analysis

The first, second and third date of formation of microtuber was carefully followed and recorded. 50% days to set microtuber was recorded and used for analysis. The number of microtubers produced by each explant was counted before harvest. The diameter (in millimeter) of each microtuber was measured by Digital Caliper. Immediately after harvest, each microtuber was weighed on sensitive balance to get the mean microtuber weight in gram. After 15 days of light exposure, the microtubers were treated with gibberellic acid (GA3) and incubated in the dark before planting in the green house. The number of the microtubers germinated and established was counted to get their percent survival under in vivo. The data were subjected to the analysis of variance (ANOVA) at 5% level of significance using SAS

*Corresponding author:Miheretu Fufa, Adami Tullu Agricultural Research Center, Plant Biotechnology Team, P.O. Box 35, Zeway, Ethiopia, E-mail: miheretufufag@gmail.com

statistical software Version 9.2 [11]. The REGWQ multiple comparison procedure was used for separating significant means.

Results and Discussion

Analysis of variance (ANOVA) revealed that sucrose and variety interaction had very highly significant effect (α=5%) on days to Tuberization and on the average number, diameter (millimeter) and weight (gram) of microtubers (Table 1). This implied that there is interdependence of sucrose and genotype on *in vitro* Tuberization of potato. Thus, the response of genotypes to a given level of sucrose was not the same.

Effects of sucrose on days to Tuberization, mean microtuber number and diameter

At 40 gram liter[-1] sucrose, both varieties did not produce microtubers. However, when 60 gram liter[-1] sucrose was added to growth media, 'Hunde' variety produced microtubers in 36 days, which is significantly earlier than that of 'Ararsa' variety (43 days). Increasing concentration of sucrose from 60 to 80 gram liter[-1] delayed microtuber formation in both varieties, but more pronounced on 'Ararsa' variety. This might be due to the marked variation in the responses of plant gene to changing sucrose status. Some genes are induced, some are repressed, and others are minimally affected [12].

Moreover, in both varieties, microtuber number and size get reduced as the concentration of sucrose increased from 60 gram liter[-1] to 80 gram liter[-1]. At 120 gram liter[-1] of sucrose, both genotypes did not produce microtubers (Table 1). The absence of microtuber formation at high sucrose concentration might be due to the effect of super optimal level of sucrose that can result in an unfavorable osmotic condition for water uptake, and thus affecting microtuber formation of the seedlings.

Effects of sucrose on mean microtuber weight

In both varieties, a decreasing trend in mean weight (gram) of microtuber was observed as the level of sucrose increased (Figure 1). This might be again due to the effect of high sucrose level on osmotic condition of the culture that affect cell turgidity [12], and hence, microtuber weight.

In both varieties, Murashige and Skoog (MS) medium supplemented with 60 gram liter[-1] sucrose exhibited a better response than the other concentrations in mean values of microtuber number, diameter and weight, and was found optimum. Accordingly, this medium gave an average value of (1.97 ± 0.02) microtuber number, (3.60 ± 0.04 mm) microtuber diameter, and (0.08 ± 0.002 g) weight of microtuber in the variety 'Ararsa' variety after 42.57 ± 0.58 days of culture. On the other hand, it gave mean value of (2.90 ± 0.031) microtuber number, (2.95 ± 0.01 mm) microtuber diameter, and (0.06 ± 0.001 g) weight of microtuber in variety 'Hunde', after 35.67 ± 0.58 days of culture (Table

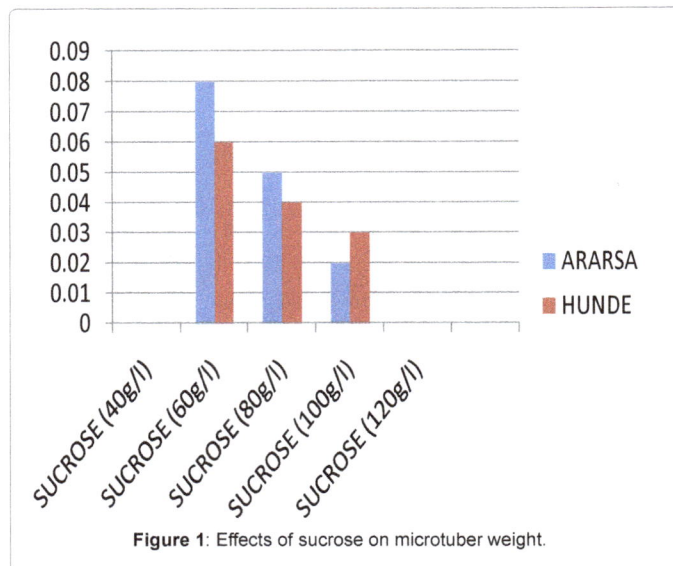

Figure 1: Effects of sucrose on microtuber weight.

1).

The present result is in agreement with that of Aslam et al. [13], who found that a medium containing 6% sucrose was optimal in terms of minimum time of induction (34), mean tuber number (1.2) and weight (0.03 gram) of microtubers per single nodal explant in cultivar Desiree. Imani et al. [14] also reported that Murashige and Skoog (MS) medium supplemented with 60 gram liter[-1] of sucrose as the best in producing the maximum number (4.20) and size (0.44 centimeter) of micro tubers. Iqbal et al. [9] also recorded similar results on mean numbers of tubers (4.8) on Murashige and Skoog (MS) medium treated with 60 gram liter[-1] sucrose. Kanwal et al. [15], on the other hand, reported that Murashige and Skoog (MS) medium supplemented with 30 and 40gram liter[-1] sucrose did not produce microtubers.

Conclusion

A protocol for *in vitro* Tuberization of potato varieties 'Ararsa' and 'Hunde' from single nodal explant has been developed. The result indicated that *in vitro* Tuberization of potato was highly dependent on sucrose and genotype interaction.

Murashige and Skoog (MS) medium supplemented with 60 gram liter[-1] exhibited fewer days to microtuber formation (35.67 ± 0.58/42.67 ± 0.58), better mean number (2.90 ± 0.031/1.97 ± 0.02), diameter (2.81 ± 0.015/3.60 ± 0.04 millimeter) and fresh weight (0.06 ± 0.001/0.08 ± 0.002 gram) of microtubers in 'Hunde' and 'Ararsa' varieties, respectively. However, microtuber production needs further improvement since the size of microtubers produced was not large

	ARARSA				HUNDE			
Sucrose (g/l)	DT	MTN	MTD	MTWT	DT	MTN	MTD	MTWT
40	0.00 ± 0.00c	0.00 ± 0.00c	0.00 ± 0.00c	0.00 ± 0.00d	0.00 ± 0.00d	0.00 ± 0.00d	0.00 ± 0.00d	0.00 ± 0.00d
60	42.67 ± 0.58b	1.97 ± 0.02a	3.60 ± 0.04a	0.08 ± 0.002a	35.67 ± 0.58c	2.90 ± 0.031a	2.95 ± 0.050a	0.06 ± 0.001b
80	45.00 ± 1.00a	1.30 ± 0.08b	3.07 ± 0.03b	0.05 ± 0.001b	40.00 ± 0.00b	2.06 ± 0.081b	2.81 ± 0.015b	0.04 ± 0.001b
100	0.00 ± 0.00c	0.00 ± 0.00c	0.00 ± 0.00c	0.02 ± 0.00d	44.67 ± 0.58a	1.30 ± 0.042c	2.56 ± 0.044c	0.03 ± 0.002c
120	0.00 ± 0.00c	0.00 ± 0.00c	0.00 ± 0.00c	0.00 ± 0.00d	0.00 ± 0.00d	0.00 ± 0.00d	0.00 ± 0.00d	0.00 ± 0.00d

Table 1: Effects of sucrose on days to tuberization, number, diameter and weight (g) of microtuber.
Means with the same letters in a column are not significantly different from each other using the Ryan-Einot-Gabriel-Welsch Multiple Range Test (REGWQ) at α= 0.05.
DT=Days to Tuberization, MTN=Microtuber Number, MTD=Microtuber diameter (mm), and MTWT=Microtuber Weight (g).

enough. Thus, trying different levels of Benzylaminopurine (BAP) in combination with sucrose and extending the time of harvesting may be helpful to improve the size of the microtuber.

References

1. Lemaga B, Kakuhenzire R, Gildemacher P, Borus D, Gebremedhin WG, et al. (2009) Current status and opportunities for improving the access to quality potato seed by small farmers in Eastern Africa. Symposium 15th triennial of the symposium of the International Society for Tropical Root Crops, Africa.

2. Ethiopian variety registers (2006) Workshop on opportunities and challenges for promotion and sustainable production and protection of the potato crop in Vietnam and elsewhere in Asia. FAO, Addis Ababa, Ethopia 25-28.

3. Badoni A, Chauhan JS (2010) Conventional Vis -a- Vis biotechnological methods of propagation in potato. A review stem cell 1: 1-6.

4. CSA (Central Statistical Authority) (2003) Ethiopian Agricultural Sample enumeration.

5. CSA (Central Statistical Authority) (2009) Agricultural sample survey. A report on area and production of crops (private peasant holdings, Meher season), Addis Ababa, Ethopia.

6. Gebremedhin W, Endale G, Berga L (edn) (2008) Root and tuber crops the untapped resources.

7. Dobranszki J, Magyar-Tabori K, Hudakl (2008) In vitro tuberization in hormone-free systems on solidified medium and dormancy of potato microtubers. Fruit, vegetable and cereal science and Biotechnology 2: 82-94.

8. Nistor A, Campeanu G, Atanasiu N, Chiru N, Karacsonyi D (2010) Influence of potato genotypes on "in vitro" production of microtubers. Romanian Biotechnol Lett.

9. Iqbal H, Zubeda C, Aish M, Rehana A, Naqvi S, Hamid R (2006) Effect of chlorocholine chloride, sucrose and BAP in vitro tuberization in potato (Solanum tuberosum L. cv. cardinal). Pak J Bot 38: 275-282.

10. Naik PS, Karihaloo JL (2007) Micropropagation for production of quality potato seed in Asia-Pacific. Asia-Pacific consortium on agricultural biotechnology, New Delhi, India.

11. SAS Institute Inc (2008) SAS/STAT 9.2 user's guide. SAS Institute Inc, Cary, NC, USA.

12. George EF, Hall MA, De Klerk GJ (2008) Plant propagation by tissue culture: Plant Tissue Culture Procedure. (3rd Edn), Springer, AA Dordrecht, The Netherlands.

13. Aslam A, Ali A, Naveed NH, Saleem A, Iqbal J (2011) Effect of interactionof6-benzyl aminopurine (BAP) and sucrosefor efficient microtuberization of two elitepotato (Solanum tuberosu L.) cultivars, Desiree and Cardinal. Afr J Biotechnol 10: 12738-12744.

14. Imani AA, Qhrmanzadeh R, Azimi J, Janpoor J (2010) The effect of concentrations of 6-Benzylaminopurine (BAP) and sucrose on in vitro potato (Solanum tuberosum L.) microtuber induction. J Agric Environ Sci 8: 457-459.

15. Kanwal A, Ali A, Shoaib K (2006) In vitro microtuberization of potato (Solanum tuberosum L.) Cultivar Kuroda: A new variety in Pakistan. Int J Agric Biol 3: 337-340.

37

Trait Associations in Some Durum Wheat (*Triticum durum L.*) Accessions among Yield and Yield related Traits at Kulumsa, South Eastern Ethiopia

Tesfaye Wolde[1], Firdisa Eticha[2], Sentayehu Alamerew[1], Ermias Assefa[1,3]*, Dargicho Dutamo[4] and Birhanu Mecha[5]

[1]*College of Agriculture and Veterinary Medicine Gelalcha, Jimma University, Ethiopia*
[2]*Ethiopian Institute of Agricultural Research, Kulumsa Research Center, Ethiopia*
[3]*Southern Agricultural Research Institute, Ethiopia*
[4]*College of Agriculture and Natural Resources, Mizan-Tepi University, Ethiopia*
[5]*College of Agricultural Sciences, Wachemo University, Ethiopia*

Abstract

Wheat (*Triticum durum L.*) is the leading source of cereal and vegetable protein in human food, having higher protein content than either maize (corn) or rice, the other major cereals. The objective was to study the association among grain yield and related traits and identify traits those have the most direct and indirect effects on grain yield. Sixty eight durum wheat accessions tested in an augmented block design. Twelve yield and related traits were recorded in the investigation. Analysis of variance revealed highly significant differences among accessions for all traits. Grain yield showed positive and significant correlation with grain filling period, productive tillers per plant, number of grains per spike, spikelets per spike, 1000 grain weight, biological yield and harvest index at both phenotypic and genotypic levels. This implies that through improving these traits possible to improve grain yield. Path coefficient analysis showed that harvest index and biological yield revealed high positive direct effect and positively high significant correlations with grain yield. Indicated these two characters can be considered for direct selection. However the present outcome is only an indication and we cannot reach a definite conclusion. Therefore, it is sensible to conduct a follow up study by incorporating more characters over different years and locations.

Keywords: Durum wheat accessions; Yield related traits; Ethiopia

Introduction

Durum wheat (**Triticum durum** L.) is a monocotyledonous plant of the **Gramineae** family of the Triticeae tribe that belongs to the genus **Triticum**. For commercial production and human consumption, it is the second most important **Triticum** species, next to common wheat (**Triticum aestivum** L.) (Garcia et al. and Simane et al.) [1,2]. It is the only tetraploid (AABB, 2n=4x=28) species of wheat which has commercially a great importance and carries raw material of numerous foods such as macaroni and semolina in alimentation of world population and it is a promising and viable alternative crop for farmers Shewry [3].

The nutritional composition indicated that 100 g of Durum wheat provides 339 calories and it consisted carbohydrate 71 g, protein 14 g, fat 2.5 g, minerals 2 g and considerable proportions of vitamins (thiamine and vitamin-B) and minerals (zinc, iron). It is also a good source of traces minerals like selenium and magnesium, nutrients which are essential to good health Belay [4]. In Ethiopia durum wheat is used to prepare local food recipes: kitta (unleavened bread), Injera (spongy flat unleavened bread), Kinche (boiled coarse-ground wheat), Nifro (boiled whole grain), Kollo (roasted grain), dabo-kollo (ground and seasoned dough) etc. [5].

Grain Yield of wheat is multifaceted quantitative trait that outcome to the actions and interactions of various component traits [6]. Thus, analysis of Correlation coefficient could be used as an vital tool to give the interrelationship of yield and yield contributing traits. It is also used to measure the strength and the direction of the relationships among the different characters and grain yield [7]. Correlation and path coefficient analysis could be used as important tools to bring information about appropriate cause and effect relationship between yield and some yield components [8]. Understanding the interrelationship between component characters will help in determining which character to be selected and when it is improved in relation to the desired characters.

The knowledge of the genetic relationship between grain yield and its components can help the breeders to improve the success and efficiency of the selection. In durum wheat to select genotypes of good yielder and the traits which contribute for the yield efficiency, it is important to study the relationships among the traits and find out their direct and indirect effects on grain yield [9]. A number of research finding showed that, the correlation coefficient measures the mutual relationship between various plant characters and determines the component characters on which selection can be based for the improvement in associated complex characters and yield [10-12]. Assessment of correlation coefficients between various yield traits helps to develop wheat cultivars with best combinations of yield attributes in wheat for obtaining higher yield [13]. In effective plant breeding program, identifying characters contributing to grain yield is important as it amplify breeding effectiveness. Therefore, the objective of this study was to assess the association among yield and yield contributing traits and identify traits those have the most direct and indirect effects on grain yield.

Materials and Methods

Description of the study area

The field experiment was conducted at Kulumsa Agricultural

---

*Corresponding author: Ermias Assefa, College of Agriculture and Veterinary Medicine, Jimma University, Ethiopia
E-mail: ethioerm99@gmail.com

Research Center (8° to 8° 2' northern latitude and 39° 07' to 39° 10' eastern longitude) which is located 160 km southeast of Addis Ababa and it is located at 8 km north of Asela town. The altitude of the center is 2200 masl with annual average rainfall of 832 mm. The annual average temperature of the study area is 16.65°C with maximum and minimum temperature of 22.8°C and 10.5°C respectively. With the soil type classified as clay loam soil with a pH of 6 (Source: Secondary data from KARC).

Experimental material

The entries for this study consisted of 68 durum wheat (*T. durum* L.) (64 land races and 4 released variety).

Experimental design and trial management

The field experiment was laid down in an Augmented Block Design having 4 blocks where each block contains 16 test entries and 4 checks (randomly allocated) with the total of 20 accessions in each blocks and accession was sown in 2 rows of 1.25 meter long and 20 cm apart. The recommended fertilizer rate of 100/100 kg/ha N/P_2O_5 in the forms of Urea and DAP was applied to each plot in the shallow furrow depths and mixed with soil at the same time during sowing. Other agronomic practices such as weeding were followed uniformly to all plots according to the recommendation of the location. No herbicide was applied to control weeds.

Data collected on-plant basis

The data on plant basis were taken for each accession from ten randomly taken plants as follows:

➤ **Plant height (cm)**: The plant height for randomly taken samples was measured in centimeter from ground level of the plant to the base of the spike excluding awns.

➤ **Spike length (cm)**: Spike length was recorded from the bath of the spike to the tip and it was recorded from randomly taken plants and measured in centimeters.

➤ **Number of spikelets per spike**: Total number of spikelets on main spike of all ten plants was counted at the time of maturity and the average was recorded.

➤ **Number of productive tillers per plant**: The number of tillers per plant bearing productive heads was counted and average was recorded for the ten randomly taken plants

➤ **Number of grains per spike**: The spike of the selected plants was threshed manually and numbers of grains per spike was counted for each accession at the time of harvest.

On plot basis

➤ **Days to heading**: The number of days taken from the date of sowing to the day on which the main ear comes out of the flag leaf completely in 50 per cent of the plants.

➤ **Days to maturity**: The numbers of days from date of sowing to a stage at which 75% of the plants have reached physiological maturity or 75% of the spikes on the plots turned golden yellow color were calculated.

➤ **Grain Filling Period (GFP)**: The grain filling period in days was computed by subtracting the number of days to heading from the number of days to maturity.

➤ **1000 grain weight (TGW) (g)**: The weight (g) of 1000 grain

from randomly sampled seeds per plot measured by using sensitive balance.

➤ **Grain Yield (GY) (kg/ha)**: The grain yield per plot was measured in grams using sensitive balance after moisture of the seed is adjusted to 12.5%. Total dry weight of grains harvested from the two rows was taken as grain yield per plot and expressed in kg/ha.

➤ **Biomass Yield (BMY) (kg/ha)**: It was recorded by weighing the total above ground biological yield harvested from all the plots of each accession at the time of harvest and converted to kg/ha.

➤ **Harvest Index (HI) (%)**: It was estimated by dividing grain yield per plot to biological yield per plot. It is the ratio of grain yield to the above ground biomass yield.

Statistical analysis

The analysis of variance for each character was performed using the SPAD software developed by IASRI, New Delhi, India. Phenotypic and genotypic correlation and path coefficient analysis GENRES Statistical Soft ware or Package [14].

Correlation coefficient (r): Phenotypic and genotypic correlations between yield and yield related traits were estimated using the method described by Miller et al. [15] and tested for statistical significance against the correlation table values at 5 and 1% levels of significance. The statistical procedures were as follows:

$$rp_{xy} = \frac{covp_{xy}}{\sqrt{vp_x \ vp_y}}$$

Where: rp_{xy} is phenotypic correlation coefficient between character x and y

$covp_{xy}$, phenotypic covariance between character x and y

vp_x, phenotypic variance for character x

vp_y phenotypic variance for character y

$$rg_{xy} = \frac{covg_{xy}}{\sqrt{vg_x \ vg_y}}$$

Where: rg_{xy}, genotypic correlation coefficient between character x and y

$covg_{xy}$, genotypic covariance between character x and y

vg_x, genotypic variance for character x

vg_y, genotypic variance for character y

The correlation coefficients were worked out to determine the degree of association of a character with yield and among the yield components. where 'n' is the number of genotypes. To test the significance of correlation coefficients, the following formula was adopted [16]:

T=r/SE(r)

Where, $SE(r) = \frac{1-r2}{\sqrt{n-2}}$

Where, r is correlation coefficient; n is number of genotypes. To test the significance of correlation coefficient, the calculated t-value can be compared with tabulated t-value at (n-2) degree of freedom at 0.05 and 0.01 levels of probability [17].

Path coefficient analysis: The path coefficient analysis was carried out using GENRES Statistical Software Package [14] to study the direct and indirect contributions of the traits to the associations. A measure of direct and indirect effects of each character on grain yield was estimated using a standardized partial regression coefficient known as path coefficient analysis, as suggested by Dewey and Lu [18]. Thus, correlation coefficient of different characters with grain yield was partitioned into direct and indirect effects adopting the following formula.

$$r_{ij} = p_{ij} + \sum rik + pkj$$

Where: r_{ij}=mutual association between the independent variable 'i' and the dependent variable 'j' as measured by correlation coefficient.

p_{ij}=components of direct effects of the independent variable 'i' on dependent variable 'j' as measured by the path coefficient.

$\sum rik\ pkj$ = Summation of component of indirect effects of independent trait 'i' on the given dependent trait 'j' via all other independent variable 'k'.

The contribution of the remaining unknown factor was measured as the residual factor. It was calculated by:

Residual effect=$\sqrt{(1-R^2)}$ where: R= $\sum pij\ rij$

Results and Discussion

Analysis of variance (ANOVA)

The analysis of variance for the 12 characteristics is presented in Table 1. It revealed that significant (P<0.1) differences were observed among the genotypes for all characters studied y. This gives an ample opportunity to plant breeders for improvement of these characters through selection. Similarly, the research finding of Abinasa [19] reported highly significant differences among durum wheat genotypes for days to heading, days to maturity, number of productive tillers per plant, plant height, spike length, spikilets per spike, thousand grain.

Correlation of grain yield with other traits: The phenotypic and genotypic correlation coefficients among the characters were presented in Table 2. The results showed that generally the genotypic correlation coefficients (rg) were higher than the phenotypic correlation coefficients (rp) indicating the association among these characters were under genetic control and indicating the preponderance of genetic variance in expression of characters. Grain yield was positively and significantly correlated with grain filling period (0.37**, 0.32**), number of grain yield per spike (0.42**, 0.41**), thousand grain weight (0.43**, 0.42**), biomass yield (0.34**, 0.33**), harvest index (0.59**, 0.57**) and productive tillers per plant (0.27*, 0.23*) at both genotypic and phenotypic levels respectively. It implies that improving one or more of the above characters could result in high grain yield. Similarly, Tsegaye et al. [20] reported that grain yield had positive and significant phenotypic and genotypic correlations with number of tillers per plant, thousand grain weigh, biological yield and harvest index.

Yousaf et al. [21] reported that yield per plant had highly significant and positive genotypic and phenotypic correlations with number of productive tillers per plant, number of spikelets and number of grains per spike had significant positive correlations with spike length at both levels. Tila et al. [22] found that the biological yield and harvest index had positive and significant genotypic and phenotypic correlations with grain yield. Ahmad et al. and Akcura (Ali and Shakor and Peymaninia et al.) [10,23-25] also reported strong positive correlation of biomass yield and harvest index on grain yield in wheat.

Correlation among characters

Phenotypic and genotypic correlation: Days to heading showed positive and significant correlation with days to maturity (0.46**, 0.51**) at both phenotypic and genotypic levels (Figures 1 and 2; Table 3). Similarly, Khan et al. [8] reported days to maturity are positively and significantly correlated with days to heading at both phenotypic and genotypic levels. However, it displayed negative and significant association with grain filling period at both phenotypic and genotypic levels (-0.29*, -0.32**) indicating that the shorter days to heading have the longer grain filling period but, it had positive and highly significant association with days to maturity at both phenotypic and genotypic levels these showed the difficulty in simultaneous improvement of these traits. It had also revealed positive and highly significant correlation with spikelets per spike (0.51**, 0.62**) at both phenotypic and genotypic levels while, significant and positive correlation with spike length at genotypic level (0.24*). It implies that the longer flowering period will enhance the possession of sufficient spikelets per spike accompanied by long spike length. On the contrary, it was found negative and significant correlation with harvest index (-0.24*, -0.26*) at both phenotypic and genotypic levels.

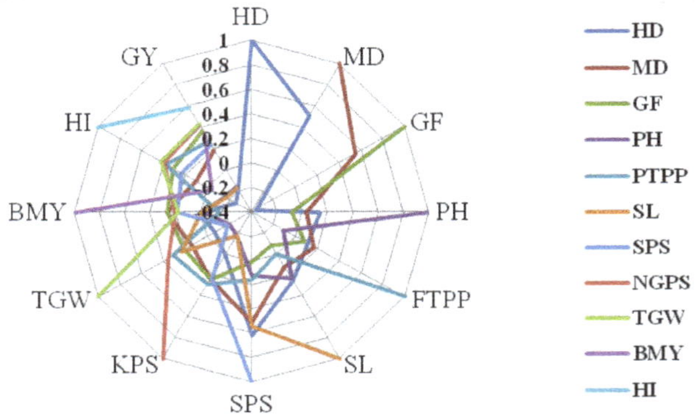

DH=Days to heading, MD=Days to maturity, GF=Grain filling period, PH=Plant height, PTPP=Productive tillers per plant, SL=Spike length, SPS=Spikeletes per spike, NGPS=Number of grains per spike, TGW=Thousand grain weight, BMY=Biomass yield, HI=Harvest index and GY=Grain yields

Figure 1: Graphic presentation of genotypic correlations of different traits with grain yield.

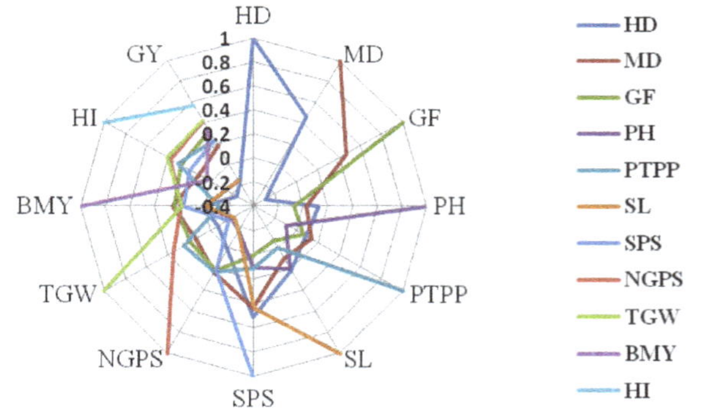

DH=Days to heading, MD=Days to maturity, GF=Grain filling period, PH=Plant height, PTPP=Productive tillers per plant, SL=Spike length, SPS=Spikeletes per spike, NGPS=Number of grains per spike, TGW=Thousand grain weight, BMY=Biomass yield, HI=Harvest index and GY=Grain yields

Figure 2: Graphic presentation of phenotypic correlations of different traits with grain yield.

Character	Block (adj) (df=3)	Error (df=9)	TG (adj) (df=67)	SC (df=3)	AC (df=63)	AC vs. SC (df=1)	CV (%)
Days to heading (days)	1.22	4.95	38.23**	118.22**	32.48**	161.02**	3.37
Days to maturity (days)	5.16	2.83	52.64**	36.66**	47.88**	400.51**	1.51
Grain filling period (days)	4.22	12.11	57.15**	273.22**	30.79*	1069.45**	7.67
Plant height (cm)	1.66	1.13	304.07**	57.01**	61.35**	16336.32**	0.92
No. of productive tillers plant	0.68	0.62	2.12*	0.50ns	1.64*	37.46**	16.55
Spike length (cm)	0.68	0.50	3.01**	0.85ns	1.33*	115.80**	8.25
Number of spikelets spike-1	2.64	1.15	5.1*	3.18ns	4.14*	71.25**	5.23
Number of kernels spike-1	4.72	4.11	118.43**	283.22**	46.75**	4140**	5.53
Thousand kernels weight (g)	8.05	3.30	142.99**	300.60**	38.28**	6267.21**	7.43
Grain yield plot-1 (g)	4.91	9.75	295.13**	128.41**	146.51**	10158.77**	10.12
Biomass yield plot-1 (g)	6.22	11.17	1336.67**	1050.89**	1264.31**	6752.81**	2.07
Harvest index (%)	3.08	3.744	89.02**	2.74ns	58.16**	2292.15**	10.16

Df=Degrees of freedom; *, ** Significant at 5% and 1% probability level respectively; CV=Coefficient of Variation; TG=Total genotypes; SC=Standard checks; AC=Accessions.
Table 1: Analysis of variance (Mean squares) for the 12 characters of 68 durum wheat accessions.

	HD	MD	GF	PH	FTPP	SL	SPS	KPS	TGW	BMY	HI	GY
HD	1.0	0.51**	-0.36**	0.14	0.11	0.25*	0.62**	0.014	-0.07	0.012	-0.26*	-0.19
MD	0.46**	1.0	0.55**	0.03	0.17	0.11	0.51**	0.24*	0.14	0.27*	0.11	0.19
GF	-0.29*	0.47**	1.0	-0.08	0.08	-0.09	0.002	0.24*	0.21	0.27*	0.32**	0.37**
PH	0.13	0.03	-0.07	1.0	-0.11	0.23*	0.12	-0.17	-0.2	-0.02	-0.15	-0.16
PTPP	0.09	0.14	0.06	-0.09	1.0	-0.007	0.16	0.29*	0.31**	-0.15	0.38**	0.27*
SL	0.21	0.1	-0.08	0.19	-0.005	1.0	0.55**	-0.18	0.24*	-0.005	-0.16	-0.17
SPS	0.51**	0.44**	0.002	0.1	0.12	0.44**	1.0	0.23*	-0.18	0.19	0.24*	0.24*
NGPS	0.013	0.23*	0.21	-0.17	0.23*	-0.16	0.21	1.0	0.36**	0.2	0.4**	0.42**
TGW	-0.07	0.14	0.19	-0.2	0.25*	-0.22	-0.17	0.35**	1.0	0.18	0.43**	0.43**
BMY	0.01	0.26*	0.24*	-0.02	-0.12	-0.005	0.17	0.19	0.18	1.0	-0.04	0.34**
HI	-0.24*	0.11	0.28*	-0.14	0.31**	-0.14	0.2	0.38**	0.41**	0.03	1.0	0.59**
GY	-0.18	0.18	0.32**	-0.16	0.23*	-0.15	0.23*	0.41**	0.42**	0.33**	0.57**	1.0

T=0.23 (p<0.05) and T=0.302 (p<0.01) for df=n-2 where, n=number of genotypes, DH=days to heading, MD=days to maturity, GF=grain filling period, PH=plant height, PTPP=productive tillers per plant, SL=spike length, SPS=spikeletes per spike, NGPS=number of grains per spike, TGW=thousand grain weight, BMY=biomass yield, HI=harvest index and GY=grain yields.
Table 2: Estimates of phenotypic (below diagonal bold) and genotypic (above diagonal not bold) correlation coefficients among yield and yield components in 68 durum wheat accessions tested at Kulumsa (2013/2014).

Characters	HD	MD	GF	PH	FTPP	SL	SPS	KPS	TGW	BIOY	HI	GYg
HD	0.072	-0.054	-0.035	-0.009	0.002	-0.010	0.019	0.001	-0.0004	0.005	-0.217	-0.19
MD	0.037	-0.105	0.056	-0.002	0.004	-0.005	0.008	0.017	0.0008	0.128	0.108	0.19
GF	-0.027	-0.063	0.093	0.006	0.002	0.004	0.022	0.018	0.001	0.134	0.320	0.37**
PH	0.015	-0.006	-0.012	-0.044	-0.004	-0.014	0.003	-0.019	-0.002	-0.015	-0.208	-0.16
NPTPP	0.007	-0.016	0.007	0.007	0.024	0.0002	0.012	0.018	0.002	-0.059	0.314	0.27*
SL	0.020	-0.014	-0.011	-0.017	-0.0002	-0.035	0.029	-0.015	-0.002	-0.003	-0.173	-0.17
SPS	0.041	-0.051	0.025	-0.007	0.003	-0.021	0.026	0.0003	-0.0004	0.034	-0.108	0.24*
NGPS	0.001	-0.034	0.031	0.015	0.008	0.009	0.023	0.054	0.003	0.117	0.490	0.42**
TGW	-0.007	-0.022	0.030	0.020	0.009	0.014	-0.012	0.037	0.004	0.119	0.575	0.43**
BMY	0.0008	-0.029	0.027	0.002	-0.003	0.0002	0.0021	0.014	0.001	0.452	-0.033	0.34**
HI	-0.019	-0.014	0.037	0.011	0.009	0.007	-0.0032	0.033	0.003	-0.018	0.810	0.59**

Residual=0.195; DH=Days to heading, DM=Days to maturity, GFP=Grain filling period, PH=Plant height (cm), NPTPP=No. of productive tillers plant-1, SL=Spike length (cm), NSPS=No. of spikelets spike-1, NGPS=No. of grains spike-1, TKW=1000 kernel weight (g), GY=Grain yield ha-1, BMY=Biomass yield tons ha-1 and HI=Harvest index.
Table 3: Estimate of direct effect (bold face and diagonal) and indirect effects (off diagonal) at genotypic level in 68 durum wheat accessions tested at Kulumsa (2013/14).

The association of maturity date with grain filling period displayed positive and highly significant at both phenotypic and genotypic levels (0.47**, 0.55**), indicating that the given durum wheat accessions required sufficient maturity time to posses more grain. It had also showed positive and significant association with spikelets per spike (0.44**, 0.51**), number of kernels per spike (0.23*, 0.24*) and biomass yield (0.26*, 0.27*) at both phenotypic and genotypic levels, indicating to improve these traits it needs long maturation time. This implies that to improve yield contributing traits in the study require enough maturation periods. The result is in agreement with Gashaw et al. [7].

Grain filling period had positive and significant correlation with biomass yield (0.24*, 0.27*) and harvest index (0.28*, 0.32**) at both phenotypic and genotypic level while, positive and significant with number of kernels per spike (0.24*) only at genotypic level. Plant height revealed positive and significant correlation with spike length (0.24*) at genotypic level where as insignificant and positive at phenotypic level. Iqbal et al. [26] found similar results in their studies. It indicated the tall wheats may possess long spike and accompanied to hold adequate amount of spikelets. Thus, the result exhibited the accessions of durum wheat have tall nature and susceptible to lodging. It is in agreement with Abinasa et al. [19] reported that negative and non significant correlation was observed for plant height with kernels number perspike and grain yield and similar finding also presented by Majumder et al. [27]. However, it contradicted the reports of Gashaw et al. [7] displayed that plant height had positive and significant correlation with number of kernels per spike, biomass yield, and thousand kernel weight at both phenotypic and genotypic levels.

The association of productive tillers per plant exhibited positively and highly significant to significant with number of grains per spike (0.23*, 0.29*), thousand grain weight (0.25*, 0.31**) and harvest index (0.31**, 0.38**), indicating that selection based on productive tillers per plant may lead to promotion of these traits. It contradicted the finding of Iftikhar et al. [28]. Spike length exhibited positively and highly significant with spikelets per spike (0.44**, 0.55**) at both phenotypic and genotypic levels. This showed the longer spike length may possess more numbers of spikelets and able to produce sufficient yield, so that selection based on this traits would improve the grain yield in durum wheat. The result is in agreement with Iftikhar et al. and Bilgin et al. [28,29]. It had also showed positive and significant correlation with thousand grain weight at genotypic level (0.24*).

Spikelets per spike revealed positive and significant correlation with number of kernels per spike (0.23*) and harvest index (0.24*) at genotypic level. The result is partially in agreement with Ali et al. [30]. The association of number of grains per spike showed positively and highly significant with thousand grain weight (0.35**, 0.36**) and harvest index (0.38**, 0.40**) at both phenotypic and genotypic levels, suggested that high grain numbers per spike required for the improvement of both thousand grain weight and harvest index. The result is in accordance with Akcura and Iftikhar [23,28] but, it was contradicted the finding of Yousaf and Dutamo [21,31]. On the other hand it revealed positive and insignificant association with biomass yield at both levels. Thousand grain weight showed positively and highly significant correlation with harvest index (0.41**, 0.43**) at both genotypic and phenotypic levels whereas positive and insignificant with biomass yield at both levels. It is partially in accordance with the finding Abinasa et al. [19]. The association of biomass yield exhibited insignificant and positive with harvest index at phenotypic level while, negative and insignificant at genotypic level.

Path coefficient analysis

Direct and indirect effects: The genotypic direct and indirect effects of different characters on grain yield are presented in Table 3. The results of path coefficient analysis of the study revealed that almost all characters exhibited positive direct effect except days to maturity, plant height and spike length. The maximum positive direct effect was exhibited on grain yield by harvest index (0.81**). It had also positive and highly significant genotypic correlation with grain yield makes the direct selection of this character is effective for yield improvement. The result confirms the finding of Abinasa et al. and Izzat et al. [19,32]. The indirect effects of this character were positive with grain filling period and number of kernels per spike. But, it had negative and negligible indirect effects for the rests of the traits. Therefore, harvest index, grain filling periods and number of kernels per spike should be considered in durum wheat breeding program designed to simultaneously improve grain yield.

Biomass yield had positive direct effects as well as significant positive correlation with grain yield (0.45). The indirect effects via maturity date, productive tillers per plant, and harvest index are negative whereas, indirect effects via days to heading, grain filling period, plant height, spike length, spikelets per spike, number of kernels per spike and thousand grain weight were positive. The result suggested that the direct selection of biomass yield will improve grain yield effectively. On the other hand the indirect selection of grain filling period, plant height, spike length, spikelets per spike, number of kernels per spike and thousand grain weight via biomass yield will improve grain yield. It confirms the finding of Ali [24] and Similarly, Hannachi et al. [33] who reported that biomass and harvest index exerted positive direct effect on grain yield.

Days to maturity exhibited negative direct effects on grain yield, even though it had positive correlation, indicating that indirect effects to be cause of correlation. Thus, the negative direct effect of days to maturity was counter balanced by its positive indirect effects via days to heading, grain filling period, biological yield and harvest index as well as rendered the positive genotypic correlation coefficient. Grain filling period showed positive direct effect and positive genotypic correlation on grain yield, it indicated that this trait should be considered in further selection procedures for higher grain yield. The indirect effects of this trait were positive and high to moderate with harvest index and biomass yield respectively but, it had negative indirect effects with days to heading and maturity date however, negligible positive indirect effects for the rest traits. Days to heading had positive direct effects with grain yield. Oğuz et al. [34] reported similar results in durum wheat. However, the correlation coefficient it had with grain yield was negative this negative relationship between early flowering accompanied with sufficient grain filling period characters and grain yield is desirable if stresses conditions such as terminal heat and drought are expected during growing season. This suggestion can be justified by earlier report of Gelalcha and Hanchinal [35] in which correlation between days to flowering and days to maturity with grain yield was negative but, this result is weak agreement with maturity date. The indirect effect of days to heading was negative and high with harvest index, days to maturity and grain filling periods whereas positive and weak effects with the rest of the traits.

Number of grains per spike revealed positive direct effects and positively high significant correlation with grain yield, it reflect that this characteristic is essential to improve grain yield directly. The indirect effects were showed positive for all characteristics except for days to maturity. It exhibited high positive indirect effects with harvest index, biological yield and grain filling period. Thus, the indirect selection of all these traits via number of grains per spike may evolve high yielding genotypes. This result is in agreement with Dogan [36] who reported direct positive effect of grain number per spike and 1000-grain weight on yield.

Productive tillers per plant manifested positive direct effect and significant positive correlation with grain yield. Hence, selection for yield improvement can be done on the basis of productive tillers per plant. The indirect effects of this trait were positive with days to heading, grain filling period, plant height, spike length, spikelets per spike, number of kernels per spike, thousand grain weight and harvest index. It indicated that the improvement of one of these traits via productive tillers per plant will improve grain yield in durum wheat. However, it showed negative indirect effect via biomass yield and maturity date. Spike length showed negative direct effects and negative correlation with grain yield this indicated that the direct selection of this trait is not crucial unless together with indirect effects. The indirect effect of this characteristic was highly significant positive correlation with spikelets per spike indicated not only the length of spike but also incorporated with spikelet number to improve grain yield under selection. On the other hand it exhibited negligible and negative indirect effects for the of rest traits rather than days to heading.

Thousand grain weight revealed weak positive direct effects and highly significant positive genotypic correlation with grain yield. The result indicated that even though it had weak positive direct effects, the correlation was strong because the positive and high indirect effects with harvest index, biomass yield, number of kernels per spike and grain filling period renders its correlation to be stronger. Residual effect in the present study was 0.195 (Table 3) which indicated the traits in the path analysis expressed the variability in grain yield by 80.5%.

Conclusion

The genotypic and phenotypic correlation of grain yield was positive and significant with grain filling period, number of productive tillers per plant, number of spikelet's per spike, number of grains per spike, thousand grain weight, biomass yield and harvest index. Indicated that these characters contributed positively towards yield, and should be considered when selecting for high grain yield. Path coefficient analysis showed that biological yield and harvest index showed positive direct effect and highly significant positive correlation. Therefore much attention should be given to them as there traits are helpful for direct selection.

Acknowledgements

The authors would like to thank Jimma University College of Agriculture and Veterinary Medicine (JUCAVM) and Kulumsa Agricultural Research Center (KARC) for their financial support.

References

1. Garcia MLF, Ramos JM, Garcia MMB, Jimenez TP (1991) Ontogenetic approach to grain production in spring barley basedon path coefficient analysis. Crop Science 31: 1179-1185.

2. Simane B, Struik PC, Nachit MM, Peacock JM (1993) Ontogenic analysis of field components and yield stability of durum wheat in water limited environments. Euphytica 71: 211-219.

3. Shewry P (2009) Increasing the health benefits of wheat. FEBS Journal 276: 71-71.

4. Belay G (2006) Triticum turgidum L. In: Cereals and pulses/Céréales et légumes secs, PROTA, Wageningen, Netherlands.

5. Badebo A, Gelalcha S, Ammar K, Nachit MM, Abdalla O (2009) Overview of durum wheat research in Ethiopia: Challenges and prospects Paper presented at the Proceedings of Borlaug Global Rust Initiative technical workshop, Cd. Obregon, Sonora, Mexico.

6. Singh SP, Dewivedi VK (2002) Character association and path analysis in wheat. Agric Sci Digest 22: 255-257.

7. Gashaw A, Mohammed H, Singh H (2007) Selection criterion for improved grain yields in Ethiopian durum wheat genotypes. Afr Crop Sci J 15: 25-31.

8. Khan AS, Ashfaq M, Asad MA (2003) A correlation and path coefficient analysis for some yield components in bread wheat. Asian J of Plant Sci 2: 582-584.

9. Flexas J, RibasCarbo MIQ, DiazEspejo ANT, Galmes J, Medrano H (2007) Mesophyll conductance to CO2: current knowledge and future prospects. Plant Cell and Enviro 31: 602-621.

10. Ahmad B, Khalil IH, Iqbal M, Hidayat UR (2010) Genotypic and phenotypic correlation among yield components in bread wheat under normal and late plantings. Sarhad J Agric 26: 259-265.

11. Leilah AA, Al-Khateeb SA (2005) Statistical analysis of wheat yield under drought conditions. Elsevier 61: 483-496.

12. Zecevic V, Kenezevic D, Micanovic D (2004) Genetic correlations and path coefficient analysis of yield and quality components in wheat (Triticum aestivum L). Genetika 36: 13-21.

13. Tahir MHN, Sadaqat HA, Bashir S (2002) Correlation and path coefficient analysis of morphological traits in sunflower (Helianthus annuus L) populations. Int J Agric Biol 4: 341-343.

14. GENRES statistical software (1994) Data entry module for pascal intl soft ware solution.

15. Miller PA, Williams JC, Robinson HF, Comtsock RF (1958) Estimation of genotypic environmental variances and co variances in upland cotton and their implication in selection. Agronomy journal 50: 126-131.

16. Sharma JR (1998) Statistical and biometrical techniques in plant breeding. New Age International Publication, New Delhi, p: 432.

17. Snedecor GW, Cochran WG (1989) Statistical Methods. 8th edn. Ames: Iowa State Press.

18. Dewey DR, Lu RH (1959) A correlation and path coefficient analysis of components of crested wheat grass seed production. Agron J 51: 515-518.

19. Abinasa M, Ayana A, Bultosa G (2011) Genetic variability, heritability and trait associations in durum wheat (Triticum turgidum L var durum) genotypes. African Journal of Agricultural Research 6: 3972- 3979.

20. Tsegaye D, Tadesse D, Yigzaw D, Getnet S (2012) Genetic ariability correlation and path analysis in durum wheat germplasm (Triticum durum Desf). African Journal of Biotechnology 1: 107-112.

21. Yousaf A, Atta BM, Akhter J, Monneveux P, Lateef Z (2008) Genetic variability, association and diversity studies in wheat (triticum aestivum L.) germplasm. Pak J Bot 40: 2087-2097.

22. Tila M, Sajjad H, Muhammad A, Muhammad IK, Roshan Z (2005) path coefficient and correlation studies of yield and yield associated traits in candidate bread wheat (triticum aestivuml) lines. Suranaree J Sci Technol 13: 175-180

23. Akcura M (2011) The relationships of some traits in Turkish winter bread wheat landraces. Turk J Agric For 35: 115-125.

24. Ali IH, Shakor EF (2012) Heritability, variability, genetic correlation and path analysis for quantitative traits in durum and bread wheat under dry farming conditions. Mesoptamia J Agric 40: 27-39.

25. Peymaninia Y, Valizadeh M, Shahryari R, Ahmadizadeh M (2012) Evaluation of morpho physiological responses of wheat genotypes against drought stress in presence of a leonardite derived humic fertilizer under greenhouse condition. The journal of animal and Plant Sciences 22: 1142-1149.

26. Iqbal N, Aqsa T, Hameed A, Arshad R (2007) Evaluation of Pakistani wheat germplasm for bread quality based on allelic variation in HMW glutenin subunits. Pak J Bot 43: 1735-1740.

27. Majumder DAN, Shamsuddin AKM, Kabir MA, Hassan L (2008) Genetic variability, correlated response and path analysis of yield and yield contributing traits of spring wheat. J Bangladesh Agric Univ 6: 227-234.

28. Iftikhar R, Khaliq I, Ijaz M, Rashid MA (2012) Association Analysis of Grain

Yield and its Components in Spring Wheat (Triticum aestivumL.). American Eurasian J Agric & Environ Sci 12: 389-39.

29. Bilgin O, Korkut KZ, Başer İ, Dağlioğlu O, Öztürk İ, et al. (2008) Determination of variability between grain yield and yield components of durum wheat varieties (Triticum durum desf.) in Thrace region. Journal of Tekirdag Agricultural Faculty 5: 101-109.

30. Ali Y, Atta BM, Akhter J, Monneveux P, Lateef Z (2008) Genetic variability, association and diversity studies in wheat (Triticum aestivum L.) germplasm. Pak J Bot 40: 2087- 2097.

31. Dutamo D, Alamerew S, Eticha F, Assefa E (2015) Path Coefficient and Correlation Studies of Yield and Yield Associated Traits in Bread Wheat (Triticum aestivum L.) Germplasm. World Applied Sciences Journal 33: 1732-1739.

32. Izzat SA, Tahir ISA, Abdalla BE, Abu EH, Ibrahim S, et al. (2000) Genetic improvement in grain yield and associated changes in traits of bread wheat cultivars in the Sudan. In: CIMMYT 2000 The Eleventh Regional Wheat Workshop for Easter Central and Southern Africa. Addis Ababa Ethiopia: CIMMYT, p: 60-66.

33. Hannachi A, Fellahi ZEA, Bouzerzour H, Boutekrabt A (2013) Correlation, Path Analysis and Stepwise Regression in Durum Wheat (Triticum Durum Desf.) under Rainfed Conditions. Journal of Agriculture and Sustainability 3: 122-131.

34. Oğuz B, Kayihan ZK, İsmet B, Orhan D, İrfan Ö (2011) Genetic variation and interre-lationship of some morpho-physiological traits in durum wheat (triticum durum L.) desf. Pak J Bot 43: 253-260.

35. Gelalcha S, Hanchinal RR (2013) Correlation and path analysis in yield and yield components in spring bread wheat (Triticum aestivum L.) genotypes under irrigated condition in Southern India. African Journal of Agricultural Research 8: 3186-3192.

36. Dogan R (2009) The correlation and path coefficient analysis for yield and some yield components of durum wheat (Triticum turgidumvar. durumL.) in west Anatolia conditions. Pak J Bot 41: 1081-1089.

The Benefits of Crop Rotation Including Cereals and Green Manures on Potato Yield and Nitrogen Nutrition and Soil Properties

Adrien N'Dayegamiye[1], Judith Nyiraneza[2], Michèle Grenier[1], Marie Bipfubusa[1] and Anne Drapeau[1]

[1]Research and Development Institute for the Agro-Environment (IRDA), Complexe Scientifique, Canada
[2]Agriculture and Agro-Food Canada, Crops and Livestock Research Centre, 440 University Avenue, Charlottetown, Canada

Abstract

Soil quality decline is a common concern in potato production systems. Including cereals and green manures (GMs) in potato rotation could improve soil productivity and sustain potato (*Solanum tuberosum* L.) yields and quality. This experiment initiated in Québec, eastern Canada, assessed the effects of two cycles of 2-yr potato rotations with cereals and GMs on soil properties and potato yield and quality, and on disease incidence from 2008 to 2011. Three cereals [corn (*Zea mays*), barley (*Hordeum vulgare*), and oat (*Avena sativa*)] seeded in spring, three summer GMs [mustard (*Sinapsis alba*), japanese millet (*Setaria italica*), and pearl millet (*Pennisetum glaucum*)], four fall GM crops [oat, mustard, wheat (*Triticum aestivum*), and rye (*Secale cereale*)], and continuous potato (*Solanum tuberosum* L.) were grown on main plots in 2008 and 2010. In following years (2009 and 2011), each main plot was split, and five N fertilizer rates (0, 50, 100, 150 and 200 kg N ha^{-1}) were applied to potato. After two rotation cycles in 2011, soils under cereals and summer and fall GMs had higher soil water-extractable organic C and N contents, nitrate levels, soil respiration, urease and dehydrogenase activities, and a larger proportion of soil macro aggregates (0.25 to 2 mm), compared with the continuous potato. Cereals and summer GMs increased marketable potato yield and specific gravity, whereas fall GMs increased tuber yield but reduced tuber quality. In addition, fall GMs favored the incidence of common scab and black scurf. Summer GMs had higher potato N uptake and N efficiency compared to cereals and fall GMs. Although fall GMs produced higher net returns than cereals and summer GMs, they may not represent a viable long term option to sustain potato production and to enhance soil quality. Results indicated that growing cereals or summer GMs in rotation with potato is an interesting alternative to improve soil properties while sustaining potato yield and quality. A fall GM with a better growth during the fall season may sustain potato yield and quality, if included in longer potato rotation, but this remains to be determined.

Keywords: Cereals in potato rotation; Summer and fall green manures; Potato yield and N efficiency; Disease incidence; Net benefit; Soil properties changes

Introduction

Potato is an important crop worldwide. In the last 50 years, yields of potato have been relatively constant in spite of the increase in chemical inputs [1]. Nitrogen inputs are often higher than potato requirements to ensure acceptable tuber yield and quality [2]. However, excess N can delay maturity and senescence of plants and decrease yields and specific gravity of the tubers [3]. Moreover, residual N is prone to be leached causing environmental issues [4] especially on sandy soils [5], which are well adapted to potato production because of their low water-holding capacity. The potential of N losses is also exacerbated by the low potato N use efficiency due to its shallow and poorly developed root system [6]. Overall, the potato crop returns low amount of organic residues to the soil and frequent potato cropping often leads to soil structural degradation due to the depletion of soil organic matter [1,7].

Management practices that add organic matter such as crop rotations and GMs have proved to be efficient in enhancing the sustainability of intensive potato production [8], with positive effects on potato yields and soil organic matter content, structure, microbial biomass, and fertility [1,7,8]. Once incorporated into the soil, green manure crops may contribute to following potato nutrition because much of the nutrients retained by GMs become available through decomposition and mineralization processes [9]. According to Weinert et al. [10], GMS can provide to the following crop between 20% and 55% of their N fertilizer requirements [10]. However, the ability of GMs to release N depends on their chemical composition, including the C: N ratio, and lignin and polyphenol contents, and on environmental conditions such as soil moisture and temperature [11,12]. Furthermore, the N supply

from GMS varies depending on its yield and N content [3,12]. To maximize the N use efficiency while minimizing the N loss, the amount and the timing of N release from GMS must be synchronized with the subsequent potato N needs for maximum tuber yields [13].

Compared to non-legumes, legume crops are more important sources of N for subsequent potato crop since they can biologically fix significant amounts of N$_2$ [13]. In some potato-based crop rotations, different legume crops have been found to supplying 25 to 260 kg N ha^{-1}[14]. However, legumes are not popular green manure crops in potato based rotations because farmers are concerned with the potential increase in diseases such as common scab [7,14] representing the fourth most economically important disease for potato production in North America [15]. Moreover, establishment costs of legume crops can also be 10 times higher than grasses species [14]. Non-legume green manure crops could have high potential as alternative rotation crops for potato and provide significant quantity of N to subsequent crops by capturing soil nutrients. When grown in the fall after harvest of the main cash crop, GMS can be used to retain residual fertilizer N and other

*Corresponding author: Adrien N'Dayegamiye, Research and Development Institute for the Agro-Environment (IRDA), 2700 Einstein, Complexe Scientifique, D 1 110, Quebec, G1P 3W8, Canada
E-mail: adrien.ndaye@irda.qc.ca

nutrients that could otherwise be leached out of the cropping systems. Typical non-legume species used as green manure include cereals such as rye, millet, wheat, oat, and buckwheat (*Fagopyrum esculentum*) and *Brassicaceae* such as mustard, canola, and radishes [10,12,13]. Crop rotations including GMs may also enhance potato yields and quality by breaking the cycle of weeds, pests, and diseases, although their action mechanisms are varied and often unknown [16]. *Brassica* species are often considered the most effective to suppress soil-borne diseases, likely because of their high contents in glucosinolates which breakdown in toxic compounds such as isothiocyanates [14,17]. Incorporation of GMs may also stimulate other microorganisms that help biological control of potato diseases through competition [16,18].

Potato has a high value by unit surface cultivated and, thus, potato producers prefer short rotation cycles [7]. Despite its potential multiple benefits on soil quality and potato productivity, growing GMs during the regular cropping season usually results in a loss of income [14]. On the other hand, GMs planted in the fall must have rapid growth, great biomass production and nutrient scavenging ability. In eastern regions of Canada, continuous potato cultivation has been progressively replaced by crop rotations with corn and spring cereals such as oat, barley, and rye [19] that contribute to add significant quantities of organic matter to the soil. However, choosing the best crop rotation for potato is still challenging. Potato producers prefer to include cash crops in potato rotation, or to grow fall GMs after potato harvest. The impact on soil properties and potato yield and quality may depend on the crop specie included in rotation. This 4 year study (two successive cycles of a 2 y potato crop rotations) aimed to compare the different potato rotations with such as cereals, including oat, barley, and corn, and with GMs seeded either in summer (mustard, Japanese millet and pearl millet) or in the fall (oat, mustard, wheat and rye) on potato yield, net income, N nutrition, disease incidence and on soil properties changes.

Materials and Methods

Site description and experimental design

A 4 yr study was conducted from 2008 to 2011 at the research station of the Institut de Recherche et Développement en Agroenvironnement located in Deschambault, Québec, eastern Canada (46°41'27"N; 71°58' 18" O). The soil was a fine sandy loam of the Batiscan series (sandy over clayey, mixed, non-acid, frigid, Typic Humaquept) with on Average 490 g kg^{-1} sand, 300 g kg^{-1} silt and 210 g kg^{-1} clay. The initial soil pH (water)

was 6.3, and the soil total C and N contents were 40.0 g kg^{-1} C and 1.9 g kg^{-1} N. The soil available P and K contents were 81.0 and 46.0 mg kg^{-1}, respectively. Cereal and green manure crops were grown in 2008 and 2010, while potato was grown in 2009 and 2011. During the four cropping seasons (May-September), mean air temperatures were 14.2, 13.2, 15.6 and 15.5°C in 2008, 2009, 2010 and 2011, respectively and total precipitation was 749.6, 681.6, 517.2, and 468.3 mm, respectively (Table 1). In the four years of the experiment, potato crop was not irrigated.

This research was conducted to determine the effect of various previous crops and GMs on potato yield, disease incidence and soil properties in two cycles of a two-year potato rotation. Previous cash crops (cereals), summer GMs, fall GMs, and continuous potato were compared. In the years following cereals and GMS cultivation as main treatments, the plots were split to accommodate 5 fertilizer N rates. The experiment was conducted over 4 years using a randomized complete block design in years with the crop rotations, and a split-plot design in years when N rates were tested.

The experiment was laid out in a complete randomized block design with three replications. Each block was divided in 11 experimental units of 30 by 6 m. Treatments established in 2008 and 2010 on the same field consisted of a control with continuous potato (*Solanum tuberosum* L.), cereal-potato rotations [corn (*Zea mays*), barley (*Hordeum vulgare*) and oat (*Avena sativa*)], summer GMs [mustard (*Sinapsis alba*), japanese millet (*Setaria italica*) and pearl millet (*Pennisetum glaucum*)], and potatoes followed by fall GMs [oat, mustard, wheat (*Triticum aestivum*) and rye (*Secale cereale*)].

The periods of seeding and harvest of the different crops are summarized in Table 2. Cereals and summer GMs were seeded in May 2008 and 2010, whereas potato was planted at the same period in plots for the fall GMS and control treatments. Fall GMs were established after potato harvest at the end of August in 2008 and in the mid-September in 2010 due to heavy rainfall that delayed potato harvest. To avoid the use of a same cultivar in the four years of the experiment, potato (cv. Norland) which is a common cultivar in eastern Canada and is tolerant to common scab, was planted at 43000 pl ha^{-1} in the years of crop rotation establishment (2008 and 2010). Corn hybrid used in the experiments were Dekalb-343-2550 corn heat units (CHU) at 80,000 plants ha^{-1}, and oat (cv Rigodon) and barley (cv. Myriam) were seeded at 120 and 150 kg ha^{-1}, respectively. Mustard, pearl millet and Japanese millet as summer GMs were seeded at 21,40 and 30 kg ha^{-1}, respectively. Fall GMs (oat, mustard, wheat, and rye) were seeded at 120, 21,160 and 120 kg ha^{-1}, respectively. Potato and corn received N fertilizer at a rate of 170 kg N ha^{-1} as calcium ammonium nitrate (27%). Phosphorus and K were applied at a rate of 50 kg P$_2$O$_5$ ha^{-1} as super-phosphate and K at a rate of 150 kg K$_2$O ha^{-1} as KCl. Nitrogen fertilizers for oat, barley and summer GMs were applied at a rate of 50 kg N ha^{-1}, 50 kg P$_2$O$_5$, and 150 kg K$_2$O ha^{-1} with the same fertilizer types as above. Fall GMs were seeded after potato harvest and were not fertilized as they may use residual fertilizer N that could otherwise be leached out of the field.

Cereal grain yields were determined by harvesting in mid-August a surface of 1.5 by 6 m in the center of each plot with a combine. Corn was harvested in late October in the two central rows from each plot. After harvest, cereal straws and corn stalks were collected, weighed, and organic residues left on the soil. Summer GMs were harvested in August at the flowering stage. Total aboveground biomass was determined and the biomass was returned to the soil and incorporated by disc harrowing to a depth of 10 cm. In both years (2008 and 2010), the pearl millet biomass was chopped in August and left on the soil, and the plants were left to regrow in the fall. Potato in the check plots and in soils with fall

Temperature(°C)				
Months	2008	2009	2010	2011
May	9.4	9.4	12.3	14.8
June	16.4	13.6	15.8	19.3
July	18.2	17.3	20.4	21.2
August	17.3	15.2	18.2	17.9
September	12.8	12.3	13.6	14.7
October	10.9	11.6	12.5	10.8
Rainfall (mm)				
May	59.4	97.5	68.9	92.3
June	180.9	76.2	74.6	52.3
July	167.7	151.9	92.8	44.2
August	106.1	140.5	77.7	91.8
September	103.4	88.8	90.8	80.8
October	132.3	126.7	112.4	106.2

Table 1: Temperature and rainfall at Deschambault Research Station (2008-2011).

Crops	Seeding Rates	Seeding Periods	Harvest/growth termination Times 2008
Potato	43000 pl ha⁻¹	21 May	25-Aug
Oat	120 kg ha⁻¹	25 May	17-Sep
Barley	150 kg ha⁻¹	26-May	17-Sep
Corn	80000 grains	30-May	30-Oct
Mustard	21 Kg ha⁻¹	25-May	14-Aug
Japanese Millet	40 kg ha⁻¹	25-May	14-Aug
Pearl Millet	30 kg ha⁻¹	26-May	21-Aug
Potato/	43000 pl ha⁻¹	21-May	25-Aug
Oat	120 kg ha⁻¹	27-Aug	04-Nov
Potato/	43000 pl ha⁻¹	21-May	25-Aug
Mustard	21 Kg ha⁻¹	27-Aug	04-Nov
Potato/	43000 pl ha⁻¹	21-May	25-Aug
Wheat	165 kg ha⁻¹	27-Aug	04-Nov
Potato/	43000 pl ha⁻¹	21-May	25-Aug
Rye	120 kg ha⁻¹	27-May	04-Aug

Crops	Seeding Rates	Seeding Periods	Harvest/growth termination Times 2010
Potato	43000 pl ha⁻¹	21 May	25-Aug
Oat	120 kg ha⁻¹	25 May	17-Sep
Barley	150 kg ha⁻¹	26-May	17-Sep
Corn	80000 grains	30-May	30-Oct
Mustard	21 Kg ha⁻¹	25-May	14-Aug
Japanese Millet	40 kg ha⁻¹	25-May	14-Aug
Pearl Millet	30 kg ha⁻¹	26-May	21-Aug
Potato/	43000 pl ha⁻¹	21-May	25-Aug
Oat	120 kg ha⁻¹	27-Aug	04-Nov
Potato/	43000 pl ha⁻¹	21-May	25-Aug
Mustard	21 Kg ha⁻¹	27-Aug	04-Nov
Potato/	43000 pl ha⁻¹	21-May	25-Aug
Wheat	165 kg ha⁻¹	27-Aug	04-Nov
Potato/	43000 pl ha⁻¹	21-May	25-Aug
Rye	120 kg ha⁻¹	27-May	04-Aug

Table 2: Crops in potato rotation in 2008 and 2010, and seeding and harvest periods. Potato was seeded as subsequent crop in the same fields in 2009 and 2011.

GMs was harvested at the end of August in 2008 and in mid-September in 2010 on 6 m row sections of the two middle rows of each plot to measure tuber yields. Pearl millet and fall GMs were hand harvested in late November on three 6 m, and the total aboveground biomass was determined and returned to the soil. Thereafter, all plots of the experiment were tilled to a depth of 15 cm. At each crop harvest, organic residues and vegetative biomass were sampled to determine dry matter and N contents.

In 2009 and 2011, the main plots with the rotations crops were divided in five subplots to which five N fertilizer rates for potato were randomly assigned: 0, 50, 100, 150, and 200 kg N ha⁻¹. The experiment

design was a split-plot with crop rotations seeded in the same fields in 2008 and 2010 as main plots, and N fertilizer rates as subplots in 2009 and 2011. Each subplot was 6 m long and 6 m wide, and consisted of 6 rows with 0.90-m row spacing. Nitrogen fertilizer was applied as calcium ammonium nitrate (27%) and was split-applied in two fractions: the first 50 kg N ha⁻¹ as a starter fertilizer and the remaining N fertilizer was added at the tuber initiation development stage. Phosphorus and K fertilizers were applied at seeding at rates of 50 kg P_2O_5 and 150 kg K_2O ha⁻¹ as triple superphosphate and KCl, respectively. Potato (cv Snowden) is the most popular cultivar in potato chips industry, and is susceptible to common scab. To avoid the disease incidence, potato (cv. Norland) which is tolerant to common scab was planted in 2008 and 2010 in the treatments with fall GMs and with continuous potato as control. In spring 2009 and 2011, potato (cv. Snowden) was planted at 43000 plants ha⁻¹ Titan treatments were applied against potato beetles, Quadris against black scurf (*Rhizoctonia solani* Kühn), Allegro 500 EF against potato fungal diseases (*Phytophtora infestans* and *Alternaria solani*), and pentachloronitrobenzene (PCNB) for common scab (*Streptomyces scabies*)

Soil and plant sampling and analysis

Soil samples were collected in spring 2008 at 20 cm depth in the experimental field before initiation of the experiment. Ten soil cores were taken randomly with a 2 cm diam. stainless auger (Oakfield model B, Oakfield Apparatus Co. Oakfield, WI), and bulked to make one composite soil sample. Twenty soil samples were then obtained, air-dried and sieved to pass a 2-mm sieve to determine soil pH, soil texture, and available P and K contents. Sub-samples were ground to pass a 0.25-mm sieve for total C and N analyses. Soil pH was measured in 1:1 soil/water solution. Extractable soil P and K contents were determined in a Mehlich III solution [20] and measured on inductively coupled plasma optical emission spectrometer (Perkin Elmer 4300 DV, Boston, MA). The soil C and N contents of the whole soil were determined by dry combustion using an automated analyzer (Leco C-N 1000, LECO, St. Joseph, MI). Particle size analysis (texture) was performed using the pipette method after the destruction of organic matter with H_2O_2 and dispersion with sodium hexametaphosphate [21].

In 2011, soil samples were collected before potato seeding and fertilizer application, in main plots with previous rotational crops. Ten soil cores were taken from each plot at 0-30 cm depth and pooled to make one composite soil sample, air-dried and sieved to pass a 2 mm sieve for pre-plant soil nitrate test (PPNT). A sub-sample was ground to pass a 0.25 mm sieve for C and N analyses. At the potato tuber initiation growth stage, soil samples were collected only in the 0 kg N ha⁻¹ N rate treatment of each previous crop at 0-30 cm depth to determine changes in soil physical and biological properties and soil pre-sidedress (PSNT) contents following two cycles of 2 yr potato rotations with cereals and green manure crops. Ten soil cores were taken between rows in each plot, pooled to make one composite soil sample, sieved to pass a 2 mm sieve and then stored at 4°C until analysis. A portion of the moist soil samples was used to assess various microbial activities (respiration, urease, alkaline phosphatase, and dehydrogenase).

Soil respiration was measured according to the soil incubation method as described by Anderson [22]. Briefly, wet soil samples (100 g) were incubated in triplicate in 1-L Erlenmeyer flasks at 30° for 20 days. The quantity of CO_2 released was trapped using 10 mL of 1.0 M NaOH and determined by titration after adding 1.5 M $BaCl_2$ in excess. Dehydrogenase activity was determined by colorimetric measurement of triphenyltetrazolium formazan (TPF) produced by the reduction of 2,3,5-triphenyltetrazolium chloride (TTC) according to the method of Casida et al. [23]. Alkaline phosphatase activity was determined by colorimetric measurement of the p-nitrophenol released when 1 g of field-moist sample was incubated with 4 mL of buffered (pH 11) sodium p-nitrophenyl phosphate solution, 0.2 mL of toluene and 1 mL of p-nitrophenol phosphatase at 37°C for 1 h [24]. Urease activity was determined on field-moist soil samples (2.5 g) incubated at 37°C for 2 h and the amount of NH^+ produced was determined by N spectrophotometry at 636 mM using indophenol blue [24]. The concentrations of soil nitrate before seeding (PPNT) and at post-seeding (PSNT) were determined with 2M KCl [25]. The soil C and N contents of the whole soil were determined by dry combustion using an automated analyzer (Leco C-N 1000, LECO, St. Joseph, MI). Water-extractable organic C (WEOC) and N (WEON) were extracted with cold water on air-dried soil ground to 0.25 mm following the method of Curtin et al. [26]. The concentration of WEOC was determined by wet combustion (Model TOC, 5000, Shimadzu Corp., Kyoto, Japan) and the concentration of WEON with an automated colorimeter (Model AA II, Technicon Instruments, Tarrytowmn, NY).

Three intact soil blocks of about 600 g (0-30 cm depth) were taken between rows in the 0 kg N ha^{-1} N rate treatment of each previous crop with a spade, to assess the soil water-stable aggregates. Soil blocks were sieved at 5 mm in the field and kept at 4°C until analysis. Water-stable soil macroaggregates were determined by the wet-sieving method. Forty grams were put on the top of a series of sieves (5,2,1, and 0.25 mm), which were immersed in water and shaken for 10 min. The soil fractions recovered on each sieve were dried at 65°C for 24 h, corrected for sand content and expressed as a percentage of total dry soil [27]. Aggregate mean weight diameter (MWD) was calculated according to Haynes and Beare [28].

Potato yield, quality and N efficiency determination

In 2009 and 2011, potato was harvested on 10 m row sections of the two middle rows of each plot to measure tuber yields. The tubers were weighed and tuber size categories and marketable (47-114 mm diam.) yield determined. A representative 4 kg sub-sample of tubers was taken from each plot to determine tuber dry matter, specific gravity, N concentration, and presence of diseases. Ten representative tubers from each plot were quartered along the long axis, and one quarter from each tuber was sliced into strips of 1 cm × 1 cm, weighed, oven dried at 55°C, and re-weighed to determine dry matter content. The tuber samples were ground to pass a 0.15 mm sieve for total N determination. Potato specific gravity was determined from three samples per plot by using the "weight in air/weight in water" method. Randomly selected tubers from each plot were rinsed with water and rated for common scab (*Streptomyces scabiei*) and black scurf (*Rhizoctonia solani* Kühn) occurrence. Disease severity was determined using a rating scale of zero to six based on the percentage of tuber surface covered with scab or black scurf lesions, i.e., 0, 0%; 1, 1%-5%; 2, 6%-15%; 3, 16%-25%; 4, 26%-35%; 5,36%-60% and 6.61%-100% [29]. The C and N concentration of potato tubers and green manures were determined by dry combustion (Leco C-N 1000, LECO, St. Joseph, MI, USA).

Potato nitrogen fertilizer efficiency (NFE) was determined as described by Wortman et al. [30]: NFE (kg tuber kg^{-1} N applied)=(Tuber yield from N fertilized plot-Tuber yield from unfertilized plot)/N applied × 100; Nitrogen recovery efficiency (NRE) was calculated as per Zvomuya et al. [31]: NRE=(N_{treat}-$N_{control}$)/$N_{applied}$ × 100, where N_{treat} represents the amount of nitrogen stored in the tubers of a given fertilizer treatment, $N_{control}$ is the amount of N stored in the tubers of the control plot.

Economic analysis

Net returns were determined annually for each crop rotation system as the difference between gross margins and total variable costs, both of which were determined using the local market prices published in economic references [32]. Variable costs included inputs costs (seeds, fertilizers, limestone, pesticides, fuel, oil, and lubricants), field operation costs (tillage, harrowing, sowing, pulverizing, harvest), sale costs related to drying grain (propane gas and electricity, dryer maintenance) and crop insurance. Gross margins were generated by sale of grains, straw, and income from the local crop insurance program (Programme d'assurance de stabilisation des revenus agricoles, ASRA, Financière Agricole du Québec). Gross margins were computed per hectare by multiplying the yield of each crop by its local market price.

Statistical methods

Data were analyzed separately for each year of the experiment (2009 and 2011) and analysis of variance performed using PROC MIXED of SAS [33] treating potato preceding crops and N fertilizer rates as fixed effects and block as random. Type III F tests were performed for the fixed factors and a priori contrasts were used to compare treatment means when the preceding crop effect was found significant at the 0.05 probability level. Different crop groups were compared: 1. continuous potato, 2. cereal crops, 3. summer GMS, and 4. fall GMs. When the N rate effect was significant (P<0.05), orthogonal polynomial contrasts were performed to assess if effect of the N fertilizer effect was linear or quadratic. For soil properties changes measured in 2011, a one-way ANOVA was used to compare the effect of the preceding crops on the soil parameters. Mixed procedure of SAS was used to conduct statistical analysis treating the preceding crops as fixed factors and blocks as random. Crop groups as above were compared using contrasts when the preceding crop effect was significant at 0.05 probability level. A one-way ANOVA was also used to compare physical and chemical characteristic of rotational crops in 2008 and 2010. A priori contrasts were also used to compare the different preceding crops.

Results and Discussion

Green manure biomass and nitrogen accumulation

Quantities of biomass, carbon and nitrogen incorporated in soil from different previous crops: For summer GMs, the amounts of aboveground biomass varied from 4.1 to 8.4 Mg dry matter ha^{-1} in 2008 and from 5.3 to 8.8 Mg ha^{-1} in 2010, and were the lowest for yellow mustard (Tables 3 and 4). The amounts of C returned to the soil following the incorporation of summer GMs varied from 1558 to 3336 kg C ha^{-1} in 2008 and from 2067 to 3520 kg C ha^{-1} in 2010. The amounts of N returned to the soil varied between 78 and 95 kg N ha^{-1} in 2008 and between 124 and 138 kg N ha^{-1} in 2010, and were highest for summer GMs over two years. Despite its low biomass yield, yellow mustard supplied similar amounts of N as the other summer GMs, likely due to its greater N scavenging ability as compared with Japanese and pearl millet. In both years, fall GMs produced lower biomass yields (1.2-2.0 Mg ha^{-1}) and supplied less C (444-840 kg C ha^{-1}) and N (47-77 kg N ha^{-1}) to the soil than summer GMs (Tables 3 and 4). This could be explained in part by the shorter period of time elapsed between planting and harvesting of fall GMs, as compared with summer GMs. Average air temperatures are likely lower in the fall than in the summer, and the days are shorter with less daily global radiation. Fall GMs aboveground biomass, C and N contents returned to the soil were lower in 2010 than in 2008 because heavy rainfall in August and September 2010 altered crop growth. Corn stalk yields were 4.3 and 3.3 Mg ha^{-1} in 2008 and

Crops	Dry matter Mg ha⁻¹	C kg ha⁻¹	N kg ha⁻¹	C g kg⁻¹	N g kg⁻¹	C/N
Summer GMs£						
Pearl millet	8.3	3336	83	402	12	34
Yellow mustard	4.1	1558	95	380	24	16
Japanese millet	7.2	2786	78	387	12	32
Fall GMs						
Oat	2	840	53	420	28	15
Wheat	1.9	798	52	420	27	16
Mustard	2	744	77	372	39	10
Rye	2.1	873	59	416	30	14
Cereals						
Corn	4.3	1892	21	440	4.9	90
Barley	2.9	1282	16	442	5.6	79
Oat	2.6	1139	17	438	6.4	68
Analysis of Variance (P value)						
Crop effect	<0.0001	<0.0001	<0.0001	ns	<0.0001	<0.0001
Contrasts						
1,2,3 vs 4,5,6,7	<0.0001	<0.0001	<0.0001	ns	<0.0001	<0.0001
1,2, 3 vs 8,9,10	<0.0001	<0.0001	<0.0001	ns	<0.0001	<0.0001
4,5,6.,7 vs 8,9,10	<0.001	<0.0001	<0.0001	ns	<0.0001	<0.0001

£GMs: green manures; ns, not significant.

Table 3: Dry matter, carbon and nitrogen contents, and C/N of different crops in 2008.

2010, respectively and barley and oat straw yields were 2.7 and 2.3 Mg ha⁻¹, respectively. The N concentrations of corn stalk and cereal straws were low (4.9 to 7.5 g kg⁻¹) and their N input in the soil varied from 15 to 21 kg ha⁻¹ (Tables 3 and 4). In 2008 and 2010, the C returned to the soil by cereal residues varied from 913 to 1892 kg C ha⁻¹.

The C/N ratio for summer and fall GMs manures varied from 10 to 34, indicating that these organic residues may have a high mineralization rate, except for Japanese millet with a C/N of 34 which could induce a slight N immobilization. Organic mineralization rate is mainly controlled by their biochemical composition, such as C/N ratio and polyphenol, lignin, and nitrogen contents [34,35]. Fall GMs had lower C/N ratios than summer GMs (Tables 3 and 4), which suggests that they may decompose and release N more rapidly. The C/N ratio was 88 for corn stalk and 63 and 74 for oat and barley, showing that these organic residues with high C/N ratios may decompose slowly in soil, and their influence on soil microbial activity and aggregation is low [36]. The author observed that potato tuber yield declined following oat cover crop with a high C/N ratio due to its low rate of N mineralization. Results showed that the biochemical composition of the studied crops was different, therefore, their effects on soil N availability and on soil properties changes may differ.

Effect of cereals and green manures on soil properties soil total and dissolved carbon and nitrogen contents: Total soil C and N contents varied from 21 to 23 g C kg⁻¹ and from 1.6 to 1.8 g N kg⁻¹ (Table 5). Soil total C and N contents were not significantly increased after two cycles of crop rotations with cereals (barley, oat and corn) and summer and fall GMs. This finding is in accordance with other studies reporting that changes in soil organic matter content are gradual and not detectable in short-term rotations [37]. Contradictory results, however, were obtained by other investigators [12,38]. For instance, N'Dayegamiye and Tran [12] found that cultivating GMs, such as

millet, mustard, and canola for one year significantly increased C and N contents of a silt loam soil. These results may be related to different conditions under which the experiments were conducted, including the use of different GMs species, different amounts and quality of green manure biomass returned to the soil, climatic conditions, and different soil types. It is well documented that clay soils retain more organic matter than sandy soils [39], because organic matter associated with clay minerals benefit more physical and chemical protection against decomposition than that associated with sand [35,40]. Potato is mostly cultivated on sandy soils and the decline of soil organic matter is of much concern. The low soil organic matter and N accumulation observed in the present study indicates that the total quantities of organic residues (3 to 16 Mg ha⁻¹) and C (1000 to 3800 kg C ha⁻¹) incorporated in soils in 2008 and 2010 were not sufficient to induce an increase in soil organic matter. N'Dayegamiye [35] showed that three paper-mill sludges applications at 40 to 60 Mg ha⁻¹ yr⁻¹ significantly increased the soil C content. Moreover, GMs as used in the study also contain readily mineralizable organic materials as suggested by their low C/N ratios (Tables 3 and 4), and that they were likely mineralized more rapidly in the sandy soil of the present study than in the silt loam studied by N'Dayegamiye et Tran [12] and N'Dayegamiye [35]. Soil water extractable C and N (WEOC, and WEON) concentrations varied from 110 to 140 mg C kg⁻¹, and from 15 to 24 mg N kg⁻¹, respectively (Table 5). By contrast to soil total C and N contents, soil WEOC and WEON concentrations were both significantly (p<0.01) increased by the preceding crops in comparison with continuous potato. While WEOC did not vary significantly among the preceding crops, contrast analysis showed that WEON concentrations were larger in soils previously cropped to cereals and summer GMs than in those previously cropped to fall GMs or under continuous potato (Table 5). On average, soil WEOC contents were 17% higher in soils under the cereal-potato rotations and following summer and fall GMs than in

Crops	Dry matter Mg ha⁻¹	C kg ha⁻¹	N kg ha⁻¹	C g kg⁻¹	N g kg⁻¹	C/N
Summer GMs£						
Pearl millet	8.8	3520	133	400	15	26
Yellow mustard	5.3	2067	138	390	25	16
Japanese millet	8.3	3237	131	390	15	26
Fall GMs						
Oat	1.2	504	48	420	40	11
Wheat	1.2	504	53	420	45	9
Mustard	1.2	444	47	370	36	10
Rye	1.3	559	53	430	40	10
Cereals						
Corn	3.3	1468	17	445	5.1	87
Barley	2.5	1102	15	441	6.2	70
Oat	2.1	913	16	435	7.5	58
Analysis of Variance (P value)						
Crop effect	<0.0001	<0.0001	<0.0001	ns	<0.0001	<0.0001
Contrasts						
1,2,3 vs 4,5,6,7	<0.0001	<0.0001	<0.0001	ns	<0.0001	<0.0001
1,2, 3 vs 8,9,10	<0.0001	<0.0001	<0.0001	ns	<0.0001	<0.0001
4,5,6.,7 vs 8,9,10	<0.0001	<0.0001	<0.0001	ns	<0.0001	<0.0001

Table 4: Dry matter, carbon and nitrogen contents, and C/N of different crops in 2010.

those under continuous potato. These findings are in agreement with those reported by Asmar et al. [41] where the addition of fresh organic material to soil increased the amount of water-soluble organic matter.

Proceeding Crops	C g kg^{-1}	N g kg^{-1}	WEOC mg kg^{-1}	WEON mg kg^{-1}
Continuous potato	21	1.6	110	15
Barley	23	1.7	140	22
Corn	23	1.8	130	24
Oat	22	1.8	130	20
Mustard	23	1.8	130	21
Japanese Millet	21	1.8	120	16
Pearl Millet	23	1.9	120	28
Potato/mustard	23	1.8	140	17
Potato/wheat	22	1.8	140	17
Potato/Rye	23	1.9	120	18
Potato/Oat	23	1.8	120	21
Analysis of variance and contrasts (P value)				
Effect of the preceding crops	ns	ns	0.0038	0.0019
Summer GMs£ vs. fall GMs	ns	ns	ns	0.0453
Summer GMs vs. cereals	ns	ns	ns	ns
Fall GMs vs. cereals	ns	ns	ns	0.0289
Green manures + cereals vs. continuous potato	ns	ns	0.0523	0.0472

¥ 2, 3 and 4: cereal-crop rotations; 5, 6 and 7: summer GMs; 8, 9, 10 and 11: fall GMs cropped after potato harvest.

£ GMs: green manures; ns, not significant.

€ Soil sampling was done in 2011 in plots that received no N fertilizer (0 kg N ha)$^{-1}$.
Table 5: Effects of different preceding crops on soil total carbon and nitrogen, and water-extractable organic carbon (WEOC) and nitrogen (WEON) concentrations in 2011€.

These results also corroborate other studies showing that WEOC was a better indicator than total C to assess short-term changes in soil organic matter quality [42,43]. Although it accounts for a small proportion of soil organic matter, WEOC is considered the most bioavailable fraction [44] and is considered as a substrate for soil microbes [45]. In other studies, WEOC was found to be strongly correlated with macroaggregate stability, and it was suggested that it is involved in aggregates formation [46,47]. Thus, one might expect that cereal straw and corn stalks in the cereal-potato rotations and GMs which have increased water-soluble C might promote soil microbial activities and soil aggregation in the studied soil.

Soil microbial activities

Soil respiration, urease and dehydrogenase activities were higher in soils previously cropped to cereals and GMs compared with those under continuous potato (Table 6). Soil alkaline phosphatase activity was not significantly increased by the preceding crops. Contrast analysis showed that soil respiration and urease activity were higher in soils previously cropped to cereals than with GMs, and did not differ between summer and fall GMs treatments (Table 6). In comparison with continuous potato, soil respiration and urease activity were on average 43 and 53% greater in soils previously cropped with cereals, and 16 and 22% in those with GMs. Cereal straws and corn stalks were left on the soil and the C quantities applied to the soil were half of those added with summer GMs. The larger increase in microbial activities found in soils previously cropped to cereals suggests that more organic matter was still available for soil microorganisms likely because barley, oat and corn organic residues were more recalcitrant to microbial degradation than residues from the GMs and might thus lastlonger.

Our findings agree with others studies supporting that soil respiration, microbial biomass, and soil enzyme activities respond more quickly to changes in crop management than soil total C [35,48]. Increased soil respiration generally reflects an increase in soil

Proceeding Crops	PPNT mg NO^{-3}	PSNT N kg^{-}	Soil Respiration mg CO_2	Alkaline Phosphatase µg PNP g^{-1}	Urease µg N-NH4 g^{-1}	Dehydrogenase µg TPF^{-1}
Continuous potato	2	6	326	136	29	22
Barley	3	9	434	151	43	27
Corn	2.9	11	467	175	47	48
Oat	2.6	9.7	494	156	43	44
Mustard	2.8	12.6	387	143	36	32
Japanese Millet	2.3	10.3	378	131	34	39
Pearl Millet	3.6	11.7	434	162	40	49
Potato/mustard	2.9	6.8	374	133	32	24
Potato/wheat	2.5	6.7	369	139	37	41
Potato/Rye	2.9	8.3	361	153	33	42
Potato/Oat	3.4	11	354	135	35	28
Analysis of variance and contrasts (P value)						
Effect of the preceding crops	0.0479	0.0335	0.0313	ns	0.0452	0.0497
Summer GMs£ vs. fall GMs	ns	0.0492	ns	ns	ns	ns
Summer GMs vs. cereals	ns	ns	0.0041	ns	0.0067	ns
Fall GMs vs. cereals	ns	0.0518	<0.0001	ns	<0.0001	ns
Green manures+cereals vs. continuous potato	0.004	0.0417	0.0051	ns	0.0458	0.0384

¥2, 3 and 4: cereal-crop rotations; 5, 6 and 7: summer GMs; 8, 9, 10 and 11: fall GMs cropped after potato harvest.

£ GMs: green manures; ns, not significant.

€ Soil sampling was done in 2011 in plots that received no N fertilizer (0 kg N ha)$^{-1}$

Table 6: Effects of different preceding crops on pre-plant soil nitrate (PPNT), pre-sidedress soil nitrate (PSNT), soil respiration, and enzyme activities in 2011€.

microbial biomass [49] and, thereby, the greater respiration in soils previously cropped to cereals and GMs indicates that these preceding crops stimulated soil microbial growth and activity by supplying labile organic matter, as shown by soil WEOC (Table 5). The overall increases in soil respiration and enzyme activities should enhance the decomposition of soil organic matter and incorporated plant residues, thereby promoting soil aggregation and nutrient availability to the following potato crop [43].

Soil macroaggregates and mean weight diameter of aggregates

The proportion of soil macroaggregates (>0.25 mm diam.) varied between 31% and 62% while MWD varied from 0.78 and 1.41 mm (Table 7). Soil macroaggregates >0.25 mm represented more than 83% in silt loam which received paper mill sludge application [35] and 70% in clay loam soils with dairy cattle manure application [50]. In the present study, the preceding crops increased the proportions of soil macroaggregates 0.25-1 mm and 1-2 mm by 48% and 23%, respectively, compared with continuous potato (Table 7). This finding is consistent with other studies where the use of GMs in potato crop rotations improved water-stable aggregates [38,51]. Contrast analysis showed that the proportion of soil macroaggregates (>0.25 mm diam.) was not significantly different between the crop systems with cereals and GMs. Results showed that the preceding crops did not increase the proportion of soil macroaggregates>2 mm and MWD in studied sandy loam, contrary to N'Dayegamiye [35] who found that organic residues rapidly increased soil macroaggregates>2 mm and MWD in a silt loam. This difference may be attributed to the different soil types between studies. For example, MacRae and Mehuys [37] reported that crop rotations improved soil aggregation and bulk density of a clay soil, but had no effect on a sandy loam. N'Dayegamiye [35] found that increases of soil C contents, water-stable aggregates and MWD following application of mixed paper mill sludge and dairy cattle manure were lower in sandy loam than in clay loam soils.

Soil pre-plant test and pre-sidedressed nitrate test

Soil nitrate concentrations were measured on a 0-30 cm soil layer. Pre-Plant soil nitrate (PPNT) before potato seeding in 2011 varied between 2.0 and 3.4 mg kg^{-1} and Pre-Sidedress soil nitrate (PSNT) at mid-season varied between 6.0 and 12.6 mg kg^{-1} (Table 8) and were both significantly (P<0.05) increased following cereals and GMs in comparison with continuous potato. Soils previously cropped to cereals and GMs had PPNT values 47% higher than under continuous potato. Soil nitrate contents at mid-season (PSNT) were at least threefold higher than PPNT, indicating that mineralization rate of soil organic N and incorporated crop residues increased as the potato growing season progressed. As for PPNT, the lowest PSNT values were recorded in soils under continuous potato. In comparison with continuous potato, PSNT values were 36 to 78% greater in soils previously cropped to cereals and summer GMs, respectively. Contrast analysis showed that soil PSNT contents were higher following summer GMs and cereals than following fall GMs. The highest PSNT content in soils previously cropped to summer GMs is attributed to their high N input (Tables 3 and 4). The increase in soil nitrate following cereals was not expected as their residues generally add small quantities of N to soil, compared with summer and fall GMs and may induce soil N immobilization due to their high C/N ratios (Tables 3 and 4). By improving soil aggregation, the cereal-potato rotations probably favored soil organic matter mineralization. Soil nitrates measured at mid-season were supplied by soil organic matter or organic residues mineralization [52,53]. Increases in NO$_3$N at mid-season in soils previously cropped to cereals and GM occurred at the period of greatest crop N needs, corresponding to the tuber bulking growth stage, which likely enhanced potato yield and N nutrition, as reported by Zebarth and Rosen [54].

Potato Yields and Tuber Specific Gravity

In 2009, total and marketable potato yields were significantly (p<0.01) affected by the preceding crops and by N fertilizer application rate, and showed no significant interaction between these main effects (Table 9). Contrast analysis showed that conventional rotations with cereals and summer or fall GMs enhanced total and marketable potato yields compared with continuous potato, and did not show significant difference among them. Total and marketable tuber yields were 32% and 34% (10.1 and 9.3 Mg ha^{-1}, respectively) greater in soils previously cropped to cereals and GMs than in soils under continuous potato. Our results are in agreement with numerous studies that found positive effects of crop rotations and GMs on potato production [53,55,56]. In all cases, total and marketable potato yields increased linearly with increasing N fertilizer rates (Table 9). This showed that after only a one-year rotation with cereals or GMs, soil N availability remained a limiting factor for potato yields.

Total and marketable potato yields were generally lower in 2011 than in 2009 (Tables 9 and 10), due to lower precipitations and drought that occurred in summer 2011 (Table 1). In 2011, total and marketable potato yields ranged from 23.2 and 42.3 Mg ha^{-1} and from 18.9 to 38.4 Mg ha^{-1} (Table 10), respectively, and were significantly increased (p<0.01) in soils previously cropped to cereals and GMs, and by N fertilizer addition. Data also showed a significant interaction between the preceding crops and N rates (p<0.05). Total and marketable potato yields increased in the following order: summer GMs>cereal-potato rotations>fall GMs> continuous potato. Increases in potato yields following GMs in 2011 were greater than those obtained in 2009 after

Proceeding Crops	>5 mm	5-2 mm	2-1 mm	1-0,25 mm	MWD
Continuous potato	8	18.8	10.1	10.4	1.3
Barley	6.9	13.8	10.9	12.8	1.3
Corn	5.6	18.7	16.1	17	1.4
Oat	2.7	19.4	11.5	17.6	1
Mustard	3.5	15.2	11.9	20.4	1
Japanese Millet	6.2	20.5	12.4	14.3	1.3
Pearl Millet	8.7	21.6	16	15.8	1.6
Potato/mustard	2.4	10	7.8	10.9	0.8
Potato/wheat	9.6	17.5	14.8	17.6	1.7
Potato/Rye	7.8	12.6	11	10.9	1.3
Potato/Oat	3.2	15.3	12.2	17.7	0.91
Analysis of variance and contrasts (P value)					
Effect of the preceding crops	ns	ns	0.0471	0.0523	ns
Summer GMs£ vs. fall GMs	ns	ns	ns	ns	ns
Summer GMs vs. cereals	ns	ns	ns	ns	ns
Fall GMs vs. cereals	ns	ns	ns	ns	ns
Green manures + cereals vs. continuous potato	ns	ns	0.0452	0.0322	ns

¥ 2, 3 and 4: cereal-crop rotations; 5, 6 and 7: summer GMs; 8, 9, 10 and 11: fall GMs cropped after potato harvest.

£ GMs: green manures; ns, not significant.

€ Soil sampling was done in 2011 in plots that received no N fertilizer (0 kg N ha)$^{-1}$.

Table 7: Effects of different preceding crops on soil water-stable macroaggregate size distribution and mean weigh diameter (MWD) in 2011€.

Proceeding Crops	Total yield (Mg kg⁻¹)	Marketable yield (Mg kg⁻¹)	Specific gravity (g cm⁻³)	Black scurf incidence %
Continuous potato	31.4	27.2	1.0923	8
Barley	43	38	1.1003	6
Corn	37.2	29.9	1.0991	9
Oat	45.7	41.9	1.0951	11
Mustard	44.7	39.9	1.0972	2
Japanese Millet	40	33.7	1.0998	6
Pearl Millet	40.5	37.1	1.0983	4
Potato/mustard	33.3	28.1	1.0945	10
Potato/wheat	43.8	39.5	1.0853	14
Potato/Rye	42.2	37.9	1.0826	17
Potato/Oat	44.6	39.1	1.0819	12

Analysis of variance and contrasts (P value)

Effect of the preceding crops	0.0051	0.0019	0.0274	0.0041
Effect of N fertilizer	0.0023	0.0037	ns	0.0452
Preceding crops × Nitrogen Fertlizer	ns	ns	ns	0.039
Linear effect of N	0.0055	0.0023	0.31	0.0274
Quadratic effect of N	ns	ns	ns	0.0419

Contrasts (Pr>F)

Summer GMs£ vs. fall GMs	ns	ns	0.0061	0.0032
Summer GMs vs. cereal - potato rotations	ns	ns	ns	0.0029
Fall GMs vs. cereal-potato rotations	ns	ns	0.0028	ns
Green manures + cereal-potato rotations vs continuous potato	0.0043	0.0027	ns	0.0389

ns, not significant; Incidence of common scab was not detected in 2009.
£ GMs, green manures.

Table 8: Effects of cereal-potato rotation and green manures on potato yields and tuber specific gravity, and black scurf incidence in 2009.

only one rotation cycle. The highest overall total and marketable yields were achieved when potato followed pearl millet (12.2 and 14.4 Mg ha⁻¹)and mustard (9.0 and 13.2 Mg ha⁻¹ These findings provide evidence that pearl millet and mustard grown as summer GMs were the best preceding crops for maximizing potato productivity. Although japanese millet returned similar amounts of N and the residues had C: N ratio similar to pearl millet, it had a lower impact on potato yields, suggesting that factors other than N supply may explain the high performance of potato grown after pearl millet and mustard.

Lowest potato yields were obtained in 2011 in soils with fall GMs, compared with other rotation crops (Table 7) Fall GMs produced lowest biomass quantity and returned the least amounts of N to the soil particularly in 2010 where they provided less than 50% of those returned by summer GMs (Table 4). Dryer climatic conditions in 2011 also has not favored the fall green manure residues decomposition as they were incorporated in soil in late fall 2010 whereas summer GMs were incorporated in mid-August 2010. It has been shown that the rate and timing of organic N mineralization depends on soil moisture

[57,58].

In 2011, potato yield response to N fertilizer application was quadratic (Table 10) contrary to 2009 where the N fertilizer effect was linear (Table 9), showing that N supply from the preceding crops to the following potato was more important in 2011 than in 2009, likely because the soil N supply was greater after the second cycle of rotation. This is consistent with many reports which showed that only a fraction of N from organic amendments becomes plant-available in the first year after incorporation [12] while their residual effects could last for 2-3 years due to the mineralization of accumulated N in soils [59]. Creamer and Baldwin [60] estimated that, in temperate regions, 9 to 29% of the added residues N is mineralized in the following crop cycle. Those findings demonstrate that using appropriate GMs and crop rotations could reduce or avoid N fertilizer inputs on subsequent potato crop, leading tolower N fertilizer costs as well as to lower risk of NON losses.

Positive effects on potato yields may also be attributed to soil properties improvement (Tables 5-8). Results of the present study suggest that benefits of cereal-potato rotations and GMs on potato yields could be related to increased N availability through crop residue mineralization, stimulation of soil microbial activity, and improved soil aggregation (Tables 6 and 7), which favored potato growth and N uptake. However, they also may be attributable to other factors not assessed in the present study. For example, Gasser et al. [5] related the positive effects of yearly fall rye and barley grown every 3 yr on the following potato yields and tuber specific gravity to improvement in soil water content at the critical flowering stage.

Tuber specific gravity

Potato specific gravity averaged between 1.0819 and 1.1003 g cm⁻³ in 2009, and from 1.0878 to 1.0945 g cm⁻³ in 2011 (Tables 9 and 10). In 2009, tuber specific gravity was significantly (p<0.05) affected by the preceding crops but not by N fertilizer application. Contrast analysis showed that tuber specific gravity was the lowest forpotato grown after fall GMs, while it did not differ among cereal-potato rotations and summer GMs (Table 9). In 2011, tuber specific gravity was significantly affected by the previous crops, N fertilizer, and their interaction (Table 10). As in 2009, the lowest tuber specific gravity values were observed for potato grown after fall GMs, while they were the highest for potato grown in cereal-potato rotations and following summer GMs. It could be inferred that N was provided later in 2011 cropping season by fall green manure decomposition which delayed potato maturity, resulting in lower tuber specific gravity [61]. Our results corroborate earlier studies in eastern Canada [55] and other investigations that reported that excess N decreases potato specific gravity, particularly during the tuber bulking growth stage [62].

Nitrogen uptake, nitrogen recovery, and nitrogen fertilizer efficiency

Potato N uptake and N fertilizer efficiency were in general lower in 2011 than in 2009, while apparent N recovery was similar in both years (Table 11). This is attributable to drier climatic conditions in 2011 than in 2009. However, the data indicate that potato N uptake was significantly increased by cereal-potato rotations and GMs, compared with continuous potato. In 2009, potato N uptake varied in the following order: summer and fall GMS>cereal-potato rotations>continuous potato. Potato N uptake increased linearly with increasing N fertilizer rates likely due to low N supply from the incorporated plant residues after one crop rotation cycle. There was a significant interaction between the preceding crops and N fertilizer rate on potato N uptake,

Proceeding Crops	Total yield	Marketable yield	Tuber Specific gravity (g cm⁻³)	Common Scab incidence %
Continuous potato	29.6	23.1	1.0916	3.49
Barley	32.1	26.4	1.0917	0.64
Corn	30.6	28.7	1.0945	0.46
Oat	32.8	28.4	1.0915	0.63
Mustard	38.4	38.4	1.0922	0.07
Japanese Millet	37.3	32.4	1.0924	0.42
Pearl Millet	42.3	35.4	1.0934	0.6
Potato/mustard	23.2	22.4	1.085	2
Potato/wheat	33.4	28.2	1.093	2
Potato/Rye	28.3	29.2	1.089	2.17
Potato/Oat	33.4	18.9	1.0878	1.83
Analysis of variance and contrasts (P value)				
Effect of the preceding crops	0.0025	0.0012	0.0036	0.0013
Effect of N fertilizer	0.0023	0.0062	0.0451	ns
Preceding crops × Nitrogen	0.0469	0.0281	0.0423	ns
Fertlizer Linear effect of N	0.0019	0.0037	0.0066	0.8
Quadratic effect of N	0.0026	0.0016	ns	ns
Contrasts (Pr>F)				
Summer GMs£ vs fall GMs	0.0025	0.0038	0.004	0.0325
Summer GMs vs cereal - potato rotations	0.0028	0.0059	ns	ns
Fall GMs vs cereal-potato rotations	0.0465	0.0033	0.0099	0.0468
Green manures + cereal-potato rotations vs continuous potato	0.0358	0.0472	ns	0.0048

£ GMs, green manures.
Incidence of black scurf was not detected in 2011
ns, not significant

Table 9: Effects of cereal-potato rotation and green manures on potato yields and tuber specific gravity, and common scab incidence in 2011.

demonstrating that these cropping systems helped to increased N fertilizer use by potato, likely due to improved soil conditions (Tables 5-8), which favored potato growth and N uptake. This finding is in accordance with other studies that reported synergistic beneficial effects of organic amendment and mineral N inputs on potato yields [56,63].

In 2011, potato N uptake was significantly increased by the preceding crops and N fertilizer application and no significant interaction between these two main effects was observed (Table 11). These results indicate that two rotation cycles with GMs provided significant amounts of available N to the following potato, whereas cereal-potato rotations increased potato N uptake probably due to their positive impact on soil properties (Tables 4 and 5), which likely favored soil organic N mineralization and availability (Table 4). Contrast analysis showed that higher N uptake occurred when potato succeeded summer GMs than cereals or fall GMs. Earlier studies showed that slow release of N from green manure residues are better synchronized with plant N uptake than inorganic N fertilizer [9,12,64], and could result in increasing N uptake efficiency. Therefore, discrepancies between preceding crops in this study may reflect poorer or better synchrony between N supply from residue decomposition with active potato N uptake. Under the cold temperate climate of Quebec (Canada), N'Dayegamiye and Tran [12] showed that incorporation of GMs such as millet and mustard to the soil in late summer or in early fall of the preceding year resulted in good synchrony between N release and timing of wheat N needs in the subsequent cropping season.

Potato apparent N fertilizer recovery averaged from 23% to 58% and from 29% to 57% in 2009 and 2011, respectively (Table 11). These values are consistent with other studies in eastern Canada where apparent N recovery in whole potato plant ranged from 29% to 77% [19,65]. Potato N use efficiency varied from 193 to 258 kg yield kg⁻¹ N applied in 2009 and from 87 to 109 kg yield kg⁻¹ N applied in 2011. In both years (2009 and 2011), potato apparent N recovery and N use efficiency were significantly increased by the preceding crops and N fertilizer application, and the interaction of the preceding crops and N fertilizer was significant (Table 11). This finding is consistent with other studies that reported enhanced potato apparent N recovery when combining GMs or other organic amendments such as manures or composts with mineral N fertilizer [56,64].

In 2009, potato apparent N recovery and N use efficiency decreased linearly with increasing N fertilizer rates, while the effect was quadratic in 2011, which again indicates a greater N supply from crop residues after the second cycle of rotation. Furthermore, increases of apparent N recovery and N use efficiency may reflect better synchronization between N release from green manure residues and potato N demands and thus, lower risk of N leaching losses [14,56]. In both years of the study, potato N uptake, apparent N recovery and N use efficiency were higher in soils previously cropped to summer GMs than cereals or fall GMS. This impact of summer GMS on N supply and N use efficiency may probably be related to greater N supply and to improved soil conditions which likely favored potato growth and N uptake.

Incidence of common scab and black scurf

Incidence of black scurf (*Rhizoctonia solani*) on potato varied between 2 and 17% in 2009 (Table 8), but was not detected in 2011.

Proceeding Crops	2009			2011		
	N Uptake kg N ha⁻¹	Apparent N Recovery %	NFE kg yield kg⁻¹ N	N Uptake kg N ha⁻¹	Apparent N Recovery %	NFE kg yield kg⁻¹ N
Continuous potato	119.3	33	193	97.5	29	89
Barley	135.6	40	228	105.6	29	87
Corn	140.3	49	202	98.4	48	102
Oat	144.1	35	216	116.7	34	92
Mustard	142.8	45	226	152.9	50	119
Japanese Millet	146.8	58	266	124	56	106
Pearl Millet	132.2	38	258	137	41	109
Potato/mustard	87.2	23	195	82.3	54	97
Potato/wheat	134.6	28	230	120.7	40	92
Potato/Rye	139.7	31	231	124.6	46	96
Potato/Oat	131.4	53	254	82.4	57	84
Analysis of variance and contrasts (P value)						
Effect of the preceding crops	0.0053	0.0398	0.0472	0.0013	0.0038	0.0046
Effect of N fertilizer	0.001	0.0453	0.0023	0.0021	0.0082	0.0061
Preceding crops × Nitrogen	0.0461	0.0463	0.0493	ns	0.0469	0.0457
Fertlizer Linear effect of N	0.0026	0.0383	0.0412	0.0031	0.0031	0.0018
Quadratic effect of N	ns	ns	ns	0.0057	0.0057	0.0013
Contrasts (Pr>F)						
Summer GMs£ vs. fall GMs	ns	0.0369	0.0458	0.0052	ns	0.0081
Summer GMs vs. cereal - potato rotations	0.0083	0.0312	0.0015	0.0019	0.0493	0.0048
Fall GMs vs. cereal-potato rotations	0.0425	ns	ns	ns	0.0484	ns
Green manures + cereal-potato rotations vs continuous potato	0.0319	0.0495	0.0037	0.0459	0.0488	0.0057

£ GMs, green manures.
ns, not significant

Table 10: Effects of cereal-potato rotation and green manures on potato N uptake, apparent N recovery, and N use efficiency (NFE).

Conversely, common scab (*Streptomyces scabies*) was not detected in 2009 while its incidence varied from 0.13 to 3.49% in 2011 (Table 9). These opposite trends for common scab and black scurf between the two cycles of rotations were probably related to different environmental conditions since R. solani was favored by low temperature in the cropping season of 2009 (Table 1) as showed by Harrison [66], while S. scabies was promoted by high temperature in 2011 and low levels of precipitation (Table 1) as recorded by Loria et al. [67]. The incidence of black scurf was significantly influenced by the preceding crops and N fertilizer application, and the interaction was significant (Table 8). Orthogonal contrast analysis showed that N fertilizer application had a significant linear effect (P<0.05) on the incidence of black scurf on potato. Contrast analysis showed that the incidence of black scurf in 2009 increased in the following order: fall GMs> continuous potato=cereal-potato rotations>summer GMs (Table 8). In 2011, the incidence of common scab was significantly (P<0.01) affected by the preceding crops, and was higher for fall GMs, compared to other rotational crops (Table 9).

Data obtained in the present study are in accordance with previous investigations which showed that 2 or 3 yr cereal- potato rotations and GMs can reduce the occurrence of potato diseases, including black scurf [17,68,69] and common scab [16]. In the present study, the lowest incidence of black scurf and common scab was recorded for potato grown after summer GMs (mustard, japanese millet, and pearl millet). The suppressive effects of these green manure crops on potato black scurf and/or common scab have been demonstrated in various studies reporting the control of potato diseases with pearl millet [70], japanese millet [68], and mustard [15,17]. As shown in the present study, these green manures also increased soil microbial activity (Table 6) which may result in increased competition for existing nutrients [18,71] or increased populations of antagonistic organisms against S. scabies [71,72] or R. solani [15]. The suppressive effects of mustard and other Brassica spp. are often attributed to their high contents in glucosinolates [15-17]. Cereals used in the present study included barley, corn and oat which have been recognized to suppress potato diseases [73]. Our results revealed that the cereals significantly reduced the incidence of common scab and black scurf, compared with fall GMs and continuous potato. Increases of black scurf incidence and common scab in response to fall GMs suggest that these organic materials incorporated late in the season were still decomposing during the following season, and likely stimulated soil microorganisms including R. solani and *Rhizoctonia solani*.

Economic returns

Gross margins and net returns (Table 11) were maximized in potato cropping systems including fall GMs, whereas they were the lowest for rotation systems with cereals and summer GMs. In the 2008-2009 rotation cycle, fall GMs increased total net returns by 11% to 28%, representing $1775 to $4567 ha⁻¹, compared to those generated by continuous potato. The overall net benefit however doubled in the

Rotation	2008-2009			2010-2011		
	Crop Value	Total Variable Cost	Total net return	Crop Value	Total Variable Cost	Total net return
Continuous potato	16070	5544	10526	15203	6154	9049
Barley-potato	9788	3714	6074	9988	3906	6082
Corn-potato	8617	4179	4438	12534	4148	8386
Oat-potato	10895	3702	7193	10877	3884	6993
Mustard-potato	8213	3671	4542	12600	3839	8761
Japanese millet-potato	10146	3697	6449	13316	3796	9520
Pearl millet-potato	10732	3741	6991	13748	3644	10105
Potato/mustard-potato	18190	5653	12538	18741	6493	12248
Potato/wheat-potato	17845	5613	12233	19230	6694	12536
Potato/rye-potato	20637	5709	14929	22497	6830	15667
Potato/oat-potato	20107	5630	14478	22104	6488	15616

Gross margins were calculated based on crop yields means in each treatment

Table 11: Economic metrics for two consecutive 2-yr potato rotation cycles with cereal and green manures.

2010-2011 rotation cycle, when fall GMs increased net returns by +23% to +48%, representing a gain of $3199 to $6618 ha^{-1} over those generated by continuous potato. In both crop rotation cycles, potato rotations including rye and oat as fall GMs were the most economically profitable. The high profitability of potato rotations with fall GMs is related to potato being harvested on all years, contrary to the systems including cereals or summer GMs.

By contrast, rotation systems with cereals and summer GMs reduced total net benefit by 32% to 57%, representing $5175 to $7453 ha^{-1} compared with continuous potato. Cereal-potato rotations were less profitable than continuous potato due to the low value of cereal products compared with that of potato. The low net returns obtained with summer GMs were explained by the lack of cash crop during the first year of each 2 y rotation cycle. Our results agree with finding by Snapp et al. [14] who showed that growing green manure crops during regular cash crop season reduce the profitability due to the lack of marketable output on that particular year. However, net returns obtained with summer GMs were increased in 2011 compared with the first rotation cycle (2008-2009). In 2009 and 2011, net returns with summer GMs represented 50 and 70% of those obtained with fall green manures, respectively. Although net returns for fall GMs remained higher than for the other cropping systems tested, including fall GMs may be non- profitable as it was demonstrated that it reduced tuber specific gravity and increased the incidence of common scab and black scurf (Tables 9 and 10). As fall GMs are cultivated on all years after potato harvest, this crop system may also deteriorate soil properties and productivity. After two consecutive 2 yr rotation cycles (2008-2011), potato yields obtained with fall GMs were lower than with summer GMs. Therefore, rotation systems with summer GMs may become more profitable than continuous potato with or without fall GMS on the longer term.

Conclusion

Including cereals and GMs in potato rotation is expected to improve soil properties and boost potato yields and quality. Crop rotation systems including cereals (oat, barley and corn), summer GMs (mustard, Japanese millet and pearl millet) and fall GMs (mustard, wheat, rye and oat) increased total and marketable potato yield, compared to continuous potato. Summer GMs and cereals as preceding crops supported higher potato yield and greater N nutrition than fall GMs, which was attributed to higher N input by summer GMs and to

improvement of soil properties (soil NO_3^- concentration, microbial and enzymatic activities, soil structure). Even if cereals and fall GMs also improved soil properties significantly, their impact on potato yields and N nutrition was less than summer GMs because of their low N input. It remains challenging to design crop rotations to build soil health and sustain potato productivity, but growing cereals or summer GMs in rotation with potato is an interesting alternative to improve soil properties while sustaining potato yield and quality. Including fall GMS in potato rotation will have a small impact on potato yield and soil properties, compared to summer GMs and cereals as preceding crops. Moreover, fall GMs produced potato with lower specific gravity and they induced common scab and black scurf incidence disease. Fall GMs permit more frequent potato cultivation, compared to summer GMs and cereals, but they cannot restore soil properties to the same extent. In the cold humid temperate regions of eastern Canada, fall GMs produce lower biomass and add less N to soil, which contributes less to potato yield and N nutrition, compared to summer GMS. Results showed that continuous potato production with fall GMs even though profitable may not represent a viable option to sustain potato production and soil quality on the longer term. Since farmers prefer to include cash crops in the potato rotation or plant a fall GM after potato harvest, attention must be paid to the choice of fall GM species and consider a longer potato rotation to avoid frequent cultivation of potato, which causes a deterioration in soil properties and increases disease incidence.

Acknowledgements

This research was supported by the Research Institute for the Agri environment (IRDA), Québec, CANADA.

References

1. Mallory EB, Porter GA (2007) Potato yield stability under contrasting soil management strategies. Agr J 99: 501-510.

2. Munoz F, Mylavarapu RS, Hutchison CM (2005) Environmentally responsible potato production systems: A review. J Plant Nutr 28: 1287-1309.

3. Plotkins JMB (2000) The effects of green manure rotation crops on soils and Potato Yield and Quality. Thesis, University of Maine, p: 214.

4. Snapp SS, Fortuna AM (2003) Predicting nitrogen availability in irrigated potato systems. Hor Technology 13: 598-604.

5. Gasser O, N'Dayegamiye A, Laverdière MR (1995) Short-term Effects of crop rotations and wood-residue amendments on potato yields and soil properties of a sandy loam soil. Can J Soil Sci 75: 385-390.

6. Tyler KB, Broadbent FE, Bishop JC (1983) Efficiency of nitrogen uptake by

potatoes. Amer Potato J 60: 261-269.

7. Po EA, Snapp SS, Kravchenko A (2009) Rotational and Cover Crop Determinants of Soil Structural Stability and Carbon in a Potato System. Agron J 101: 175-183.

8. Stark C, Condron LM, Stewart A, Di HJ, O'Callaghan M (2007) Influence of organic and mineral amendments on microbial soil properties and processes. Appl Soil Ecol 35: 79-93.

9. Stute JK, Posner JL (1995) Synchrony between legume nitrogen release and corn demand in the upper Midwest. Agron J 87: 1063-1069.

10. Weinert TL, Pan WL, Moneymaker MR, Santo GS, Stevens RG (2002) Nitrogen Recycling by Nonleguminous Winter Cover Crops to Reduce Leaching in Potato Rotations. Agr J 94: 365-372.

11. Wagger MG, Cabrera ML, Ranells NN (1998) Nitrogen and carbon cycling in relation to cover crop residue quality. J Soil Water Conserv 53: 214-218.

12. N'Dayegamiye A, Tran TS (2001) Effects of green manures on soil organic matter and wheat yields and N nutrition. Can J Soil Sci 81: 371-382.

13. Cherr CM, Scholberg JMS, Mcsorley R (2006) Green manure approaches to crop production. Agron J 98: 302-319.

14. Snapp SS, Swinton SM, Labarta R, Mutch D, Black JR, et al. (2005) Evaluating cover crops for benefits, costs, and performance within cropping system niches. Agron J 97: 322-332.

15. Larkin RP, Griffin TS (2007) Control of soilborne diseases of potato using Brassica green manures. Crop Prot 26: 1067-1077.

16. Larkin RP, Griffin TS, Honeycutt CW (2010) Rotation and cover crop effects on soilborne potato diseases, tuber yield, and soil microbial communities. Plant Disease 94: 1491-1502.

17. Sexton P, Plant A, Johnson SB, Jemison JJ (2007) Effect of a mustard green manure on potato yield and disease incidence in a rainfed environment. Crop Management, p: 6.

18. Olanya OM, Lambert DH, Porter GA (2006) Effects of pest and soil management systems on potato diseases. Am J Potato Res 83: 397-408.

19. Zebarth BJ, Drury CF, Tremblay N, Cambouris AN (2009) Opportunities for Improved Fertilizer Nitrogen Management in Production of Arable Crops in Eastern Canada: A Review. Can J Soil Sci 89: 113-132.

20. Mehlich A (1984) Mehlich 3 soil test extractant: a modification of Mehlich 2 extractant. Communications in Soil Science & Plant Analysis 15: 1409-1416.

21. Gee GW, Bauder JW (1986) Particle-size analysis. In: Klute A (ed.) Methods of Soil Analysis. 2nd edn. Agron Monogr Part 1, ASA and SSSA, Madison, p: 384-411.

22. Anderson JPE (1982) Soil respiration. In: Miller ALRH, Keeny DR (eds.), Methods of soil analysis Part 2, 2nd edn. ASA and SSSA, Madison, p: 841-845.

23. Casida LE, Klein DA, Santro T (1964) Soil dehydrogenase activity. Soil Sci 98: 371-376.

24. Tabatabai MA (1994) Soil enzymes. In: Weaver R (ed.), Methods of soil analysis Part 2 Microbiological and biochemical properties. SSSSA Book Ser 5, Madison, p: 903-947.

25. Bremner JM (1965) Total nitrogen. In: Black CA (ed.), Methods of soil analysis. Agron Monogr Part 2. ASA and SSSA, Madison, USA, pp: 1149-1178.

26. Curtin D, Wright CE, Beare MH, McCallum FM (2006) Hot-water extractable nitrogne as an indicator of soil nitrogen availability. Soil Sci Soc Am J 70: 1512-1521.

27. Kemper WD, Rosenau RC (1986) Aggregate stability and size distribution. In: Page AL (eds.) Methods of Soil Analysis Part 1 Physical and Mineralogical Methods. Agron Monogr 9, ASA and SSSA, Madison, pp: 425-442.

28. Haynes RJ, Beare MH (1996) Aggregation and organic matter storage in mesothermal, humic soils. In: Carter MR, Stewart BA (eds.) Structure and organic matter storage in agricultural soils. CRC Press, Boca Raton, pp: 213-262.

29. Conn KL, Lazarovits G (1999) Impact of animal manures on virticillium wilt, potato scab, and soil microbial populations. Can J Plant Pathol 21: 81-92.

30. Wortman CS, Tarkalson DD, Shapio CA, Dobermann AR, Frguson RB, et al. (2011) Nitrogen use efficiency of irrigated corn for three cropping systems in Nebraska. Agron J 103: 76-84.

31. Zvomuya F, Rosen CJ, Russelle MP, Gupt SC (2003) Nitrate leaching and nitrogen recovery following application of polyolefin-coated urea to potato. J Environ Qual 32: 480-489.

32. Agricultural Income Stabilization Insurance Program, ASRA, Financière Agricole du Québec (2011) Agdex 740/825.

33. SAS Institute (2003) The MIXED Procedure. In: SAS/STAT User's Guide, Ver 9.1, SAS Institute, Cary, NC, pp: 2664-2844.

34. Mellilo JM, Aber JD, Linkins AE, Ricca A, Fry B, et al. (1989) Carbon and nitrogen dynamics along the decay continuum: Plant litter to soil organic matter. Plant Soil 115: 189-198.

35. N'Dayegamiye A (2009) Soil properties and crop yields in response to mixed paper mill sludges, dairy cattle manure, and inorganic fertilizer application. Agron J 101: 826-835.

36. Neeteson JJ (1989) Effects of legumes on soil mineral nitrogen and response of potatoes to nitrogen fertilizer. In: Vos J, et al. (eds.) Effects of crop rotation on potato production in the temperate zones. Kluwer Academic Publishers, Dordrecht, pp: 88-93.

37. McRae RJ, Mehuys GR (1988) The effect of gren manuring on the physical properties of temperate-area soils. Adv Soil Sci 3: 71-94.

38. Grandy AS, Porter GA, Erich MS (2002) Organic amendment and rotation crop effects on the recovery of soil Organic Matter and Aggregation in Potato Cropping Systems. Soil Sci Soc Am J 66: 1311-1319.

39. Sorensen, Lasse H (1974) Rate of decomposition of organic matter in soil as influenced by repeated air drying- rewetting and repeated additions of organic material. Soil Biol Biochem 6: 287-292.

40. Kandeler E, Tscherko D, Spiegel H (1999) Long-term monitoring of microbial biomass, N mineralization and enzyme activities of a Chernozem under different tillage management. Biology & Fertility Soils 28: 343-351.

41. Asmar F, Eiland F, Nielsen NE (1994) Effect of ex- tracellular-enzyme activities on solubilization rate of soil organic nitrogen. Biol Fert Soils 17: 32-38.

42. Baldock JA, Kay BD, Schnitzer M (1987) Influence of cropping treatments of the monosaccharide content of the hydrolysates of a soil and its aggregate fractions. Can J Soil Sci 67: 489-499.

43. Haynes RJ (2008) Soil organic matter quality and the size and activity of the microbial biomass: their significance to the quality of agricultural soils. In: Huang Q, Huang PM, Violante A (eds.) Soil mineral- microbe-organic interactions: theories and applications. Springer, pp: 201-230.

44. Marschner B, Kalbitz K (2003) Controls of bioavailability and biodegradability of dissolved organic matter in soils. Geoderma, 113: 211-235.

45. Carter MR (2002) Soil quality for sustainable land management. Agron J 94: 38-47.

46. Angers DA, Mehuys GR (1989) Effects of cropping on carbohydrate content and water-stable aggregation of a clay soil. Can J Soil Sci 69: 373-380.

47. Veum KS, Goyne KW, Kremer R, Motavalli PP (2012) Relationships among water stable aggregates and organic matter fractions under conservation management. Soil Sci Soc Am J 76: 2143-2153.

48. Angers DA, N'Dayegamiye A, Côté D (1993) Tillage-induced differences in organic matter of particle- size fractions and microbial biomass. Soil Sci Soc Am J 57: 512-516.

49. Anderson JPE, Domsch KH (1978) A physiological method for the quantitative measurement of microbial biomass in soils. Soil Biol Biochem 10: 215-221.

50. Nyiraneza J, Chantigny MH, N'Dayegamiye A, Laverdière MR (2009) Dairy cattle manure improves soil productivity in low residues systems. Agron J 101: 207-214.

51. Porter GA, Opena GB, Bradbury WB, McBurnie JC, Sisson JA (1999) Soil management and supplemental irrigation effects on potato: I. Soil properties, tuber yield, and quality. Agron J 91: 416-425.

52. Nyiraneza J, N'Dayegamiye A, Gasser MO, Giroux M, Grenier M, et al. (2010) Soil and crop parameters related to corn nitrogen response in Eastern Canada. Agron J 102: 1478-1490.

53. Bélanger G, Walsh JR, Richards JE, Milburn PH, Ziadi N (2001) Critical nitrogen curve and nitrogen nutrition index for potato in eastern Canada. Am J Potato Res 78: 355-364.

54. Zebarth BJ, Rosen CJ (2007) Research perspective on nitrogen BMP development for potato. Amer J Potato Res 84: 3-18.

55. Nyiraneza J, Snapp S (2007) Integrated management of inorganic and organic nitrogen and efficiency in potato system. Soil Sci Soc Am J 7': 1508-1515.

56. Essah SYC, Delgado JA (2009) Nitrogen management for maximizing tuber yield, quality and environmental conservation. In: Yanful EY (ed.), Appropriate technologies for environmental protection in the developing world. Springer, pp: 317-325.

57. Cassman KG, Munns DN (1980) Nitrogen mineralization as affected by soil moisture, temperature and depth. Soil Sci Soc Am J 44: 1233-1237.

58. Myers RJK, Campbell CA, Weier KL (1982) Quantitative relationship between net nitrogen mineralization and moisture content of soils. Can J Soil Sci 62: 111-124.

59. N'Dayegamiye A, Nyiraneza J, Whalen JK, Grenier M, Drapeau A (2012) Growing soybean prior to corn increased soil nitrogen supply and N fertilizer efficiency for corn in cold and humid conditions of Eastern Canada. Sustainable Agriculture Research 1: 257- 267.

60. Creamer NG, Baldwin KR (2000) An evaluation of summer cover crops for use in vegetable production systems in North Carolina. Hort Science 35: 600-603.

61. Santerre CR, Cash JN, Chase RW (1986) Influence of cultivar, harvest-date and soil nitrogen on sucrose, specific gravity and storage stability of potatoes grown in Michigan. Am J Potato Res 63: 99-110.

62. Sparrow LA, Chapman KSR (2003) Effects of nitrogen fertiliser on potato (Solanum tuberosum L., cv. Russet Burbank) in Tasmania 1 Yield and quality. Aust J Exp Agr 43: 631-641.

63. Essah SYC (2012) Potato tuber yield, tuber size distribution, and quality as impacted by preceding green manure cover crops. In: He Z, Larkin R, Honeycutt W (eds.) Sustainable Potato production: Global Case Studies, pp: 99-115.

64. Warman PR, Rodd AV, Hicklenton PR (2011) The effect of MSW compost and fertilizer on extractable soil elements, tuber yield, and elemental concentrations in the plant tissue of potato. Potato Research 54: 1- 11.

65. Båth B (2000) Matching the availability of N mineralized from green-manure crops with the N-demand of field vegetables. Doctoral thesis, Agraria 222, Swedish University of Agricultural Sciences, Uppsala, Sweden, pp: 1401-6249.

66. Harrison JG, Serale RJ, Williams NA (1997) Powdery scab disease of potato-a review. Plant pathology. 46: 1-25.

67. Loria R, Bukhalid RA, Fry BA, King RR (1997) Plant pathogenicity in the genus STREPTOMYCES. Plant Disease. 81: 836-846.

68. Specht LP, Leach SS (1987) Effects of crop rotation on Rhizoctonia disease of white potato. Plant Disease 71: 433-437.

69. Honeycutt CW, Clapham WM, Leach SS (1996) Crop rotation and nitrogen fertilization effects on growth, yield and disease incidence in potato. Amer Potato J 73: 45-61.

70. Bélair G, Dauphinais N, Fournier Y, Dang OP, Clément MF (2005) Effect of forage and grain pearl millet on Pratylenchus penetrans and potato yields in Quebec. Journal of Nematology 37: 78-82.

71. Kinkel LL, Schlatter DC, Bakker MG, Arenz BE (2012) Streptomyces competition and co-evolution in relation to plant disease suppression. Research in Microbiology 163: 490-499.

72. Millard WA, Taylor C (2008) Antagonisms of microorganisms as the controlling factor in the inhibition of scab by green manuring. Annals of Applied Biology 14: 202-216.

73. Davis JR, Huisman OC, Everson DO, Nolte P, Sorensen LH, et al. (2010) Ecological relationship of Verticillium wilt suppression of potato by green manure. Am J Potato Res 87: 315-326.

Permissions

List of Contributors

Mahendra Kumar Trivedi, Alice Branton, Dahryn Trivedi and Gopal Nayak
Trivedi Global Inc., Henderson, USA

Mayank Gangwar and Snehasis Jana
Trivedi Science Research Laboratory Pvt. Ltd., Bhopal, Madhya Pradesh, India

Shahzad Imran, Muhammad Arif, Wasif Shah and Abdul Latif
The University of Agriculture, Peshawar, Khyber Pakhtunkhwa, Pakistan

Arsalan Khan and Muhammad Ali Khan
Agriculture Research Institute (ARI) Tarnab Peshawar, Pakistan

Amare Seyoum and Sentayehu Alamerew
College of Agriculture and Veterinary Medicine, Jimma University, Jimma, Ethiopia

Dagne Wegary
CIMMYT–Ethiopia, ILRI Campus, Addis Ababa, Ethiopia

S.K. Sahoo and B. Singh
Department of Entomology, Punjab Agricultural University, Ludhiana-141004, Punjab, India

Avinash Kumar Gautam and Shrivastava AK
Department of Farm Machinery and Power Engineering, College of Agricultural Engineering, JNKVV, Jabalpur, Madhya Pradesh, India

Kator P E
Department of Agricultural Technology, Delta State Polytechnic, P.M. B. 005, Ozoro, Delta State, Nigeria

Adaigho. P
Department of Agricultural Extension Management, Nigeria

Dereje Shimelis
Ethiopian Sugar Corporation Research and Training Division, Variety Development Directorate, Biotechnology Research Team, Wonji Research Center, Wonji, Ethiopia

Kassahun Bantte
Jimma University College of Agriculture and Veterinary Medicine, Jimma, Ethiopia

Tilaye Feyissa
Addis Abeba University, Science and Technology faculty, Addis Abeba, Ethiopia

Abera Degefa, Mengistu Bosie, Yohannes Mequanint, Endris Yesuf and Zeleke Teshome
Sugar Corporation, Research and Development Center, Wonji, Ethiopia

Mijena Bikila
Ethiopian Sugar Corporation, Research and Training Division, Sugarcane Production Research Directorate, Agronomy and Protection Research Team, Finchaa Research Center, Finchaa, Ethiopia

Nigussie Dechassa and Yibekal Alemayehu
Haramaya University, College of Agriculture, Haramaya, Ethiopia

Azeb Hailu
Tigray Agricultural Research Institute, Mekelle Agricultural Research Center, Mekelle, Ethiopia

Sentayehu Alamerew
Colleges of Agriculture and Veterinary Medicine, Jimma University, Ethiopia

Ermias Assefa
Colleges of Agriculture and Veterinary Medicine, Jimma University, Ethiopia
Southern Agricultural Research Institute, Bonga Agricultural Research Center, Bonga, Ethiopia

Mandefro Nigussie
Oxfam America, Horn of Africa Regional Office, Ethiopia

Eba Muluneh Sorecha and Birhanu Bayissa
School of Natural Resources Management and Environmental Sciences, Haramaya University, Dire Dawa, Ethiopia

Ambachew Dametie, Abiy Fantaye and Zeleke Teshome
Sugar Corporation, Research and Training, Wonji, Ethiopia

Abrham Shumbulo
Department of Horticulture, College of Agriculture, Wolaita Sodo University, Ethiopia

Mandefro Nigussie
Oxfam America, Horn of Africa Regional Office, Addis Ababa, Ethiopia

Sentayehu Alamerew
Jimma University College of Agriculture and Veterinary Medicine, Jimma, Ethiopia

Tibebu Belete
Department of Plant Sciences and Horticulture, College of Dry Land Agriculture, Samara University, Samara, Ethiopia

Firew Mekbib
Department of Plant Science, College of Agricultural Sciences, Haramaya University, Ethiopia

Million Eshete
Debre Zeit Agricultural Research Center, Debre Zeit, Ethiopia

Sajid Khan, Ahmad Khan, Haris Khan and Said Badshah
Department of Agronomy, The University of Agriculture, Peshawar, Pakistan

Maaz Khan
Department of Soil and Environmental Sciences, The University of Agriculture, Peshawar, Pakistan

Fazal Jalal
Department of Agriculture, Abdul Wali Khan University, Mardan, Pakistan

Spandana B and Nagalakshmi S
Institute of Biotechnology (IBT), Acharya NG Ranga Agricultural University (ANGRAU), Guntur, Andhra Pradesh, India

DSRS Prakash
Department of Human Genetics, Andhra University, Visakhapatnam, Andhra Pradesh, India

Mawia A. Musyoki, Wambua F. Kioko, Ngugi P. Mathew, Agyirifo Daniel, Nyamai D. Wavinya, Matheri Felix, Lagat R. Chemutai, Njagi S. Mwenda, Mworia J. Kiambi and Ngari L. Ngithi
Department of Biochemistry and Biotechnology, School of Pure and Applied Sciences, Kenyatta University, Nairobi, Kenya

Karau G. Muriira
Molecular Laboratory, Kenya Bureau of Standards, Kenya

Abiyot Lemma
EIAR, Worer Agricultural Research Center, Addis Ababa, Ethiopia

Getaneh Woldeab
EIAR, Ambo Plant Protection Research Center, Ambo, Ethiopia

Selvaraj T
Ambo University, Ambo, Ethiopia

Y.M.A.M. Wijerathna
The Division for International Studies,Robert H. Smith Faculty of Agriculture, Food and Environment,The Hebrew University of Jerusalem. Rehovot 76100, Israel

Mathew Piero Ngugi, Oduor Richard Okoth and Mgutu Allan Jalemba
Department of Biochemistry and Biotechnology, Kenyatta University, Nairobi, Kenya

Omwoyo Richard Ombori and Cheruiyot Richard Chelule
Department of Plant and Microbial Sciences, Kenyatta University, Nairobi, Kenya

Njagi Joan Murugi
Department of Environmental Health, Kenyatta University, Nairobi, Kenya

Ravi Sharma
Eco-physiology Laboratory Department of Post-graduate Studies and Research in Botany K R College Mathura Formerly Head Department of Botany K R College, Mathura and Ex-Principal ESS ESS College, Agra (Dr B R Ambedkar University formerly Agra University, Agra) 281 001 UP India

Belay Tolera
Ethiopian Sugar Corporation, Research and Training Division, Variety Development Directorate, Biotechnology Research Team, Wonji Research Center, Wonji, Ethiopia

Mulugeta Diro
Capacity Building for Scaling up of Evidence-based Best Practices in Agricultural Production in Ethiopia (CASCAPE) Project, Addis Ababa, Ethiopia

Derbew Belew
Jimma University, College of Agriculture and Veterinary Medicine, Jimma, Ethiopia

Miheretu Fufa
Oromia Agricultural Research Institute, Adami Tullu Agricultural Research Center, Plant Biotechnology Team, Zeway, Ethiopia

Mulugeta Diro
Southern Agricultural Research Institute, Ethiopia

Abiyot Lemma, Hadush Hagos and Amrote Tekle
Ethiopian Sugar Corporation Research and Training Wonji, Ethiopia

Yohannes Zekarias
Bayer Trade Representative Office Addis Ababa, Ethiopia

Messgo Moumene S, Laidani M, Saddek D and Houmani Z
Laboratory for Research on Medicinal and Aromatic plants, Science and life Faculty, University of Blida1, BP. 270, Soumaa road, 09100, Blida, Algeria

Olubunmi OF
Department of Crop Protection and Environmental Biology, University of Ibadan, Nigeria

Bouznad Z
Laboratory of Phytopathology and Molecular Biology, National graduate school of Agronomy El Harrach, Algeria

Fufa Miheretu
Sinana Agricultural Research Center, Horticulture division, Ethiopia

Endale Hailu and Getaneh Woldeab
Ethiopian Institute of Agricultural Research, Ambo Plant Protection Research Center, Ambo, West Shewa, Ethiopia

Amare Tesfay, Mohammed Amin and Negeri Mulugeta
Department of Plant Science, College of Agriculture and Veterinary Science, Ambo University, Ethiopia

Hamad Saeed
Department of Agriculture, Wiltshire College, Lacock, Chippenham, Wiltshire, UK

Ivan G. Grove, Peter S. Kettlewell and Nigel W. Hall
Crop and Environment Research Centre, Harper Adams University College, Newport, Shropshire, UK

Ian J. Fairchild and Ian Boomer
School of Geography, Earth and Environmental Sciences, University of Birmingham, UK

Katja Kempe, Anastassia Boudichevskaia, Robert Jerchel, Dmitri Pescianschi, Renate Schmidt and Mario Gils
Leibniz Institute of Plant Genetics and Crop Plant Research (IPK), Germany

Martin Kirchhoff and Ralf Schachschneider
Nordsaat Saatzucht GmbH, Böhnshauser Straße 1, D-38895 Langenstein, OT Böhnshausen, Germany

Suma R
University of Horticultural Sciences, Bagalkot, Karnataka, India

Savitha CM
KVK, V.C. Farm, Mandya, Karnataka, India

Tessema Tamirat, Sentayehu Alamerew and Temesgen Menamo
Jimma University, Collage of Agriculture and Veterinary Medicine, Jimma, Ethiopia

Getaneh Woldeab, Endale Hailu and Teklu Negash
Ethiopian Institute of Agricultural Research, Ambo Plant Protection Research Center, Ambo, West Shewa, Ethiopia

Teklay Ababa
Tigray Agricultural Research Institute, Alamata Agricultural Research center, Alamata, Ethiopia

Szczepanek M and Grzybowski K
Department of Agrotechnology, UTP University of Science and Technology

Miheretu Fufa
Adami Tullu Agricultural Research Center; Plant Biotechnology Team, Zeway, Ethiopia

Mulugeta Diro
Cascape Project Coordinator, Addis Ababa, Ethiopia

Tesfaye Wolde and Sentayehu Alamerew
College of Agriculture and Veterinary Medicine Gelalcha, Jimma University, Ethiopia

Ermias Assefa
College of Agriculture and Veterinary Medicine Gelalcha, Jimma University, Ethiopia
Southern Agricultural Research Institute, Ethiopia

Firdisa Eticha
Ethiopian Institute of Agricultural Research, Kulumsa Research Center, Ethiopia

Dargicho Dutamo
College of Agriculture and Natural Resources, Mizan-Tepi University, Ethiopia

Birhanu Mecha
College of Agricultural Sciences, Wachemo University, Ethiopia

Adrien N'Dayegamiye, Michèle Grenier, Marie Bipfubusa and Anne Drapeau
Research and Development Institute for the Agro-Environment (IRDA), Complexe Scientifique, Canada

Judith Nyiraneza
Agriculture and Agro-Food Canada, Crops and Livestock Research Centre, 440 University Avenue, Charlottetown, Canada

Index